X International Scientific Practical Conference – Innovative Technologies in Engineering 2019

IOP Conference Series: Materials Science and Engineering
Volume 582

Yurga, Russia
23 – 25 May 2019

ISBN: 978-1-5108-9813-4
ISSN: 1757-8981

Printed from e-media with permission by:

Curran Associates, Inc.
57 Morehouse Lane
Red Hook, NY 12571

CURRAN ASSOCIATES INC.
proceedings.com

Some format issues inherent in the e-media version may also appear in this print version.

Institute of Physics
Dirac House, Temple Back
Bristol BS1 6BE UK

Phone: 44 1 17 929 7481
Fax: 44 1 17 920 0979

techtracking@iop.org

Additional copies of this publication are available from:

Curran Associates, Inc.
57 Morehouse Lane
Red Hook, NY 12571 USA
Phone: 845-758-0400
Fax: 845-758-2633
Email: curran@proceedings.com
Web: www.proceedings.com

Table of contents

Volume 582

X International Scientific and Practical Conference - Innovative Technologies in Engineering

23–25 May 2019, Yurga, Russian Federation

Accepted papers received: 2 July 2019
Published online: 17 October 2019

Preface

Papers

IOP Conf. Series: Materials Science and Engineering **582** (2019) 011001 doi:10.1088/1757-899X/582/1/011001

PREFACE

Innovative Technologies in Engineering Conference Proceedings follow the same-name 10th International Scientific Practical Conference held at Yurga Institute of Technology (branch) of National Research Tomsk Polytechnic University from 23th to 25th May, 2019. The Proceedings contain the selected scientific reports submitted to the Conference. Having started in 1996 the scientific conference at Yurga Institute of Technology (branch) of National Research Tomsk Polytechnic University acquired its international status in 2010. This year scholars from Russia, Kazakhstan, Ukraine, Azerbaijan and China have submitted their papers to the Conference.

The scientific reports published in these Proceedings have been revised and approved by the editorial team of the Conference. All of the reports exhibit clear, concise, and precise expositions that appeal to a broad international readership interested in mechanical engineering, welding, metallurgy, material science, as well as in computer-aided manufacturing. The reports present original ideas or results of general significance supported by clear reasoning and compelling evidence, and employ methods, theories and practices relevant to the research. The authors state clearly the questions and the significance of their research to theory and practice, describe how the research contributes to new knowledge; and provide tables and figures that meaningfully add to the narrative.

Organizing Committee of the Conference thank all the participants of the Conference for their fruitful work and personal contribution to the development of these Conference Proceedings.

Dmitry A. Chinakhov, PhD,
Conference Convenor,
Director, Yurga Institute of Technology
of National Research Tomsk Polytechnic University

IOP Conf. Series: Materials Science and Engineering **582** (2019) 011002 doi:10.1088/1757-899X/582/1/011002

Peer review statement

All papers published in this volume of *IOP Conference Series: Materials Science and Engineering* have been peer reviewed through processes administered by the proceedings Editors. Reviews were conducted by expert referees to the professional and scientific standards expected of a proceedings journal published by IOP Publishing.

IOP Conf. Series: Materials Science and Engineering **582** (2019) 012001 doi:10.1088/1757-899X/582/1/012001

Stress-strain State of Circular Plates with Concentric Inserts under Tension

N L Velikanov[a], S I Koryagin[b] and O V Sharkov[c]

Immanuel Kant Baltic Federal University, Kaliningrad, 236016, Russia

E-mail: [a]nvelikanov@kantiana.ru, [b]skoryagin@kantiana.ru, [c]osharkov@kantiana.ru

Abstract. The article discusses the method of determining stresses in circular plates with concentric absolutely rigid inserts. The plate is stretched by a uniformly distributed load. A design scheme is proposed and an analysis of mathematical expressions describing the deformation process at the interface of the plate and rigid insert is performed. Mathematical dependences are obtained for the calculation of normal and tangential stresses, allowing to determine their distribution at the design stage. It was established that the zone of influence on the stress distribution of a rigid insert does not exceed the diameter of the insert. The use of various welding technologies allows changing the resultant field of normal and tangential stresses and increasing the load capacity of the permanent joint.

1. Introduction

In the context of scientific and technological progress, the development of science, technology and production is especially important. When developing complex metal structures, techniques and algorithms are necessary for calculating the strength characteristics [1–6], including the characteristics of permanent joints.

This type of joints include metal welding, which in many industries is one of the main factors determining the pace of technological progress, and has a significant impact on the efficiency of social production. There is hardly any mechanical engineering industry, which doesn`t use metal welding (Figure 1) [7–13].

Residual stresses are those that exist in the structure or in its individual elements in the absence of external force, thermal and other actions. In engineering to designate residual stresses the names of the technological processes are also used: welding stresses, quenching stresses, deformation stresses, straightening stresses or finishing stresses.

Residual (preliminary) stresses have a significant role in engineering [14–22]. Knowledge of the internal stress state in materials and structures makes it possible to calculate the structure more accurately. There are plenty of examples of damage caused by large technological stresses. One of the most frequent is the destruction of pipelines in which cracks can develop up to tens of kilometers in length.

The welding of different types of metal structures made it possible to most effectively use workpieces obtained by rolling, bending, punching, casting and forging, as well as metals with different physicochemical properties. Welded structures are more light-weighted and less labor intensive. Permanent joint of almost all metals and alloys of various thicknesses are obtained by welding.

To determine the stresses in a circular plate with a rigid insert under external load, installed by welding, one can use the solution of the plane elasticity problem for annular plates in polar coordinates. The polar coordinates of the points are the polar radius r and the polar angle θ, and the stress state is determined by the tangents τ, normal radial σ_r and tangential σ_θ stresses [23–25].

2. Determination of stresses and strains

2.1. Stress-strain state in the absence of external load

Let us determine residual stresses in a circular plate of radius R_1 with an absolutely circular rigid insert of radius R (Figure 2) in the absence of external forces (p =0).

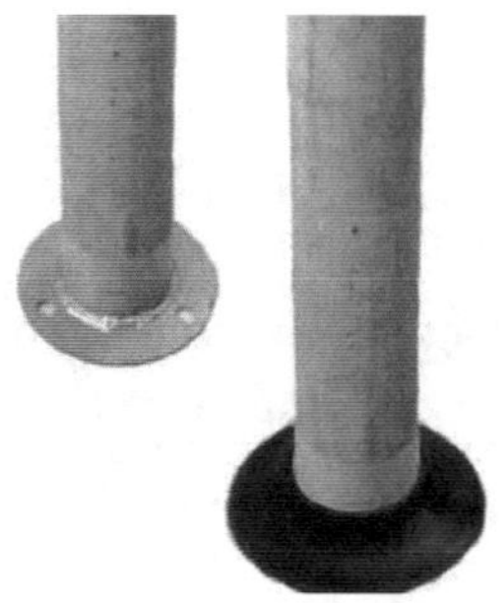

Figure 1. Welded structures with circular welded insert.

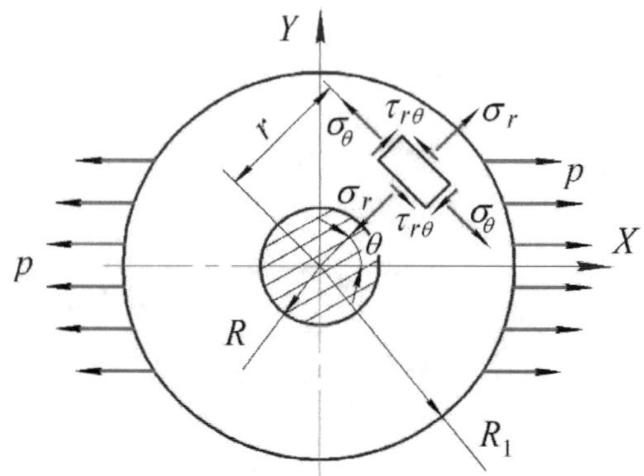

Figure 2. Circular plate with rigid insert under uniaxial tension with uniform load.

In this case, the displacements of the points of the plate on the boundary with the insert in the Cartesian coordinate system are

$$g_1 = u_1 + iv_1, \tag{1}$$

where $u_1 = \Delta_1 \cos\theta$ – the displacements of points in the longitudinal (along the axis OX) direction; $v_1 = \Delta_2 \sin\theta$ – the displacements of points in the transverse (along the axis OY) direction; Δ_1, Δ_2 – the longitudinal and transverse shrinkage of the welded seam [23].

Representing radial u and tangential v displacements in polar coordinates as

$$\begin{cases} u = u_1 \cos\theta + v_1 \sin\theta; \\ v = -u_1 \sin\theta + v_1 \cos\theta. \end{cases} \tag{2}$$

we obtain the following

$$\begin{cases} u = 0.5(\Delta_1 + \Delta_2) + 0.5(\Delta_1 - \Delta_2)\cos 2\theta; \\ v = -0.5(\Delta_1 - \Delta_2)\sin 2\theta. \end{cases} \tag{3}$$

Boundary conditions on the contour $r = R$ will be written as

$$\begin{cases} u(R,\theta) = 0.5(\Delta_1 + \Delta_2) + (\Delta_1 - \Delta_2)\cos 2\theta; \\ v(R,\theta) = -(\Delta_1 - \Delta_2)\sin 2\theta. \end{cases} \tag{4}$$

Boundary conditions on the contour $r = R_1$ will be written as

$$\begin{cases} \sigma_r(R_1,\theta) = 0; \\ \tau(R_1,\theta) = 0. \end{cases} \tag{5}$$

For annular plates the general solution is known in the form of an infinite trigonometric series in $\sin(n\theta)$ and $\cos(n\theta)$ (n – is the series term number).

Since the boundary conditions (4, 5) are homogeneous or contain only expansion terms with numbers n=0 and n=2, then in general expressions for stresses it is sufficient to leave the terms with the indicated numbers and only those that determine the expansion of displacements u by cosines, and v -by sinuses.

Therefore, displacements can be determined by the following expressions:

$$\begin{cases} u(r,\theta)=\dfrac{2a_0r(1-\mu)}{E}-\dfrac{a_1(1+\mu)}{Er}+\dfrac{\left(-a_2r(1+\mu)+\dfrac{a_3(1+\mu)}{3r^3}-\dfrac{2a_4r^3\mu}{3}+\dfrac{2a_5}{r}\right)\cos 2\theta}{E}; \\ v(r,\theta)=\dfrac{-(1+\mu)(R^2-R_1^2)}{Er}+\dfrac{\left(a_5r(1+\mu)+\dfrac{a_3(1+\mu)}{3r^3}+\dfrac{a_4r^3(\mu+3)}{3}-\dfrac{a_5(1-\mu)}{r}\right)\sin 2\theta}{E}. \end{cases} \quad (6)$$

where E – modulus of elasticity (Young's modulus); μ – coefficient of transverse deformation (Poisson's ratio); a_0, a_1, a_2, a_3, a_4, a_5 – unknown coefficients to be determined.

Stresses can be determined by the following expressions:

$$\begin{cases} \sigma_r=\dfrac{a_1}{r^2}+2a_0-\left(a_2+\dfrac{a_3}{r^4}+\dfrac{2a_5}{r^2}\right)\cos 2\theta; \\ \sigma_\theta=-\dfrac{a_1}{r^2}+2a_0+\left(a_2+2a_4r^2+\dfrac{a_3}{r^4}\right)\cos 2\theta; \\ \tau_{r\theta}=\dfrac{(R^2-R_1^2)}{r^2}+\left(a_2-\dfrac{a_3}{r^4}+a_4r^2-\dfrac{a_5}{r^2}\right)\sin 2\theta. \end{cases} \quad (7)$$

Substituting displacements (1) and stresses (2) into the boundary conditions at $r=R$ and $r=R_1$, we determine the unknown coefficients, equating the free terms to coefficients at $\cos 2\theta$, $\sin 2\theta$. Substituting the coefficient values into expression (2), we obtain the following expressions for the residual stresses when welding absolutely rigid insert of radius R into the plate of radius R:

$$\sigma_{r1}=\frac{E(\Delta_1+\Delta_2)R}{2[(1-\mu)R^2+(1+\mu)R_1^2]}\left(1-\frac{R_1^2}{r^2}\right)-\frac{(\Delta_1-\Delta_2)ERR_1^4[(1+\mu)R_1^6+(3-\mu)R^6]}{A+B}\times$$
$$\times\left(\frac{3R_1^4(1+\mu)(R^2-R_1^2)-[(1+\mu)R_1^6+(3-\mu)R^6]}{2R_1^2[(1+\mu)R_1^6+(3-\mu)R^6]}-\right. \quad (8)$$
$$\left.-\frac{3R_1^2R^2[(1+\mu)R_1^4+(3-\mu)R^4]}{2r^4[(1+\mu)R_1^6+(3-\mu)R^6]}+\frac{2}{r^2}\right)\cos 2\theta.$$

$$\sigma_{\theta 1}=\frac{E(\Delta_1+\Delta_2)R}{2[(1-\mu)R^2+(1+\mu)R_1^2]}\left(1+\frac{R_1^2}{r^2}\right)-\frac{(\Delta_1-\Delta_2)ERR_1^4[(1+\mu)R_1^6+(3-\mu)R^6]}{A+B}\times$$
$$\times\left(\frac{3R_1^4(1+\mu)(R^2-R_1^2)-[(1+\mu)R_1^6+(3-\mu)R^6]}{2R_1^2[(1+\mu)R_1^6+(3-\mu)R^6]}-\right. \quad (9)$$
$$\left.-\frac{3R_1^2R^2[(1+\mu)R_1^4+(3-\mu)R^4]}{2r^4[(1+\mu)R_1^6+(3-\mu)R^6)]}-\frac{6(1+\mu)(R^2-R_1^2)r^2}{(1+\mu)R_1^6+(3-\mu)R^6}\right)\cos 2\theta.$$

$$\tau_{r\theta}=\frac{(\Delta_1-\Delta_2)ERR_1^4[(1+\mu)R_1^6+(3-\mu)R^6]}{A+B}\times$$
$$\times\left(\frac{3R_1^4(1+\mu)(R^2-R_1^2)-[(1+\mu)R_1^6+(3-\mu)R^6]}{2R_1^2[(1+\mu)R_1^6+(3-\mu)R^6]}-\right. \quad (10)$$
$$\left.-\frac{3R_1^2R^2[(1+\mu)R_1^4+(3-\mu)R^4]}{2r^4[(1+\mu)R_1^6+(3-\mu)R^6]}-\frac{3(1+\mu)(R^2-R_1^2)r^2}{(1+\mu)R_1^6+(3-\mu)R^6}-\frac{1}{r^2}\right)\sin 2\theta.$$

where the following notation is used

$$A = 3(1+\mu)(R^2 - R_1^2)R^4[(1+\mu)R_1^4 + (3-\mu)R^4];$$

$$B = [(1+\mu)R_1^6 + (3-\mu)R^6][(1+\mu)R_1^6 + (3-\mu)R^6][(1+\mu)(R^2 - R_1^2)(R_1^2 - 3R^2) + 4R_1^4].$$

When radius R_1 tends to infinity in expressions for σ_{r1}, $\sigma_{\theta 1}$, $\tau_{r\theta 1}$, we pass to expressions for residual stresses in an infinite plate with a welded absolutely rigid circular insert.

2.2. Stress-strain state under the action of external load

In the case of tension of the plate along the OX axis at the edge $r = R_1$ by the normal load p (Figure 2), additional stresses are introduced into the plate. It can be assumed that at the edge points $r = R_1$, the stress field in the plate is determined in Cartesian coordinates by the components $\sigma_x = p$; $\sigma_y = \tau_{xy} = 0$, and in polar coordinates according to the formulas:

$$\begin{cases} \sigma_{r2} = 0{,}5p(1+\cos 2\theta); \\ \tau_{\theta 2} = -0{,}5p\sin 2\theta. \end{cases}$$

The insert is considered to be non-deformable or absolutely rigid and firmly connected to the plate along the line of their connection. So the radial and tangential displacements of the points of the insertion edge $r = R$ are equal to zero. The expressions for the boundary conditions are as follows: on the contour $r = R$ – $u_2(R,\theta) = 0$ and $v_2(R,\theta) = 0$; on the contour $r = R_1$ – $\sigma_{r2}(R_1,\theta) = 0{,}5p(1+\cos 2\theta)$ and $\tau_{\theta 2}(R_1,\theta) = -0{,}5p\sin 2\theta$.

Boundary conditions on the contour $r = R$ will be written as

$$\begin{cases} (R,\theta) = 0; \\ v(R,\theta) = 0. \end{cases} \tag{11}$$

Boundary conditions on the contour $r = R_1$ will be written as

$$\begin{cases} \sigma_r(R_1,\theta) = 0.5(\Delta_1 + \Delta_2) + (\Delta_1 - \Delta_2)\cos 2\theta; \\ \tau(R_1,\theta) = -(\Delta_1 - \Delta_2)\sin 2\theta. \end{cases} \tag{12}$$

Expressions for stresses by analogy with the welded insert are taken as (2), and for displacements as (1), given that the boundary conditions contain only terms with numbers n=0 and n=2. Substituting these expressions into boundary conditions, we determine the unknown coefficients.

Considering these coefficients in expression (2), we obtain the following formulas for determining stresses under uniaxial tension of the circular plate having an absolutely rigid insert firmly connected to the plate

$$\sigma_{r3} = \frac{pR_1^2}{2[(1-\mu)R^2 + (1+\mu)R_1^2]}\left(1+\mu+\frac{(1-\mu)R^2}{r^2}\right) + \\ + \left\{\frac{p}{2} + C\left[\frac{(3-\mu)R^6 + (1+\mu)R_1^4[3(R^2 - R_1^2) - R_1^2]}{R_1^2\, 2[(3-\mu)R^6 - (1+\mu)R_1^6]} + \right.\right. \\ \left.\left. + \left(\frac{3R_1^2R^2[(3-\mu)R^4 - (1+\mu)R_1^4]}{2r^4[(1+\mu)R_1^6 + (3-\mu)R^6]} + \frac{2}{r^2}\right)\right]\right\}\cos 2\theta. \tag{13}$$

$$\sigma_{\theta 3}=\frac{pR_1^2}{2[(1-\mu)R^2+(1+\mu)R_1^2]}\left(1+\mu-\frac{(1-\mu)R^2}{r^2}\right)$$
$$-\left\{\frac{p}{2}+C\left[\frac{(3-\mu)R^6+(1+\mu)R_1^4[3(R^2-R_1^2)-R_1^2]}{R_1^2\,2[(3-\mu)R^6-(1+\mu)R_1^6]}+\right.\right. \tag{14}$$
$$\left.\left.+\left(\frac{3R_1^2R^2[(3-\mu)R^4-(1+\mu)R_1^4]}{2r^4[(1+\mu)R_1^6+(3-\mu)R^6]}+\frac{6(1+\mu)(R^2-R_1^2)r^2}{[(1+\mu)R_1^6+(3-\mu)R^6]}\right)\right]\right\}\cos 2\theta$$

$$\tau_{r\theta 3}=(-\frac{p}{2}-C\left\{\frac{p}{2}+C\left[\frac{(3-\mu)R^6+(1+\mu)R_1^4[3(R^2-R_1^2)-R_1^2]}{R_1^2\,2[(3-\mu)R^6-(1+\mu)R_1^6]}-\right.\right. \tag{15}$$
$$\left.\left.-\left(\frac{3R_1^2R^2[(3-\mu)R^4-(1+\mu)R_1^4]}{2r^4[(1+\mu)R_1^6+(3-\mu)R^6]}+\frac{3(1+\mu)(R^2-R_1^2)r^2}{[(1+\mu)R_1^6+(3-\mu)R^6]}+\frac{1}{r^2}\right)\right]\right\}\sin 2\theta.$$

where

$$C=pR_1^4R^2(1+\mu)[(1+\mu)R_1^6+(3-\mu)R^6][R^2(1+\mu)(R^2-R_1^2)[R^2(3-\mu)(R_1^4-3R^4)-$$
$$-2R_1^4(3(1+\mu)R_1^2-2\mu R^2)]-[(3-\mu)R^6-(1+\mu)R_1^6][(1+\mu)(R_1^4-3R^4)-$$
$$-2R_1^4][2R_1^2-(1+\mu)R^2]]]^{-1}.$$

When the radius R_1 tends to infinity in the expressions for σ_{r3}, $\sigma_{\theta 3}$ $\tau_{r\theta 3}$ we come to the expressions for the tension stresses of the infinite plate with the rigid circular insert [23].

The stresses in the stretched plate with the welded circular rigid insert are determined by the overlapping of the stresses of the two cases considered above, i.e.

$$\begin{cases}\sigma_{r4}=\sigma_{r1}+\sigma_{r3};\\ \sigma_{\theta 4}=\sigma_{\theta 1}+\sigma_{\theta 3};\\ \tau_{r\theta 4}=\tau_{r\theta 1}+\tau_{r\theta 3}.\end{cases} \tag{16}$$

If the circular plate with the non-deformable insert firmly connected to it is under external pressure p (Figure 2), then the boundary conditions in the problem of determining the stresses in the plate are expressed as follows: on the contour $r=R$ – $u_5(R,\theta)=0$ and $v_5(R,\theta)=0$; on the contour $r=R_1$ – $\sigma_{r5}(R_1,\theta)=-p$ and $\tau_{r5}(R_1,\theta)=0$.

Substituting the expressions for displacements (1) and stresses (2) into these boundary conditions, we obtain that the coefficients, a_2, a_3, a_4, a_5 are zero. Having determined the other two coefficients a_0 and a_1 and substituting them into formulas (2), we calculate the stresses in the plate with non-deformable insert under external pressure:

$$\sigma_{r6}=\frac{p}{(1-\mu)R^2+(1+\mu)R_1^2}\left(\frac{R^2R_1^2}{r^2}-(1+\mu)R_1^2-(2-\mu)R^2\right). \tag{17}$$

$$\sigma_{r3}=-\frac{p}{(1-\mu)R^2+(1+\mu)R_1^2}\left(\frac{R^2R_1^2}{r^2}+(1+\mu)R_1^2+(2-\mu)R^2\right). \tag{18}$$

$$\tau_{r\theta 6}=0. \tag{19}$$

When the external pressure acts on a stretched plate with the welded rigid insert, the stresses in the plate are determined by the formulas

$$\begin{cases} \sigma_{r7} = \sigma_{r4} + \sigma_{r6}; \\ \sigma_{\theta 7} = \sigma_{\theta 4} + \sigma_{\theta 6}; \\ \tau_{r\theta 7} = \tau_{r\theta 6}. \end{cases} \quad (20)$$

3. Calculation results and discussion

Formulas (1)–(20) determine the mathematical basis of the calculation method of the residual stress field in the annular plate with the absolutely rigid concentric insert under tension by external load.

Figure 3 shows the stresses at the boundary of the rigid insert. Figure 4 shows the residual welding stresses in the circular plate with the welded rigid insert. It was assumed that the stresses at the weld boundary reached the yield strength of the plate material. It can be seen that both radial σ_{r1} and tangential stresses $\sigma_{\theta 1}$ decay at a distance equal to two radii from the edge of the hole (r/R=2).

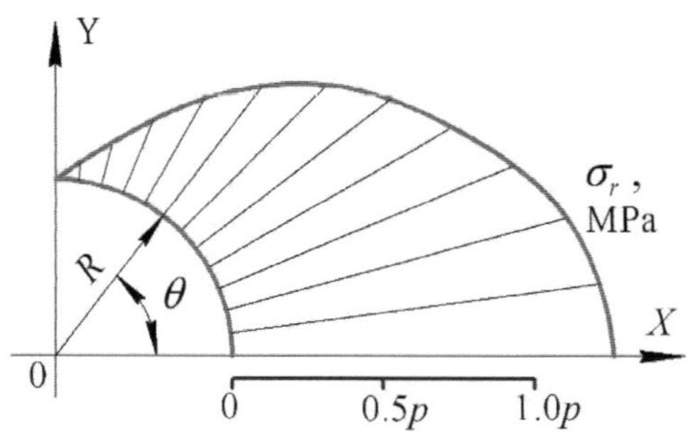

Figure 3. Distribution of radial stresses along the boundary of the rigid insert in the circular plate under uniaxial tension by a uniform load.

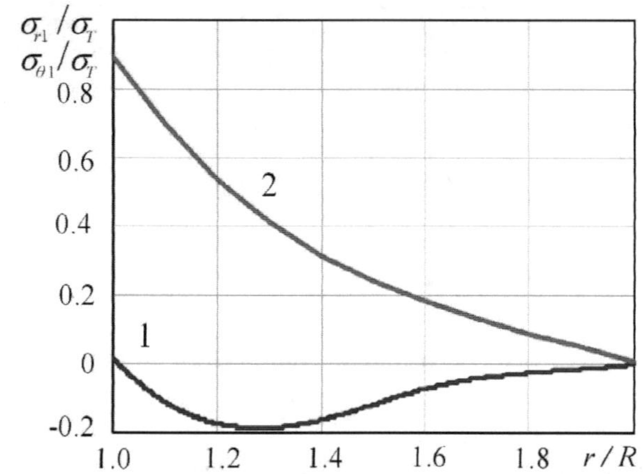

Figure 4. Distribution of residual welding stresses in the circular plate with welded rigid insert: 1 – σ_{r1}/σ_T ; 1 – $\sigma_{\theta 1}/\sigma_T$.

4. Conclusion

It is established that the width of the zone of significant influence of rigid insert on the distribution of residual stresses depends on the size of the insert and does not exceed the diameter of the insert. The field of residual stresses can be changed by the variation of the welding parameters. The ability to determine the residual stresses by calculation can improve the strength characteristics of complex metal structures at the design stage.

References

[1] Qu S Z, Wu X H and Sun Q 2018 *Engineering structures* **165** 457–470
[2] Lu Y, Yao C, Xuan W, Mengyue L and Youzhen Z 2019 *Journal of Constructional Steel Research* **152** 246–260
[3] Jiang J, Zhang J, Liu J, Chiew S P and Lee C K 2018 *Journal of Constructional Steel Research* **151** 238–252
[4] Velikanov N L, Koryagin S I and Sharkov O V 2016 *IOP Conference Series: Materials Science and Engineering* **124** 012094.
[5] Macdonald K 2011 *Fracture and Fatigue of Welded Joints and Structures* (Philadelphia: Woodhead Publishing) p. 360
[6] Xiaoyi L and Tak-Ming C 2019 *Structures* **17** 58–65
[7] Dubrovskiy A, Aliukov S, Dubrovskiy S and Alyukov A 2018 *Engineering Letters* **26(4)** 526–534
[8] Fu Y H, Zhang H O, Wang G L and Wang H F 2017 *Journal of materials processing technology* **250** 220–227

[9] Agyenim-Boateng E, Huang S, Sheng J, Yuan G, Wang Z W and Zhou J Z 2017 *Surface & coatings technology* **328** 44–53
[10] Hicks J 2000 *Welded design. Theory and practice* (Cambridge: Abington Publishing) p. 141
[11] An K, Yuan L, Dial L, Spinelli I, Stoica A D and Gao Y 2017 *Materials & design* **135** 122–132
[12] Gornostajev D, Aryassov G and Penkov I 2016 *International Review of Mechanical Engineering* **10(2)** 115–124
[13] Junaid M, Khan F N , Rahman K and Baig M N 2017 *Optics and laser technology* **97** 405–419
[14] Huang H, Tsutsumi S, Wang J D, Li L Q and Murakawa H 2017 *Finite elements in analysis and design* **135** 1–10
[15] Ramos H M E, Tavares S M O and de Castro P M S T 2018 *Science and Technology of Materials* **30(1)** 6–15
[16] Seo S, Huang E W, Woo W and Lee S Y 2017 *International journal of fatigue* **104** 408–415
[17] Kainuma S, Yang M Y, Jeong Y S, Inokuchi S, Kawabata A and Uchida D 2017 *Engineering failure analysis* **79** 520–537
[18] Sabzi M and Dezfuli S M 2018 *Journal of Manufacturing Processes* **33** 74–85
[19] Yildirim H C 2017 *International journal of fatigue* **101(2)** 408–420
[20] Yukhymets P, Gopkalo A, Zecheru G and Mihovski M 2016 *International journal of offshore and polar engineering* **26(3)** 272–277
[21] Madyira D M, Kumbaa T and Kaymakcia A 2017 *Procedia Manufacturing* **8** 665–672
[22] Vasileiou A N, Smith M C, Balakrishnan J, Francis J A and Hamelin C J 2017 *Nuclear engineering and design* **323** 309–316
[23] Velikanov N L, Koryagin S I and Sharkov O V 2018 *Journal of Physics: Conference Series* **1015** 052028
[24] Maceri A 2010 *Theory of Elasticity* (New York: Springer Science & Business Media) p. 716
[25] Sadd M H 2014 *Elasticity: Theory, Applications, and Numerics* (Amsterdam-Tokyo: Academic Press) p. 600

IOP Conf. Series: Materials Science and Engineering **582** (2019) 012002 doi:10.1088/1757-899X/582/1/012002

Influence of the Thermodynamic Characteristics of the Combustion Process on Vehicle Emissions in Spark-Ignition Engines

N M Smolenskaya[a], V V Smolenskii[b]

Togliatti State University, Togliatti, Russian Federation

E-mail: [a]Nata_smolenskaya@mail.ru, [b]Biktor.cm@mail.ru

Abstract. The paper deals with the influence of the thermodynamic characteristics of the combustion process on toxicity in spark-ignition engines using the example of the UIT-85 single-cylinder unit. The studies were conducted on compressed natural gas and compressed natural gas with the addition of hydrogen gas. Also shown is the relationship of exhaust toxicity with the flame conductivity characteristic, reflecting the intensity of the combustion process in the flame front.

1. Introduction

Thermodynamic characteristics of the combustion process reflect in an integral form all its main manifestations. An analysis of the presence or absence of a relationship between the various thermodynamic characteristics and the main toxic components of the engine exhaust gases shows us how best to influence the workflow to ensure the necessary toxicity reduction. This analysis allows you to better identify those ways of influencing the workflow that will bring the greatest effect of reducing toxicity. This analysis was performed for compressed natural gas and compressed natural gas (CNG) with hydrogen additives. These fuels are currently considered the most likely alternative to liquid hydrocarbon fuels [1,2].

2. Experimental technique

Experimental studies were carried out on a single-cylinder UIT-85 installation (Figure 1a). Include information about the geometric parameters of the engine UIT-85: number of cylinders – 1; working volume – 0.652 liter; compression ratio – 7; diameter of the cylinder – 85 mm; piston stroke – 115 mm; length of connecting rod – 266 mm; rotational speed – 900 rpm; ignition – spark plug, at a fixed ignition advance angle of 13 BTDC for a more visual comparison of the results obtained.

a)

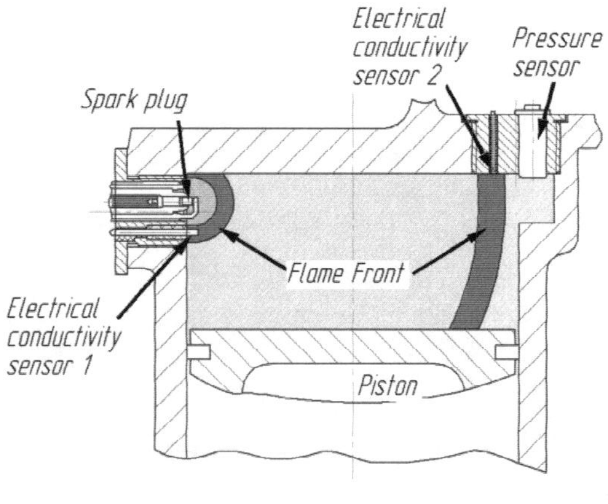

b)

Figure 1. Single-cylinder research unit UIT-85: (a) general view of the installation; (b) combustion chamber with installed sensors.

The portable gas analyzers «META» Avtotest-01.03P and Avtotest-02.03P determined the concentration of toxic components (NO_x, CH and CO) in the exhaust gases. The measurement error on the passport of the gas analyzer is 5%. The pressure in the engine cylinder was measured with a Kistler pressure sensor. Local conditions of propagation of the flame front was determined by means of sensors of ionization.

3. Results and Discussion

The most obvious picture of toxicity is given by characteristics not relative ("ppm" or "%"), but absolute [3]. For example, the amount of toxic components in one four-stroke cycle. Therefore, figures 2 and 3a show the toxicity characteristics in mg / cycle when working on CNG and CNG with hydrogen. The characteristics in Figures 2 and 3a are given to better understand the influence of the thermodynamic characteristics of the combustion process on toxicity in spark ignition engines [4,5].

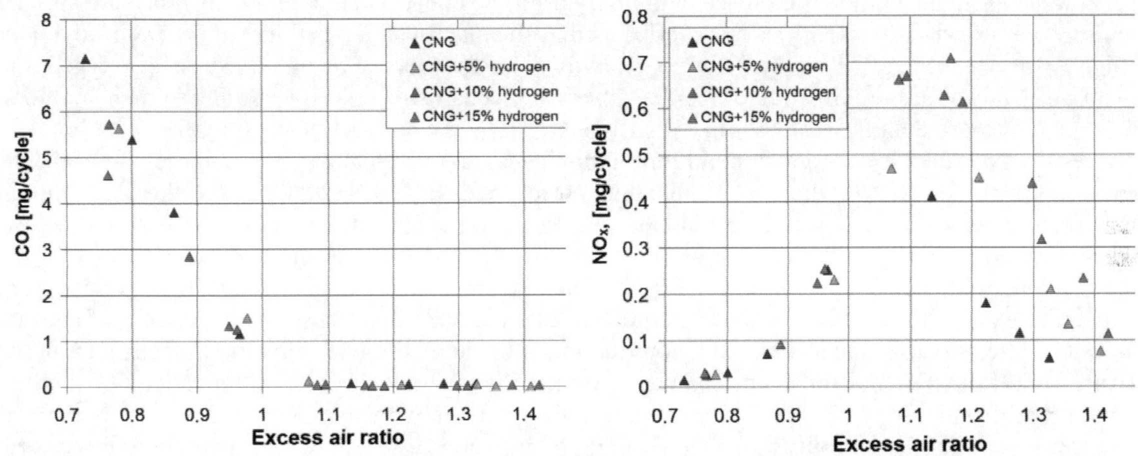

a) b)

Figure 2. The concentration of toxic components in the exhaust gases for UIT-85 when operating on CNG and CNG with hydrogen: (a) carbon monoxide (CO); (b) nitrogen oxides (NO_x).

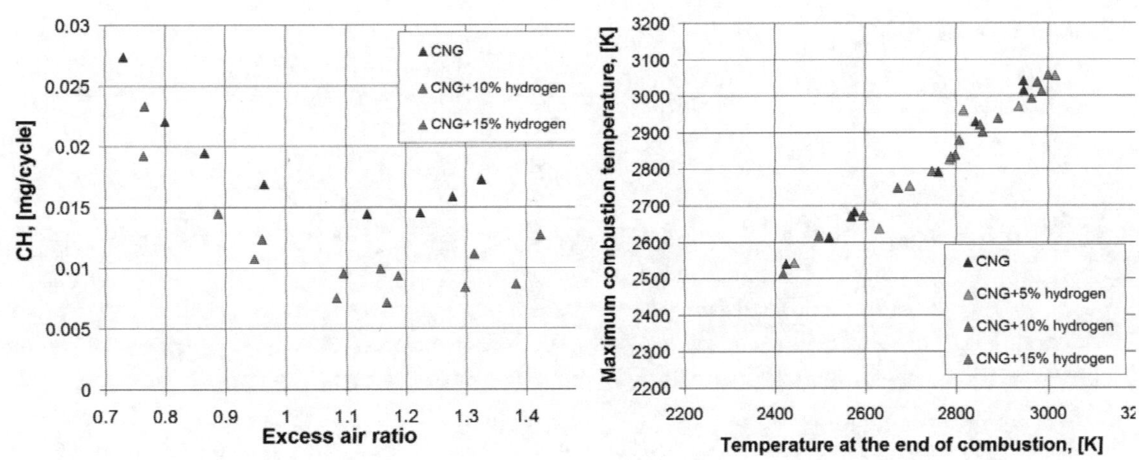

a) b)

Figure 3. Characteristics of toxicity and temperature in UIT-85 when operating on CNG and CNG with hydrogen: (a) unburned hydrocarbons (CH); (b) the relationship of the temperature of the end of combustion with the maximum temperature of the cycle.

Also for clarity, Figure 3b shows the linear relationship between the maximum temperature of the cycle and the temperature at the end of the combustion process. These temperatures reflect two main problem points actively influencing the formation of toxic components in the combustion process. Namely: the maximum temperature reflects the maximum rate of oxidative reactions, and the temperature of the end of combustion reflects the thermodynamic conditions at the completion of active oxidative reactions due to the completion of the combustion process.

When analyzing the experimental data, it was found that both temperatures considered have an impact on the toxicity of exhaust gases. Figures 4 and 5a show graphs for those temperatures whose effect is more linear. Assessing the relationship of carbon monoxide (CO) toxicity with the temperature of the end of combustion (Figure 4a), we note that for mixtures rich in oxygen, the amount of CO is minimal. For mixtures with a small lack of oxygen, an increase in CO toxicity is observed, but it is associated with a lack of oxygen, and no significant temperature effect was detected. The same applies to mixtures with a significant lack of oxygen, where there is an increase in CO toxicity with decreasing temperature. The decrease in temperature occurs due to an excessive decrease in the rate of heat generation when the mixture is enriched with fuel, starting with excess air coefficients less than 0.9. The toxicity of nitrogen oxides (NO_x) is more influenced by the maximum temperature in the engine cylinder (Figure 4b). Here the effect of temperature is obvious. So for mixtures with an excess of oxygen, the main mechanism for the formation of NO_x is a thermal mechanism, which, according to many authors, accounts for more than 90% of NO_x in exhaust gases [6,7]. We see that this assumption also holds in our results. NO_x toxicity has a maximum for mixtures with an excess of oxygen in the region of the air excess factor of 1.05 – 1.1, where the temperature corresponds to 2900 – 2950K. Then there is a drop in temperature, which leads to a decrease in NO_x. At maximum cycle temperatures of 2600K and below, or 2450K and below, a sharp decrease in the effect of temperature on NO_x toxicity is observed for the temperature of the end of combustion [8,9]. This effect suggests that starting from a maximum temperature of 2600K and 2450K for the temperature of the end of combustion, the amount of thermal NO_x decreases sharply, and a significant contribution is made by the NO_x formation mechanism in the flame front (fast NO_x) [9,10]. This mechanism is practically the only one in the combustion of a mixture with a significant lack of oxygen. This is observed for mixtures with an air excess ratio less than 0.94. NO_x formed in the flame front is affected by the rate of chemical reactions and their energy intensity, which is reflected by an increase in NO_x with increasing combustion temperature. It should also be noted that both mechanisms have approximately equally affecting the formation of NO_x in mixtures with a slight lack of oxygen [10]. The formation of NO_x is determined by the temperature characteristics and the amount of oxygen, and the chemical composition of the fuel has no fundamental effect on the formation of NO_x.

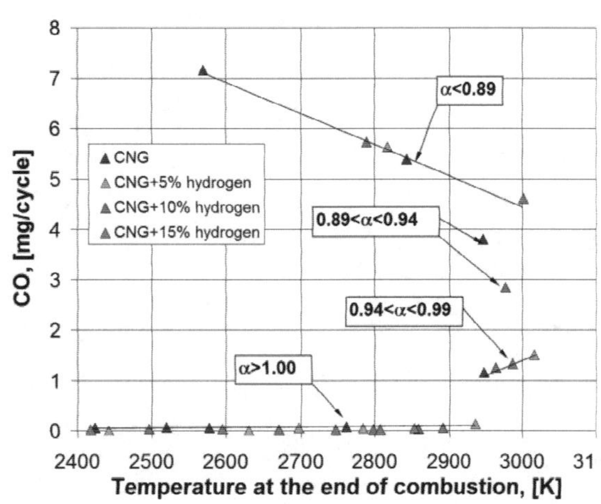

a)

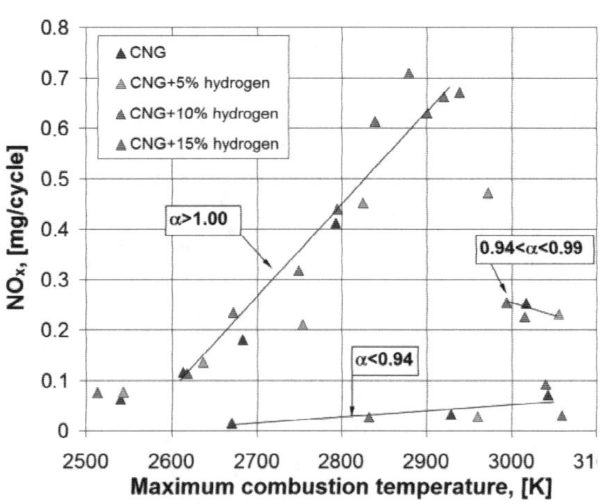

b)

Figure 4. The relationship of temperature with toxic components in the exhaust gases: (a) the connection of carbon monoxide (CO) with the temperature of the end of combustion; (b) the relationship of nitrogen oxides (NO_x) with the maximum temperature cycle.

Consider the effect of temperature on the formation conditions of unburned hydrocarbons (CH) in an engine cylinder (Figure 5a). From figure 5a, it can be seen that CH is practically independent of temperature during combustion. Their content is more dependent on the adequacy of oxygen for complete combustion and reduction of the total amount of burning hydrocarbons due to the replacement of hydrocarbon fuel (natural gas) with hydrogen, which is completely burned, taking oxygen with it. Therefore, for mixtures with a significant lack of oxygen, we cannot see a decrease in toxicity due to a lower CNG concentration with the addition of hydrogen. Considering the mechanisms of formation of CH in the engine cylinder, they mainly emit the mechanism of formation of CH in volume due to lack of oxygen or with a significant reduction in the rate of combustion. This leads to the breakthrough of unburned CH into the zone of combustion products, where oxidation processes are much less active. This mechanism, in addition to the composition of the mixture, is also reflected in the temperature of the combustion process [11]. It is seen that in our case, the combustion temperature exceeds 2400K. Therefore, a significant reduction in the rate of combustion in the studies was not observed. The second mechanism of CH formation is the mechanism associated with the breaking of the burning chains in the vicinity of the walls. Due to the active reduction of the temperature of the mixture at the wall due to heat sink into the wall [11]. The same mechanism takes place when the flame is attenuated in clamped volumes, which is, in our case, the area between the piston and the cylinder to the first compression ring. In our case, the design of the engine in the process of experiments has not changed. Therefore, we will not be able to track the effect of this mechanism. Consider the influence of the area of the walls of the combustion chamber at the moment of the end of combustion on the CH toxicity (Figure 5b). Figure 5b clearly shows the significance of increasing the wall area with an increase in the duration of the combustion process for CH toxicity. Taking into account that when working on mixtures with an excess of oxygen, we have complete combustion in the flame front. Therefore, the main source of CH will be the near-wall frozen layer, an increase in the area of which linearly affects the toxicity of CH. With a lack of oxygen, we also observe a linear relationship between the toxicity of CH and the area of the wall layer at the end of the combustion process. The obtained linear dependences show that CH formed in the volume due to the lack of oxygen is added to CH remaining in the wall layer. The influence of the fuel composition is determined only by the fact that the rate of combustion increases with the addition of hydrogen. This leads to a decrease in the area of the near-wall layer. No other toxicity reduction factors with the addition of hydrogen in CNG were detected during detailed thermodynamic analysis of the combustion process. This shows that although the addition of hydrogen and reduces the concentration of hydrocarbon fuels, in the process of combustion, hydrogen is completely burned taking oxygen. A more slowly burning hydrocarbon fuel, with a lack of oxygen in the mixture, it is also not enough. This leads to the fact that behind the flame front the amount of bulk hydrocarbons is determined by the general lack of oxygen and does not depend on the proportion of hydrogen in the fuel. Thus, we can say that hydrogen as a chemical element does not have an activating effect on the formation of the toxic components in question. Its influence affects the thermodynamic characteristics of the combustion process, which affect the conditions for the formation of toxic components.

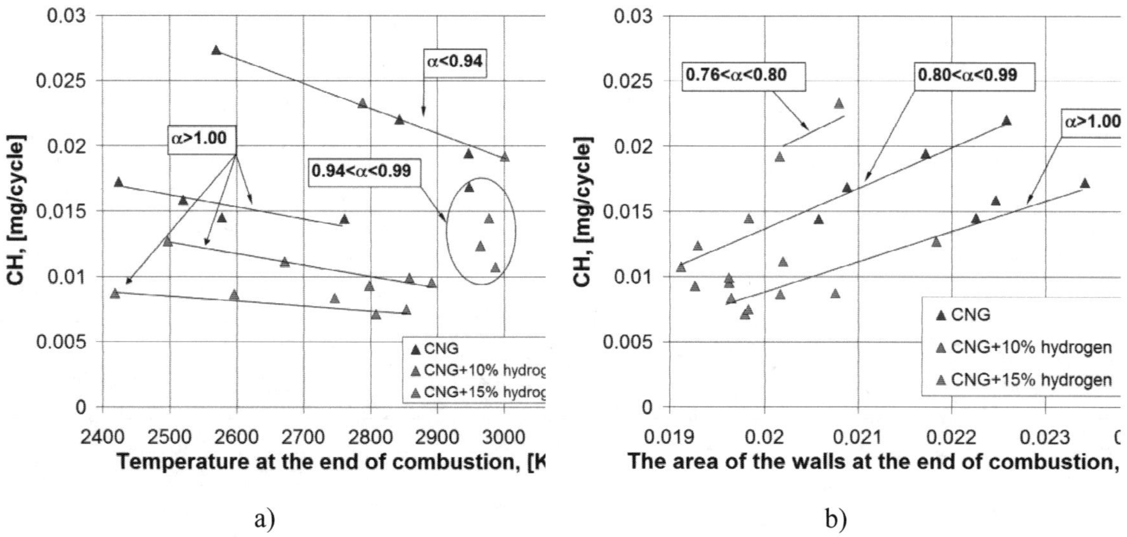

Figure 5. The relationship of unburned hydrocarbons (CH) with the characteristics of the end of combustion: (a) the temperature of the end of combustion; (b) wall area at the end of combustion.

When analyzing the factors affecting the toxicity of CH, the dependence of the amount of CH in the exhaust gases with energy introduced with the fuel into the engine cylinder was detected (figure 6a). The results obtained in Figure 6a correlate well with the conclusions made above. Thus, an increase in the energy introduced into the cylinder with fuel with a lack of oxygen leads to a linear increase in the toxicity of CH. However, there is no separation of the results due to the addition of hydrogen to CNG [12]. This confirms the statement that hydrogen is completely burned, and the entire lack of oxygen falls on CNG. Therefore, the influence of the type of fuel in figure 6a for mixtures with a lack of oxygen was not detected. For mixtures with an excess of oxygen, a decrease in CH toxicity is due to faster burning in a smaller volume, which leads to the obtained results.

In addition to analyzing the influence of the thermodynamic characteristics of the combustion process on the toxicity of exhaust gases, an assessment was made of the possibility of determining the toxicity of exhaust gases of a spark-ignition engine using the flame conductivity characteristics. The main characteristic of the electrical conductivity of the flame is the magnitude of the amplitude of the ion current signal. The greatest correlation was shown by the amplitude of the ion current from the sensor installed 80 mm from the spark plug. In Figure 1b, it is designated as a electrical conductivity sensor 2. The best correlation of electrical conductivity on the sensor remote from the spark plug is explained by the fact that it is the characteristics of the completion of combustion that largely determine the toxicity of exhaust gases. Therefore, the comparison is made exactly by the electrical conductivity sensor 2. A comparison of the toxicity characteristics with the magnitude of the ion current signal amplitude is shown in Figures 6a, 7a, and 7b. Let us begin the analysis of the relationship between toxicity and electrical conductivity from Figure 6a, which shows CH toxicity. It may be noted that there is a clear correlation between CH toxicity and the amplitude of the ion current signal for mixtures that do not have an oxygen deficiency. At the same time, combustion with a lack of oxygen occurs at high rates of combustion, which leads to an increase in the ion current. But oxygen deficiency increases the amount of bulk CH, thereby making monitoring of CH toxicity over the signal amplitude for these compounds not possible.

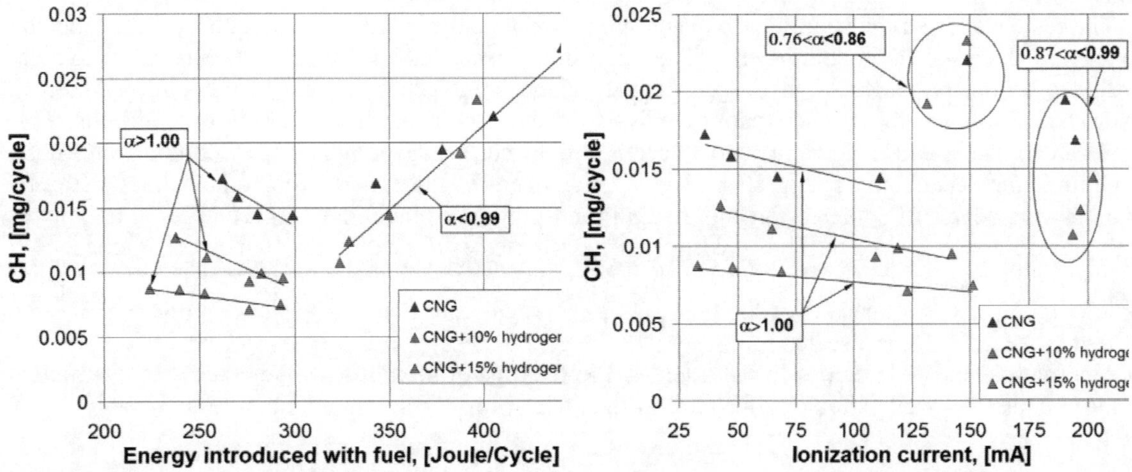

a) b)

Figure 6. The relationship of CH toxicity to the characteristics of the combustion process: (a) the relationship with the energy introduced into the cylinder with the fuel; (b) a relationship with the electrical conductivity of the flame at the end of the main phase of the combustion process.

For the association of NO_x toxicity with the amplitude of the ion current signal (Figure 7a), a greater correlation is observed than with CH toxicity. So there is a clear correlation in the zone of mixtures with an excess of oxygen, where the thermal mechanism of NO_x formation dominates. The characteristics of NO_x toxicity in the zone of mixtures with a significant lack of oxygen are also well correlated, where the mechanism of NO_x formation in the flame front is the main one. In the zone of excess air coefficients of 0.94 – 0.99, the electrical conductivity of the flame is at its maximum and varies slightly with the addition of hydrogen to CNG, which makes it difficult to assess toxicity in this range for all toxic components studied.

Considering the connection of CO toxicity with the amplitude of the ion current signal (Figure 7b), we can note the possibility of correlation only for mixtures with air excess factors less than 0.94. Since in the zone of excess air coefficient of 0.94 – 0.99, the correlation is difficult for the previously noted reasons. And in the zone of mixtures with an air excess factor of more than 1, CO toxicity has minimal values and varies by the level of gas analyzer error. Thus, an analysis of CO toxicity by flame conduction has no practical meaning.

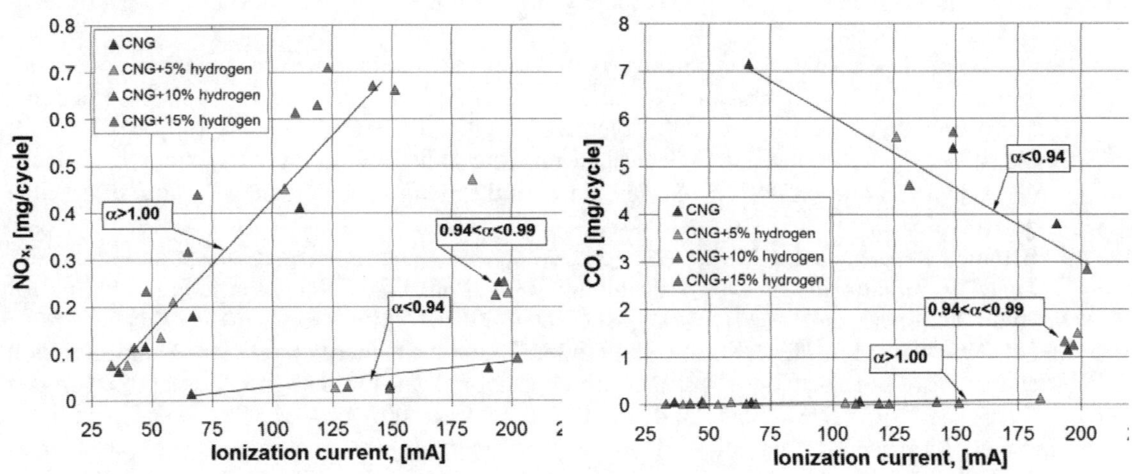

a) b)

Figure 7. Relation of toxicity with flame electrical conductivity characteristics at the end of the main phase of the combustion process: (a) connection with nitrogen oxides (NO_x); (b) association with carbon monoxide (CO).

The results showed the significance of thermodynamic analysis to assess the potential for reducing toxicity of the main toxic components. The analysis showed that the perceived decrease in toxicity attributed to the properties of hydrogen activating the combustion process is not entirely correct. Hydrogen, as shown by thermodynamic analysis, leads to a change in combustion conditions. This results in an increase in temperature and a decrease in the surface layer at the end of combustion. At the same time, the data obtained suggests that having achieved the same thermodynamic characteristics by changing the degree of compression and the ignition timing, we obtain the same changes in toxicity as with the addition of hydrogen. Thus, hydrogen affects toxicity only within the framework of changes in the thermodynamic characteristics of the combustion process and is not a catalyst for the formation of toxic components. An analysis of the possibilities of regulating toxicity by the amplitude of the ion current in the flame front revealed some regularities. So, by the amplitude of the ion current, it is possible to effectively evaluate toxicities for NO_x and CH in the area of mixtures with an excess air coefficient of more than 1, as well as toxicity for NO_x in the area of mixtures with an air excess factor less than 0.94.

4. Conclusion

1. The effect of temperature in the combustion process on the mechanisms of the formation of NO_x and CO is shown, taking into account the ratio of air to fuel.

2. It was found that by temperature and wall area at the end of the completion of combustion, it is possible to evaluate CH toxicity, taking into account the air-to-fuel ratio.

3. It was revealed that hydrogen has a significant effect on the thermodynamic characteristics of the combustion process, thereby affecting the engine's toxicity for CH, NO_x and CO. In this case, the effect of replacing the main hydrocarbon fuel to reduce toxicity as a consequence of a decrease in the concentration of hydrocarbon fuel in the engine cylinder was not detected. On the contrary, it is shown that the addition of hydrogen in CNG under the same thermodynamic conditions does not affect the toxicity of CH in the parietal layer or in the concentration of CH behind the flame front with the same lack of oxygen.

4. An assessment of the possibility of determining toxic components by the amplitude of the ion current showed the possibility of this approach in assessing toxicity of NO_x and CH in the area of mixtures with an excess air coefficient of more than 1, as well as toxicity of NO_x in the area of mixtures with an excess air factor of less than 0.94.

References

[1] Wallington T. J., Kaiser E. W., Farrell J. T. Automotive fuels and internal combustion engines: a chemical perspective, The royal society of chemistry, 35, 2006. pp. 335–347. DOI: 10.1039/B410469M

[2] Das L., Gulati R., Gupta P. A comparative evaluation of the performance characteristics of a spark ignition engine using hydrogen and compressed natural gas as alternative fuels, International Journal of Hydrogen Energy. 2000; 25(8): 783-93 doi:10.1016/S0360-3199(99)00103-2

[3] Smolenskaya N.M., Smolenskii V.V. Influence of the Type of Fuel on the Toxicity of Exhaust Gases in SI Engines, IOP Conf. Series: Earth and Environmental Science 224 (2019) 012044 doi:10.1088/1755-1315/224/1/012044

[4] Mohammed S.E., Baharom M.B., Aziz A.R. Analysis of engine characteristics and emissions fueled by in-situ mixing of small amount of hydrogen in CNG, International Journal of Hydrogen Energy 36(6) (2011) 4029-4037. DOI: 10.1016/j.ijhydene.2010.12.065}

[5] Ma, F. Wang Y., Liu H. Q., Li Y. Experimental study on thermal efficiency and emission characteristics of a lean burn hydrogen enriched natural gas, International Journal of Hydrogen Energy 32 (2007) 5067–5075. DOI: 10.1016/j.ijhydene.2007.07.048}

[6] Unich A., Morrone B., Mariani A. The impact of natural gas – hydrogen blends on internal combustion engines performances and emissions, SAE paper. 2009. 2009-24-0102. DOI: 10.4271/2009-24-0102

[7] Futko, S.I. Analysis of Nitrogen-Oxide Formation Mechanisms in Filtration Combustion of Methane–Air Mixtures, Combustion, Explosion, and Shock Waves (2003) 39: 627. https://doi.org/10.1023/B:CESW.0000007674.81766.e9

[8] Bowman C.T. (2000) Gas-Phase Reaction Mechanisms for Nitrogen Oxide Formation and Removal in Combustion. In: Vovelle C. (eds) Pollutants from Combustion. NATO Science Series (Series C: Mathematical and Physical Sciences), vol 547. Springer, Dordrecht. DOI https://doi.org/10.1007/978-94-011-4249-6_7

[9] Merryman E. L., Levy A. Nitrogen oxide formation in flames: The roles of NO2 and fuel nitrogen, Symposium (International) on Combustion, Volume 15, Issue 1, 1975, Pages 1073-1083. https://doi.org/10.1016/S0082-0784(75)80372-9

[10] England G.C., Heap M.P., Pershing D.W. et al. Mechanisms of NOx formation and control: Alternative and petroleum-derived liquid fuels, Symposium (International) on Combustion, Volume 18, Issue 1, 1981, Pages 163-174. https://doi.org/10.1016/S0082-0784(81)80021-5

[11] Matula R.A. (1973) Mechanism of Hydrocarbon Formation in Combustion Processes. In: Springer G.S., Patterson D.J. (eds) Engine Emissions. Springer, Boston, MA. **DOI** https://doi.org/10.1007/978-1-4684-1983-2_3

[12] Dmitrenko, Y.M., Zhdanok, S.A., Klyovan, R.A. et al. Methane-to-hydrogen conversion in the filtration-combustion wave of rich methane-air mixtures, J Eng Phys Thermophys (2007) 80: 304. https://doi.org/10.1007/s10891-007-0040-x

IOP Conf. Series: Materials Science and Engineering **582** (2019) 012003 doi:10.1088/1757-899X/582/1/012003

Methods for Assessing the Thermodynamic Characteristics of the Combustion Process Using the Indicator Diagram in Spark-Ignition Engines

V V Smolenskii[a], N M Smolenskaya[b]

Togliatti State University, Togliatti, Russian Federation

E-mail: [a]Biktor.cm@mail.ru, [b]Nata_smolenskaya@mail.ru

Abstract. The paper deals with the evaluation of the thermodynamic characteristics of the combustion process according to the experimentally obtained indicator diagram of pressure in spark ignition engines. The proposed method allows to determine the main characteristics of the combustion process (change in internal energy, heat generation characteristic, speed and intensity of combustion processes in the flame front, volume of the burned mixture, volume of unburned mixture) using the experimental indicator diagram of pressure obtained for a spark ignition engine.

1. Introduction

The indicator diagram of the pressure in the engine cylinder carries a significant amount of information about the course of thermal processes in compression, combustion and expansion [1, 2]. The indicator diagram is an integral characteristic of thermal processes. Her detailed analysis allows to determine the following local parameters of the combustion process: the average temperature of the burned and unburned mixtures, the volume of the burned and unburned mixtures, the average propagation velocity of the flame front at each moment of the combustion process, the flame front area, the intensity of the combustion processes in the flame front, the average flame front thickness, heat dissipation characteristic [3]. All these data allow us to better assess the impact of various engine operating parameters on environmental and effective performance. It should also be noted the complexity of the calculations of the combustion process in modern software products with a significant proportion of the dilution of the fuel-air mixture with residual gases that have an unstable chemical composition. Therefore, at present, in addition to the use of modern software products for the calculation of internal combustion engines, experimental studies are also actively used [4]. In such studies, research on engine performance at low loads and idling play a large role. These characteristics significantly affect the performance of the car in a city. In this regard, an increase in reliable data when analyzing the results of experimentally obtained indicator pressure diagrams for engine operation modes with significant dilution with residual gases is an important task of scientific research.

2. Theoretical analysis

The combustion chamber is a thermodynamic system that is limited to the cylinder head, valves, cylinder liner and piston. There are losses of thermal energy through the walls, which can be determined by the equations proposed by Voshni [5]. Since the temperatures in the combustion zone, when operating at idle and low loads, barely exceed 2000 K, there are practically no dissociation effects. Pressure on these modes usually does not exceed 4 – 5 MPa. This allows us to consider the leakage through the piston rings insignificant and neglected in the calculations.

The indicator diagram of pressure from the first law of thermodynamics, you can determine the amount of heat released during combustion in the flame front. The total character of which is a characteristic of heat dissipation and describes the energy release due to combustion (1):

$$\Delta Q_C = \Delta U - \Delta Q_{walls} + p \cdot \Delta V \quad (1)$$

where: ΔQ_C – the amount of heat released in the combustion process; ΔU – change the internal energy of the system; ΔQ_{walls} – loss of heat energy through the walls; $p \cdot \Delta V$ – thermodynamic system operation.

In equation (1), heat losses to the cooling system through walls are calculated in accordance with the equation proposed by Voshni [5]. The operation of the thermodynamic system is determined by the indicator diagram and the characteristic motion of the piston for each section of the indicator diagram. The size of the section is adopted so that the pressure change from the beginning to the end of the section can be considered straight, in our case it is 0.2724 degrees of the angle of rotation. Earlier, the authors obtained a method for estimating the internal energy from the parameters of the indicator diagram. The method allows to more reliably determine this parameter under conditions of dilution with residual gases, including during idling and low load [6]

$$\Delta U = \frac{(C_{V1} + C_{V2})}{2} \cdot \frac{(V_1 + V_2)}{2} \cdot \frac{(P_2 - P_1)}{R_\mu} \cdot \frac{2}{[Z_r(p,T)_1 + Z_r(p,T)_2]} \quad (2)$$

where: C_{V1} and C_{V2} – average molar heat capacities with a constant volume at the beginning and at the end of the considered section; V_1 and V_2 – volumes of the thermodynamic system at the beginning and at the end of the considered section; P_1 and P_2 – pressure on the indicator diagram at the beginning and at the end of the section under consideration; R_μ = 8,314 J/(mol·K) – the universal gas constant; $Z_r(p,T)_1$ and $Z_r(p,T)_2$ – compressibility factors of real gas at the beginning and at the end of the considered section.

Also, the amount of heat released in the flame front is determined from the following expression (3):

$$\Delta Q_C = H_u \cdot \Delta m_{\text{combustion}}, \quad (3)$$

where: H_u – net calorific value of the fuel; $\Delta m_{\text{combustion}}$ – mass of burnt fuel in the area under study.

Therefore, the proportion of burnt mixture in the area is determined from the expression (4):

$$\Delta \chi = \frac{\Delta m_{\text{combustion}}}{m_{\text{fuel}} - m_{\text{incomplete combustion}}}, \quad (4)$$

where: m_{fuel} – the mass of fuel entering the engine cylinder; $m_{\text{incomplete combustion}}$ – mass of fuel not participating in the combustion process due to lack of oxygen when working on rich fuel-air mixtures.

Thus, we can determine the heat release characteristic from the expression (5):

$$\chi = \sum \Delta \chi, \quad (5)$$

As a result, we have all the necessary data regarding the thermodynamic system in which combustion takes place. The thermodynamic system, where the combustion process takes place, can be divided into several zones. The dual-zone approach is often used to simulate combustion [7,8]. The combustion chamber is divided into a hot zone with a burnt fuel-air mixture and a cold zone with an unburned fuel air mixture. These zones are separated by a thin flame front (middle line passing through the middle of a real flame front) in which combustion and heat release occurs (Figure 1). It is also assumed that inside the combustion chamber the pressure propagates at the speed of sound, and the rate of combustion is significantly lower than it. Therefore, the pressure in the burned and unburned mixtures is considered the same [9].

We present the equations for the two-zone model of the combustion process. The equations of state for the two-zone combustion process model (6, 7):

$$p \cdot V_{\text{burn}} = m_{\text{burn}} \cdot R_{\text{burn}} \cdot T_{\text{burn}}, \quad (6)$$

$$p \cdot V_{\text{unburn.}} = m_{\text{unburn.}} \cdot R_{\text{unburn.}} \cdot T_{\text{unburn.}}, \quad (7)$$

where: p – pressure in the burned and unburned mixtures; V_{burn} and $V_{\text{unburn.}}$ – volumes of burned and unburned mixtures; m_{burn} and $m_{\text{unburn.}}$ – mass of burned and unburned mixtures; R_{burn} and $R_{\text{unburn.}}$ –

gas constants for burned and unburned mixtures; T_{burn} and $T_{\text{unburn.}}$ – temperature burned and unburned mixtures.

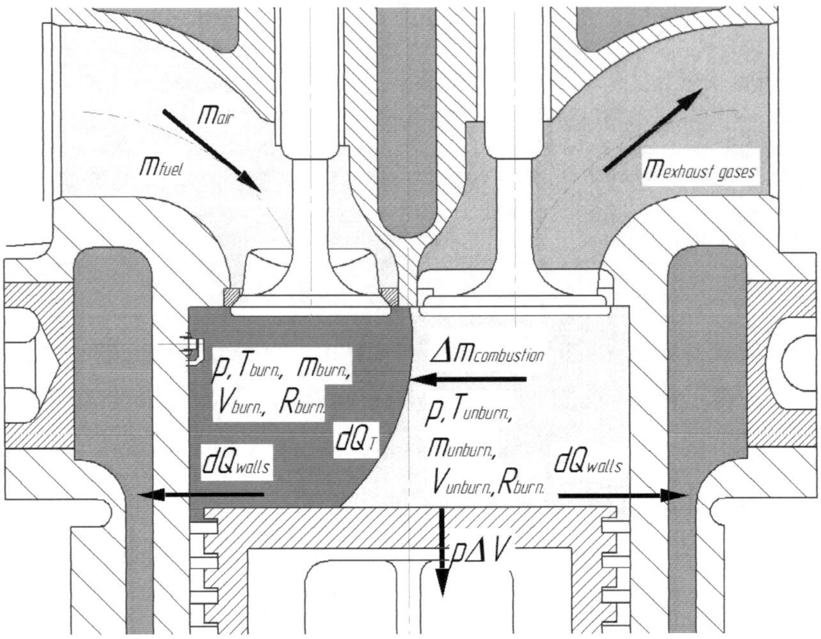

Figure 1. Two-zone simulation of the combustion chamber system

Express the mass of the burned and unburned mixtures through the mass of the working mixture (mixture of fuel, air, residual gases) and the characteristics of heat generation:

$$m_{\text{burn}} = m_{\text{work}} \cdot \chi_i ; \tag{8}$$

$$m_{\text{unburn.}} = m_{\text{work}} \cdot (1 - \chi_i) . \tag{9}$$

where: χ_i – current value of heat dissipation characteristic.

Express from equations (6) and (7) the temperature of the burned and unburned mixtures:

$$T_{\text{unburn.}} = \frac{p \cdot V_{\text{unburn.}}}{m_{\text{unburn.}} \cdot R_{\text{unburn.}}} = \frac{p \cdot V_{\text{unburn.}}}{m_{\text{work}} \cdot (1 - \chi_i) \cdot R_{\text{unburn.}}} ; \tag{10}$$

$$T_{\text{burn}} = \frac{p \cdot V_{\text{burn}}}{m_{\text{burn}} \cdot R_{\text{burn}}} = \frac{p \cdot V_{\text{burn}}}{m_{\text{work}} \cdot \chi_i \cdot R_{\text{burn}}} . \tag{11}$$

Suppose that the temperature of the thermodynamic system in the engine cylinder can be determined through the calculated expression for the temperature of the gas mixture without taking into account compressibility. Since it is already assumed that the pressure in the burned and unburned part of the combustion chamber is the same. Express the mass of burned and unburned mixtures through the mass of the working mixture as it is done in equations (8) and (9):

$$T = \frac{\frac{m_{\text{burn}}}{m_{\text{work}}} \cdot C_{\text{burn}} \cdot T_{\text{burn}} + \frac{m_{\text{unburn.}}}{m_{\text{work}}} \cdot C_{\text{unburn.}} \cdot T_{\text{unburn.}}}{\frac{m_{\text{burn}}}{m_{\text{work}}} \cdot C_{\text{burn}} + \frac{m_{\text{unburn.}}}{m_{\text{work}}} \cdot C_{\text{unburn.}}} = \frac{\chi_i \cdot C_{\text{burn}} \cdot T_{\text{burn}} + (1 - \chi_i) \cdot C_{\text{unburn.}} \cdot T_{\text{unburn.}}}{\chi_i \cdot C_{\text{burn}} + (1 - \chi_i) \cdot C_{\text{unburn.}}} . \tag{12}$$

where: C_{burn} and $C_{\text{unburn.}}$ – average molar heat capacity with a constant volume of burned and unburned mixtures.

The average molar heat capacity at a constant volume of the burned and unburned mixtures is determined by the well-known thermophysical formulas.

Express from equation (12) the temperature of the burned and unburned mixtures, taking into account expressions (10) and (11):

$$T_{\text{burn}} = \frac{T \cdot \left(\chi_i \cdot C_{\text{burn}} + (1-\chi_i) \cdot C_{\text{unburn.}}\right) - C_{\text{unburn.}} \cdot \dfrac{p \cdot V_{\text{unburn.}}}{m_{\text{work}} \cdot R_{\text{unburn.}}}}{\chi_i \cdot C_{\text{burn}}} ; \quad (13)$$

$$T_{\text{unburn.}} = \frac{T \cdot \left(\chi_i \cdot C_{\text{burn}} + (1-\chi_i) \cdot C_{\text{unburn.}}\right) - C_{\text{burn}} \cdot \dfrac{p \cdot V_{\text{burn}}}{m_{\text{work}} \cdot R_{\text{burn}}}}{(1-\chi_i) \cdot C_{\text{unburn.}}} . \quad (14)$$

The volume of the thermodynamic system in the engine cylinder at the current time is determined by the expression (15):

$$V_i = V_{\text{comb. chamb.}} + \frac{\pi \cdot D^2}{4} S_i ; \quad (15)$$

where: $V_{\text{comb. chamb.}}$ – volume of the combustion chamber; D – cylinder diameter; S_i – piston stroke relative to the top dead center.

Assuming that the zones are separated by a thin flame front (middle line passing through the middle of the real flame front), we can express the volume of the thermodynamic system through the volumes of burned (16) and unburned (17) mixtures:

$$V_{\text{burn}} = V_i - V_{\text{unburn.}} , \quad (16)$$

$$V_{\text{unburn.}} = V_i - V_{\text{burn}} . \quad (17)$$

Let us express from the equations (6) and (7) the volumes of the burned and unburned mixtures taking into account (8) and (9) respectively:

$$V_{\text{burn}} = \frac{m_{\text{work}} \cdot \chi_i \cdot R_{\text{burn}} \cdot T_{\text{burn}}}{p} , \quad (18)$$

$$V_{\text{unburn.}} = \frac{m_{\text{work}} \cdot (1-\chi_i) \cdot R_{\text{unburn.}} \cdot T_{\text{unburn.}}}{p} . \quad (19)$$

Consider the solution of expressions (18) and (19) separately. Let us substitute into expression (18) the formula (13) in which we express the volume through expression (16):

$$V_{\text{burn}} = \frac{m_{\text{work}} \cdot \chi_i \cdot R_{\text{burn}} \cdot \dfrac{T \cdot \left(\chi_i \cdot C_{\text{burn}} + (1-\chi_i) \cdot C_{\text{unburn.}}\right) - C_{\text{unburn.}} \cdot \dfrac{p \cdot [V_i - V_{\text{burn}}]}{m_{\text{work}} \cdot R_{\text{unburn.}}}}{\chi_i \cdot C_{\text{burn}}}}{p} \quad (20)$$

In formula (20) one unknown quantity is the volume of the burned mixture. Simplify the formula (20) and expressing from it the volume of the burnt mixture, we obtain the calculation formula (21):

$$V_{\text{burn}} = \frac{m_{\text{work}} \cdot R_{\text{burn}} \cdot R_{\text{unburn.}} \cdot T \cdot \left(\chi_i \cdot C_{\text{burn}} + (1-\chi_i) \cdot C_{\text{unburn.}}\right) - p \cdot R_{\text{burn}} \cdot C_{\text{unburn.}} \cdot V_i}{p \cdot \left(R_{\text{unburn.}} \cdot C_{\text{burn}} - R_{\text{burn}} \cdot C_{\text{unburn.}}\right)} \quad (21)$$

Let's carry out similar actions with the formula (19) to determine the volume of the unburned mixture and get the calculation formula (22)

$$V_{\text{unburn.}} = \frac{m_{\text{work}} \cdot R_{\text{unburn.}} \cdot R_{\text{burn}} \cdot T \cdot \left(\chi_i \cdot C_{\text{burn}} + (1-\chi_i) \cdot C_{\text{unburn.}}\right) - p \cdot R_{\text{unburn.}} \cdot C_{\text{burn}} \cdot V_i}{p \cdot \left(R_{\text{burn}} \cdot C_{\text{unburn.}} - C_{\text{burn}} \cdot R_{\text{unburn.}}\right)} \quad (22)$$

Determining by the formula (21) or (22) the volume of the burned or unburned mixture, we can determine all the remaining volumes by the formulas (16) and (17). And using formulas (10) and (11), we can calculate the temperature of the zone of the burned mixture and the zone of the unburned mixture.

By formulas (8) and (9) we know the mass of the zone of the burned mixture and the mass of the zone of the unburned mixture.

3. Experimental technique

Experimental studies were carried out on a single-cylinder UIT-85 installation. Include information about the geometric parameters of the engine UIT-85: number of cylinders – 1; working volume – 0.652 l; compression ratio – 7; diameter of the cylinder – 85 mm; piston stroke – 115 mm; length of connecting rod – 266 mm; rotational speed – 900 rpm; ignition – spark plug, at a fixed ignition advance angle of 13 BTDC for a more visual comparison of the results obtained.

4. Results and Discussion

Let us give an example of the analysis of the pressure indicator diagram obtained on the UIT-85 single-cylinder research unit using the proposed method. As an example, the mode of operation of the plant on compressed natural gas is chosen. Natural gas is fed through a calibrated nozzle into the intake manifold. The mixture in the engine cylinder is significantly diluted with residual gases, due to the low compression ratio and valve overlap. This allows you to evaluate the performance of the proposed method on the mode of operation of a single-cylinder unit simulating engine operation at low loads.

Figure 2 shows the graphs of the obtained thermodynamic characteristics of the combustion process according to the indicator diagram in the UIT-85. For example, the mode of operation on compressed natural gas is selected with an excess air ratio of 1.005 and a residual gas ratio of 0.85 [3]. The results obtained in Figure 2 show the applicability of the proposed method for analyzing the thermodynamic characteristics of the combustion process according to the experimentally obtained indicator diagram of pressure in spark ignition engines. The obtained characteristics of changes in the volume of burned and unburned mixtures are explained by the shape of the UIT-85 combustion chamber. In Figure 2a it can be seen that the active increase in the volume of the burnt mixture starts at 370 degrees BTDC. And at 395 degrees BTDC the activity of combustion and the growth of the volume of the burned mixture is reduced. This is confirmed by the above parameters of the characteristics of heat generation and changes in the volume of burned and unburned mixtures in Figure 2b. An analysis of the shape of the pressure indicator chart with the characteristics of changes in the volume and mass of the burned mixture (Figure 2a and 2b) shows the accuracy of the above calculation.

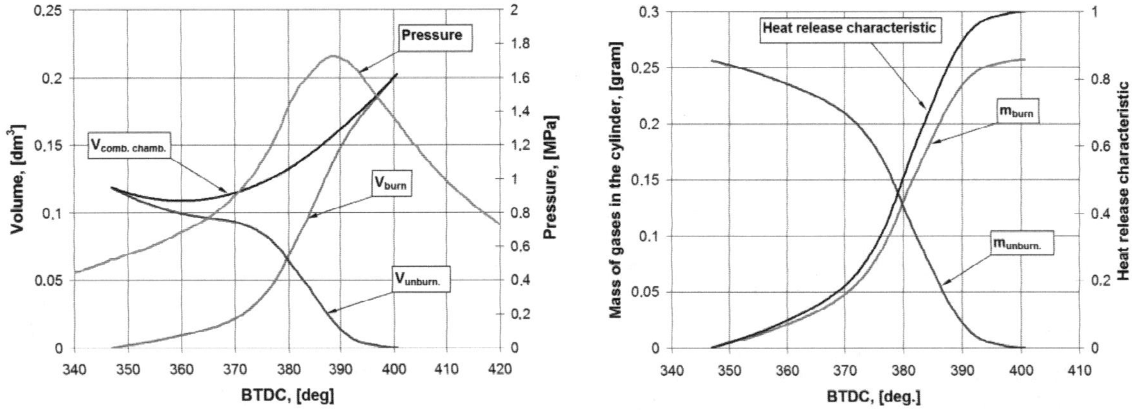

a) b)

Figure 2. Graphics thermodynamic characteristics of the combustion process depending on the angle of rotation of the crankshaft: (a) indicator pressure, volume of the combustion chamber, volumes of burned and unburned mixtures; (b) the characteristic of heat generation is given, and the change is the mass of the zone of the burnt and unburned mixtures

Having obtained the necessary data on the volume change during the combustion process, one can proceed to the analysis of the change in the velocity of propagation of the flame front. To do this, we construct a three-dimensional model of the thermodynamic system under consideration, which is a combustion chamber with a moving piston. Using the 3D model, it is possible to determine, for each moment of time under consideration, the position of the spherical flame front (Figure 3a). The propagation parameters of the spherical flame front, taking into account the position and volume of the burned mixture, were determined in increments of 5 degrees BTDC. The results are shown in Figure 3b. Figure 3b shows the characteristics of the average speed of propagation of a spherical flame front from the moment of the spark discharge to the corresponding position of the flame front to the angle. The speed was considered from the change in the radius of the spherical flame front. Also shown in Figure 3b is the change in speed between the test sites, and two points are shown, obtained from experimental data. These points correspond to the appearance of a signal on the ionization sensors. The first ionization sensor is mounted 7 mm from the spark plug electrode. The second ionization sensor is installed 80 mm from the spark plug electrode.

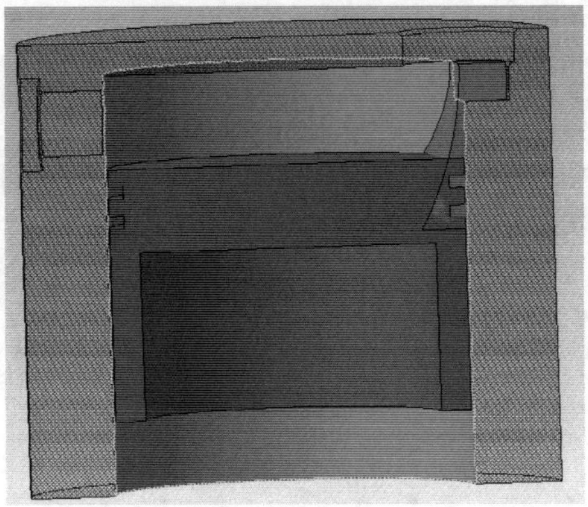

a)

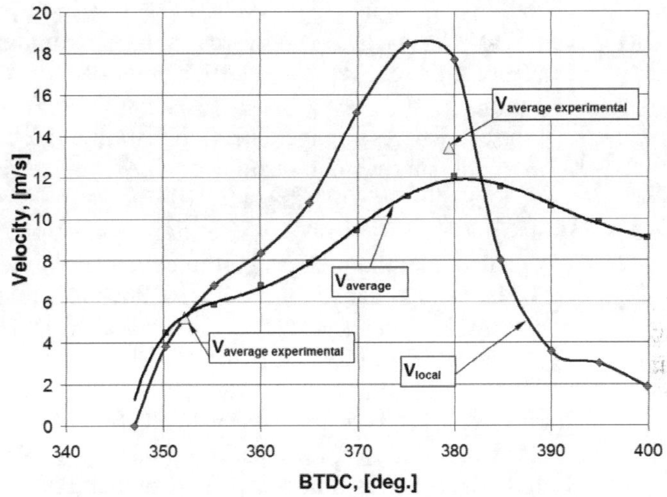

b)

Figure 3. Analysis of the change in the speed of propagation of the flame front: (a) a three-dimensional model of the combustion chamber, cylinder and piston UIT-85 to determine the parameters of the combustion process; (b) Characteristics of the change in the flame propagation velocity of the flame front.

The results obtained correspond to the general picture of the combustion process and are fairly well correlated with the experimental data obtained for the initial combustion phase (first ionization sensor) and the beginning of the final combustion phase (second ionization sensor).

Therefore, assuming that on average the flame front extends from the spark plug spherically, you can determine the exact coordinates of its front border. This will determine the speed of propagation of the flame front. You can also determine the flame front area, from where, knowing the current mass burning rate and the average density of the unburned mixture, you can determine the average flame front thickness at a given time [10, 11, 12].

The presented results make it possible to simplify the process of analyzing the thermodynamic parameters of the combustion process, on operating conditions with increased emissions of toxic components. This may allow a fresh look at the possibilities of optimizing the workflow for the creation of modern low-toxic engines.

5. Conclusion

1. The possibility of evaluating the main characteristics of the combustion process is shown by analyzing the indicator diagram of pressure in the cylinder of a spark-ignited engine.

2. The proposed method allows to determine the main thermodynamic characteristics of the combustion process according to experimentally obtained indicator diagram of pressure, including when the engine is running at idle and low loads. In these modes there is a significant amount of residual gases that significantly complicate the modeling process for these conditions, which requires the mandatory use of bench experimental studies. The proposed method allows to expand the understanding of the

conditions for the formation of toxic substances in spark ignition internal combustion engines. This is necessary to identify opportunities to improve the environmental performance of the engine at low loads and idling.

References

[1] Li Y., Jia M., Chang Y., Kokjohn S.L., Reitz R.D. Thermodynamic energy and exergy analysis of three different engine combustion regimes, Applied Energy, 180 (2016) 849-858. DOI: 10.1016/j.apenergy.2016.08.038.

[2] Zhao Z., Wang S., Zhang S., Zhang F. Thermodynamic and energy saving benefits of hydraulic free-piston engines, Energy, 102 (2016) 650-659. DOI: 10.1016/j.energy.2016.02.018.

[3] Teh K.-Y, Miller S L , Edwards C F, Thermodynamic requirements for maximum internal combustion engine cycle efficiency. Part 1: Optimal combustion strategy, International Journal of Engine Research, 9(6) (2008) 449-465. Information on https://www.researchgate.net/publication/245395230_Thermodynamic_requirements_for_maximum_internal_combustion_engine_cycle_efficiency_Part_1_Optimal_combustion_strategy DOI: 10.1243/14680874JER01508.

[4] Maurya, R.K.; Mishra, P. Parametric investigation on combustion and emissions characteristics of a dual fuel (natural gas port injection and diesel pilot injection) engine using 0-D SRM and 3D CFD approach. Fuel 2017, 210, 900–913. https://doi.org/10.1016/j.fuel.2017.09.021

[5] Warrier G. R., Dhir V. K. Heat Transfer and Wall Heat Flux Partitioning During Subcooled Flow Nucleate Boiling, Journal of Heat Transfer 128(12). 2006. p.1243-1256. DOI: 10.1115/1.2349510

[6] Smolenskaya N.M., Smolenskii V.V., Korneev N.V. Increase in the thermodynamic efficiency of the working process of spark-ignited engines on natural gas with the addition of hydrogen, IOP Conf. Series: Earth and Environmental Science 121 (2018) 052009 doi :10.1088/1755-1315/121/5/052009.

[7] Hongsheng Liu, Maozhao Xie, Dan Wu Simulation of a porous medium (PM) engine using a two-zone combustion model, Applied Thermal Engineering, Volume 29, Issues 14–15, October 2009, Pages 3189-3197. https://doi.org/10.1016/j.applthermaleng.2009.04.021

[8] La Xiang, Enzhe Song and Yu Ding, A Two-Zone Combustion Model for Knocking Prediction of Marine Natural Gas SI Engines, Energies 2018, 11, 561; doi:10.3390/en11030561

[9] Rakopoulos, C.D.; Rakopoulos, D.C.; Kyritsis, D.C. Development and validation of a comprehensive two-zone model for combustion and emissions formation in a DI diesel engine. Int. J. Energy Res. 2003, 27, 1221–1249. https://doi.org/10.1002/er.939

[10] Aziz, A.R.; Anbese, Y.T.; Hagos, F.Y.; Heikal, M.R. Characteristics of Early Flame Development in a Direct-Injection Spark-Ignition CNG Engine Fitted with a Variable Swirl Control Valve. Energies 2017, 10, 964. https://doi.org/10.3390/en10070964

[11] Mahmood, H.A.; Adam, N.M.; Sahari, B.B.; Masuri, S.U. New Design of a CNG-H2-AIR Mixer for Internal Combustion Engines: An Experimental and Numerical Study. Energies 2017, 10, 1373. https://doi.org/10.3390/en10091373

[12] Schießl, R.; Maas, U. Analysis of endgas temperature fluctuations in an SI engine by laser-induced fluorescence. Combust. Flame 2003, 133, 19–27. https://doi.org/10.1016/S0010-2180(02)00538-2

IOP Conf. Series: Materials Science and Engineering **582** (2019) 012004 doi:10.1088/1757-899X/582/1/012004

Development of Methods for Assessing the Operational Effectiveness of Sets of Personal Protective Equipment for Fighters and Rescue Units

M Rodicheva[1,a], A Abramov[1,b], E Surzhenko [2,c], E Gneusheva[1,d], T Vetrova[1,d]

[1] Orel State University named after I.S.Turgenev, Naugorskoe shosse 29, Orel, Russian Federation, 302020

[2]Saint Petersburg State University of Industrial Technologies and Design, Bolshaya Morskaya 18, Saint Petersburg, Russian Federation, 191186

E-mail: [a] rodicheva.unpk@gmail.com, [b] Ant-lin88@mail.ru, [c] esurzh@mail.ru, [d] emign@yandex.ru

Abstract. The article discusses modern approaches to the design of special clothing for the technical support of rescue operations. Some problems of using modern thermoregulating textile materials have been analyzed. The author's method of estimating heat and mass transfer in packages of textile materials is considered, some research results are presented.

1. Introduction. The composition of the technical support of rescue operations includes sets of personal protective equipment for fighters, the formation of which must take into account the requirements for the level of ergonomic properties. Among them is the ability to protect the human body during changing cooling effect. Introducing products with "active" heat-shielding properties into a clothing set can meet this requirement [15]. Clothing with "active" heat-shielding properties is developed based on smart thermoregulating textiles. Such materials change their hygroscopicity and total thermal resistance during changes of intensity of the cooling effect of the environment due to the introduction nano-fillers or electronic components into fibers structure [13].

Nano-fillers of modern textile canvases are introduced into the structure of fibers. With a decrease in the cooling effect, these fillers melt, absorbing the heat corresponding to the phase transition, and when cooled, they polymerize, releasing the previously accumulated heat. Figure 1 shows the cut of the fiber "Outlast" with heat-accumulating fillers thermocules [1].

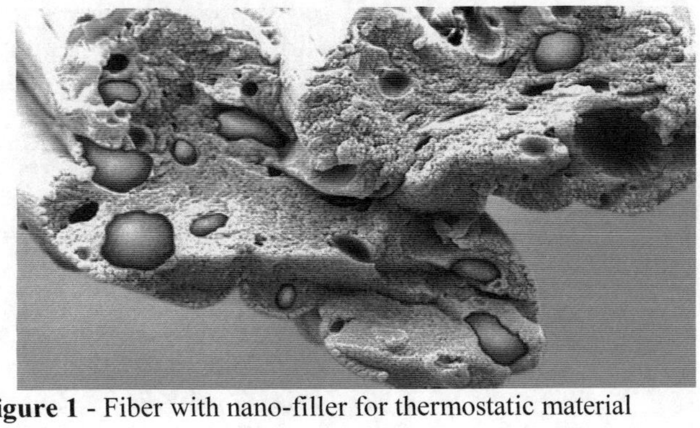

Figure 1 - Fiber with nano-filler for thermostatic material

When using textile materials with electronic components, in the structure of a set of clothes, the transmission of signals and thermal energy is carried out. Integrated microcircuits into the structure of the canvas measure the parameters of microclimate under clothes and, if necessary, transmit control signals to the electronic heating components, thereby adjusting the parameters of the microclimate under clothes [2].

For the introduction of conductive elements into the structure of the textile package were proposed (Figure 2):

- methods of applying conductors to the surface of the fibers (figure 2 a) [5, 8, 16];
- methods for integrating conductors into the structure of yarns (Figure 2b, 2c) [11];
- methods of integrating conductors into the structure of textiles (figure 2 g) [2, 4, 6, 10-12].

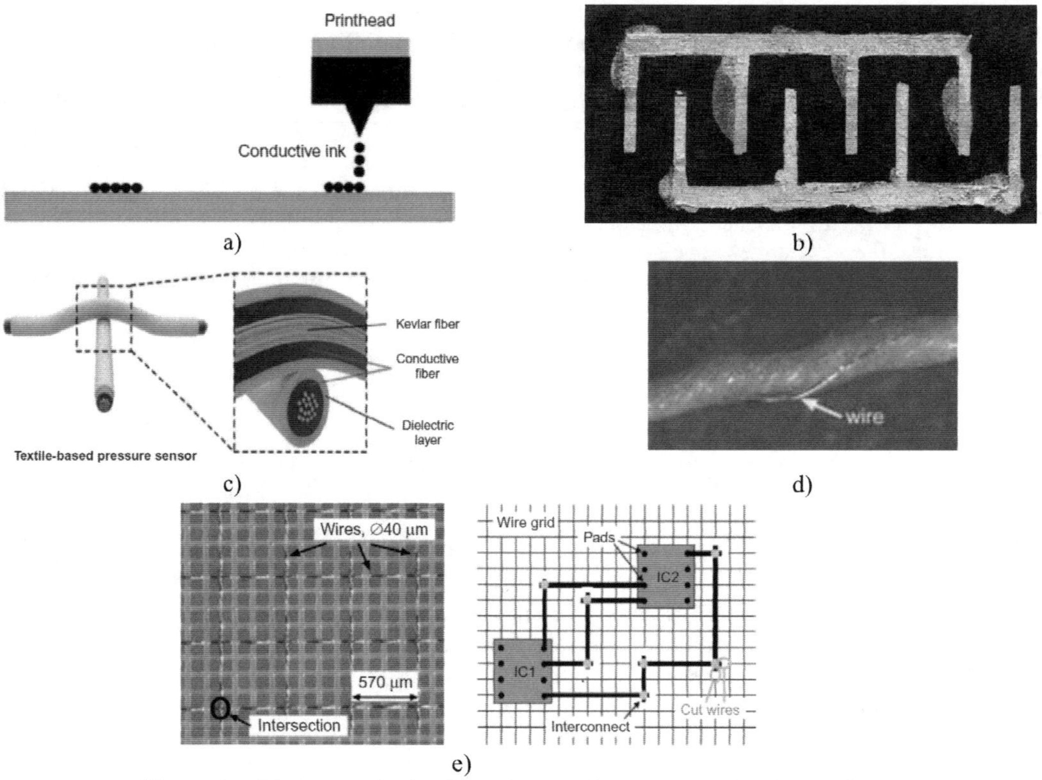

Figure 2 – Modern methods of creating textiles with portable electronics

On the basis of numerous "temperature-controlled" materials, clothing systems are created, which consisted of a limited number of products that can be worn by a person in different sequences. Replacing one product in a set with another allows you to significantly change the properties of the entire set of personal protective equipment. Thus, a limited set of products allows you to create a set of special clothes for the effective protection of a fighter in various climatic conditions.

This approach became widespread in the development of numerous systems of materials, in particular, the Extended Cold Weather Clothing System III (ECWS) for the US Army, which is based on a large amount of physiological research. By summarizing their results, specialists have developed fairly simple and intuitive graphic diagrams illustrating the principles of the formation of clothing sets depending on the degree of physical activity and operational features. An example of such a scheme is shown on Figure 3.

The development of such schemes for the creating personal protective equipment for fighters of rescue units is an urgent task. However, taking into account the specifics of the rescuers' tasks, the mechanical transfer of recommendations developed for the ECWS material system can lead to serious design errors. The use of sufficiently laborious and costly methods of physiological research in this case is also not difficult. The selection of a system of materials, as well as the development of recommendations for their acquisition are based on the results of experimental studies. In this regard, the urgent task is to develop adequate and affordable methods for studying the operational effectiveness of textile package, including "thermostatically controlled", for personal protective equipment for emergency rescue equipment.

Experience shows, the tasks of forming sets of clothing for rescuers are often solved using television and calorimetric methods of research [9, 10]. The result of such studies are: the distribution of the temperature field on the surface of a set of clothes or indirect parameters, such as the TRF-indicator of the quality of materials. For example, the distribution of the temperature field on the surface of a set of traditional and modern materials is presented on Figure 4 [9].

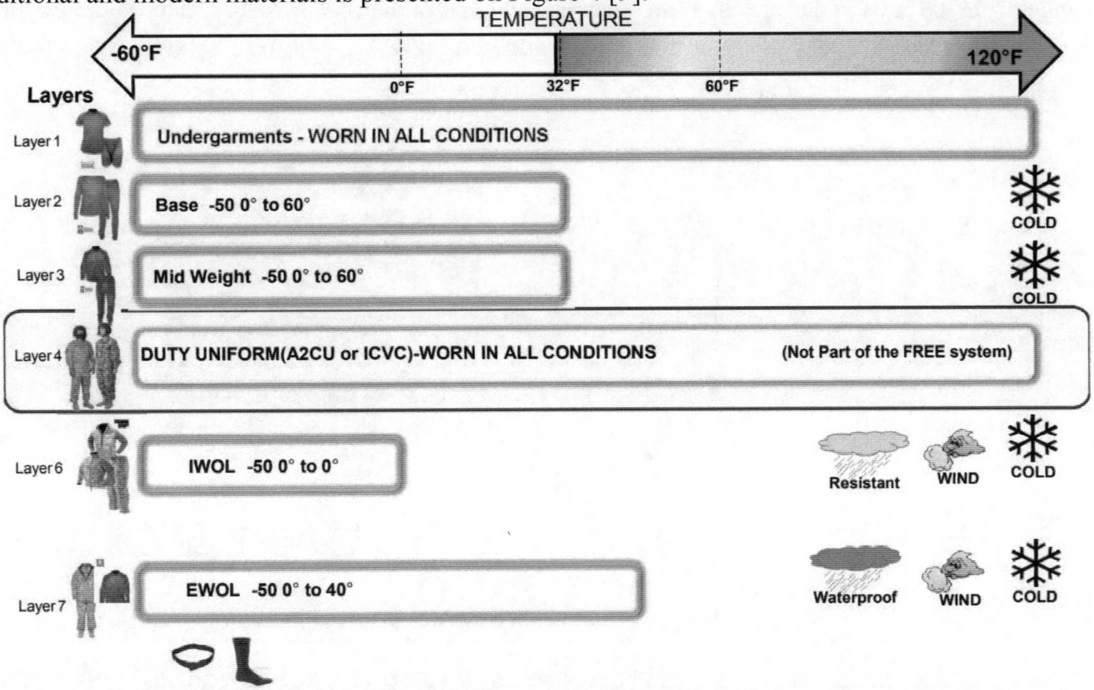

Figure 3 – An example of a graphic scheme with recommendations for the formation a set of clothes based on the ECWS material system

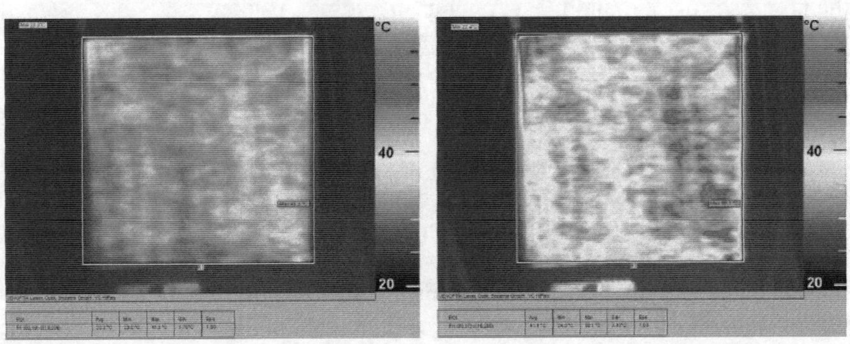

a) b)

Figure 4 – Some results of television studies of the surface of sleeves made of traditional and thermoregulatory textile

The main disadvantage of these and similar methods is the impossibility of a quantitative assessment of the dynamics of the temperature field and heat flows in the structure and on the surface of clothing packages.

2. Formulation of the problem. An urgent task during designing of personal protective equipment for fighters of special units is the development of methods for the experimental evaluation of the operational effectiveness of sets of clothes with inner thermal regulatory control, which should:

- ensure the possibility of accurate quantitative assessment of indicators of the properties of innovative materials;

- allow to investigate the influence of the original properties of "thermostatic" materials on the operational efficiency of clothing;
- to ensure the replication of the test bench.

To date, a sufficient theoretical apparatus has been developed that makes it possible to develop methods for researching the thermal properties of innovative materials that meet these requirements.

3. Theoretical part. It is known that the human body can be approximated by a system of embedded cylinders. On the basis of this presentation, numerous models of thermal processes in the "man-clothing-environment" system have been proposed. For example in Figure 5, a shows the representation of the human body . To maintain the constancy of temperature the mechanisms for changing the level of heat production and sweating should be implemented in the body element.

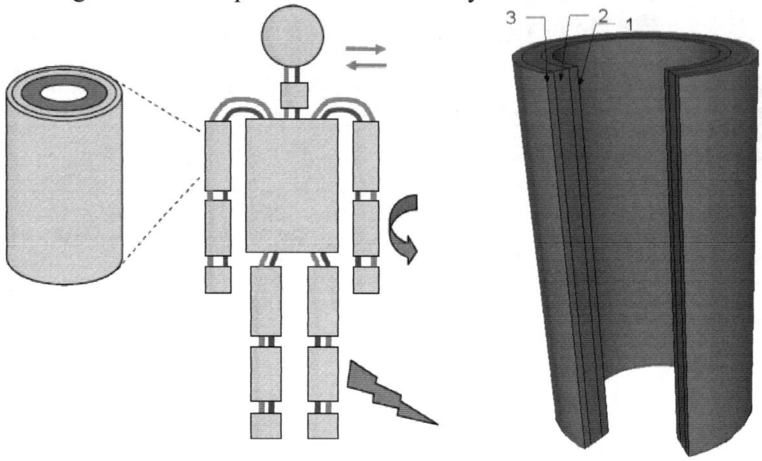

a) b)

Figure 5 – Theoretical bases of representation of the person's body and a package of materials

The package of clothes in this case can be replaced by a set of nested cylindrical elements (Figure 5, b) [12]. Depending on the degree of freedom of the clothes, air gaps can form between the layers or the contact between the layers of clothing can be quite dense.

To study the processes of heat and mass transfer in a textile package, it is necessary to simulate the operating conditions under which heat and moisture flows are dispersed through the structure of a package of clothes into the environment. For this, the authors propose an experimental setup, presented in Figure 6.

The installation contains:

1. The physical model of the body element (Figure 6, a) in the form of a cylinder filled with water (pos. 1, Figure 6, b). Thermal control functions are implemented by an automated system (pos. 4, 5), which controls the temperature of the cylinder by means of a meter (pos. 2) and, if necessary, controls the operation of the electric heater (pos. 3);
2. Moisturizing device (Figure 6, c), which allows you to supply model moisture on the surface of the element of the human body.

During the experiment, the package of materials is located on the surface of the thermal model of segment of a human body (pos. 1, figure 6, d). A distributor (pos. 2) was installed at the upper end of the model, and an air intake (pos. 3) with temperature, velocity (pos. 4) and humidity (pos. 5) transducers were placed above it.

In the process of research, various conditions of heat and mass transfer are reproduced, which are observed in the "man-clothing-environment" system during rescue operations. The study is conducted during five phases, in each phase evaluates the dynamics of the heat flow density (Figure 7, a) and moisture flows (Figure 7, b) in the structure and on the surface of the textile package (Figure 7) [12].

In the first phase, the conditions of heat exchange of the body element with the environment are modeled with a moderate level of physical exertion, when the effect of sweat on the intensity of heat and mass transfer can be neglected. In the second phase, a sharp increase in physical exertion is reproduced, at which the device that moistens the package of materials due to sweat is activated. For this, moisture is supplied to the surface of the element of the human body. During this phase, the heat flux density (q_{pack}, W/m^2) in the package and the value of its moisture content (u_{pack}, g/g) increase. Quantitative values of these parameters characterize the ability of a textile package to absorb moisture, while maintaining heat-shielding properties.

By the beginning of the third phase of the experiment, stabilization of heat and moisture fluxes in the structure of the package of materials is observed, which indicates their saturation.

In the fourth phase (drying phase) the moisturizing device is turned off, which simulates the conditions for reducing the physical load of the fighter. Under these conditions, an important indicator is the intensity of moisture removal from the structures of the textile package, which results in a decrease in the duration of the fourth phase of the experiment. By the beginning of the fifth phase of the experiment, the conditions of stationary heat exchange are restored. The dynamics of heat and moisture flows during the five phases of the experiment illustrates the intensity of heat and mass transfer processes in a textile package for heat-shielding clothing.

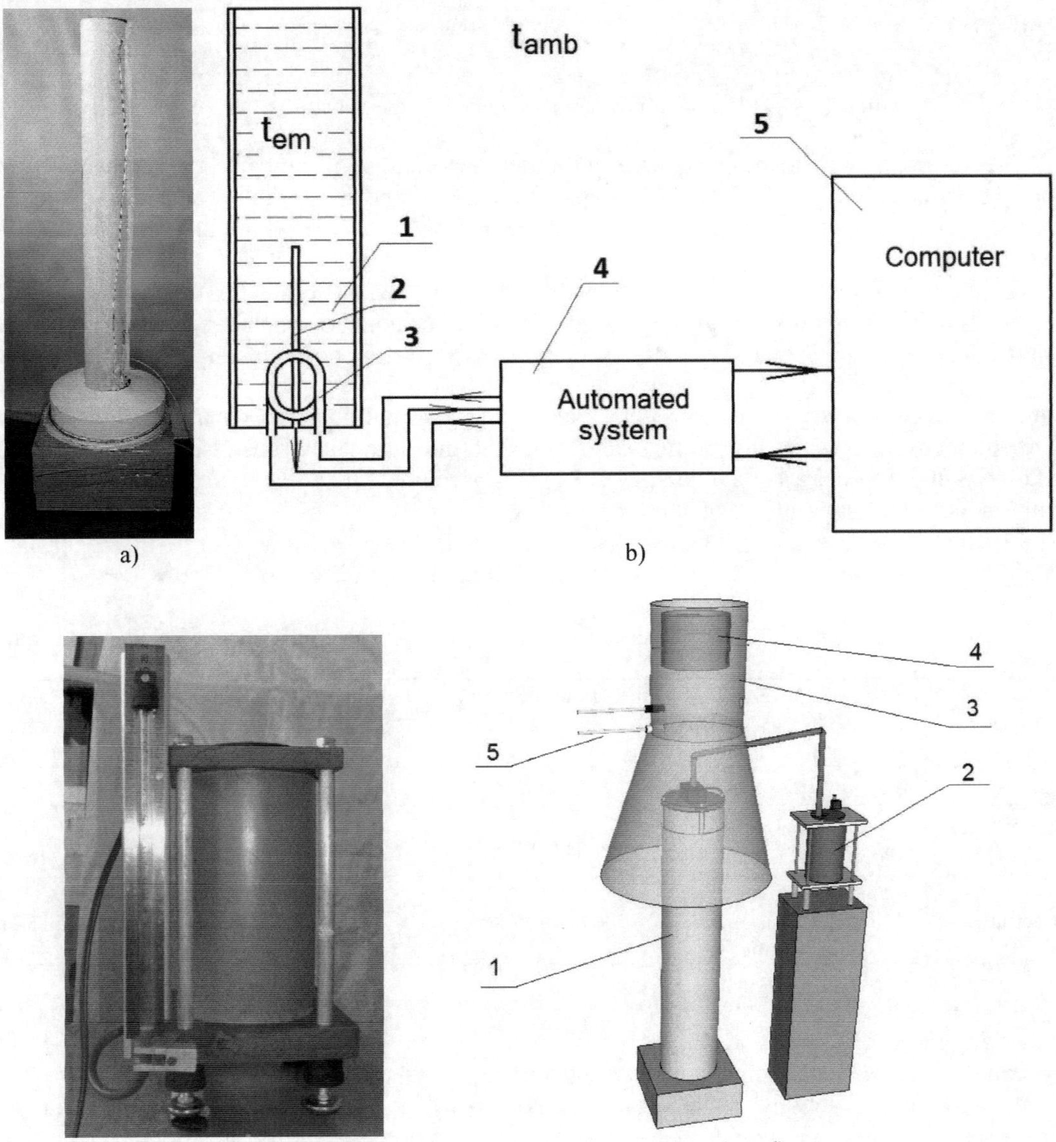

a) b)

c) d)

Figure 6 – The scheme of the experimental setup

IOP Conf. Series: Materials Science and Engineering **582** (2019) 012004 doi:10.1088/1757-899X/582/1/012004

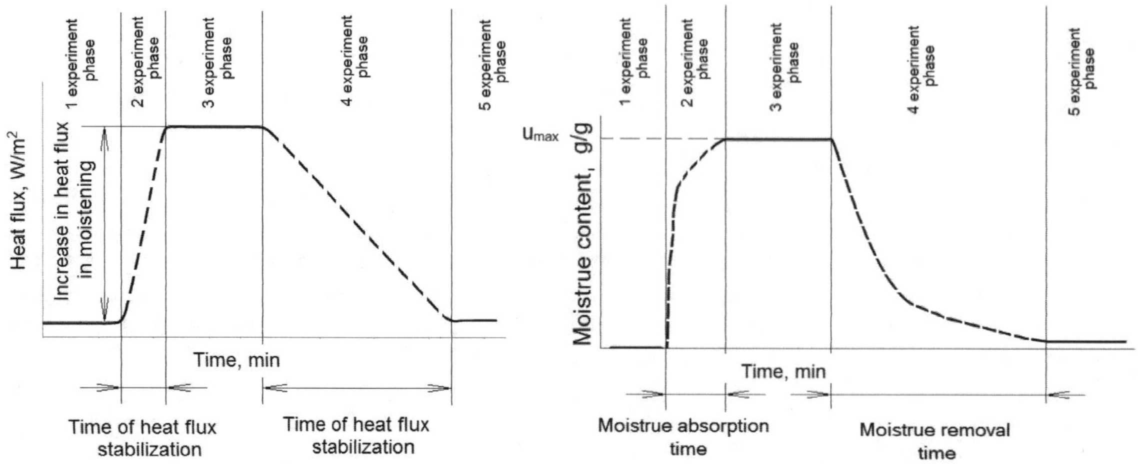

a) b)

Figure 7 – Heat flow and moisture flow curves in the structure of the textile package

The results obtained are compared with the parameters of human heat production when performing different work (q_{cell}, W/m^2). Discrepancies between the density of the heat flow in the textile package and heat production allows us to estimate the thermal state of a person under the conditions of operation.

Comparing the results of research obtained for different variants of the textile package, as well as these results with the data of physiological studies, you can make graphical diagrams illustrating recommendations for the formation of a set of clothes based on the chosen system of materials.

4. Some research results. Let us consider an example of results obtained for a textile package with "active" protective properties, including samples of "thermostatic" materials (Table 1). Now clothing based on this and similar variants of textile packages is used in the manufacture of clothing for outdoor activities at ambient temperatures of about t = + 10 °C.

The results of the research will be compared with similar results obtained for textile packages that are traditionally used for the production of special clothing of fighters of rescue teams.

Table 1. The composition of the package "temperature-controlled" materials of clothing for outdoor activities

Textile package's layer	Type of fabric	Fiber composition	Density, g/m^2	Thickness, мм	Weave structure
Underwear	Woven fabric «Outlast»	90% cotton	240	0.6	Plain
Interim layer	Woven fabric «Polartech Power Dry»	100% polyamide	360	2.4	Purl-Stretched
Inner layer	Fabric laminated to reactive polymer membrane	100% polyester	430	0.45	Rip-Stop

During conducting experimental studies, the conditions of human heat transfer were simulated with the following levels of physical activity:

- medium level: the value of human heat production q_{people} = 180 W/m^2, sweating is absent;
- increased physical activity: human heat production q_{people} = 350 ÷ 500 W/m^2, active sweating;
- decrease in physical activity: decrease in heat production to q_{cell} = 180 W/m^2, cessation of sweating.

The results of the study are presented in Figure 8.

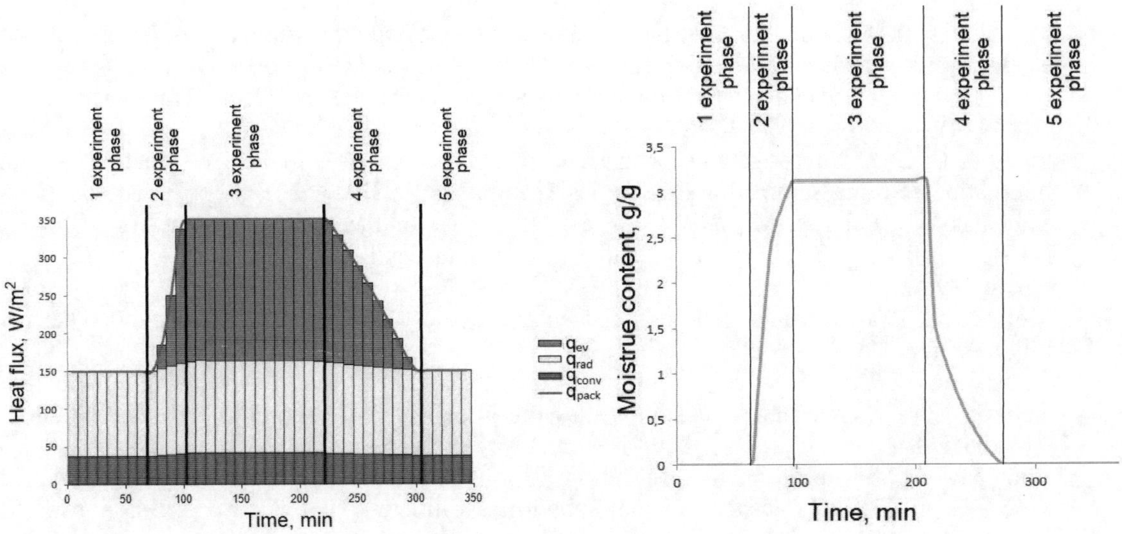

a) b)

Figure 8 The heat flow curve in the material package structure according the Table 1

In the first phase of the experiment, the heat flux density in the textile package is q_{1f} = 150 W/m^2, which allows maintaining a person's heat balance with an average level of physical activity. According to the results of preliminary studies, in the packages of traditional materials of similar configuration, the thickness and weight of which is higher, this indicator is q_{1f}= 160 ÷ 190 W/m^2. This testifies to the positive effect of "thermostatic" materials on the heat-shielding properties of clothing.

When moisture is absorbed in the second phase of the experiment to values u = 2.3 g/g, the heat flux density in the package of textile with "active" heat-shielding properties is q_{2f} = 340 W/m^2. In packages of traditional textiles of similar configuration, the value is q_{2f}= 420 ÷ 450 W/m^2. Thus, part of the heat expended in the evaporation of moisture in the structure of the textile package is compensated by nano-fillers of the Outlast material. During the fourth phase of the experiment, in textile packages with active heat-shielding, a more intensive drying process is observed in comparison with packages of traditional textiles. In spite of this, the density of the heat flux in the structure of such packages is 15 ÷ 25 % lower than in packages of traditional materials. The presented experimental data are consistent with known calculated results [3].

5. Conclusion. As can be seen, the author's research method allows one to obtain comparative results characterized the influence of the original properties of new textile on the intensity of heat and mass transfer processes in the "man-clothing-environment" system. In the future, these results will allow to develop a system of materials, as well as recommendations on the formation of a set of clothes within the selected system.

References

[1] Chandrasekhar P. Conductive Polymers, Fundamentals and Applications, A Practical Approach // Kluwer Academic Press, – Boston, – USA, – 1999, – P. 43-76.

[2] Decaens J., Vermeersch O. Specific testing for smart textiles chapter: Advanced Characterization and Testing of Textiles A volume in The Textile Institute Book Series // CTT Group, Saint-Hyacinthe, – Canada, – 2017, – P. 351 – 374.

[3] Fan J., Cheng X.Y. Heat and moisture transfer with sorption and phase change through clothing assemblie // Textile Research Journal, – 2005, – Vol. 75 (3) – P. 187-196.

[4] Kallmayer, C., Simon, E. Large area sensor integration in textiles // IOP Conf. Series: Earth and Environmental Science, – 50, – 2017, – P.49-55.

[5] Knittel D., Schollmeyer E. Conductive Polymer based Electro-conductive Textiles for Novel Applications // Technical Textiles, – №1, – 2015, – P. 16-18

IOP Conf. Series: Materials Science and Engineering **582** (2019) 012004 doi:10.1088/1757-899X/582/1/012004

[6] Lee, J., et al. Conductive fiber-based ultrasensitive textile pressure sensor for wearable electronics. // Advanced Materials, – 2015, – № 27 (15), – p. 2433–2439.
[7] Locher, I. Design and characterization of purely textile patch antennas // IEEE Trans. Adv. Pack, – №29 (4), – 2006, – P. 777–788
[8] Ionescu, C. 2012. Investigations on current capabilities of PEDOT:PSS conductors // 35th International Spring Seminar on Electronics Technology (ISSE), – 2012, – p. 59–64.
[9] Michalak, M., Felczak, M., & Więcek, B. Evaluation of the thermal parameters of textile materials using the thermographic method // Fibres and Textiles in Eastern Europe, – №17(3), – p. 84–89.
[10] Mondal, S. Phase change materials for smart textiles–An overview // Applied Thermal Engineering, – 2008, – №28(11), – p. 1536–1550.
[11] Numakura, D. Advanced screen printing "Practical approaches for printable & flexible electronics" // Microsystems, Packaging, Assembly & Circuits Technology Conference, – 2008, – p. 205–208.
[12] Rodicheva M. Advanced Technique and the Results of a Research of a Heat-Mass-Exchange Processes in Clothes Packages in the Subnormal Climate // IOP Conf. Series: Earth and Environmental Science. – 2017. – №66. – 9 P.
[13] Stoppa M., A. Chiolerio Testing and evaluation of wearable electronic textiles and assessment thereof // Performance Testing of Textiles: Methods, Technology and Applications – 2014, – P.11957 – 11992.
[14] Wallace G.G. e.t.c. Conductive Electroactive Polymers, Intelligent Materials Systems // 2nd ed., CRC Press, – USA, – 2003.
[15] Wang, F., et al. A review of technology of personal heating garments // International Journal of Occupational Safety and Ergonomics, – 2010, –№16(3), – p. 387–404.

IOP Conf. Series: Materials Science and Engineering **582** (2019) 012005 doi:10.1088/1757-899X/582/1/012005

Forecasting the Price of Products Using Neural Networks

G N Boyarkin, I V Revina[a], O G Shevelyova

Omsk state technical university, 11, Mira pr, Omsk, 644050, Russia

E-mail: [a] revina_i@rambler.ru

Abstract. When developing new products, the justification of the price of future products is an important aspect of product competitiveness. Therefore, price assessment methods are paid great attention to. The article offers an information model for forecasting the price of pumping equipment using neural networks. The main technical characteristics of the pumping equipment used in the computational experiment were chosen. The developed and trained neural network showed a high level of correlation between predicted and real price values. Thus, price prediction using neural networks is a convenient tool in managing business processes. This is especially relevant at the stage of developing new products when comparing prices for similar products from different manufacturers.

1. Introduction

Price is an important parameter of product competitiveness. The price of products is a monetary expression of value. The cost is determined by the cost of production and sales of products, product benefits for the consumer [1].

Production costs, in accordance with the economic content, include the following groups:

- labour costs;
- tangible costs (for raw materials, consumables, parts, etc.);
- depreciation charge;
- allocations for social needs;
- other allocations.

Costing defines different types of costs: technological, shop, production and complete. The total cost is the basis for the pricing of manufactured products [2].

There is a law of supply and demand along with the law of value in a market economy. The law of demand (supply) reflects the relationship between the price of goods and the amount of demand (supply) for goods.

When choosing a pricing strategy, the following approaches are distinguished:

- calculated (method of full costs - adding of a certain size to the cost of goods, ensuring break-even);
- standard-parametric (specific indicators method, regression analysis method, aggregate method, price calculation method taking into account consumer effect, etc.);
- methods of market pricing (method of determining prestigious prices, the method of following market prices, a method with a focus on demand, etc.) [3].

Thus, many factors are to be considered while determining the price.

When forecasting the price of technical products, a correlation is often used between the price and one or several technical parameters determining the consumer properties of technical products [4,5].

Using statistical methods of data analysis (regression analysis, maximum likelihood method, etc.) leads to values averaging, and, consequently, to the loss of a certain information content.

Neural networks are one of the data mining tools. Neural networks are a powerful distributed computing process consisting of simple elements operating according to certain rules. The choice of network architecture, its parameters to solve a specific problem is an informal and difficult formalized procedure [6].

If there is a large amount of data, incomplete information, redundant information, and data with a high level of noise, etc., it is most advisable to use neural networks. They solve the task of finding patterns effectively. Therefore, neural networks are actively used in studies of economic processes [7, 8, 9], in forecasting [10, 11, 12], in pricing [13, 14].

The purpose of this article is to choose a neural network to forecast prices for pumping equipment.

2. The problem statement

Pumps vary in design, use, and type [15]. Centrifugal pumps were selected for pumping wastewater in this article. The main characteristics of the pumps of this group are:

- productivity (x_1);
- pressure head (x_2);
- electric motor power (x_3);
- nominal diameter of the discharge port (x_4);
- nominal diameter of the impeller (x_5);
- efficiency (x_6);
- mass (x_7);
- rotational speed (x_8);
- rated current (x_9).

The specified basic parameters (x_i, i=1…9) will be considered as input when simulating. Price (Y) is output parameter simulation. Thus, a neural network can be used to forecast the price of pumps (Figure 1).

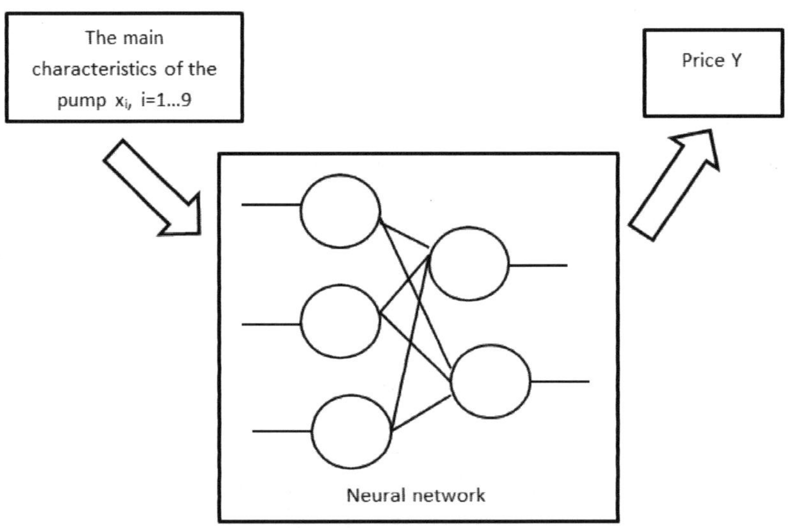

Figure 1. Data processing using neural network

So it is necessary:

- to develop a neural network, where the input data will be the main characteristics of the pumps, and the output value of the price;
- to train the neural network using known algorithms;
- to assess the accuracy of the trained neural network;
- to test a neural network by known combinations of characteristics – price;
- to complete price forecasting using neural network.

3. Neural network model of prices forecasting

Simulating neural networks is time consuming, and ready-made software products are used for its implementation. The first packets of neural networks appeared in the late 80s of the twentieth century. To date, the number of such packages is very extensive. These may be add-ons to software products to support neural network simulating: Deep Learning Toolbox (former Matlab Neural Network Toolbox), Statistica Automated Neural Network, a set of libraries to implement some data processing: Excel

Neural Package, specialized neural network packets: NeuroSolutions, Neural Work Professional, Neuro Shell, Mem Brain, Neuro Pro and so on. The analysis of these software products by such criteria as functionality, data processing functionality, documentation, speed of operation showed that the most preferred is Matlab Neural Network Toolbox. This package will be used in this research.

The initial (input) data for the simulation were taken from open Internet sources; the initial database was formed on their basis. It included 520 pump items, 500 of them were training samples and 20 were control samples. A fragment of the database is shown in Figure 2.

Figure 2. Database fragment for simulation

Titles	Productivity, Q m³/h	Head pressure	El.motor power, kW	Nominal diameter of the discharge port	Nominal diameter of the impeller	Efficiency ,%	Mass, kg	Rotational speed	Rated current, A	Price
PF2 50/120.110 - 1,1/2 - 016	10	15	1.1	50	120	45	42	2730	2.6	93279
PF2 50/120.110 - 1,1/2 - 036	10	15	1.1	50	120	47	42	2730	2.6	111451
PF2 50/140.138 – 3/2 - 016	25	20	3	50	140	48	45	2940	7	98058
PF2 50/140.138 – 3/2 - 036	25	20	3	50	140	50	45	2940	7	116230
PF2 50/150.145 – 3/2 - 016	16	27	3	50	150	65	70	2940	7	98 058
PF2 50/150.145 – 3/2 - 036	16	27	3	50	150	67	70	2940	7	116 230
PF2 50/150.155 – 3/2 - 016	16	32	3	50	150	66	70	2940	7	98058
PF2 50/200.185 – 7,5/2 - 016	25	40	7.5	50	200	44	155	2895	20	213998
PF2 50/200.180 – 11/ 2 - 016	45	34	11	50	200	60	160	2895	21	231187
PF2 65/135.125 – 3/2 - 016	50	11	3	65	135	68	65	2940	7	99100
PF2 65/135.130 – 3/2 - 036	50	13	3	65	135	70	65	2940	7	116230

The article shows the simulation used neural networks of direct recognition, consisting of two layers, one of which is hidden. The choice of the network structure was made on the grounds that insufficient complexity can lead to a significant error, and excessive complexity can lead to the perception of noise as data for simulation, i.e. deprives the neural network of important properties of generalization.

Training of the neural network was carried out by the method of back propagation of error. While training, the values of the weights of the network are determined on the basis of examples (database) that form a training set, which has the form:

$$\langle \tilde{X},\tilde{Y} \rangle = \{(\tilde{x}_k,\tilde{y}_k)\}_{k=1,...500} \tag{1}$$

where $\tilde{x}_k \in R^n$ is input vector from the *k*-th sample (n=9 is number of inputs), $\tilde{y}_k \in R^m$ is a vector of required values (teacher's instructions) ($m=1$ is number of outputs). Network training is with the teacher.

For each input set there is a pair set, specifying the required output values. These sets form a training pair together.

$w \in R^W$ is a current vector of neural network weights, determining the proximity of the network response vector y_k at the *k*-th sample and the corresponding vector of teacher instructions $\tilde{y}_k$. W is the number of weights in the direct propagation neural network.

The quality of training, which determines the degree of compliance with the data from the training set is determined by a function of the form:

$$E(w) = \sum_{k=1}^{500} \left(y_k(w) - \tilde{y}_k\right)^2 = \sum_{k=1}^{500} \left(F(x_i, w) - \tilde{y}_k\right)^2 \tag{2}$$

The purpose of training a neural network is to determine such a value of the weight vector w^*, so that function (2) takes the minimum value. Thus, the process of training a neural network is reduced to solving the problem of unconditional optimization of the form:

$$w^* = \underset{w \in R^W}{arg\ min}\, E(w) \tag{3}$$

There are many methods for solving (3). Several algorithms to be solved can be selected in software Deep Learning Toolbox (former Matlab Neural Network Toolbox). The Levenberg-Marquardt algorithm has several advantages: a fairly fast convergence; minimum computation time; no negative effect of local minima. The Levenberg-Marquardt algorithm represents a combination of the least squares method with the gradient descent method [16]. The analysis of the Levenberg-Marquardt algorithm showed that it achieves the smallest error in neural networks [17].

Algorithm for creating and configuring a neural network simulation in Deep Learning Toolbox (former Matlab Neural Network Toolbox) consisted of several stages:

1. determination of a training set;
2. selection of the network structure;
3. the training sample is divided into three subsets: "Training", "Validation", "Testing" (training is done according to "Training", the training will be stopped according to "Validation", and the total network operation will be determined according to the "Testing" data set);
1. determine the number of neurons in the hidden layer (change to adjust the network);
2. choose the training algorithm;
3. run the network;
4. obtain and analyze the results (if the results obtained are not satisfactory, the modeling steps are repeated from stage 3).

4. Results and discussion

According to the results of computational experiments, the two-layer structure of the neural network is optimal, consisting of 25 neurons in the hidden layer and one neuron in the open layer. The network has 9 inputs pump characteristics and 1 output price pumps. The structure of the neural network is shown in Figure 3.

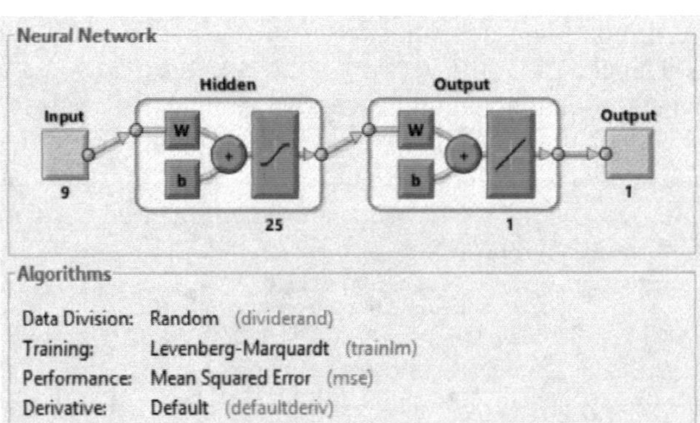

Figure 3. The structure of the neural network

Network training takes place in 39 iterations, i.e. training stops when the network training ability stops improving on the data set "Validation". The accuracy of the training (the dependence of the mean square error on the training iteration) is presented in Figure 4, errors for three data sets "Training", "Validation", "Testing" are shown. Training is terminated when an error on the set "Validation" does not decrease. "Best" index shows the best training outcome. "Test" shows the generalizing ability of the network.

The approximation of data by a neural network and a network error is shown in Figure 5. It is considered satisfactory for research purposes. A regression dependence plot is shown in Figure 6. "Fit"

IOP Conf. Series: Materials Science and Engineering **582** (2019) 012005 doi:10.1088/1757-899X/582/1/012005

line is a regression line based on empirical data, and "Y=T" line is a line constructed on the basis of the values calculated after the training ("T" array represents the data used for training, and "Y" array of output data calculated from the neural network training and serves to verify the accuracy of the mode).

R=0.96432 is the correlation coefficient that suggests a strong relationship between variables. Thus, the constructed neural network has high accuracy.

The results are saved in "Script" (Figure 7). The saved "Script" performs data modeling, not involved in the training process of the neural network.

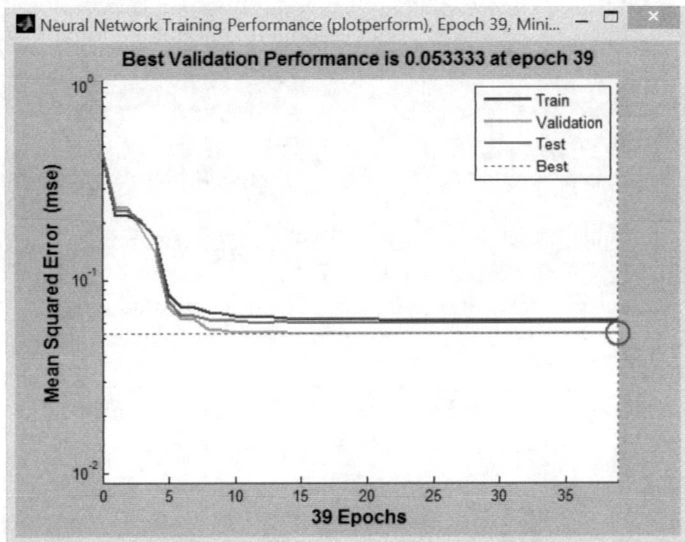

Figure 4. Accuracy of training of the neural network (the dependence of the standard error on the iteration of training)

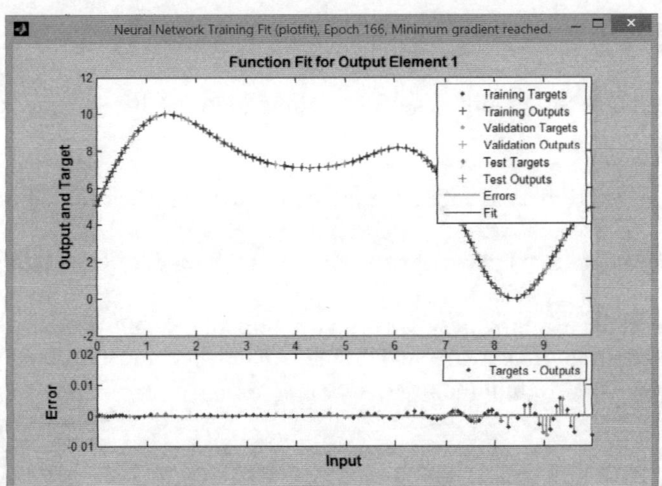

Figure 5. Neural network data approximation and network error

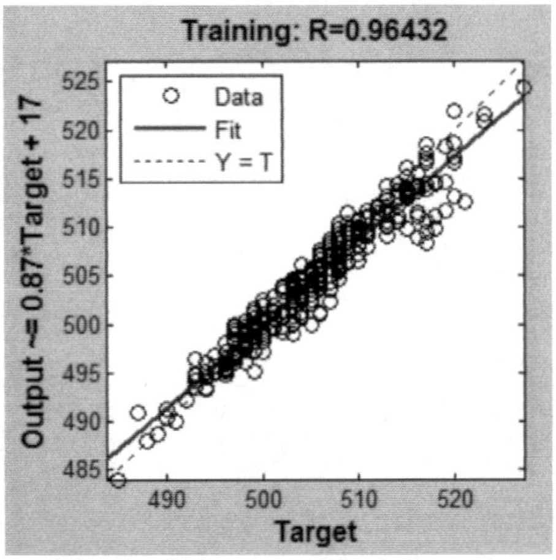

Figure 6. Regression graph

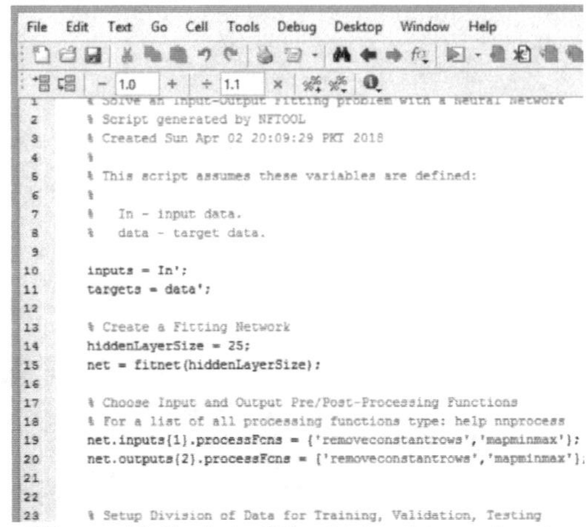

Figure 7. Script of the trained neural network

To check the work of the trained neural network, one selects data not involved in the training process (Figure 2, see the fill) and run the saved "Script". The results of the simulation are presented in Figure 8 a), and the initial value is presented in Figure 8 b). The price offered by the neural network is 228091, and the price of the source data is 231187. The obtained results suggest the possibility of using neural networks for forecasting

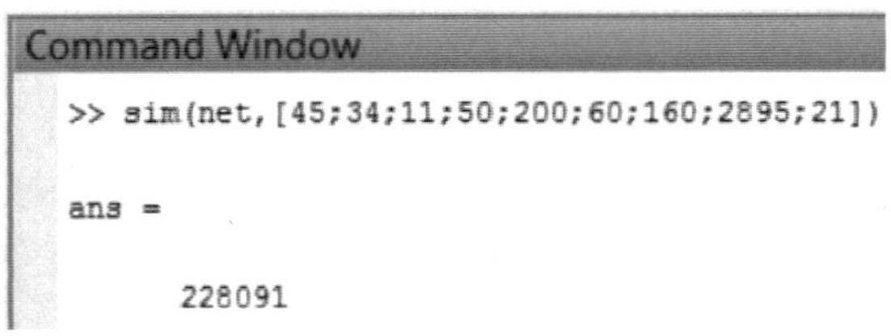

a

45.0	43.0	11.0	50.0	200.0	60.0	160.0	2895.0	21.0	231187

b

Figure 8. Simulation results: a – simulation results; b – initial value

5. Conclusion

The article discusses an information model developed to forecast the price of pumping equipment using neural networks.

The performed computational experiment showed a high level of correlation between predicted and real price values.

Thus, price forecast using neural networks is a convenient tool in managing business processes. Such forecasting is especially important in the development of new technical products, when comparing prices for products, analogues of different manufacturers.

References

[1] Fatkhutdinov R A 2000 *Organization of production* (Moscow: INFRA-M)

[2] Gorfinkel V Ya, Schwander V A 2004 *Economics of Organizations (Enterprises)* (Moscow: YUNITI)

[3] Vasyukhin O V 2010 *Basics of pricing* (St. Petersburg: St. Petersburg State University ITMO)

[4] Killingsworth E C, Wolz R R 1990 *AIAA-1990-2038, SAE, ASME, and ASEE, 26th Joint Propulsion Conf.* Requirements for business jet Aircraft 10–16

[5] Tseytltna T O, Balashov V V, Dunaevsky A I, Smirnov A V 2009 *Proceedings of MIPT* Predicting the price of a promising supersonic administrative aircraft **1(3)** 160–170

[6] Haykin S 2006 *Neural networks* (New Jersey: Prentice-Hall)

[7] Xiaobing H, Xiaolian L, Yuanqian R 2018 *Cognitive Systems Research* Enterprise credit risk evaluation based on neural network algorithm **52** 317–324

[8] Stupak A A 2017 *Investment and Innovation Management* Inventory Management using neural networks **3** 95–102
[9] Kurnikov D S, Petrov S A 2017 *Juvenis scientia (@jscientia)* The use of neural networks in the economy **6** 10–12
[10] Pengju Z 2018 *Cognitive Systems Research* Quantitative analysis of portfolio based on optimized BP neural network **52** 709–714
[11] Fedotov D V, Semenkin E S. 2014 *Vestnik SibSAU* On predicting economic indicators using neuroevolutionary models **5(57)** 299–304.
[12] Omarov A V, Repnikov A I, Gryaznov I E 2018 *Izvestiya VolSTU* Predicting Problems Using Fuzzy Neural Networks in the MATLAB Package **2(212)** 80–82
[13] Dongqing Z, Guangming Z, Jing L, Kaiping M, Huan L 2018 *Computers and Electronics in Agriculture* Prediction of soybean price in China using QR-RBF neural network model **154** 10–17
[14] Bugorsky V N, Nikitin N A 2006 *Applied Informatics* Neural Networks in Retail Management **2** 34–41
[15] Karelin V I , Minaev A V 1986 *Pumps and pumping stations* (Moscow: Stroiizdat)
[16] MATLAB – book1, see http://matlab.exponenta.ru/neuralnetwork/book1/index.php.
[17] Hagan M.T. Menhaj M 1994 *IEEE Transactions on Neural Networks* Training Feedforward Networks with the Marquardt Algorithm **5(6)** 989–993

IOP Conf. Series: Materials Science and Engineering **582** (2019) 012006 doi:10.1088/1757-899X/582/1/012006

Structure and Electrical Erosion Resistance of An Electro-Explosive Coating of the ZnO-Ag System

D A Romanov[a], S V Moskovskii, K V Sosnin

Department of natural science disciplines of Professor V. Finkel, Siberian State Industrial University, 42 Kirov street, Novokuznetsk, 654007, Russia

E-mail: [a] romanov_da@physics.sibsiu.ru

Abstract. For the first time, electroerosion-resistant coatings of the ZnO-Ag system were obtained by electro-explosive spraying. The coating structure was investigated by atomic force microscopy and scanning electron microscopy. The formed coating is a homogeneous composite material consisting of a silver matrix and ZnO inclusions located in it. Tests of sprayed coatings for electrical erosion resistance under arc erosion conditions were carried out. A formation mechanism of hierarchical levels of the structure of electro-explosive coatings was suggested. The mechanism is based on theoretical concepts of the formation and evolution of fractal aggregates with diffusion-limited and cluster-cluster aggregations. Based on the electrical properties of silver and zinc oxide, the dependence of electrical contact resistance on the number of on/off cycles in the process of testing for electrical erosion resistance is substantiated.

1. Introduction

Power systems safety, reduction of energy losses and saving material resources is mainly determined by the reliability of electrical contacts. Reliability of electrical contacts determines the operation efficiency of energy and industrial equipment, and by and large the global energy system. Due to poor-quality contacts, they are heated resulting in an increase of power losses and emergency situations. According to the National Association of Rescue and Environmental Organizations, 50% of fires in industrial enterprises and in the residential stock are caused by electrical equipment malfunction, while 50% of electrical equipment fires are caused by electrical contact failures.

The use of powder metallurgy methods allows implementing in one material a diverse and contradictory complex of properties that an electrical contact material should have [1]. Currently, a large number of electro-contact materials have been developed for their use in various operating conditions. As a rule, they include a matrix with high electrical conductivity and a high-melting component (filler) with high wear and electro-erosion resistance. [2]. The most promising for manufacturing contacts are composite materials based on silver, copper, copper-nickel, aluminum matrix and refractory filler. These include the following systems: W-Cu, Mo-Cu, W-C-Cu, Mo-C-Cu, Ti-B-Cu, TiB2-Cu, TiB2-Al, W-Ni-Cu, Mo-Ni-Cu, Cr-C-Cu, Cr-Cu, CdO-Ag, SnO2-Ag, ZnO-Ag, ZnO-Ag, W-Ag, Mo-Ag, W-C-Ag и Mo-C-Ag etc. [3–8].

Since the destruction process of the material begins with its surface, for a number of practical applications, for example, hardening the contact surfaces of medium and heavy-duty switches and switching devices, the formation of protective coatings is promising, since in this case the electrical erosion resistance of the contact surface, and not the entire volume, is important. It is economically and technically reasonable to develop an approach to materials creation, which ensures the mechanical strength of a component by the use of cost-efficient substrates, and the special surface properties are provided by continuous or local formation of composite coatings, the properties of which correspond to operating requirements. This approach can give 90% savings. Expert assessments confirm this trend. One of the priority areas of condensed matter physics is the development of methods for improving the operating characteristics of various materials. With this in mind, the hardening of the surface of arc-resistant electrical contacts is a crucial task for the development of contemporary technologies.

The aim of this work is to study the structure and the phase morphology of electro-explosive coating of the ZnO-Ag system, using scanning and atomic force microscopy, as well as conducting tests for electrical erosion resistance.

2. Materials and methods

The subject of research was copper contacts of the KPV-604 contactor, on the contact surfaces of which the ZnO-Ag system was formed by the electro-explosive method [9]. A 250 mg silver foil was used as an explosive conductive material on the surface of which a 50 mg CdO powder sample was placed in the zone of the explosion. The exposure time of plasma on the sample surface was ~100 μs, the absorbed power density at the jet axis ~5,5 GW/m2, the pressure in the shock-compressed layer near the irradiated surface ~12,5 MPa, the residual gas pressure in the working chamber ~100 Pa; the plasma temperature at the nozzle exit ~104 K, the thickness of the heat affected zone ≈50 μm. The structure and morphology of the coating and the adjacent substrate layer were analyzed by scanning methods (Carl Zeiss EVO50) with an EDAX X-ray spectrometer and atomic-force microscopy (Solver NEXT). Tests for electrical erosion resistance of coatings under conditions of arc erosion were carried out at the contacts of the CJ20 electromagnetic actuators with alternating current and inductive load in accordance with the requirements of the AC-3 test mode for switching wear resistance when operating in a three-phase circuit with a low voltage value of 400/230 V, frequency 50 Hz for currents up to 320 A and $\cos\varphi = 0.35$ and the number of switching cycles 6000.

3. The results and discussion

The structure and elemental composition of the section of the coating/substrate system formed by the electro-explosion method were studied by scanning electron microscopy. A typical image of the cross-sectional structure of the ZnO-Ag composition is shown in Fig. 1.

The analysis of the results obtained shows that the formed coating is a homogeneous composite material (Fig. 1, a, b). According to the structural morphology and the etching contrast, the forming coating consists of a light silver matrix and dark ZnO inclusions with dimensions varying from 0.3 to 0.5 μm (Fig. 1, b). The elemental composition of the coating was analyzed by X-ray microanalysis methods. The research results are given in Fig. 1 c-e. Analyzing the results presented in Fig. 1 c-e, it can be noted that the concentrations of copper (Fig. 1 c), zinc (Fig. 1 d) and silver (Fig. 1 e) in the coating vary slightly in its thickness. This fact also indicates the structural homogeneity of the coating obtained.

Atomic force microscopy was performed in the coating layer located at a distance of 10 μm (Fig. 2) from the coating surface, as well as at the interface between the coating and the copper substrate (Fig. 3).

Since the electro-explosive coating is formed by a silver matrix and ZnO powder particles located in it, small particles of ZnO powder can be crumbled out of the matrix during the preparation of thin sections. In this case, pores, i.e. dark areas in Fig. 2, a, b 30 to 100 nm deep and 2 to 5 nm wide (Fig. 2, c) are formed at the site of fallen particles. ZnO particles are dispersed to 2 ... 5 nm in the process of an electric explosion during the formation of a pulsed plasma jet of products of the electric explosion of conductors.

Separate large particles of various shapes with sizes ranging from 10 to 15 nm are also detected. These ZnO particles do not crumble out of the silver matrix during the preparation of the thin section; they are sharply highlighted in color, i.e. they are lighter than the matrix. They are randomly located in a silver matrix. As can be seen from Fig. 2, a, b large particles have a complex structure. They are composed of spheres (globules) with a diameter of 2 to 5 nm (these are small spherical particles described above). The ratio of the silver matrix, large and small particles of ZnO powder is 0.6: 0.15: 0.25. Taking into consideration the fact that large ZnO particles consist of smaller globular ZnO particles then the ratio of the silver matrix and the inclusions of ZnO powder is 0.6: 0.4. This ratio is proportional to the content of ZnO powder and silver foil used for electro-explosive coating. The average roughness of the surface profile of the ZnO-Ag system coating is 100 nm.

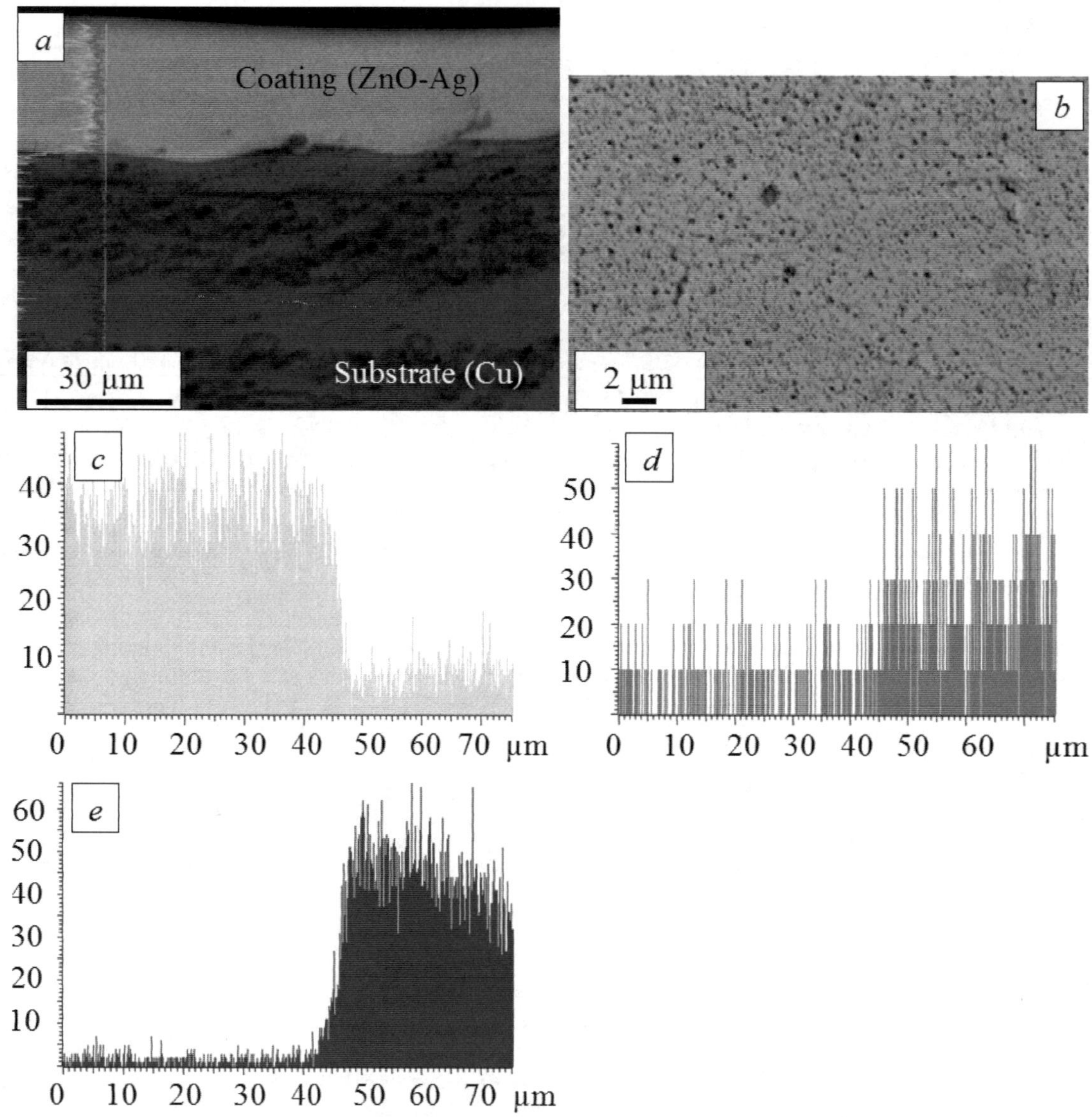

Figure 1. Structure of electro-explosive coating of the Cu-Ag system (a, b) and concentration profiles of copper (c), zinc (d) and silver (e) atoms along the line indicated on (a). The copper concentration is determined by the characteristic Kα1 X-rays; zinc concentration is determined by Kα1; silver concentration is determined by Lα1.

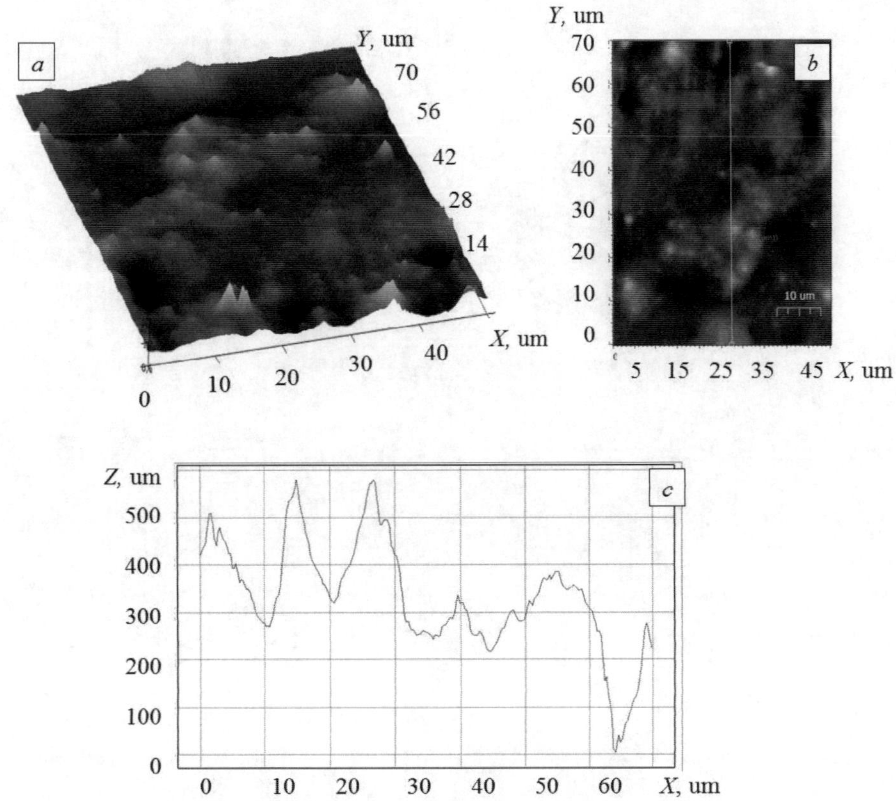

Figure 2. The coating structure of the ZnO-Ag system, detected by atomic force microscopy
a – 3D distribution of uneven relief in height, b – secant position (top view), c – roughness distribution along the base length

Thus, it was possible to identify an important structural element - the ZnO globule, i.e. a spherical particle with a diameter of 2 to 5 nm. There is a multi-level hierarchical structure of the ZnO-Ag coating system, which is based on uniform spherical ZnO particles with a diameter of 2 to 5 nm. The fact that the ZnO inclusions, located in the silver matrix are made of a single structural unit is a very important argument in favor of the fractal mechanism for the formation of an electro-explosive coating. Such particles constitute the first hierarchical level of the structure of the electro-explosive coating of the ZnO-Ag system. The second hierarchical level consists of globules -large particles of various shapes with sizes ranging from 10 to 15 nm, which, in turn, form the sediment of micron-sized particles of irregular shape, detected by scanning electron microscopy. At the boundary between the coating and the copper substrate (Fig. 3, a, b) there are visible dark depressions ranging in size from 10 to 15 nm. Aforementioned large ZnO particles crumbled out of them. In addition, surface periodic structures appear at the coating/substrate boundary in a silver matrix. The secant held perpendicular to these structural formations (Fig. 3, b, c) suggests that the wavelength in them is on average 3 nm. The structures are the residual nanorelief of the surface. After the end of the impact of a pulsed plasma jet of products of the electrical explosion of conductors on a substrate and cooling the surface, the induced relief is fixed in the form of surface periodic structures. They can be formed due to evaporation, melting of the surface and displacement of the melt by excess vapor pressure, thermocapillary phenomena and thermochemical reactions, thermal deformations, the emergence and development of various instabilities such as Rayleigh-Taylor [10], Kelvin-Helmholtz [11], Marangoni [12] and others. In general, the phenomenon is universal and is an example of self-organization in a system with no initially selected directions and structures [13]. The energy regimes for obtaining surface periodic structures correspond to heating the material to a temperature approximately equal to the melting point (lower limit), but not higher than the temperature of developed evaporation. [13]. This regime was used in electro-explosive coating in the present work.

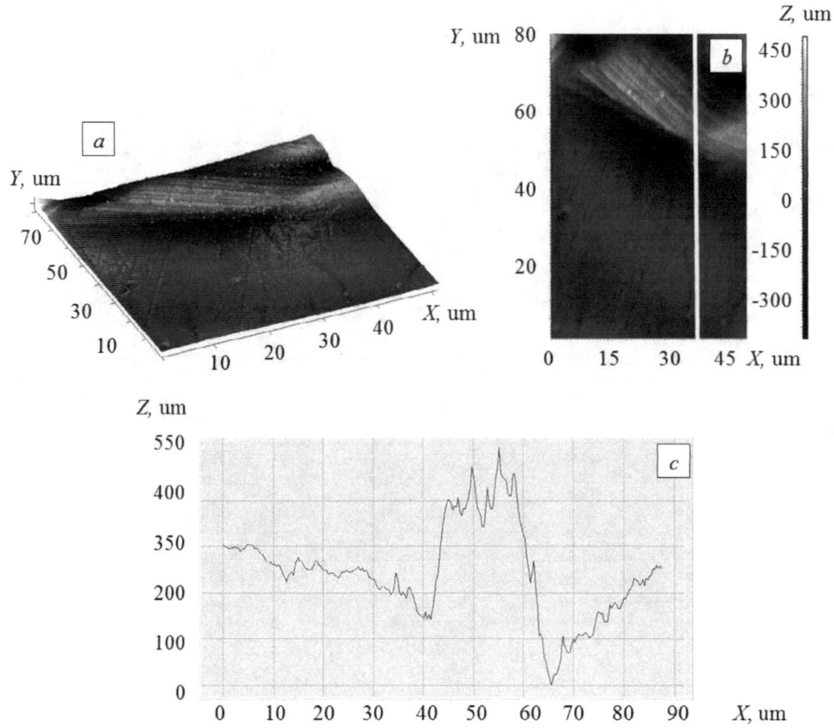

Figure 3. Structure on the border of the ZnO-Ag coating with a copper substrate, revealed by atomic force microscopy
a – 3D distribution of uneven relief in height, b – secant position (top view), c – roughness distribution along the base length

Surface profilometry in Fig. 3a showed that the roughness parameter of the electro-explosive coating of the ZnO-Ag system is 73 nm. In this case, the maximum profile protrusion reaches 536.85 nm, and the depth - 497.5 nm. Comparing the roughness parameters at the electro-explosive coating/substrate boundary (Fig. 3) with the roughness parameter in the coating layer located at a distance of 10 μm (Fig. 2), it can be noted that they differ by 28%. In Fig. 3a 35% of the photo is taken by the substrate and 65% -by the coating. It strongly suggests that the substrate roughness is lower than the coating roughness. This is natural since due to precipitations of ZnO inclusions from the silver matrix, the roughness parameter of the coating increases.

The assembly mechanism of such a complex multi-level ZnO in the silver matrix can be represented as follows. The uniform size r1 of spherical ZnO particles of the first hierarchical level the can be explained within the framework of the model of diffusion-limited aggregation by the mechanism of diffusion-limited aggregation “particle-cluster” [14]. Plasma components of a multiphase jet of products of the electrical explosion of conductors distributed in a certain effective volume on a certain effective interface space of the substrate of the same size are assembled into a cluster, which will then turn into a sphere with a diameter of 2 to 5 nm. [14]. It should be noted that the size of the effective volume in which the formation of blanks of spherical particles of the first hierarchical level takes place will depend on the composition of the explosive conductors and the temperature. The formation of ZnO clusters occurs only at the initial stage of the impact of a pulsed plasma jet on the substrate. At the next growth stage, the particle-cluster aggregation mechanism ceases to operate, but the cluster sizes increase. Growth begins simultaneously and ends synchronously when the ZnO is in a plasma state, the clusters are transformed into particles of almost the same size (under the conditions discussed, these are spheres with a diameter of 2 to 5 nm). Particle sizes are not large enough for gravity force to dominate. In other words, the spheres with a diameter of 2 to 5 nm are maintained in a suspended state mainly due to Brownian motion and intermolecular interaction forces with other components of the molten metal. Under the action of Brownian forces, the spheres continue the chaotic movement that stimulates their agglomeration. Self-assembly of particles of the first hierarchical level leads to the formation of a second hierarchical level of globules with r2 size from 10 to 15 nm. The formation of this hierarchical level occurs mainly through the cluster-cluster mechanism. Further, such globular particles can be enlarged due to conglutinating into micrometer particles.

Let us analyze the dependence of the contact resistance (R) on the number of on / off cycles (N) when testing ZnO-Ag coatings for electrical erosion resistance under arc erosion conditions (Fig. 4).

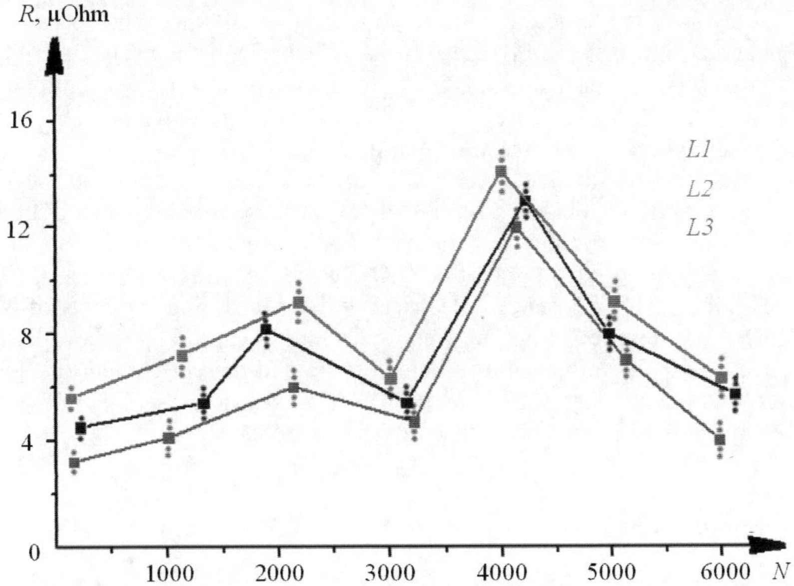

Figure 4. The dependence of the contact resistance (R) on the number of on / off cycles (N) when testing electro-explosion coatings of the ZnO-Ag system for electro-erosion resistance under conditions of arc erosion.

At the beginning of the experiment, the value of R contacts is 11, 7.5 and 9.5 μOhm for phases L1, L2, and L3, respectively. After the first on / off cycles, R begins to decrease from the initial values in linearity to the following values.: for phase L1, the value of R is 6.3 μOhm with N = 3059; for phase L2, the value of R is 4.7 μOhm at N = 2956 and is the smallest among all phases; for phase L3, the value of R is 5.4 μOhm with N = 3211. A gradual decrease in R when N is varied from 0 to 3211 indicates that a gradual burn-in of contacts occurs on this test part, which results in a stable surface roughness. In the next section of the curve, with increasing N, R increases to maximum values: for the L1 phase, the value R is 15.6 μOhm at N = 5015; for phase L2, the value of R is 13.3 μOhm at N = 5123 and is the smallest; for the L3 phase, the value of R is 14.3 μOhm with N = 4967. This indicates that at the present stage of the experiment intensive evaporation of the fusible silver matrix begins under the influence of an electric arc. The contact surface is enriched with particles of ZnO powder, which has a lower electrical conductivity (10-8 S / m) if compared with silver (62.5 MS / m) [15]. For this reason, the contact resistance on this part of the graph increases. When the maximum values of R are reached on all three phases with a further increase in N, the electrical resistance decreases again. The reason is that at N ≈ 5000 due to the evaporation of silver, the contact surface is enriched with refractory ZnO particles with a low electrical conductivity. Upon reaching N ≈ 5000, mechanical accumulation of ZnO particles on the contact surface occurs. As a result, the electrical resistance decreases again, namely, for the L1 and L2 phases, the R values are 6.2 μOhm with N = 6023 and 4 μOhm with N = 5983, respectively. For L3, the value of R equals 5.7 μOhm with N = 6122. The tests have shown that the formed coatings of the ZnO-Ag system satisfy for switching wear testing [16].

Thus, the increase in electrical resistance during the tests of electro-explosive coatings of the ZnO-Ag system for switching wear resistance is caused by the evaporation of the low-melting silver matrix under the influence of an electric arc and the enrichment of the coating surface with ZnO particles. Electrical contacts, hardened by electro-explosive coatings of the ZnO-Ag system, are capable of mechanically cleaning the surface of ZnO particles. The formed coatings of the ZnO-Ag system satisfy for switching wear testing.

4. Conclusion

This work presents a mechanism for the formation of hierarchical levels of the structure of electro-explosive coatings of the ZnO-Ag system. By comparing the images of atomic force microscopy of the surface of layers of electro-explosive coatings of the ZnO-Ag system, the typical sizes of the first and second hierarchical levels of the structure were established, which, under the coating conditions used, were from 2 to 5 nm and from 10 to 15 nm, respectively. The mechanism of self-assembly of hierarchical nanostructures is based on theoretical concepts of the formation and evolution of fractal aggregates with diffusion-limited and cluster-cluster aggregations. Such electro-explosive coatings are used to create contacts of switches of high-power electrical networks.

Surface periodic structures with an average wavelength of 3 nm were found at the coating/substrate boundary. They represent the residual nano-relief of the surface, which appeared after the end of the impact of a pulsed plasma jet of products of the electrical explosion of conductors on a substrate and cooling of the surface. Formed coatings of the ZnO-Ag system satisfy the testing of contactors for switching wear resistance. The increase in electrical resistance during the tests of electro-explosive coatings of the ZnO-Ag system for switching wear resistance is caused by the evaporation of the low-melting silver matrix under the influence of an electric arc and the enrichment of the coating surface with ZnO particles. Electrical contacts, hardened by electro-explosive coatings of the ZnO-Ag system, are capable of mechanically cleaning the surface of ZnO particles.

Acknowledgments
The present work was performed within Russian Science Foundation project No. 18-79-00013.

References

[1] K. Katsuyoshi, Powder metallurgy, InTech, Croatia, 2012.
[2] F. Gerard, Advances in condensed matter and materials research, Nova Science Publisners, New York, 2005.
[3] K.A. Taiasov, V.P. Isupov, B.B. Bokhonov, Yu.A. Gaponov, B.P. Tolochko, M.R. Sharafutdinov, S.S. Shatskaya, J. Mater. Synth. Process. 8 (1) (2000) 21–27.
[4] A.M. Russell, K.L. Lee, Structure-property relations in nonferrous metals, John Wiley & Sons, Hoboken, 2005.
[5] N.M. Talijan, Zastita Materijala 52 (3) (2011) 173–180.
[6] A.H. Qureshi, S.M. Azhar, N. Hussain, J. Therm. Anal. Calorim. 99 (2010) 203–209.
[7] V. Bukhanovsky, Rudnytsky M., Grechanyuk M., Minakova R., C. Zhang, Mater. and Technol. 50 (4) (2016) 523–530.
[8] R.P.L. Nijssen, Composite Materials An Introduction, Inholland University of Applied Sciences, Netherlands, 2015.
[9] D.A. Romanov, S.V. Moskovskii, E.A. Martusevich, E.A. Gayevoy, V.E. Gromov, Metalurgija 57 (4) (2018) 299–302.
[10] M.H. Obied Allah, Allah. Gen. Math. Notes 20 (1) (2014) 67–76.
[11] E.C. Harding, J.F. Hansen, O.A. Hurricane, R.P. Drake, H.F. Robey, C.C. Kuranz, B.A. Remington, M.J. Bono, M.J. Grosskopf, R.S. Gillespie, Phys. Rev. Lett. 103 (2009) 045005. [12] D.A. Bratsun, A. De Wit, Phys. Fluids 16 (4) (2004) 1082– 1096.
[13] W. Tian, L. Qi, X. Chao, J. Liang, M. Fu, Compos. Part B: Eng. 162 (1) (2019) 1–10
[14] D. Asnaghi, M. Carpineti, M. Giglio, M. Sozzi, Phys. Rev. A: At. Mol. Opt. Phys. 45 (2) (1992)1018–1023.
[15] R. Venkata Krishna Rao, K. Venkata Abhinav, P.S. Karthik, Surya Prakash Singh, RSC Adv. 5 (2015) 77760–77790.
[16] B. Miedzinski, M. Habrych, J. Wandzio, N. Grechanyuk, IET Conf. Publications 2012 (2012) 78–81.

IOP Conf. Series: Materials Science and Engineering **582** (2019) 012007 doi:10.1088/1757-899X/582/1/012007

Cyber and Physical Systems Technology Classification for Production Activity of the Industry 4.0 Smart Factory

D A Zakoldaev, A G Korobeynikov, A V Shukalov, I O Zharinov

Faculty of Information Security and Computer Technologies, Saint Petersburg National Research University of Information Technologies, Mechanics and Optics, 49, Kronverksky Av., Saint Petersburg, 197101, Russia

E-mail: mpbva@mail.ru

Abstract. The task is to create a company of new type (a smart factory) for the Industry 4.0. A smart factory is for production cycle of the item designing manufacturing. A smart factory is equipped with cyber and physical equipment controlled with computerized system which is a part of the company digital infrastructure. Cyber and physical production apply innovative production technologies for the item life cycle supported in a smart factory. There is a scheme of informative technologies classification in automatic production of the Industry 4.0. The classification is done with technological operation done in the company with the technology of each type. There is a scheme of informative technologies distribution in the Industry 4.0 smart factory by components of production infrastructure. The role and place of each informative technology of the Industry 4.0 smart factory is described.

1. Introduction

To digitalize the industrial economy is a development direction of state science and technical politics. Digital economy elements [1-3] are informative technologies and means of industrial production, standards of digital certification and other. United components of digital economy implemented in production infrastructure of item designing companies created smart factories [4, 5] of the Industry 4.0 to conduct their activity with minimum part of humans.

Smart factories of the Industry 4.0 are a new type of company [6, 7] equipped with cyber and physical systems which function automatically. Cyber and physical systems (CPS) united into a production line is a deep automatizing to manufacture the item designing components.

Cyber and physical systems function automatically without humans to make technological operations. The primary human role [8, 9] in cyber and physical production is to monitor the item manufacturing process and maintenance and service of cyber and physical technological equipment.

To control cyber and physical systems in production they require [10, 11] to integrate program and technological solutions which base is digital informative technologies. Informative technologies are to support production processes completion and to form the smart factory digital infrastructure.

The Industry 4.0 smart factory digital infrastructure also permits:

- to organize joint function of cyber and physical systems which results will be the completion of the company production plans with optimal (quasi-optimal) criteria;
- to provide production process flexibility in automatic mode by implementing self-organization technologies of cyber and physical systems which is operative and adaptive resetting of technological equipment;
- to provide the system of electronic documents in company with electronic exchange of technological documentation components for the item and production data which are made after the item is done;
- to provide informative support of user (exploiters) of the item and repair services and the maintenance;

IOP Conf. Series: Materials Science and Engineering **582** (2019) 012007 doi:10.1088/1757-899X/582/1/012007

- to organize distributed system of control in company with minimal expenses after the personnel made some mistakes.

2. Progressive informative technologies of the Industry 4.0

The composition of the Industry 4.0 production company requires to implement into smart factory infrastructure (serial plants) some progressive informative technologies with item digital models and twins of cyber and physical systems [10, 11]. Informative technologies in production are to support the technological operation completion of the item manufacturing being done in physical machines level and cyber-level.

Classification scheme of progressive informative technologies of the Industry 4.0 is given in figure 1.

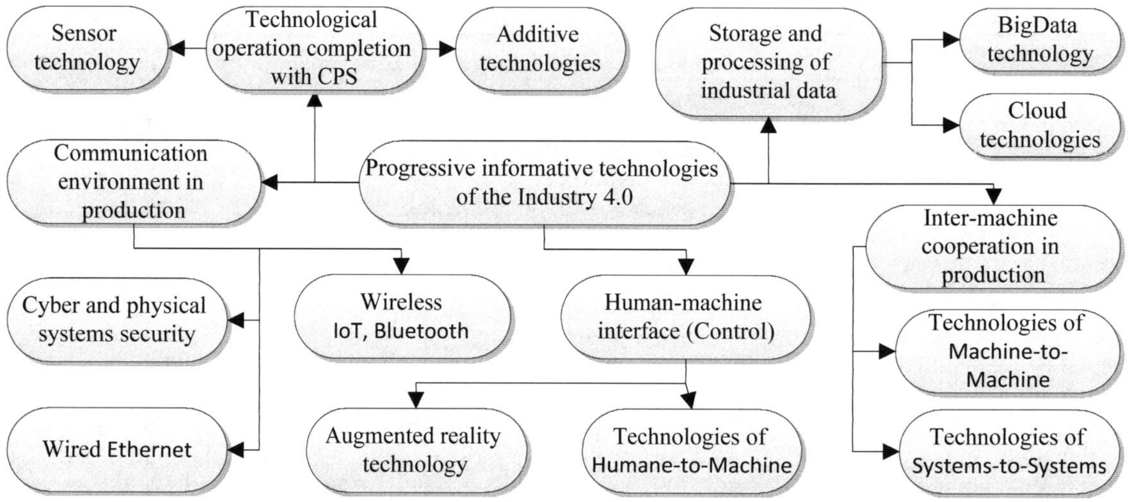

Figure 1. Classification of informative technologies being implemented in the production activity of the Industry 4.0 smart factory (IoT – Internet of Things).

Informative technologies classification of the Industry 4.0 smart factory is given in technology groups to support the following technological operations:

- to organize informative exchange (communication environment in production) based on production data transferring net protocols;
- to organize technological operations completion with cyber and physical systems;
- to storage and process vast amounts of industrial data after the technological operations;
- to organize inter-machine cooperation (cooperation of cyber and physical systems) when the technological operation is being done in production section;
- to organize human and machine interface [5] (interaction of operator in production and cyber and physical system) after the monitoring of production activity in a smart factory and the solution of the tasks of production process situation control.

These production operations are the closed loop cycle of the item designing component manufacturing with technical documentation given to the designers of the Industry 4.0 digital factory electronically (digital model).

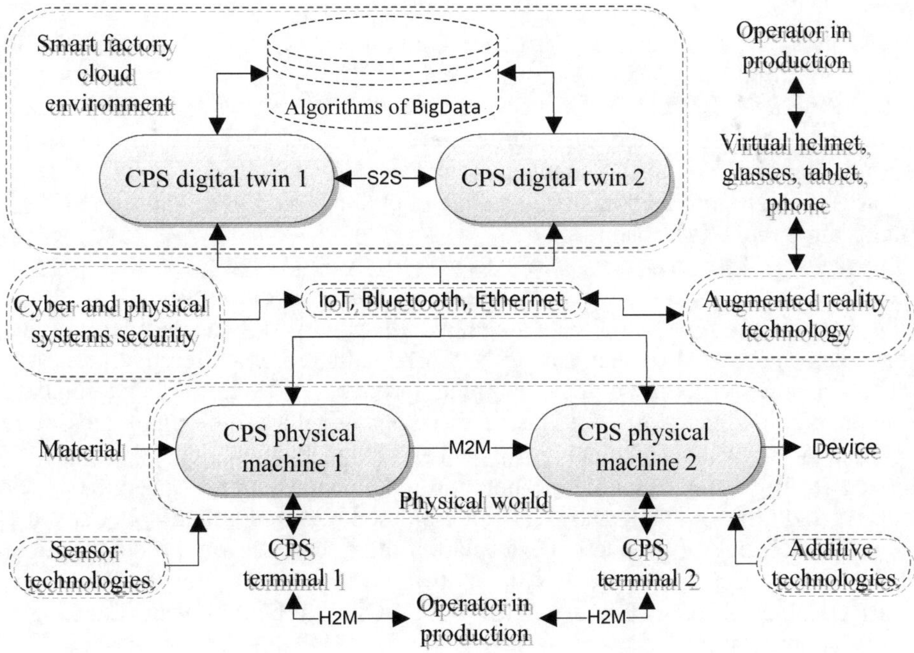

Figure 2. The distribution of progressive informative technologies by the components of cyber and physical production (S2S – Systems-to-Systems, M2M – Machine-to-Machine, H2M – Humane-to-Machine).

Informative progressive technologies distribution scheme by components of cyber and physical production of the Industry 4.0 smart factory is given in figure 2. Line with spaces are objects to define the technology type and simple lines is a place of application of those technologies in production. Having analyzed the scheme in figure 2 we see that each technology of cyber and physical production can be divided into two groups:

- technologies of direct application in cyber and physical production (for example, additive technologies being used in work chambers of cyber and physical systems);
- technologies of interaction to complete the production tasks of a smart factory (for example, technologies of IoT are being applied with cloud technologies and cyber and physical systems security technologies).

The most part of informative progressive technologies are for consolidative interaction with each other as a part of cyber and physical production of a smart factory.

3. Communication technologies of cyber and physical systems in production

Communication environment of the Industry 4.0 smart factory can be done with wired and wireless nets which user are operators in production (through the terminals) and cyber and physical systems. The most important information being circulated in the communication environment of a smart factory is:

- data of technological operations completion state in cyber and physical systems;
- material and parts consumption data which is necessary for the production;
- data of well functionality of cyber and physical systems and other.

This data through the wired Ethernet channels or wireless of Bluetooth helps the computerized control system [3] to plan the production process operatively for each cyber and physical system which is a production active agent [8, 9] connected to the Internet. This is the hybrid automatic production in a smart factory which is a single net based on the IoT technology for different cyber and physical equipment.

The IoT industrial net helps the computerized control system to control cyber and physical systems in a central way. Active properties of the agent makes the cyber and physical system to be the correction initiator of operative production plans if any emergency should arise. So IoT cyber and physical systems can be control agents which makes the system to be a non-central one. If combined central and non-central control they create the dynamic production where the technological operation are done automatically.

Quick development of technical and program communication means requires the informative security of automatic production. Your industrial data is a vital information today for a competitive company of the same activity field. To make sure the safe functionality of automatic technological equipment in a smart factory the factory communication environment [12] must have cyber and physical systems security components. Those components are machine and software solutions to cut out any unauthorized access to the production infrastructure including the remote access to the cyber and physical systems through the IoT channels.

4. Storage and processing technologies of the industrial data vast amounts

To process the vast amount of industrial data generated after the work in a smart factory there must be some new technologies of statistical analysis of the work quality criteria after given criteria. Those are technical and economic criteria (production rent, production rhythm, technological operation completion automatizing signs, cost of the item being manufactured). Technologies of informative support for production statistical analysis are BigData and cloud technologies.

The technology of Big Data is a set of horizontal program means for high-speed machine processing of non-structural data to find out their statistical value. In the Industry 4.0 the BigData technology is for mass and parallel processing of production information after some narrow production operations are completed in different cyber and physical systems to give the smart factory operator the condition completion of technological routes (for example, in the SQL (Structured Query Language) data base format of industrial processes).

Cloud technologies of the Industry 4.0 is a set of calculation resources (virtual machines, remote software, industrial data storage servers) available for automatic production operator and cyber and physical systems. Cloud technologies is for working with data amounts after the first industrial information from sensors and executive machines and from cyber and physical systems controllers is being processed with the technologies of BigData.

Implementation of cloud technologies in production infrastructure of the Industry 4.0 smart factory transferred some components and algorithms of control and data storage means from the physical world to the virtual environment to which users may access remotely. Projecting of smart factories has no restriction in territorial locations as a technological line of cyber and physical system and environments which are necessary for their application (information resources). If necessary the cyber and physical system controller through the wireless channel of Bluetooth and IoT means always may access the cloud data storage on a distant virtual server.

5. Technological operation completion in cyber and physical systems

They apply sensor and additive technologies to complete the operations with cyber and physical systems.

Sensor technologies implement in a cyber and physical system intellectual sensors for first processing the data of technological operations completion in the work chambers of technological equipment. Intellectual sensor is an advanced form of simple sensors which creates the sensor nets in a smart factory where the data is being processed with complexing mathematical algorithms. Sensor nets of cyber and physical production helps the smart factory operators who monitor the situation to receive the actual production information in each moment of time about the completion of technological operations.

Additive technologies of the Industry 4.0 smart factories is a set of technical and technological solutions to automatize production processes of part manufacturing (item) with stage by stage increasing of material on the prepared base. Unlike the CNC (computer numerical control) machines of the Industry 3.0 companies where the metal was out step by step with a cutting tool («subtractive technology»), additive production of the Industry 4.0 is more economic with raw materials (plastic, metallic, ceramic powders) when item is created and without additional mechanical treatment (soon electrotyping will be applied with additive technologies and 3D-printing).

6. Technologies of inter-machine interaction of cyber and physical systems

Conveyor manufacturing of complicated technical parts requires interaction among machines in automatic mode. Cyber and physical equipment of the Industry 4.0 use the technologies of Machine-to-Machine and Systems-to-Systems to interact. Technologies of M2M is an informative and mechanical way of data exchange and components of the item being manufactured supported with physical machines

(machines) of cyber and physical systems. For cyber and physical systems of digital twins there is the technology of S2S which helps the technological equipment to transmit to each other or to computerized system the production data which include:

- information about conditions of technological operations completion;
- information about material and parts availability which are used to manufacture the items.

In production with wireless communication nets the technologies of M2M and S2M must be done in the mobile version. Mobile production net provides data information exchange between stationed cyber and physical systems and mobile robotized transport system of a smart factory which transport blanks (parts) among work positions of cyber and physical system.

7. Human and machine interaction for cyber and physical production

Human and machine interface is necessary to organize information interaction among production operators and technological equipment (cyber and physical system). This interface has two options to be organized:

- through the cyber and physical system terminal where the operator controls directly the work modes of technological equipment (influence to the terminal itself);
- through interface of software when operator could control the working mode of cyber and physical systems remotely using the technologies of IoT and cloud services.

Cyber and physical systems control through terminal (technology of Humane-to-Machine) are used normally for repair and maintenance and during put into operation to synchronize the equipment in a single technological line.

To control cyber and physical system through the software they may use a computer of automatic work place or with a gadget of communication (tablet, phone and other). Sometimes they apply special means of communication like: virtual reality helmet or augmented reality glasses. Augmented (virtual) reality is a technology to be applied in the Industry 4.0 [6, 7] to organize the interaction of operator and cyber and physical system. In practice the augmented reality gives the user the production graphical data of cyber and physical equipment functionality.

A part of image which operator see in visual means (special glasses) is formed with the software and is applied over what the operator could really see. So operator watch the cyber and physical equipment (for example, to repair) and at the same time receives some hints of the gear being repaired which could be seen with glasses of augmented reality. Hints may include technical schemes, user instructions, recommended list of tools and other information which is necessary for the operator of the Industry 4.0 to make its job description.

Augmented reality technology is an informative support of operator activity to interact with the technologies of IoT, cloud technologies and the technologies of BigData. The place where the hint is displayed is automatically calculated according to the human position inside the cyber and physical production and the infrastructure component being watched. The hints itself are formed with artificial intelligence from the smart factory cloud environment.

8. Conclusion

To integrate informative technologies into a single infrastructure of the item designing company is a necessary condition to make the economy work with digital standards. Standards of digital economy require unique approach to organize business processes to accompany each step of the item life cycle.

Smart factory of the Industry 4.0 makes the stage of the item life cycle from production preparation, supply, item manufacturing and its quality control and later delivery of item to the exploitation point (consumer). Informative support of those stages requires from the smart factories the electronic system of control of the item engineering data with some subsystems to provide:

- to control the material purchase and parts which are necessary to manufacture the item and maintain the functionality of cyber and physical equipment;
- to control the finished product selling which are according to the technical and norm documentation (branch and state standards);
- to prepare the production technologically and processes of technological operation completion in the work chambers of cyber and physical systems;
- production planning (operative, tactical, strategical) in a smart factory and to plan the timing of the finished product delivery (graphs);

- to budget material and technical resources of the smart factory to make the manufacturing of production processes and to develop the company;
- to control the repair and maintenance of the item being manufactured in a smart factory and the items being exploited and other.

References

[1] Blunck H, Bendul J 2016 *Procedia CIRP* **57** 158-163
[2] Adenuga O T, Mpofu K, Kanisuru A M 2016 *Procedia CIRP* **57** 362-367
[3] Bakker O J, Chaplin J C, Silva L, Felli P, Sanderson D, Logan B and Ratchev S 2017 *Procedia CIRP* **63** 521-526
[4] Sormaz D N, Sarkar A, Ghosal S 2018 *Procedia manufacturing* **26** 1242-54
[5] Zheng D, Kovalenko I, Barton K and Tilbury D 2018 *Procedia manufacturing* **17** 326-333
[6] Gurjanov A V, Zakoldaev D A, Shukalov A V and Zharinov I O 2018 *IOP Conference Series: Materials Science and Engineering* **327** 022110
[7] Zakoldaev D A, Shukalov A V, Zharinov I O and Zharinov O O 2018 *Journal of Physics: Conference Series* **1015** 052033
[8] Sarkar A, Sormaz D 2018 *Procedia manufacturing* **17** 435-443
[9] Gonzalez-Briones A, Prieta F, Mohamad M S, Omatu S and Corchado J M 2018 *Energies* **11** 1928
[10] Deniaud S, Descamps P, Hilaire V, Lamotte O and Rodriguez S 2015 *Procedia computer science* **56** 520-525
[11] Zarreh A, Saygin C, Wan H, Lee Y, Bracho A and Nie I 2018 *Procedia manufacturing* **17** 1001-08
[12] Bogatyrev V A, Parshutina S A, Poptcova N A and Bogatyrev A V 2016 *Communications in Computer and Information Science* **678** 337-348

IOP Conf. Series: Materials Science and Engineering **582** (2019) 012008 doi:10.1088/1757-899X/582/1/012008

Classification of Cyber and Physical Systems of Industry 4.0

D A Zakoldaev[1], A V Gurjanov[2], A V Shukalov[1], I O Zharinov[1]

[1] Faculty of Information Security and Computer Technologies, Saint Petersburg National Research University of Information Technologies, Mechanics and Optics, 49, Kronverksky Av., Saint Petersburg, 197101, Russia
[2] Director, Stock Company «Experimental Design Bureau «Electroavtomatika» named after P A Yefimov, 40, Marshala Govorova St., Saint Petersburg, 198095, Russia

E-mail: mpbva@mail.ru

Abstract. The actual task is to create the classification system for cyber and physical technological equipment for smart factories of the Industry 4.0. Smart factories are a new type of production companies which work automatically. To project a smart factory they need to choose samples of cyber and physical systems of different purpose to unite them in automatic sections. To solve the choice task of production machine samples they need to create libraries of technical and tactical characteristics and its ontology description the base of which is the system to classify cyber and physical equipment. There is a classification scheme of industrial cyber and physical systems to work in a smart factory of item designing. The classification systems base is types of technological operations being realized in a company and also methods and technologies which the cyber and physical systems use. Cyber and physical systems functionality is described for mechanical treatment and assembly workshops of item designing company.

1. Introduction

To provide high-tech cyber and physical technological equipment is [1, 2] an actual development of economy industrial sector. The cyber and physical systems market offer today a wide nomenclature of machines which may solve the tasks of automatic production with its characteristics [3].

Each type of cyber and physical systems [4] is capable of producing only some technological operations using one or several digital technologies [5, 6]. Each types of cyber and physical systems has some machines for a particular task only with their methods [7-9] and technologies. To divide cyber and physical systems with technological operations being done and methods and technologies being used creates cyber and physical systems classification of industrial purpose.

This classification system is the base for automatic technological equipment systematization and can be developed in digital [10] production companies which are today viewed as the Industry 4.0.

The Industry 4.0 creates [11] in machine and item designing special companies to realize the item manufacturing production without human participation. Those companies of humanless and paperless technologies are called smart factories.

To synthesize [12, 13] the Industry 4.0 smart factory they need to implement criteria of company functionality and to find parameters of the production assets. So the classification system of cyber and physical equipment is an information resource to create the libraries of tactical and technical characteristics of automatic machines and its ontology to describe production processes being placed in the smart factory cloud.

2. Cyber and physical systems of industrial purpose

Cyber and physical systems are a new type of technological equipment with calculation resources to complete production tasks in automatic mode. Each cyber and physical system is for a limited number of operations only. To unite some cyber and physical systems and their cinematic interaction and agreement of their production data exchange protocols helps to create new flexible automatic productions. Flexible automatic productions (production sections) are the base for the Industry 4.0 smart factories.

The Industry 4.0 smart factories are a new type of item designing (machine designing) companies with closed loop of technological operations to produce high-tech items without humans. Self-organizing cyber and physical systems in the Industry 4.0 is for technological operations from technological maps and routes with algorithms from the smart factory clouds.

In the item designing the cyber and physical equipment is placed within two production workshops of a smart factory:

- the workshop of mechanical treatment for production operations of additive technologies to prepare the items (assembly units);
- the assembly workshop for production operations with technologies of Machine-to-Machine, Systems-to-Systems to prepare assembly units (surface montage of electric and radio components on the printed circuit boards) and to finish the item assembly.

Assembly units transportations among the workshops and within a workshop among cyber and physical systems is done with manipulator robots under the computerized system. Control algorithms and production tasks for robots and cyber and physical systems are formed in the smart factory cloud using the analysis results (BigData algorithms) for standard operative plans of technological equipment.

The classification of industrial purpose cyber and physical systems is done with technological operations from mechanical treatment and assembly workshops of the Industry 4.0 smart factory. The main classes of item designing automatic production equipment are:

- additive production equipment to complete the technological operations of 3D-printing and to cover the assembly units with lacquering and electrotyping covers;
- surface montage equipment with standard set of technological operations to place electric and radio components on printed boards, components soldering and other with subsequent washing of mounted assembly units from the flux residue, solder or other contaminations;
- transport production system equipment to load (unload) trays of cyber and physical systems to mark the items being produced to storage the items, assembly units and finished products;
- production control equipment to check the quality of humanlessly produced items with necessary technological operations.

3. Cyber and physical systems of additive production

The main purpose of cyber and physical systems in additive production is to complete technological operations of layer by layer adding (growing) of materials on the base plate. In the item designing this technological operation is done with three different types of cyber and physical systems (see figure 1):

- 3D-printers to print out the items automatically;
- production machines of lacquering which lacquers the assembly units (printed boards with mounted electric and radio components);
- electrotyping production machines to put the necessary covers of electrotyping on the made of metal items to grant them the necessary physical and chemical properties (anti-corrosivity, high electric conductivity, anti wearing out the surfaces which are necessary rugged, the surface preparation for soldering and other).

3D-printer is a complicated technological device (machine) which works automatically with numeric control (calculating device controls the process of the technological operations completion). The 3D-printer functionality is based on layer by layer creation (printing) of an item from a solid powder.

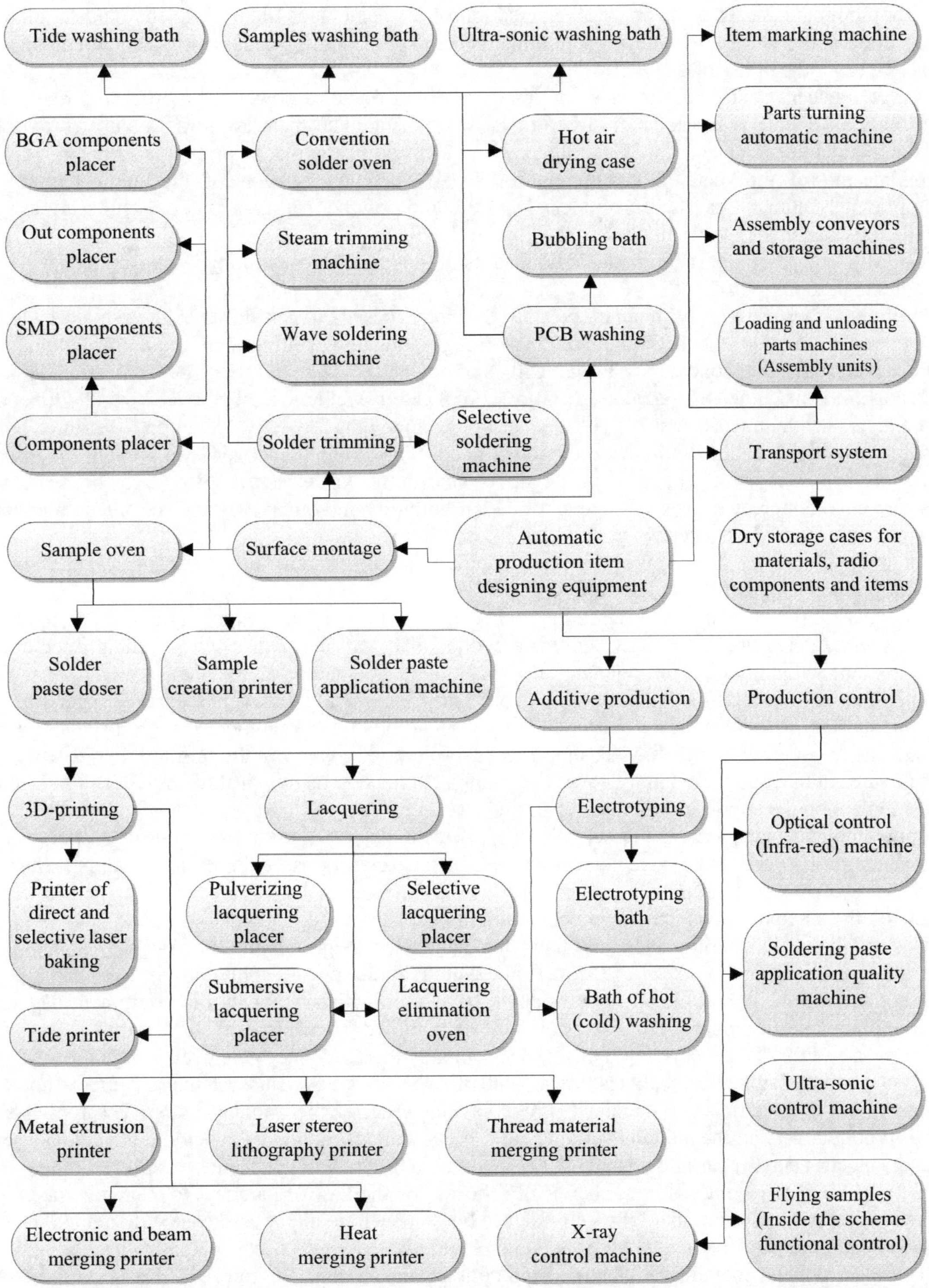

Figure 1. Cyber and physical systems classification for the Industry 4.0 item designing company.

There are several methods of 3D-printing and some additive technologies that could be supported by a 3D-printer. The most popular in practice are the following technologies:

- extrusion or layer by layer creation of a solid object (item) with the method of melt-down. The melted down material is being placed on the base with drops of small size which are rapidly cooled down and united with each other and this is how the current layer is formed (printed layer);

- photopolymerization or layer by layer creation of a solid object (item) with the method of laser stereolithography. The liquid material of polymer is being placed on the base which is cooled down with a laser beam in the ultraviolet range;

- direct (selective) sintering or layer by layer creation of a solid object (item) with the method of electron beam melt-down, heat melt-down or laser sintering. The material is melted down (metallic powder) with a heating head of a 3D-printer, with an electron beam in vacuum or a laser beam. Step by step sintering of the metallic powder is for layer by layer creation of the item. Unlike the photopolymerization where the base material is the liquid polymer (organic material) sintering use the metallic powder as the base material and the cover refusing method is the same — the influence of laser emission. And the laser emission range in both cases of those additive technologies and its output power are quite different.

In practice they use additive technologies of 3D-printers based on melt-down of gauze material under the electronic emission (the method of welding), tide printing and printing with liquid glue (gluing of a powder material) being conducted automatically and other.

Production machine of lacquering is a complicated technological device which functions automatically with numeric control. The lacquering machine puts protective cover on the printed board surface with installed radio and electronic components. This protective cover is to prevent the printed board surface being influenced with negative environmental effects (Very high or very low environmental temperature, very high level of the air humidity and other) on the electronic components installed on the board or materials from which the board is made.

There are several methods of lacquering cover putting on the printed circuit boards surface. The most popular methods in practice are the following:

- lacquering putting with pulverizing method. This kind of additive technology is used compressed air of excessive pressure in the spray and this how the pulverization is being done on the printed circuit board surface with liquid lacquering material;

- application of lacquering with the immersion. This kind of additive technology uses special bathes with lacquering material in liquid state where the printed boards are submersed with the manipulator robot. The lacquering material viscosity creates protective cover on the printed board after the submersion. In practice for a quality cover they submerse the boards two or three times (But before the next submersion the applied cover must be hardened);

- application of lacquering with selective (tidal) technology. This kind of additive technology in practice apply the lacquering cover on the limited in square sections of the printed boards (After the elements repair when the original lacquering cover was damaged in a small area) or to apply lacquering covers on the assembly units which are non-standard in construction.

The number of lacquering layers and their thickness affects significantly the assembly unit weight. In practice the maximum number of lacquering covers is three. If they apply more it may damage the final cover (In some places the air bubbles could be formed which may result in equipment damage). After application of each lacquering layer on the PCB (Printed Circuit Board) surface it is necessary to reject them and for that purpose the Industry 4.0 company they use automatic oven of rejection.

Electrotyping automatic machines are a kind of cyber and physical systems to support additive technologies of electric and chemical processes to process mechanical units. To apply good electrotyping covers on the metallic surfaces they apply work chambers of cyber and physical systems after the electric current passing through the electrolyte in the bath where the item is placed. The power (density) of current the electrolyte is chemically chosen (by the Law of Faraday) to make the molecules of metal layer not just to bombard the item surface but to penetrate the upper layer of the item. And this is how the applied electrotyping cover is secured in surface.

The most popular methods of electrotyping contain nickel, chrome, gold, silver and other metals. Electric and chemical properties of electrotyping are done with the same cyber and physical equipment. The difference how to apply the cover of different metals is based on the containing of electrolyte, its temperature and the mode of completion of some electric and chemical processes. The choice of modes, electrolyte work temperature and other details how to complete the technological operations of electrotyping is done with electrotyping cyber and physical system controller under its software placed in the cloud (libraries of technological processes completion algorithms).

The final stage of electrotyping over metallic items is the procedure of electrolyte residue elimination from the part surface. This procedure is done in special bathes of hot (cold) parts washing. The washing bathes are a kind of cyber and physical automatic equipment where the program controls the work modes of manipulator robots which submerge parts into washing liquid.

Electrotyping is a hazardous production which affects negatively the human health. To organize a production division with cyber and physical systems of electrotyping in automatic mode (without humans) may increase the quality of finished product and the level of operator personal security in a smart factory.

4. Cyber and physical systems of radio and electronic components surface montage

The surface montage of electric and radio components is done over the printed boards with cyber and physical equipment which is a part of assembly workshop of the Industry 4.0 smart factory. The surface montage includes the following technological operations:

- the application of soldering paste over the printed boards (sample printing);
- the placement of radio and electric components over the printed boards (element outputs are connected with PCB pads which were already soldered);
- solder trimming for an electric contacts of element outputs and printed boards pads and element mechanical fixture (to make the fixture stronger elements may have some ceramic bases for support, glue, mastics and other which was installed or applied before);
- PCB washing from the flux residue, solder and other contaminations which are formed after the item manufacturing technological cycle.

The solder paste is applied with special cyber and physical system:

- solder paste dozer to form a dose of solder which will be applied on the pads of the printed boards (the dosage depends on the pads linear sizes and radio and electric components output step);
- sample creation printer which produces technological masks with holes from metal which is placed over the board when the solder is being provided. The mask holes are the same with the PCB pads of the upper and lower layers of the board;
- soldering paste application machine which precisely put together the printed board and the mask (sample) to apply some doses of soldering paste over the printed board. The roller with solder presses the surface of samples with which holes the solder is being penetrated on the printed board pads.

The placement of electric and radio components over the printed board surface is done with the special type of cyber and physical equipment which can be classified with the components they install:

- the placer of output components or electric and radio items with the package type of DIP (Dual In-line Package);
- the placer of planar components or electric and radio items with the package type of SMD (Surface Mount Device);
- the placer of ball output components or electric and radio items with the package type of BGA (Ball Grid Array).

All type of placers work automatically and support one side or two sides component placement type on the printed board. Placer gears put away the electric and radio components from the belt feeder (a bobbin which cells have components already pre-arranged) with a vacuum nozzle and transfer the elements on the printed board surface (on the necessary scheme place). The element positioning precision on the printed board surface is a technical specification of cyber and physical system and may define the minimum step of elements output to work with which the system was created. The controller controls the cyber and physical system gears which memory has a program with digital description of the board geometry and the board collective scheme in its digital form.

In multi-nomenclature production technical documentation of the smart factory product (item) which is placed in company cloud and being extracted from there automatically. The transfer of technical documentation and control program into the memory of cyber and physical system according with requirements and terms which is in operative plan of the item manufacturing in company.

Solder trimming in item designing company to create a secure electric connection of electric and radio component output and PCB pad. Technological operation of solder trimming is done with the following types of cyber and physical equipment:

- convection oven of solder trimming inside of which the solder is melted down under the heat convection beam with the necessary temperature stability. The oven has work modes where the temperature beam profile can be changed with the program which depends on the board area where the components are placed (elements have different specifications as the robustness and body stability to the temperatures of soldering);

- the vapor solder trimming oven where the solder is melted down with steam. There are the technologies of soldering in the steam phase and steam phase systems of soldering. The heat transferring on the board surface helps the selective soldering of components. The selective way divide the printed board into areas and creates special modes of soldering in each area;
- wave soldering machine where the PCB bottom is submerged into a bath of prepared solder. This machine is for montage of the components which have the DIP-bodies (pin outputs for penetrating montage). After the PCB bottom is submerged into a bath of prepared solder (the prepared solder is administrated to the bottom part of the PCB as a wave) on the elements outputs there is solder residue which can penetrate to the holes between layers. With the capillary effect the solder reaches the upper surface of the board which makes a pass through solder connection which touches the pad of transfer hole and element output.

After technological operation of solder paste application, components placement and solder trimming there are some contaminations on the PCB surface which makes the item designing components to be of lower quality. To prevent a scrap to be produced after those technological operations there is the procedure of printed boards washing.

The printed boards washing is done with cyber and physical systems which differs in board and washing liquid way of interaction. There are the following types of automatic technological equipment to wash out the printed boards:

- tidal washing bath where the printed board is washed with the cleaning liquid under pressure which helps to take out all the mechanical contaminations from the board;
- ultra-sonic washing bath where the washing out of the printed boards surface is done with amplitude movements of the cleaning liquid of ultra-sonic frequency. Those movements provoke contractions of the liquid which spreads to the printed board surface;
- bubbling bath to wash out the printed board surface is done with air bubbles generated in cleaning liquid with special device.

Washing bathes are to delete contaminations from the printed board surface and from the sample surfaces which are used to apply the solder paste. When the item washing is done (PCB, sample) it is necessary to dry it out with hot air in the drying case.

5. Cyber and physical systems of smart factory transport infrastructure

Item transportation (assembly units, products) in the production division of an item designing company which functions automatically is done with robot manipulators which are a part of the Industry 4.0 smart factory transport infrastructure. The Industry 4.0 smart factory transport infrastructure has the following cyber and physical systems:

- loading (unloading) machines of parts (assembly units) with the following functions: item transferring and reception to the production machine pallets to complete technological operations (for example, components montage) and the item evacuation from the machine pallet to transport it to the next cyber and physical system;
- product marking machines put special barcodes on the items to identify the number of item being manufactured, number and containing of the technological operations done with the item and other. Barcodes are used for marking the products, radio frequency marks and other;
- machines of item turning where manipulator robots with function of linear rotation around the axe to complete some technological operations (for example, one by one montage of electric and radio components on the both surfaces of the printed boards using the equipment of one side montage);
- cases of components dry storage, parts and products (finished products) which is the part of the storage systems inside the workshop of the items being manufactured in the company.

To unite cyber and physical systems of smart factory transport infrastructure with production machines may create in item designing companies the assembly conveyors which function automatically and collective systems to organize the warehouse storage of the finished products.

6. Cyber and physical systems of production control

Production control is to monitor the quality of the item designing company product. Production control has the following types of cyber and physical systems:

- machine of optical (infra-red) control to evaluate the quality of montage and how much the given assembly units is according to its requirements, the item size and how much they correspond to the item technical documentation and other;
- ultra-sonic control machine to evaluate the quality of the given item and to find out the hidden defects which cannot be found visually with the machines of optical range (with the machines of ultra-sonic control can be found micro-fissures of metallic items produced with additive production);
- machines of X-ray control to evaluate the quality of the given item and to find out the hidden deep defects in the thickness of materials and parts (The X-ray control machines may find fissures and break-downs of printed conduits deep inside the PCB layers);
- quality control machines of the soldering paste application to evaluate the volume the paste applied on the pads of printed boards (the volume is defined after multi-frequency measurement with mathematical calculation of the paste parameters; the height is evaluated, square, volume, paste displacement from the pads and other);
- the machines of inner-schemes control (flying probes) to evaluate the quality of electric connection after the soldering electric and radio components on the printed board surface. After the verifying they may control the parameters of item functionality in different areas of the board electrical scheme. In some points defined with the controller program they apply detectors (probes).

7. Conclusion

To classify cyber and physical systems of industrial purpose is the base to systemize the equipment and technology to complete automatically technological processes of item designing components manufacturing. Given in figure 1 production machines are industrial objects which must be computerized first.

Computerized control system coordinates cyber and physical systems interaction on the algorithms of self-organization production machines. The algorithms of cyber and physical systems self-organization placed in a smart factory cloud and they function on the data analysis of:

- results of theme and production orders completion in an item designing company;
- well functionality of cyber and physical technological equipment;
- the availability of materials and parts which are necessary to complete technological processes and other.

Computerized control system in production has calculation resources of physical world and calculation resources placed in virtual (cloud) part of the smart factory. Smart factory production activity control is done with operators and interface of computerized control system interaction with the software of production processes completion automatizing.

References

[1] Nitschke C S, Williams S P 2018 *Procedia computer science* **138** 580-586
[2] Rodriguez S, Galland S, Gaud N 2015 *Procedia computer science* **56** 526-531
[3] Bogatyrev V A, Parshutina S A, Poptcova N A and Bogatyrev A V 2016 *Communications in computer and information science* **678** 337-348
[4] Wu W, Li W, Law D and Na W 2015 *Procedia engineering* **118** 1266-73
[5] Chernenkiy V, Gapanyuk Y, Terekhov V, Revunkov G and Kaganov Y 2018 *Procedia computer science* **145** 143-152
[6] Schuhmacher J, Hummel V 2016 *Procedia CIRP* **54** 19-24
[7] Ionkina K, Svistunov A, Galin I, Onykiy B and Pronicheva L 2018 *Procedia computer science* **145** 254-259
[8] Nyanga L, Van der Merwe A F, Matope S and Dewa M T 2015 *Procedia CIRP* **28** 167-172
[9] Qu S, Wang J, Govil S and Leckie J O 2016 *Procedia CIRP* **57** 55-60
[10] Yu J, Liu J, Xiang L and Zhou J 2017 *Procedia IUTAM* **22** 107-114
[11] Taratukhin V, Yadgarova Y 2018 *Procedia computer science* **123** 479-487
[12] Gurjanov A V, Zakoldaev D A, Shukalov A V and Zharinov I O 2018 *IOP Conference Series: Materials Science and Engineering* **327** 022110
[13] Zakoldaev D A, Shukalov A V, Zharinov I O and Zharinov O O 2018 *Journal of Physics: Conference Series* **1015** 052033

IOP Conf. Series: Materials Science and Engineering **582** (2019) 012009 doi:10.1088/1757-899X/582/1/012009

Determination of Rational Profile of Fine-Module Internal Ratchet Teeth When Using Cutter

O V Sharkov[a], S I Koryagin[b] and N L Velikanov[c]

Immanuel Kant Baltic Federal University, Kaliningrad, 236016, Russia

E-mail: [a]osharkov@kantiana.ru, [b]skoryagin@kantiana.ru, [c]nvelikanov@kantiana.ru

Abstract The article discusses the method for determining the length of straight sections of the working surfaces of internal fine-module ratchet teeth. A design scheme is proposed and the analysis of mathematical expressions describing the process of shaping the working edges of the ratchet teeth when cutting with a cutter by the generating method is performed. Mathematical dependences for calculation of the straight section length of the front and back edges of the ratchet teeth are obtained, allowing to determine rational profile at the design stage. It is established that the straight section on the front edge of the internal ratchet teeth can exist only at a certain value of its angle of inclination. The use of fine-module ratchet teeth of rational profile can improve the kinematic characteristics and increase the load capacity of ratchet engagement.

1. Introduction

Ratchet engagement is used in various mechanisms from ancient times to the present day. It has high load capacity, as it transmits the load by normal forces and has a relatively simple geometry and manufacturing techniques of working surfaces [1–14].

Some experts have suggested that the area of effective use of ratchet engagement is limited due to the occurrence of shock loads during the activation. Such statements are only partially valid and refer to the ratchet engagement with the traditional geometry of the working surfaces.

The reduction of the ratchet engagement module and the use of rational geometry of the working surfaces can reduce its characteristic disadvantages, while retaining the main advantages.

The paper [10] shows the possibility of manufacturing fine-module ratchet engagement m_t=0.3–0.8 mm, in which the contact of the working surfaces occurs along the straight sections. Such engagement can be used in free-wheel mechanisms.

Currently, more than fifty methods of manufacturing teeth of various profiles are used [15–26]. One of the most productive and widespread is the method of gear cutting by generating process. While cutting the internal teeth is only possible with the use of cutter.

When cutting fine-module ratchet teeth by generating method, in addition to profile straight sections there are areas that are outlined along transition curves [27].

To improve the load and kinematic characteristics of the ratchet engagement, it is necessary to produce ratchet teeth with a rational profile that ensures the greatest length of straight sections of the front l_{t1} and back l_{t2} edges of the teeth (Figure 1).

We consider a method for determining the length of straight sections of the profile of internal teeth cut by the cutter, based on the general theory of shaping cylindrical surfaces by the generating method [15–17].

2. Determination of the profile of fine-module internal ratchet teeth

2.1. Front edge of ratchet teeth profile

We introduce the following notation (Figure 2): l_{t1} and l_{t2} – the lengths of the straight sections of the front and rear edges of the ratchet teeth, r_U and r_{U1} – the radii of circles passing through the recesses and protrusions of the cutter teeth; r – the radius of the circumference of the workpiece, passing through the recesses of the internal ratchet teeth, γ_1 and γ_2 – the inclination angles of the front and back teeth edges, τ – is the pitch arc, m_t – is the ratchet teeth module, H_t – is the theoretical height of the ratchet engagement.

The extreme point of the straight section of the front edge of the internal ratchet teeth l_{t1} is profiled by the ratchet teeth peak, i.e. it lies on the circumference of the radius $r_{U1} = r_U + H_t$.

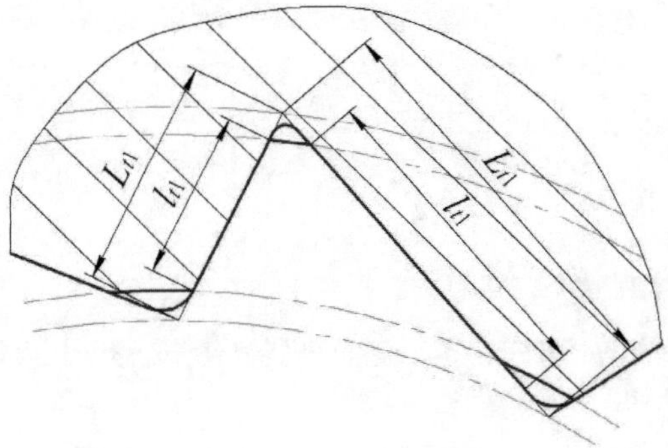

Figure 1. Scheme of straight sections on the front and back edges of the internal fine-module ratchet tooth.

The coordinates of the front edge profile of the cutter teeth in the case $x' = l_{t1}$ are determined by the system of expressions [27]

$$\begin{cases} x_{R1} = -(r - r_U)\sin\Theta'_U + r\sin(\Theta'_U - \Theta') + x'\sin(\gamma_1 + \Theta'_U - \Theta'); \\ y_{R1} = -(r - r_U)\cos\Theta'_U + r\cos(\Theta'_U - \Theta') + x'\cos(\gamma_1 + \Theta'_U - \Theta'), \end{cases} \quad (1)$$

where Θ' and Θ'_U – are the angles of workpiece and cutter rotation at the moment of shaping of the front edge of the ratchet tooth, $\Theta'_U = \Theta' r / r_U$.

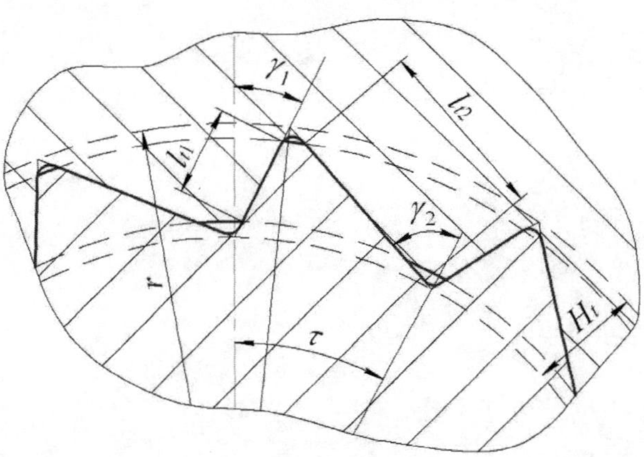

Figure 2. Design scheme for determining straight sections on the front and back edges of the internal fine-module ratchet tooth.

Then, for the extreme shaped point of the straight section of the front edge of the ratchet teeth, the condition $x_{R1}^2 + y_{R1}^2 = (r_U + H_t)^2$ is fulfilled. (2)

After substituting the system of expressions (1) into equation (2), we obtain

$$x_{R1}^2 = r^2\sin^2(\Theta'_U - \Theta') - 2r(r - r_U)\sin(\Theta'_U - \Theta')\sin\Theta'_U$$
$$+ 2rl_{t1}\sin(\Theta'_U - \Theta')\sin(\gamma_1 + \Theta' - \Theta'_U) + (r - r_U)^2\sin^2\Theta'_U$$
$$- 2l_{t1}(r - r_U)\sin\Theta'_U\sin(\gamma_1 + \Theta'_U - \Theta') + l_{t1}^2\sin^2(\gamma_1 + \Theta' - \Theta'_U);$$

$$y_{R1}^2 = r^2\cos^2(\Theta_U' - \Theta') - 2r(r - r_U)\cos(\Theta_U' - \Theta')\cos\Theta_U'$$
$$+ 2rl_{t1}\cos(\Theta_U' - \Theta')\cos(\gamma_1 + \Theta_U' - \Theta') + (r - r_U)^2\cos^2\Theta_U'$$
$$- 2l_{t1}(r - r_U)\cos\Theta_U'\cos(\gamma_1 + \Theta_U' - \Theta') + l_{t1}^2\cos^2(\gamma_1 + \Theta' - \Theta_U').$$

Therefore

$$x_{R1}^2 + y_{R1}^2 = r^2 - 2r(r - r_U)\cos\Theta' + 2rl_{t1}\cos\gamma_1$$
$$+ (r - r_U)^2 - 2l_{t1}(r - r_U)\cos(\gamma_1 - \Theta') + l_{t1}^2. \quad (3)$$

Given that $\cos(\gamma_1 - \Theta') = \cos\gamma_1 + l_{t1}/r$, the expression (3) takes the form

$$r_U^2 + 2r_U H_t + H_t^2 = r^2 - 2r_U r + r_U^2 + r^2 + l_{t1}^2 - 2r(r - r_U)\cos\Theta'$$
$$- 2l_{t1}(r - r_U)\left[\cos\gamma_1 + l_{t1}/r\right] + 2rl_{t1}\cos\gamma_1 \;. \quad (4)$$

The value of the angle Θ' is determined by the formula $\sin\Theta' = y_N/r$, where x_N are determined by the system of equations [27]. Then the expression (4) can be represented as

$$H_t(2r_U + H_t) = 2r(r - r_U) + l_{t1}^2 - 2(r - r_U)[r\cos^2\gamma_1 + l_{t1}\cos\gamma_1$$
$$- \sin\gamma_1\sqrt{r^2\sin^2\gamma_1 - 2rl_{t1}\cos\gamma_1 - l_{t1}^2}\,] - 2l_{t1}(r - r_U)[\cos\gamma_1 + l_{t21}/r] + 2rl_{t21}\cos\gamma_1$$

or

$$rH_t(2r_U + H_t) = 2(r - r_U)[r^2\sin\gamma_1 - (2l_{t1}\cos\gamma_1 + l_{t1}^2)$$
$$- r\sin\gamma_1\sqrt{r^2\sin^2\gamma_1 - (2rl_{t1}\cos\gamma_1)}\,] + r(l_{t1}^2 + 2rl_{t1}\cos\gamma_1). \quad (5)$$

We introduce the notation $2rl_{t1}\cos\gamma_1 + l_{t1}^2 = z_1$ and write the expression (5) as

$$rH_t(2r_U + H_t) = 2(r - r_U)r^2\sin^2\gamma_1 - 2(r - r_U)z_1$$
$$+ rz_1 - 2r(r - r_U)\sin\gamma_1\sqrt{r^2\sin^2\gamma_1 - z_1}$$

or

$$2r(r - r_U)r^2\sin^2\gamma_1 - rH_t(2r_U + H_t) + z_1(2r_U - r)$$
$$= 2r(r - r_U)\sin\gamma_1\sqrt{r^2\sin^2\gamma_1 - z_1}\,.$$

Finally, taking into account $H_t = m_t$ we get

$$z_1^2(2r_U - r)^2 + z_1 2r[rr_U(r - r_U)\sin^2\gamma_1 - m_t(2r_U - r)(2r_U + H_t)]$$
$$+ 2r^2 m_t(2r_U + m_t)[m_t(2r_U + H_t) - 4r(r - r_U)\sin\gamma_1] = 0. \quad (6)$$

The expression (6) can be reduced to a quadratic equation [28] $a_1 z_1^2 + b_1 z_1 + c_1 = 0$, where $a_1 = (2r_U - r)^2$;

$b_1 = 2r[2rr_U(r - r)\sin^2\gamma_1 - m_t(2r_U - r)(2r_U + m_t)]$;

$c_1 = r^2 m_t(2r_U + m_t)[m_t(2r_U + m_t) - 4r(r - r_U)\sin^2\gamma_1]$.

The roots of the quadratic equation [28] are determined as

$$z_1 = \frac{-b_1 \pm \sqrt{b_1^2 - 4a_1c_1}}{2a_1} . \quad (7)$$

Taking into account the previously introduced notation z_1, the formula for determining the straight section length of the front edge of the ratchet teeth can be represented as

$$l_{t1} = -r\cos\gamma_1 \pm \sqrt{r^2\cos^2\gamma_1 + z_1} . \quad (8)$$

The sign before the radical in formulas (7) and (8) is taken based on the results of calculations.

2.2. Back edge of ratchet teeth profile

The problem of determining the straight section length of the back edge of the internal fine-module ratchet teeth l_{t2} is solved similarly to the problem considered previously.

The coordinates of the back edge tooth profile are determined by the expression system [27] when $y' = l_{t2}$,

$$\begin{cases} x_{R2} = r\sin(\tau + \Theta''_U - \Theta'') - (r - r_U)\sin\Theta''_U - y'\sin(\gamma_2 - \tau + \Theta'' - \Theta''_U); \\ y_{R2} = r\cos(\tau + \Theta''_U - \Theta'') - (r - r_U)\cos\Theta''_U + y'\cos(\gamma_2 - \tau + \Theta'' - \Theta''_U), \end{cases} \quad (9)$$

where Θ'' and Θ''_U – are the angles of workpiece and cutter rotation at the moment of shaping of the back edge of the ratchet tooth, $\Theta''_U = \Theta'' r/r_U$.

After substituting the system of expressions (9) into equation (2), we obtain

$$\begin{aligned} x_{R2}^2 &= r^2\sin^2(\tau + \Theta''_U - \Theta'') - 2r(r - r_U)\sin(\tau + \Theta''_U - \Theta'')\sin\Theta''_U \\ &- 2rl_{t2}\sin(\tau + \Theta''_U - \Theta'')\sin(\gamma_2 - \tau + \Theta'' - \Theta''_U) + (r - r_U)^2\sin^2\Theta''_U \\ &+ 2l_{t2}(r - r_U)\sin\Theta''_U\sin(\gamma_2 - \tau + \Theta'' - \Theta''_U) + l_{t2}^2\sin^2(\gamma_2 - \tau + \Theta'' - \Theta''_U); \end{aligned}$$

$$\begin{aligned} y_{R2}^2 &= r^2\cos^2(\tau + \Theta''_U - \Theta'') - 2r(r - r_U)\cos(\tau + \Theta''_U - \Theta'')\cos\Theta''_U \\ &- 2rl_{t2}\cos(\tau + \Theta''_U - \Theta'')\cos(\gamma_2 - \tau + \Theta'' - \Theta''_U) + (r - r_U)^2\cos^2\Theta''_U \\ &+ 2l_{t2}(r - r_U)\cos\Theta''_U\cos(\gamma_2 - \tau + \Theta'' - \Theta''_U) + l_{t2}^2\cos^2(\gamma_2 - \tau + \Theta'' - \Theta''_U). \end{aligned}$$

Therefore

$$\begin{aligned} x_{R2}^2 + y_{R2}^2 &= r^2 - 2r(r - r_U)\cos(\tau - \Theta'') + 2rl_{t2}\cos\gamma_2 \\ &+ (r - r_U)^2 - 2l_{t2}(r - r_U)\cos(\gamma_2 - \tau + \Theta'') + l_{t2}^2. \end{aligned} \quad (10)$$

Given that $\cos(\gamma_2 - \tau + \Theta'') = \cos\gamma_2 + l_{t2}/r$, the expression (10) takes the form

$$\begin{aligned} r_U^2 + 2r_U H_t + H_t^2 &= r^2 + r^2 - 2r_U r + r_U^2 + l_{t2}^2 - 2r(r - r_U)[\cos\tau\cos\Theta'' - \sin\tau\sin\Theta''] \\ &+ 2rl_{t2}\cos\gamma_2 - 2rl_{t2}(r - r_U)[\cos\gamma_2 + l_{t2}/r] . \end{aligned} \quad (11)$$

The value of the angle Θ'' is determined by the formula $\cos\Theta'' = y_N/r$, where y_N are determined by the system of equations [27]. Then the expression (11) can be represented as

$$\begin{aligned} H_t(2r_U + H_t) &= 2r(r - r_U) + l_{t2}^2 - 2(r - r_U) \\ &\times[\sin\gamma_2\sqrt{r^2\sin^2\gamma_2 - 2rl_{t2}\cos\gamma_2 - l_{t2}^2} \\ &+ \cos\gamma_2(r\cos\gamma_2 + l_{t2})] + 2rl_{t2}\cos\gamma_2 \\ &- 2l_{t2}(r - r_U)(\cos\gamma_2 + l_{t2}/r) \end{aligned} \quad \text{or} \quad \begin{aligned} rH_t(2r_U + H_t) &= 2(r - r_U)[r^2 - r\sin\gamma_2 \\ &\times\sqrt{r^2\sin^2\gamma_2 - 2rl_{t2}\cos\gamma_2 - l_{t2}^2} - \\ &- r^2\cos^2\gamma_2 - 2rl_{t2}\cos\gamma_2 - l_{t2}^2] + \\ &+ (2rl_{t2}\cos\gamma_2 + l_{t2}^2)r \end{aligned} \quad (12)$$

We introduce the notation $2rl_{t2}\cos\gamma_2 + l_{t2}^2 = z_2$ and write the expression (12) as

$$rH_t(2r_U + H_t) = 2(r - r_U)[r^2\sin^2\gamma_2 - r\sin\gamma_2\sqrt{r^2\sin^2\gamma_2 - z_2} - z_2] + rz_2$$

or

$$2r(r - r_U)\sin\gamma_2\sqrt{r^2\sin^2\gamma_2 - z_2} = 2r^2(r - r_U)\sin^2\gamma_2 - 2z_2(r - r_U) + rz_2 - rH_t(2r_U + H_t).$$

Finally, taking into account $H_t = m_t$ we get

$$z_2^2(2r_U - r)^2 + z_2[4r^2(r - r_U)^2\sin^2\gamma_2 + 4r^2(r - r_U) \times (2r_U - r)\sin^2\gamma_2 - 2rm_t(2r_U - r)(2r_U + m_t)] - 4r^3m_t(r - r_U)(2r_U + m_t)\sin^2\gamma_2 + r^2m_t^2(2r_U + m_t)^2 = 0. \quad (13)$$

The expression (13) can be reduced to a quadratic equation [28] $a_2z_2^2 + b_2z_2 + c_2 = 0$, where $a_2 = (2r_U - r)^2$;

$$b_2 = 2r[2rr_U(r - r_U)\sin^2\gamma_2 - m_t(2r_U - r)(2r_U + m_t)];$$

$$c_2 = r^2m_t(2r + m_t)[m_t(2r_U + m_t) - 4r(r - r_U)\sin^2\gamma_2].$$

The roots of such quadratic equation [28] are determined as

$$z_2 = \frac{-b_2 \pm \sqrt{b_2^2 - 4a_2c_2}}{2a_2}. \quad (14)$$

Taking into account the previously introduced notation z_2, the formula for determining the straight section length of the back edge of the ratchet teeth can be represented as

$$l_{t2} = -r\cos\gamma_2 \pm \sqrt{r^2\cos\gamma_2 + z_2}. \quad (15)$$

The sign before the radical in formulas (14) and (15) is taken based on the results of calculations.

3. Calculation results and discussion

The effect of the module m_t, the angle of inclination γ_1 and the radius r on the length of the straight sections of the front l_{t1} and back l_{t2} edges of the internal ratchet teeth is shown in Figure. 3–6.

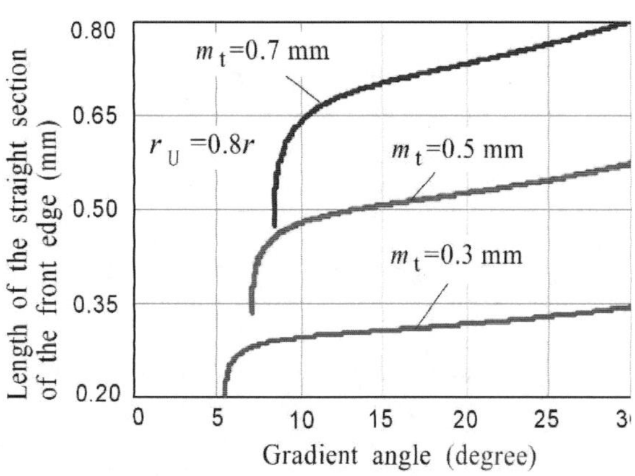

Figure 3. The effect of the angle of inclination and the module on the straight section length of the front edge of the internal ratchet tooth.

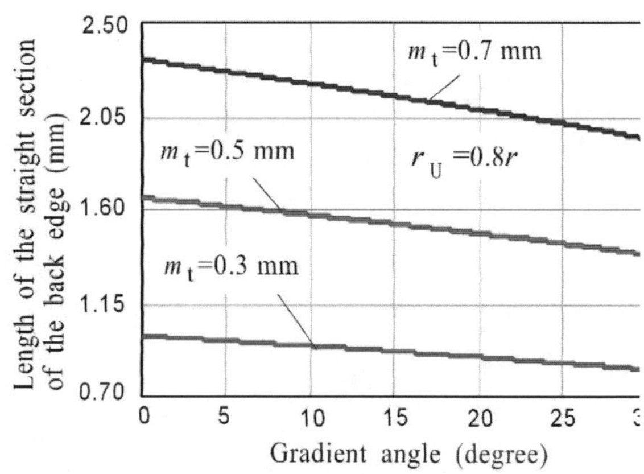

Figure 4. The effect of the angle of inclination and the module on the straight section length of the back edge of the internal ratchet tooth.

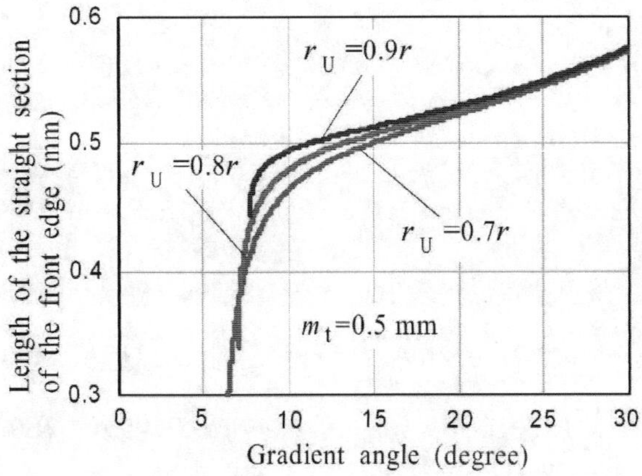

Figure 5. The effect of the cutter radius on the straight section length of the front edge of the internal ratchet tooth.

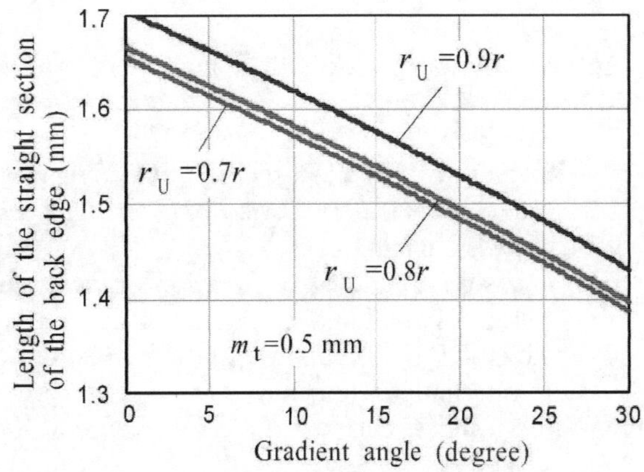

Figure 6. The effect of the angle of inclination and the module on the straight section length of the back edge of the internal ratchet tooth.

The initial data for calculations are: r =50 mm, $r_U = (0.7-0.9)r$, m_t =0.3–0.7 mm, γ_1 =0–30°. Calculations are made according to formulas (8) and (15).

The calculations performed show the following results.

The increase in the module m_t by 2.3 times leads to a linear increase in the length of the straight section of the front l_{t1} and back edges l_{t2} of the ratchet teeth, respectively, by 2.32 times.

Changing the angle of inclination γ_1 in the range from 0 to 30° leads to a nonlinear growth (1.38–1.52 times) of the straight section length of the front edge and to a linear decrease (approximately by 1.19 times) in the straight section of the back edge.

Changing the cutter radius r_U in possible range (by 1.28 times) does not have a noticeable effect on the straight section length of the front edge of the ratchet tooth.

4. Conclusion

The results show that the straight section on the front edge of the internal teeth can be obtained only starting from a certain angle γ_1, which is approximately equal to γ_1=6–8°.

The cutter radius for cutting ratchet teeth is efficiently taken from the ratio $r_U = (0.80-0.85)r$.

The parameters z_1 and z_2 defined by the dependencies (7) and (14) do not exist for all angles γ_1, because the expression under the radical becomes positive at a certain value.

It should be borne in mind that despite the front edge of the ratchet teeth is shorter, it has the greatest impact on the load capacity of engagement.

Dependences (8) and (15) can be used to assess the possibility of obtaining the rational profile of internal fine-module ratchet teeth when cutting with the cutter.

References

[1] Wu T, Wang G, Li J and Yan K 2018 *Journal of Materials Processing Technology* **259** 141–149
[2] Li W and Pang D 2018 *Mechanism and Machine Theory* **130** 523–538
[3] Jindal U C 2010 *Machine Design* (Delhi: Pearson Education India) p 892
[4] Gope P C 2012 *Machine Design: Fundamentals and Applications* (New Delhi: PHI Learning Pvt. Ltd.) p 1312
[5] Fuglede N and Thomsen J J 2016 *Journal of Sound and Vibration* **366** 447–470
[6] Wang Y, Ji D and Zhan Kai 2013 *Mechanism and Machine Theory* **70** 380–393
[7] Alyukov S V 2014 *Russian Engineering Research* **34(9)** 549–553
[8] Bondaletov V P 2008 *Russian Engineering Research* **28(9)** 845–848
[9] Neale M J 2001 *The tribology handbook* (Oxford: Butterworth-Heinemann) p 640
[10] Sharkov O V and Koryagin S I 2017 *Russian engineering research* **37(1)** 9–12
[11] Jalili N, Wagner J and Dadfarnia Mohsen 2003 *Mechatronics* **13(8–9)** 933–956

[12] Penkov I and Aleksandrov D 2011 *International Review of Mechanical Engineering* **5(7)** 1213–1218
[13] Morales F J and Benitez F G 2019 *Mechanism and Machine Theory* **134** 197–212
[14] Alyukov S V 2010 *Russian Engineering Research* **30(6)** 535-542
[15] Radzevich S P 2018 *Theory of Gearing: Kinematics, Geometry, and Synthesis* (Boca Raton: CRC Press) p 898
[16] Youssef H A and El-Hofy H 2008 *Machining Technology: Machine Tools and Operations* (Boca Raton, London, New York: CRC Press) p 672
[17] Davis J R 2005 *Gear Materials, Properties, and Manufacture* (Materials Park, USA: ASM International) p 339
[18] Fomin A A, Gusev V G, Safin R G and Safin R R 2015 *Russian Engineering Research* **35(6)** 417–420
[19] Xia L, Li D and Han J 2014 *Key Engineering Materials* **579–580** 300–304
[20] Stadtfeld H J 2011 *American Gear Manufacturers Association Fall Technical Meeting* (Alexandria, USA: AGMA) pp 1–14
[21] Gusev V G, Fomin A A and Sadrtdinov A R 2018 *Solid State Phenomena* **284** 236–241
[22] Bair B-W 2002 *Journal of Materials Processing Technology* **122(2–3)** 139–147
[23] Huang C-L, Fong Z-H, Chen S-D and Chang K-R 2009 *Mechanism and Machine Theory* **44(2)** 401–411
[24] Heinzel C and Wagner A 2013 *CIRP Annals* **62(1)** 359–362
[25] Tsay C-B, Liu W-Y and Chen Y-C 2000 *Journal of Materials Processing Technology* **104(3)** 271–279
[26] Fomin A A 2013 *Russian Engineering Research* **33(11)** 660–662.
[27] Sharkov O V, Koryagin S I and Velikanov N L 2018 *IOP Conference Series: Materials Science and Engineering* **327** 042102
[28] Korn G A and Korn T M 2013 *Mathematical Handbook for Scientists and Engineers: Definitions, Theorems, and Formulas for Reference and Review* (Mineola, New York: Dover Publication Ink) p 1152

Cyber and Physical Systems Topology for the Industry 4.0 Smart Factory

D A Zakoldaev[1], A V Gurjanov[2], A V Shukalov[1], I O Zharinov[1] and O O Zharinov[3]

[1] Faculty of Information Security and Computer Technologies, Saint Petersburg National Research University of Information Technologies, Mechanics and Optics, 49, Kronverksky Av., Saint Petersburg, 197101, Russia
[2] Director, Stock Company «Experimental Design Bureau «Electroavtomatika» named after P A Yefimov, 40, Marshala Govorova St., Saint Petersburg, 198095, Russia
[3] Department of Problem-Oriented Computing Complexes, Saint Petersburg State University of Aerospace Instrumentation, 67, Bolshaya Morskaia str., Saint Petersburg, 190000, Russia

E-mail: mpbva@mail.ru

Abstract. To project digital companies of the Industry 4.0 is a production task which requires to implement in the industrial infrastructure the elements of multi-agent environment. Multi-agent environment is components (cyber and physical systems) of two elements: the automatic one in physical level and program one (model) in the virtual level. Physical element is a digital system of automatic control. The program one is a code fragment to describe the functionality of cyber and physical system with a mathematical model. There is a scheme of physical and virtual cyber and physical system components interaction for digital automatic production of the Industry 4.0. The properties of the Industry 4.0 production company (a smart factory) depends on the interaction scheme of the digital company infrastructure. Cyber and physical systems interaction scheme is generated in a smart factory cloud. There are some options of cyber and physical systems topology which are good for the Industry 4.0 production infrastructure.

1. Introduction

The development of information technologies and progressive technological cyber and physical equipment [1-3] gives sufficient conditions to project the flexible productions of the Industry 4.0. Such automatic flexible productions are a new type of company [4, 5] which creates items without humans. The main advantage is deep automatizing of item manufacturing technological operations and also transporting (transportation) of items between cyber and physical systems (CPS). The control of cyber and physical equipment is done [6, 7] with computerized system the core of which is an artificial intelligence which is a program in a calculation cloud.

The behavior of agents which are cyber and physical systems can be described mathematically in discrete equations of automatic control with random delay. To unite agents in multi-agent [8-10] production environments helps to create multi-measure system of higher rank the functionality of which can be described with variables of condition and control which are vector and matrix. Agents interaction is for object and orientation approach using the principle of central and non-central control. Non-central control of cyber and physical production helps to avoid break-down of the smart factory production infrastructure. The main idea how to control a CPS is to grant a non-conflict access for the CPS to the informative resources of the company virtual environment [11, 12].

The ways how agents interact in a multi-agent environment are based on the technologies [13-15] of Internet of Things (IoT), cloud technologies, technologies of Systems-to-Systems (S2S), Machine-to-Machine (M2M), Humane-to-Machine (H2M), additive technologies and other technologies of the Industry 4.0 cyber and physical systems.

IOP Conf. Series: Materials Science and Engineering **582** (2019) 012010 doi:10.1088/1757-899X/582/1/012010

2. Cyber and physical system as a multi-agent environment component

Cyber and physical technological equipment is a new type of production machines to function in a hybrid line with humanless and paperless technologies. Cyber and physical equipment is a narrow thing which may complete only a set of particular technological operations. Such technological operations in automatic mode are:

- placement of radio and electronic components to the SMD (Surface Mount Device) to a PCB (Printed Circuit Board) in the item designing;
- 3D-printing of parts (additive technology of production) created from metallic or organic raw materials;
- optical, X-ray or functional quality control of the items being produced.

Cyber and physical system as an agent of a multi-agent digital production environment (encapsulated calculation system) has the following properties:

- target orientation of CPS actions to complete a particular production task (a fixed set of technological operations) as a part of the Industry 4.0 smart factory;
- communication or when an intellectual agent may interact with other agents of multi-agent environment with some defined channels of connection and interfaces to complete the production task;
- reaction or when an intellectual agent may perceive the changes of multi-agent production environment to make it safer the completion of technological operations including the ecological and industrial safety;
- proactivity or when an intellectual agent may be an active (leading) component in a multi-agent environment using the artificial intelligence to complete production tasks of cyber and physical technological equipment self-organization (adaptation and automatizing of production infrastructure).

Cyber and physical system is unlike the machines of the Industry 3.0 it has an additional component (a sub-system) which is performed in a virtual cloud. The production machine of the Industry 3.0 (computer numeric control (CNC)) have an interface terminal, mechanisms, calculator and software which together work in a single item (the machine) which is the physical level only. In the Industry 3.0 agents interact in a multi-agent environment where operator controls a CPS (technological equipment) through a terminal and the machine itself is an instrument which has or does not have localized intellectual capabilities.

The Industry 4.0 approach to organize production process in a smart factory requires the application of CPS intellectual capabilities including the components of virtual resources to organize intellectual multi-agent environment. So a cyber and physical system of the Industry 4.0 has two components (see figure 1):

- physical equipment with a work chamber to complete technological operations, gears, condition detectors of technological processes, technological equipment functionality detectors and a digital controller;
- virtual agent in a CPS cyber-level cloud with mathematical models to describe the technological operations, CPS mathematical models as a digital system of automatic control and ontology model to describe a CPS as a digital twin.

Physical level CPS are simple reflection agents to complete a particular set of technological operations. Virtual level CPS are studying agents to form effective behavior strategies for technological equipment in a multi-agent production environment of the Industry 4.0 smart factory.

Ontology is a formalized description of definitions (processes, operations and other) and their interaction for the subject field of a digital production. Ontologies are needed to create specifications of all types of CPS production which are part of the Industry 4.0 smart factory.

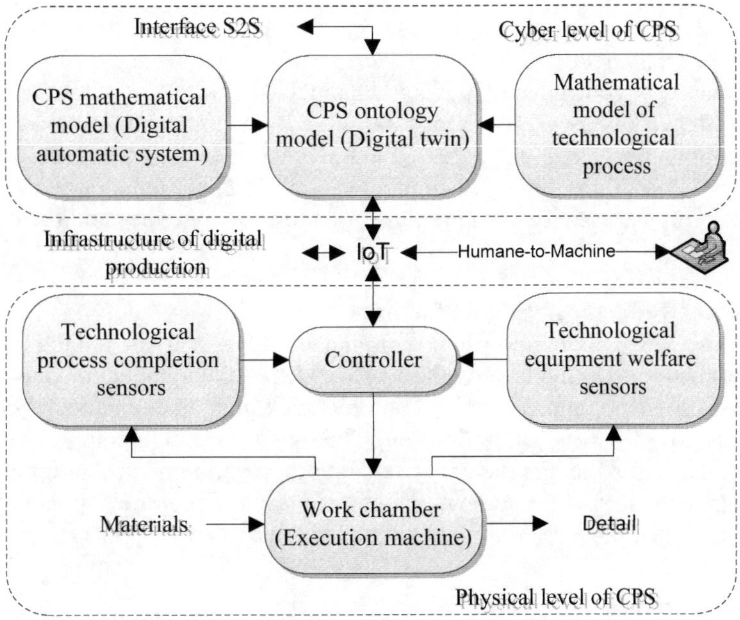

Figure 1. Cyber and physical system components for automatic digital production of the Industry 4.0.

CPS work chamber complete technological operations to transform blank (material) into an item. The technological process quality control is done periodically (the period of discrete digital system of automatic control) which proves dynamic properties of CPS. Detectors (accelerometers, pressure detectors, humidity detectors, temperature detectors and other) of technological operation completion to find out the current values of the cyber and physical production variables. The variable values to describe the technological operations completion are calculated in a cyber-level CPS cloud using mathematical models of CPS and the models of technological processes. Work parameters calculations give the needed values of the digital automatic systems as a digital twin in the cloud. Digital twin calculation is done with imitation modelling.

CPS physical level controller has a program agent (code fragment) which control the quality of technological operation inside the CPS work chamber. Controller compares the current value of technological operation completion variables and its desired values. The comparison results are the motions to the CPS work chamber gear from the controller. The executive gears are like a crank with a reducer, step engine and other automatic elements. So the detectors, controller and executive gears is a closed loop of CPS physical level in a digital system of automatic control.

Physical and virtual components CPS interaction is done with standards of the industrial Internet of Things which transfer the data wirelessly. Ontology model to describe CPS is done after the transformation of CPS description mathematical models into terms of semantic digital production united into ontology dictionaries of the Industry 4.0. So CPS ontology model is a way to describe an agent in the multi-agent system and a kind of interface to support the technology of Systems-to-Systems for the cyber and physical systems to interact.

Like the Industry 3.0, CPS and operator interaction is based on the technology of Humane-to-Machine. The H2M for interaction between operator and a CNC machine is based on controller programming (to place the control program in the machine memory which describes the order how to do technological operations of the item manufacturing which is the 3D-model) in the manual mode. The Industry 4.0 CPS is controlled by operator with the technology of H2M remotely with the industrial Internet of Things. Control program, control commands and variables is transferred by the user in the controller of the physical CPS with the protocols of IoT with guaranteed packages delivery. In this case CPS is described as digital system automatic control terms with random delay. There is a delay in the CPS channel because:

- random delays when the information is being transferred in a wireless channel from the user (or other people in digital production) to the cyber and physical system;

- technical restrictions (limited pass through of the channel) of temporal specifications in communication nets of industrial data.

A CPS has receiver and controller of the IoT channel which helps to transfer informatively the production data with other CPSs which are part the united technological line of a digital production. Cyber and physical production components united in a workshop is the production infrastructure of the Industry 4.0 smart factory. The multi-agent production environment conditions are changed because there are some results of CPS activity in each moment of time keeping in mind the dynamic properties of IoT.

3. CPS topology as the Industry 4.0 smart factory

Virtual generations of CPS dynamic configuration can be done in a smart factory based on the net architecture SDN (Software Defined Networks). This CPS self-organization has some approaches simultaneously based on: theory and methods of automatic control; mathematical theory of graphs.

Multi-agent virtual dynamic configurations of CPS have several topologies of net to complete production tasks available for the smart factory technological equipment. An almost optimal plan of item manufacturing which is good for some multi-criteria tasks. The most popular types of (see figure 2) of CPS topology of ring, bus, point to point, star, matrix and other.

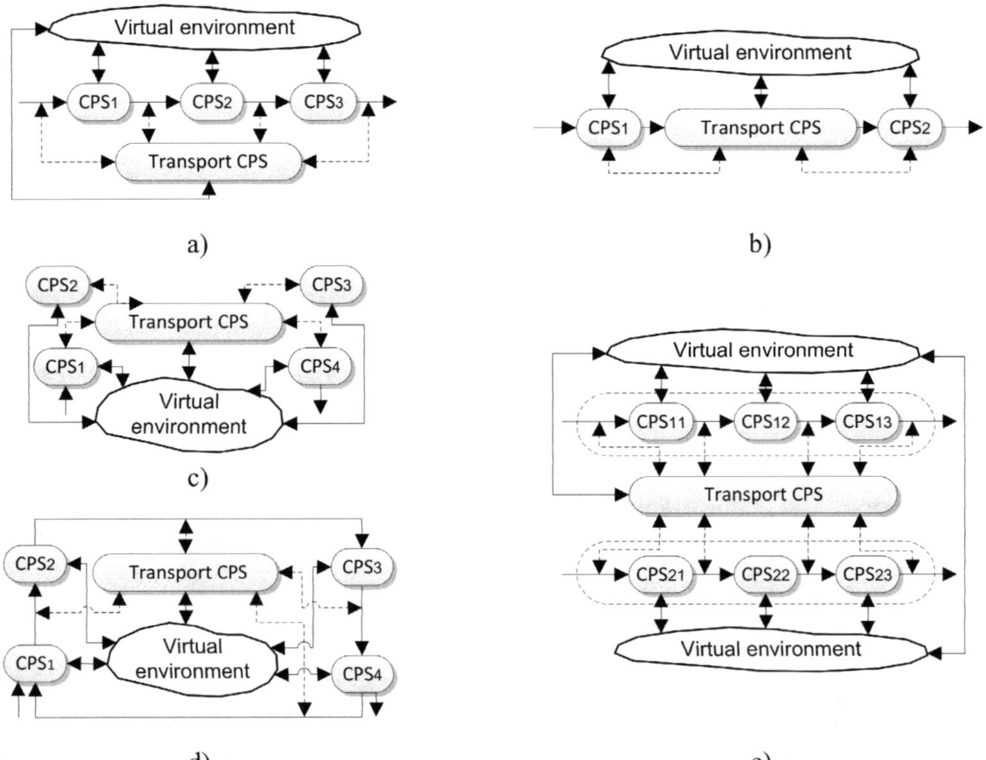

Figure 2. CPS topologies of virtual environment to organize production activity in the Industry 4.0 smart factory: a) bus topology, b) point to point topology, c) star topology, d) ring topology, e) matrix topology.

Agent interactions in the digital production virtual environment is based on the technology of Systems-to-Systems and is done automatically using semantic forms to describe production processes, components and technologies in languages of digital production ontologies. For different types of CPSs to interact (equipment of different purpose, for example, a robot manipulator which transfer a PCB to the receive area to place radio and electronic components in SMD) they use semantics from digital production ontologies placed as a resource agent of virtual environment.

Virtual environment agents interaction with physical agents of the Industry 4.0 digital production is based on digital technology of IoT where physical machines are objects of control which receives remotely the control values from cloud calculator (Industry 4.0 smart factory cloud) or from the operator. Processes of CPS in physical level and in virtual components level defines the micro-level of the smart factory multi-agent environment. In practice the principles of net control (net centering) include the agents behavior in multi-agent environment and also defines the multi-agent environment behavior in micro-level.

To project a multi-agent environment and organize CPSs interaction they use branch standards. The most popular standards and programs of industry development which are good enough to create a digital company of the Industry 4.0:

- FIPA (Foundations for Intelligent Physical Agents);
- MASIF (Mobile Agent System Interoperability Facility);
- DARPA (Defense Advanced Research Project Agency).

4. Conclusion

To implement multi-agent environments into production infrastructure of the Industry 4.0 smart factory they revised the business processes of the industrial company. The expected advantages of such agent implementation in multi-agent environments are reduced time of item manufacturing, reduced cost of production in a smart factory when at the same time the item quality gets higher.

The main developments of multi-agent environments which have cyber and physical systems in the Industry 4.0 production are the following:

- cyber and physical systems self-organization based on artificial intelligence technology may synchronize the completion of production processes for groups of cyber and physical systems in a smart factory;
- the organization of distributed calculations with the technology of BigData to collect and process the vast amounts of industrial data;
- good functionality of cyber and physical production based on the control algorithms using central and non-central principles;
- cyber and physical systems learning and multi-agent environment based on neural nets and real or model learning patterns (set of dynamic conditions) after the put into the field activity and after the real exploitation of the technological equipment.

References

[1] Leitao P, Colombo A W, Karnouskos S 2016 *Computers in industry* **81** 11-25
[2] Wang Y, Liu D, Sun Ch 2017 *Energies* **10** 267
[3] Meissner H, Ilsen R, Aurich J C 2017 *Procedia CIRP* **62** 165-169
[4] Mangharam R, Pajic M 2013 *Journal of the Indian institute of science* **93**(3) 353-388
[5] Schluse M, Priggemeyer M, Atorf L and Rossmann J 2018 *IEEE Transactions on industrial informatics* **14**(4) 1722-31
[6] Tan Y, Yang W, Yoshida K and Takakuwa S 2019 *Machines* **7** 7010002
[7] Malik A A, Bilberg A 2018 *Procedia manufacturing* **17** 278-285
[8] Colledani M, Pedrielli G, Terkaj W and Urgo M 2013 *Procedia CIRP* **7** 425-430
[9] Tolio T, Sacco M, Terkaj W and Urgo M 2013 *Procedia CIRP* **7** 25-30
[10] Terkaj W, Urgo M 2015 *Procedia CIRP* **28** 137-142
[11] Gurjanov A V, Zakoldaev D A, Shukalov A V and Zharinov I O 2018 *IOP Conference Series: Materials Science and Engineering* **327** 022110
[12] Zakoldaev D A, Shukalov A V, Zharinov I O and Zharinov O O 2018 *Journal of Physics: Conference Series* **1015** 052033
[13] Stark R, Grosser H, Beckmann-Dobrev B and Kind S 2014 *Procedia CIRP* **22** 3-14
[14] Iglesia D H, Gonzalez G V, Mendes A S, Jimenez-Bravo D M and Barriuso A L 2019 *Sensors* **19** 100
[15] Lee J, Ardakani H D, Yang Sh and Bagheri B 2015 *Procedia CIRP* **38** 3-7

Influence of Bio-organic Additives on the Ability of Yeast to Provide Biotransformation of Pesticides in Apple Must

A L Panasyuk[1], E I Kuzmina[1], L N Kharlamova[1], M V Babaeva[2] and I P Romanova[2, a]

[1] All-Russian Scientific Research Institute of Brewing, Beverage and Wine Industry - Branch of Gorbatov Research Center for Food Systems of RAS, 7, Rossolimo Str., Moscow, 119021, Russia

[2] K.G. Razumovsky Moscow State University of technologies and management (the First Cossack University), 73, str. Zemlyanoy Val, Moscow, 109004, Russia

E-mail: [a] i-p-romanova@yandex.ru

Abstract. The problem of food safety is the most important state and scientific priority, aimed at preserving and improving the health of the population, the production of high quality and safe products. The deterioration of food quality is primarily due to the chemization of agriculture in order to increase production. The use of pesticides to control weeds, diseases and pests increases yields, but these substances accumulate in the soil, and through it in crops. In addition to the negative impact on human health, there is also a negative role of pesticides in the technological processes of food production, for example, in winemaking pesticides adversely affect the activity of fermentation and the physiological state of yeast. Industrial yeast used in the technological processes of alcoholic fermentation of grape or apple musts to wines consume organochlorine pesticides as a source of carbohydrate food and lower the level of their concentration depending on the initial content. The article presents results of studies of the influence of bio-organic additives on the ability of yeast to provide biotransformation of pesticides in apple must.

1. Introduction

Ensuring food safety is the most important task of modern production. The increasing world population has therefore put a tremendous amount of pressure on the existing agricultural systems, because the new area that can be increased for cultivation purposes is very limited. That's why different pesticides (herbicides, insecticides, fungicides, nematicides, fertilizers and soil amendments) become more widely used than in the past [1, 2]. A distinctive feature of many pesticides is their high resistance to environmental factors and the ability to persist in the soil for 10 years or more and also pesticides accumulate in soil and plants [3, 4].

In addition, the most of the pesticides are non-specific and may kill the organisms that are harmless or useful for ecosystem. In general, it has been estimated that only about 0.1% of the pesticides reach the target organisms and theremaining bulk contaminates the surrounding environment. Pesticides with high persistence and toxicity are the most dangerous, first of all, it is organochlorine compounds. Consumption of food products, containing xenobiotics, poses a significant threat to human health and may be the cause of disorders in the nervous system and metabolism [5, 6].

In Russia, despite the legislative restrictions on the content of pesticides and other xenobiotics in food raw materials and products, the level of pesticide contamination remains quite high. Taking into account the increasing amount of foods containing xenobiotics, the international standard ISO 22000 has been developed. ISO 22000 contains requirements for the functioning of an effective food safety management system based on the principles of the HACCP system (Hazard Analysis and Critical Control Point) and on the implementation of programs of preconditions necessary for the production of safe food.

IOP Conf. Series: Materials Science and Engineering **582** (2019) 012011 doi:10.1088/1757-899X/582/1/012011

The basis for building a food safety system at the enterprise is the HACCP system, the most important principle of which is the analysis of dangerous factors, in particular chemical hazard - chemicals, the presence of which in food products can cause damage to the health of the consumer - pesticides, herbicides, fertilizers, insecticides, fungicides, antibiotics, growth promoters [7].

Grape and fruit wines may also contain pesticides, used in the cultivation of plants or coming from the environment. The content of some types of pesticides is significantly reduced during the process of wine production. All kinds of microorganisms, including industrial yeast of the genus *Saccharomyces* used in the technological processes of alcoholic fermentation of grape or apple musts to wines have the ability of biotransformation of pesticides. The highest sorption capacity had immobilized yeast. *Saccharomyces* yeast consume organochlorine pesticides as a source of carbohydrate food and lower the level of their concentration depending on the initial content of 60-100%. However, pesticides have a negative effect on the activity of fermentation and the physiological state of yeast and led to a decrease in the activity of yeast in 1.5-2 times [8, 9 ,10].

2. Experimental section

The main purpose of the research is to study the ability of yeast to metabolize phosphorus- and chlorine-containing pesticides and the effect of yeast activation on the intensification of the fermentation process of apple must, containing pesticides.

In order to study the ability of yeast to degrade pesticides in the fermentation of apple must, model experiments were conducted. The influence of two widely used pesticides – Bayleton (1-(4-chlorophenoxy)-3,3-dimethyl-1-(1H-1,2,4-triazol-1-yl)-2-butanone, systemic fungicide belonging to the azole group highly effective against powdery mildew, rust, and various leaf spot diseases) and Phosphamide (O,O-dimethyl S-(methylcarbamoylmethyl) ester, organophosphorus insecticide with contact and systemic action, used to combat herbivorous mites, aphids and other sucking plant pests) on two industrial species of wine yeast *Saccharomyces cerevisiae* – K-17 and Vishnevaya 33 recommended for the production of fruit wines was investigated.

Five samples of apple must were fermented periodically at a temperature of 20 °C. The first sample does not contain pesticides (control) and the other samples contain 1.5 μg/dm^3 of Phosphamide or 0.2 μg/dm^3 of Bayleton, which is more than 3 times higher than maximum permissible concentration. Active biomass of yeast cultures in the amount of 2% of must volume or 10 million cells in 1 cm^3 of must was used for must seeding.

The activity of fermentation and respiration was determined by gasometric method by the amount of carbon (IV) oxide and oxygen released during fermentation.

The mass concentration of pesticides was determined by high performance liquid chromatography on chromatograph " Shimadzu LC-4A "(Japan), equipped with spectrophotometric detector, linear flow rate measurement programmer and data processing system of CR-3A microcomputer.

The obtained results are shown in table 1.

Table 1. The influence of yeast race on the ability to metabolize phosphorus- and chlorine-containing pesticides in the fermentation of apple must

Sample	Fermentation time (days)	Pesticide residual content (% of original)	Activity	
			Fermentation (mkl/h Q CO_2)	Respiration (mkl/h Q O_2)
Apple must (control)	7	-	180	116
Apple must + Phosphamide[a]	12	9.0	103	68
Apple must + Bayleton[a]	9	23.0	112	82
Apple must + Phosphamide[b]	13	10.8	98	66
Apple must + Bayleton[b]	10	24.6	102	68

[a] *Saccharomyces cerevisiae*, K- 17
[b] *Saccharomyces cerevisiae*, Vishnevaya 33

The obtained results showed that phosphorus- and chlorine- containing pesticides have an inhibitory effect on the physiological and biochemical properties of yeast. Fermentation and respiration activity of yeast significantly decreased compared to the control in the presence of pesticides in the fermented environment, and, regardless of the yeast race, Phosphamide to a greater extent than Bayleton inhibited the processes of fermentation and respiration. The duration of fermentation of apple must containing pesticides, increased compared with the control of 1.5-2 times.

Activation of yeast *Saccharomyces cerevisiae* allows to intensify the process of fermentation of apple must containing pesticides, reduce its duration and actively carry out biodegradation of pesticides. Silicon compounds are known to play an important role in lipid metabolism in living organisms [11, 12, 13].

The use of a complex compound of silicon and carbohydrate can accelerate and simplify the process of diffusion of silicon into the yeast cell, which leads to the activation of nitrogen and carbohydrate metabolism in yeast [14].

In order to study the influence of three bio-organic additives (silicon digluconate, bio-organic supplements containing selenocysteine derivative and zinc respectively) on the fermentation activity and ability to metabolize pesticides of yeast, model experiments were conducted.

Bio-organic additives were introduced into apple must in different concentrations: an additive containing silicon digluconate in concentrations 0.5; 3.5; 7.0 ml/dm^3, an additive containing selenocysteine derivative in concentrations 0.1; 0.5; 1.0 ml/dm^3, an additive containing zinc in concentrations 0.5; 2.0; 5.0 ml/dm^3.

Nine samples of apple must were prepared using yeast biomass *Saccharomyces cerevisiae* K - 17, activated by bio-organic additives. As a control was used apple must obtained with a yeast culture without additives.

Yeast was introduced at the rate of 2% per 1 cm^3 of apple must. The accumulation of yeast biomass was carried out in stages, for several generations.

The samples were assigned the following numbering:

1 - apple must prepared using a bio-organic additive containing 0.5 ml / dm^3 silicon digluconate;

2 - apple must prepared using a bio-organic additive containing 3.5 ml / dm^3 silicon digluconate;

3 - apple must prepared using a bio-organic additive containing 7.0 ml/dm^3 silicon digluconate;

4 - apple must prepared using a bio-organic additive containing 0.1 ml / dm^3 selenocysteine derivative;

5 - apple must prepared using a bio-organic additive containing 0.5 ml / dm^3 selenocysteine derivative;

6 - apple must prepared using a bio-organic additive containing 1 ml / dm^3 selenocysteine derivative;

7 - apple must made with bio-organic additive containing 0.5 ml/dm^3 zinc;

8 - apple must made with bio-organic additive containing 2 ml/dm^3 zinc;

9 - apple must made with bio-organic additive containing 5 ml/dm^3 zinc.

Determination of the volume fraction of ethyl alcohol was carried out by distilling according to GOST 32095-2013 [15].

Determination of sugars was carried out by Bertrand method according to GOST 13192-73 [16].

Determination of titrating acids was carried out by direct titration according to GOST R 51621-00 [17].

The content of the extract was determined by the density of the aqueous solution.

The mass concentration of glycerin was determined on the liquid chromatograph "Stayer" with a refractometric detector.

The total content of anthocyanins was determined by pH-differential spectrophotometry at a wavelength of 510 nm and for fractional analysis of anthocyanins was used high performance liquid chromatography.

Determination of the content of organic acids was carried out on the liquid chromatograph "Stayer" with a spectrophotometric detector (column C18 250*4.6 mm, the volume of the injected sample is 20 μl) and by the method of high-performance liquid chromatography with column "Kromasil 5 - C18".

The chemical composition and content of volatile components were determined by gas chromatograph "Crystal 5000.1" with flame ionization detector (a capillary column with the stationary phase HP-FFAP, sample volume 1 μl).

3. Results section

The obtained results are shown in tables 2 - 4.

Table 2. The influence of activation of the yeast *Saccharomyces cerevisiae* biomass, K-17 on physico-chemical characteristics of apple must

Sample	Volume fraction of ethyl alcohol (%)	Mass concentration of sugars, in terms of invert sugar (g/dm^3)	Mass concentration of titrating acids, in terms of malic acid (g/dm^3)	Mass concentration of residual extract (g/dm^3)	Mass concentration of glycerin (g/dm^3)	Antioxidant activity, AE (mg/g)
Control	10.0	15.8	7.6	12.2	6.2	72
1	10.1	13.2	7.6	12.4	6.1	110
2	10.2	16.0	7.4	12.7	6.0	81
3	10.1	16.4	7.6	12.4	6.1	73
4	10.1	15.0	7.4	12.6	6.2	82
5	10.9	14.5	7.5	13.9	6.8	113
6	9.7	15.7	7.4	13.2	6.1	82
7	10.1	16.8	7.4	12.1	6.2	76
8	10.4	14.6	7.0	13.1	6.4	92
9	8.8	7.5	6.0	12.1	5.1	73

The alcohol content in samples 5 and 8 is higher compared to the control by 0.9% vol. and 0.4% vol. respectively. In addition, at the same time in the same samples there is an increase in antioxidant activity by 41 and 20 AE mg/g respectively.

The highest mass concentration of glycerin was observed in sample 5. The samples 6 and 9 contain the least amount of ethanol, titrating acids and glycerin in comparison with the control. Also in these samples were minimal residual extract and antioxidant activity.

Obviously, the optimal concentration of selenium and zinc necessary for the activation of yeast should not exceed the level of samples 5 and 8, that is 0.5 ml/dm^3 selenocysteine derivative and 2 ml/dm^3 zinc. A higher concentration of selenium and zinc leads to inhibition of yeast cell metabolism. In all experimental samples the content of organic acids decreases compared to the control by an average of 60 - 80%.

In addition, the mass concentration of citric acid decreased to zero in all samples, except sample 7, in which the additive containing zinc is introduced in a minimum concentration.

Table 3. The influence of activation of the yeast *Saccharomyces cerevisiae* biomass, K-17 on the composition of organic acids in apple must

Acid	Content of organic acids (g/ dm^3)									
	Control	1	2	3	4	5	6	7	8	9
Oxalic	0.319	0.006	0.008	0.007	0.011	0.008	0.012	0.015	0.014	0.014
Wine	0.778	0.011	0.012	0.018	0.041	0.041	0.044	0.057	0.049	0.075
Formic	2.963	0.290	0.202	0.241	0.315	0.303	0.294	0.435	0.284	0.362
Malic	4.285	1.892	1.721	1.912	1.854	1.895	1.832	1.939	1.86	2.066
Dairy	5.673	1.184	0.969	0.981	0.988	1.11	0.946	0.971	0.991	1.083
Acetic	5.009	0.700	0.623	0.570	1.124	0.794	0.664	0.853	0.759	1.01
Lemon	3.53	-	-	-	-	-	-	0.964	-	-
Amber	6.386	0.387	0.706	0.406	0.360	0.438	0.343	0.375	0.41	0.357

The content of volatile compounds that give the wine undesirable shades in flavor and taste and tone of oxidation (ethyl acetate, acetaldehyde, isoamyl and isobutyl alcohols) decreased in samples 5 and 8 compared to the control. In the same samples there was an increase, compared to the control, the content of ethyl esters of fatty macromolecular acids, favorably affecting organoleptic parameters (ethyl acetate, ethyl caprylate, ethyl caprate). The largest increase of aroma compounds affecting the bouquet and taste of wine, observed in the sample 6.

Table 4. The influence of activation of the yeast *Saccharomyces cerevisiae* biomass, K-17 on the residual sugar content in apple must

Sugar	Residual sugar content (g/ dm^3)									
	Control	1	2	3	4	5	6	7	8	9
Fructose	6.395	6.320	10.254	8.541	4.348	5.781	6.622	5.247	5.778	6.238
Glucose	1.832	1.738	2.067	1.736	1.439	1.360	1.429	1.343	1.315	1.426
Sucrose	2.592	2.863	3.301	2.695	2.002	2.191	2.987	2.853	2.783	3.235
Total sugar content	17.638	17.794	21.975	19.417	13.876	15.346	17.116	15.603	16.049	16.982

In order to study the effect of activated yeast biomass on organoleptic characteristics of wine materials, a working tasting was conducted. As a result of which sample 5 was recognized as the best: the wine has mild harmonious taste, the aroma is clean, with pronounced tones of apples, 9.5 points.

Thus, the best results were obtained for sample 5, so for further studies was chosen the option of yeast activation with a bio-organic additive containing 0.5 ml/dm^3 selenocysteine derivative, which corresponded to the variant of the experiment 5.

The ability of activated yeast to degrade pesticides in the fermentation of apple must was studied according to the previously described method.

The obtained results are shown in table 5.

Table 5. The influence of activation of the yeast *Saccharomyces cerevisiae* biomass, K-17 on the ability to metabolize phosphorus- and chlorine- containing pesticides in the fermentation of apple must

Sample	Fermentation time (days)	Pesticide residual content (% of original)	Activity	
			Fermentation (mkl/h Q CO_2)	Respiration (mkl/h Q O_2)
Apple must (control)	7	-	180	116
Apple must + Phosphamide[a]	12	9.0	103	68
Apple must + Bayleton[a]	9	23.0	112	82
Apple must + Phosphamide[b]	11	6.8	119	77
Apple must + Bayleton[b]	8	18.5	126	81

[a] *Saccharomyces cerevisiae*, K- 17

[b] *Saccharomyces cerevisiae*, K- 17 activated with a bio-organic additive containing 0.5 ml/dm^3 selenocysteine derivative

4. Discussion section

Thus, activation of yeast *Saccharomyces cerevisiae*, K-17 allows to intensify the process of fermentation of apple must containing pesticides, reduce its duration and actively carry out biodegradation of pesticides. In general, the activation of yeast *Saccharomyces cerevisiae*, K - 17 by using an organic additive containing selenocysteine derivative, can improve their ability to metabolize phosphorus- and chlorine- containing pesticides, as well as improve the biochemical composition of wine samples.

IOP Conf. Series: Materials Science and Engineering **582** (2019) 012011 doi:10.1088/1757-899X/582/1/012011

5. Conclusions
Thus, the use of bio-organic additives in the production of wine materials (silicon digluconate, bio-organic supplements containing selenocysteine derivative and zinc respectively) in optimal for yeast race *Saccharomyces cerevisiae*, K-17 concentrations, allows to optimize the biochemical composition and organoleptic characteristics of wine samples and increase the ability of yeast to metabolize phosphorus- and chlorine- containing pesticides by an average of 22.5%.

References
[1] Harsimran K G and Harsh G 2014 *Pesticides - Toxic Effects*, (Berlin: ResearchGate)
[2] Coppola L, Comitini F, Casucci C, Milanovic V, Monaci E, Marinozzi M, Taccari M, Ciani M, Vischetti C 2011 *New Biotechnology* **29** 99-106.
[3] Abdullah N Z, Ishaka A, Samsuddin N, Mohd Rus R, Mohamed A H 2011 *IIUM Research, Invention and Innovation Exhibition* 223.
[4] Agrawal A, Sharma B 2010 *Int. J. of Biological and Medical Research* **1** 90-104
[5] Armstrong J L, Fenske R A, Yost M G, Galvin K, Tchong-French M, Yu J 2013 *Atmospheric Environment* **66** 145-15
[6] Cothran R D, Brown J M, Relyea R A 2013 *Evolutionary Applications* **6** 832-841
[7] ISO 22000:2005 Food safety management systems. Requirements for any organization in the food chain
[8] Bokulich N A, Thorngated J H, Richardsone P M, Mills D A 2013 *Proc. Natl. Acad. Sci. U.S.A.* **111** E139–E148
[9] Jolly N P, Varela C, Pretorius I S 2014 *FEMS Yeast Res.* **14** 215–237
[10] Čuš F, Gregorčič A, Baša Česnik H, Bolta S 2010 *Food Control* **21(2)** 150-154
[11] Bityutskii N, Pavlovic J, Yakkonen K, Maksimovi V, Nikolic M 2014 *Plant Physiol. Biochem.* **74** 205–211
[12] Cornelis J T, Delvaux B, Georg R B, Lucas Y, Ranger J, Opfergelt S 2011 *Biogeoscience* **8** 89–112
[13] Deshmukh R, Vivancos J, Guerin V, Sonah H, Labbe´ C, Belzile F 2013 *Plant Mol. Biol.* **83** 303–315
[14] Mas A, Guillamon G M, Torija M J, Beltran G, Cerezo A B, Troncoso A M and Garcia-Parrilla M C 2014 *BioMed Research International* Article ID 898045
[15] State Standard 32095-2013 2013 *The alcohol production and raw material for it producing. Method of ethyl alcohol determination* (Moscow, Standartinform Publ. (In Russian))
[16] State Standard 13192-73 2015 *Wines, wine materials and cognacs. Method of sugar determination* (Moscow, Standartinform Publ. (In Russian))
[17] State Standard R 51621-2000 2009 *The alcohol production and raw material for it producing. Methods for determination of titrating acids* (Moscow, Standartinform Publ. (In Russian))

IOP Conf. Series: Materials Science and Engineering **582** (2019) 012012 doi:10.1088/1757-899X/582/1/012012

Technologies for Creating Geographic Information Resources for Monitoring the Socio-Ecological Conditions of Cities

S A Yeprintsev[a], S V Shekoyan[b], L A Lepeshkina[c] A A Voronin[d], M A Klevtsova[e],

Voronezh State University, Universitetskaya pl. 1, Voronezh, Russia, 394018

E-mail: [a]esa81@mail.ru, [b]shekoyan.syuzanna@mail.ru, [c]lilez1980@mail.ru, [d]voronin@bio.vsu.ru, [e]klevtcova@geogr.vsu.ru

Abstract. Geo-information resources have been developed for monitoring socio-ecological conditions (using the example of cities in Central Russia). The technologies of the developed GIS allow to take into account the variability of scale, the transformation of cartographic projections, changing the composition of the mapped objects, the formation of queries in numerous databases in real time, variations in the ways of displaying objects. The developed GIS system consists of hierarchically subordinate database sections and thematic cartographic tools reflecting the natural resource potential, the socio-economic and environmental-hygienic situation in the cities of Central Russia.

1. Introduction

Modern large urbanized areas are centers of acute environmental problems, which determines the high relevance of conducting monitoring studies of social and environmental conditions for the population [1, 2, 3]. At the beginning of the 21st century, a significant increase in the socio-economic potential is observed in Russia and a number of other developed countries of the world, which, in addition to positive effects, causes an increase in environmental tensions in large cities [4, 5, 6].

Among other factors that determine the environmental comfort of urbanized areas, it is necessary to highlight the natural frame surrounding this territory [7, 8].

Under the natural (ecological) framework of the territory, according to the environmental design documentation, we mean the set of the most active and interrelated in terms of the environment spatial elements (rivers and river valleys, forests, etc.), on which the ecological stability of the environment depends.Эффективное управление развитием территории основано на системе критериев качества структуры и функционирования природного каркаса.

The basic elements of the natural frame include:

• valuable natural-territorial complexes occupying a significant part of the territory of the district (as a rule, these are federal reserves and reserves, national and natural parks, large monuments of nature in area, botanical gardens) [9];

• natural-territorial complexes of the main watershed surfaces of the formation of river flows;

• large forest tracts (as a rule, these are protective forests);

• large marsh and forest natural-territorial complexes (NTC) that do not have protection status. Key elements of the natural complex are territories that have preserved unique ecological communities that are "points of ecological activity".

Thus, the natural frame of the territory, improving the quality of aeration of urbanized territories, as well as increasing the diversity of landscapes, is one of the factors determining the ecological safety of the territory [10, 11].

2. Equipment and devices used in studies

To assess the socio-ecological conditions of the urbanized areas of Central Russia, an analysis of the natural framework around the studied cities of Central Russia (Voronezh, Lipetsk, Belgorod, Kursk, Tambov) was carried out. The data of remote sensing of the Earth and the subsequent analysis of the effect of the natural framework on the ecological comfort of urbanized areas implies the need to use large data arrays, their processing and mapping. This requires the use of modern geo-information technologies that allow for the collection, analysis, processing and visualization of geodata, as well as obtaining on their basis new information about spatial-coordinated phenomena [6].

The results obtained during the interpretation of satellite images, as well as the data processing of statistical information are summarized in the GIS environment Ecological and Socio-Economic Conditions of the Urbanized territories of Central Russia, which will be used to develop a model of the effect of the natural framework on the integral value of the environmental comfort of the studied territory.

GIS offers a wide variety of approaches to the analysis of spatial data. Sometimes it is enough to use visual analysis: on the basis of the created map, see everything you need to make a decision. However, there are times when it is difficult to make a decision only on the basis of a map. Simple data visualization does not allow to make an unambiguously correct decision. When creating a map, cartographers can choose which objects are included in the map, which characters are assigned to them, which classification scheme is used, what the inscriptions look like, etc. All these cartographic elements help to understand the content of the map and determine the boundaries of the problem being analyzed, but they can also change the characteristics of the information and affect its perception and interpretation.

Created digital maps of the study area should provide accurate binding, systematization, selection and integration of all incoming and stored information (a single address space); complexity and clarity of information for decision-making; the possibility of dynamic modeling of processes and phenomena; the possibility of automated problem solving associated with the analysis of the characteristics of the territory; the possibility of operational analysis of the situation in case of emergency.

Geo-information resources being developed are considered as a certain extension of the database technology for coordinate-related information with the possibilities of organizing a database request together with the means of generating a “graphical” report, as well as analyzing spatial relationships between objects [12]. It becomes possible to display on the screen or on a hard copy only those objects or their sets that the user needs at the moment. That is, the transition from complex complex cards to a series of interconnected private cards is actually carried out. This ensures better structured information, which allows it to be used effectively (manipulation, data analysis, etc.).

3. Results and Discussion

GIS "Ecological and socio-economic conditions of urbanized areas of Central Russia" consists of a system of hierarchically subordinate sections of databases and tools for thematic mapping, reflecting the natural resource potential, socio-economic and ecological-hygienic situation of urbanized territories (fig.1).

The basis is the data of remote sensing of the Earth obtained from the Landsat-8 satellite, as well as official statistical data of environmental protection agencies and government bodies, presented in the form of textual descriptions, tabular reference data and graphic illustrations (maps, diagrams, pictures, photo slides).

The collection of information to characterize the functional planning structure and landscape-environmental conditions for model regions was carried out on the basis of special field studies, stock processing, including cartographic, material from regional integrated and sectoral environmental departments.

An array of official statistical information for a 20-year period, from 1998 to 2018, has been selected for mathematical-cartographic modeling.

GIS "Environmental and socio-economic conditions of urbanized areas of Central Russia" contain 4 sections.

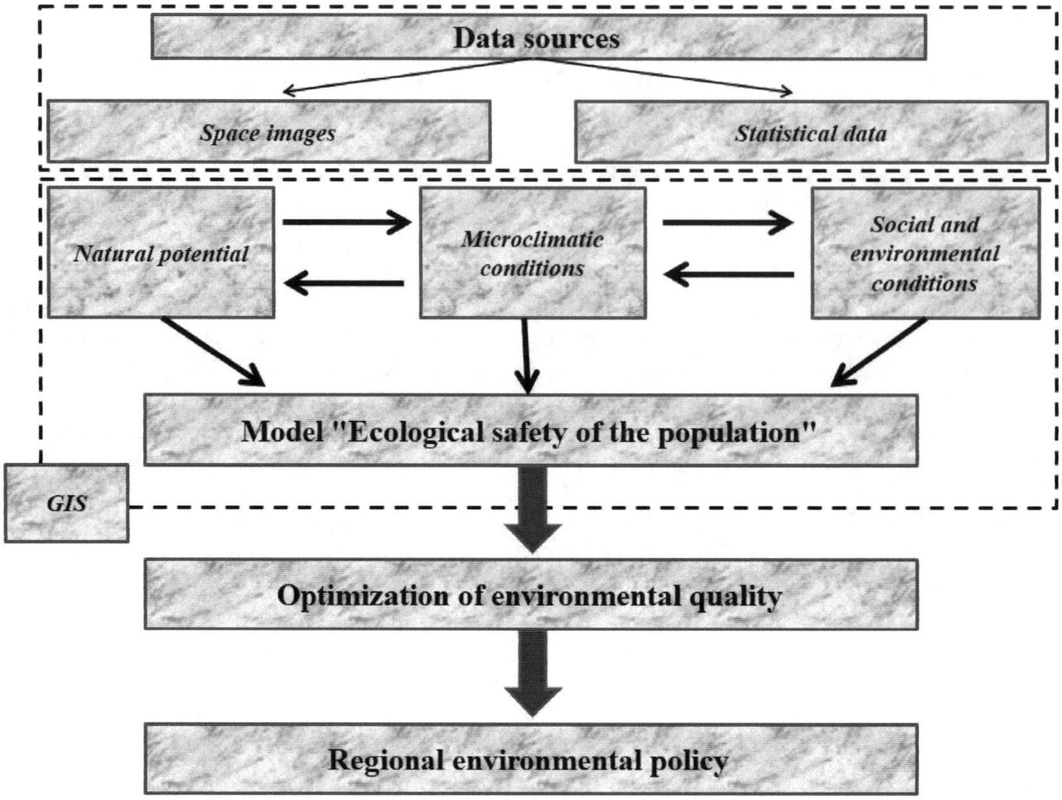

Figure 1. Logical scheme of the GIS "Environmental and socio-economic conditions of urbanized areas of Central Russia"

The first section is "Natural Potential". This section assumes the collection and synthesis of data characterizing the diversity of the landscape, which makes it more sustainable. The predominant influence of the natural frame on the quality of the environment are zones of external and internal landscaping.

Protective green areas, according to the town planning code of the Russian Federation, should be located on the windward side of the city. In urban settlements, it is necessary to provide, as a rule, an uninterrupted system of green areas and other open spaces. The share of green areas for various purposes within the building (level of greening of the building area) must be at least 40%, and within the territory of a residential area at least 25% (including the total area of greened territory of the microdistrict). The general principle of the organization of the green zone is the maximum preservation of natural green spaces, as well as the introduction of gas-dust-resistant woody plants.

The area of green areas of common use – parks, gardens, squares, boulevards, located on the residential territory of urban and rural settlements, take at the rate of 10 m^2 / person. for large, large, largest cities and 7-8 m^2 / person. for medium and small cities.

The natural potential of the territory largely influences the microclimatic conditions of the territory, as well as is itself influenced by the microclimate.

The study of the natural potential of the urbanized areas of Central Russia is carried out according to remote sensing of the earth (satellite images from the Landsat-8 satellite) using NDVI (Normalized Difference Vegetation Index) analysis [13]. As a result of the development of this section, it is assumed to calculate the coefficient of the natural potential of the territory involved in determining the integral indicator of the value of the ecological safety of the territory.

The second section is "Microclimatic conditions". This section assumes the collection and synthesis of data characterizing the microclimate of the territory - wind rose, average monthly temperatures, average wind speeds, and the meteorological potential of the atmosphere.

The microclimate of the city is formed under the influence of local conditions and urban development. Assessment of microclimatic conditions is carried out in order to identify possible ways and means of maintaining and improving the quality of the environment surrounding the building (using favorable factors and neutralizing negative impacts). Among the microclimatic factors that are essential for the population, include aeration, insolation and the degree of natural self-purification of the atmosphere.

The degree of self-purification of the atmosphere is estimated by a number of climatic indicators, and the meteorological self-purification potential of the atmosphere is considered the most common which is determined by the ratio of factors contributing to the accumulation of pollutants in the atmosphere (calm, fog), with factors contributing to the dispersion of pollutants (precipitation and wind at a speed of more than 6 m/s). For residential development, the town planning code of the Russian Federation recommends selecting areas with a favorable meteorological potential for self-cleaning of the atmosphere when the frequency of atmospheric processes contributing to the dispersion of atmospheric impurities is greater than the frequency of processes leading to accumulation of atmospheric pollution.

The forecast of the formation of the microclimate under the influence of urban development of the territory is carried out on the basis of the established efficiency of the architectural and planning means of microclimate control. To do this, you need the following information: the layout of the road network; density and number of floors of the building; nature of landscaping and improvement.

Thus, microclimatic conditions have a direct impact, and also depend on the other two blocks (natural potential and socio-ecological conditions).

As a result of the development of this section, it is assumed to calculate the coefficient of comfort of the natural conditions involved in determining the integral indicator of the value of the environmental safety of a territory.

The third section is "Social and environmental conditions". In this section, databases are maintained both on the ecological conditions of the territory and on social factors that directly or indirectly affect the ecological situation of the urbanized territory.

The main environmental factor is atmospheric pollution with man-made pollutants (nitrogen dioxide, sulfur dioxide, dust, lead, formaldehyde, phenol, carbon monoxide). These indicators are the main factors affecting the incidence of the population and the amount of environmental risk. Based on these data, it is possible to identify and quantify risk levels, as well as plan measures for organizing environmental monitoring and risk reduction in ecologically unfavorable areas.

Social factors affecting the ecological situation of an urbanized area can be attributed to - the density of roads, the number of personal vehicles among the population, the quality of housing stock, the average monthly income of the population, etc.

In compiling the database, the following annually updated information on regional statistics (Voronezh Oblast, Lipetsk Oblast, Belgorod Oblast, Kursk Oblast, and Tambov Oblast) is used, which is obtained by generating official queries on the websites of federal statistics services:

- Regional statistics (Voronezh, Lipetsk, Belgorod, Kursk, Tambov regions);
- Main public health departments of the Voronezh Lipetsk, Belgorod, Kursk, Tambov regions;
- Territorial bodies of the Federal State Statistics Service in the Voronezh, Lipetsk, Belgorod, Kursk, Tambov regions;
- Federal State Healthcare Institution "Center for Hygiene and Epidemiology in the Voronezh Region";
- Voronezh Regional Information Fund of Socio-Hygienic Monitoring (monitoring is carried out on the basis of FGUZ Center for Hygiene and Epidemiology in the Voronezh Region);
- Office of the Federal Service for Supervision of Consumer Rights Protection and Human Welfare in the Voronezh, Lipetsk, Belgorod, Kursk, Tambov regions.

Annual reports on the sanitary-epidemiological situation in the Voronezh, Lipetsk, Belgorod, Kursk, and Tambov regions:

- Annual information reports (presented on the website of the Office of Rospotrebnadzor in the Voronezh, Lipetsk, Belgorod, Kursk, Tambov regions);
- Reports on the sanitary-epidemiological situation in the Voronezh, Lipetsk, Belgorod, Kursk, Tambov regions for 2000–2018.

All of the above data sources are available via the Internet.

As a result of the operation of this section, the values of environmental risk and social comfort coefficient are calculated. These indicators will be used in determining the integral indicator of the value of ecological safety of the territory.

The fourth section of the model "Environmental safety of the population". It assumes the development of an integral indicator showing environmental comfort for the population living in the area. The integral indicator assumes that all factors of the previous three blocks are taken into account. This indicator will allow differentiating both various urbanized territories of the Voronezh, Lipetsk, Belgorod, Kursk, and Tambov regions, as well as highlight zones with different levels of environmental safety within one city.

4. Conclusion

Thus, the developed geo-information resources providing monitoring studies of the socio-ecological conditions for the population contain 4 sections: “Natural potential”, “Microclimatic conditions”, “Social-ecological conditions”, “Ecological safety of the population”

Data analysis in the GIS environment will allow to create maps reflecting the socio-ecological situation in the region, indicating environmentally hazardous man-made objects, degraded and disturbed landscapes, manifestations of exogenous processes; information on the socio-economic situation in the territory of the studied region; results of the analysis of the medico-ecological situation in the urbanized territories of the Voronezh Region.

On the basis of such maps, it is planned to develop a set of environmental design measures that would increase the integral indicator of the environmental safety of the population in a specific urbanized territory of the Voronezh Region.

These activities may include optimization of ecological-functional zoning of urbanized areas, transport networks, fuel and industrial complexes, and natural frameworks of territories.

The developed set of measures seems possible to be included in the recommendations on the optimization of territorial planning and environmental policy of the region.

The studies were carried out with the financial support of the RFFI grant (project 19-05-00660 A)

References

[1] Abanina E N Timofeev L A Agapov D A Sorokina Yu V Ustinova A N 2019 Systems of Environmental Security of Urbanized Territories Within the Framework of the Program of Ecological Development of Urbanized Territories (IOP Conference Series: Earth and Environmental Science 224(1)) pp 012031

[2] Buzyrev V Nuzhina I Zolotareva M 2018 Social-ecological priorities of town-planning activities on the urbanized territories (MATEC Web of Conferences 193) pp 01007

[3] Stossel Z Kissinger M Meir A 2015 Assessing the state of environmental quality in cities - A multi-component urban performance (EMCUP) index (Environmental Pollution 206) pp 679-687

[4] Xu, Q Zheng X Zheng M 2019 Do urban planning policies meet sustainable urbanization goals? A scenario-based study in Beijing, China (Science of the Total Environment 670) pp 498-507

[5] Liu Y Zeng C Cui H Song Y 2018 Sustainable land urbanization and ecological carrying capacity: A spatially explicit perspective (Sustainability (Switzerland) 10(9)) pp 3070

[6] Yeprintsev S A Klevtsova M A Lepeshkina L A Shekoyan S V Voronin A A 2018 Assessment of the dynamics of urbanized areas by remote sensing (IOP Conf. Series: Earth and Environmental Science 115) pp 012034

[7] Kalmanova V B 2018 Ecological planning of urbanized areas in the south of the Far East (Birobidzhan city as an example) (IOP Conference Series: Earth and Environmental Science 107(1)) pp 012130

[8] Tong H Shi P Bao S Zhang X Nie X 2018 Optimization of urban land development spatial 7 allocation based on ecology-economy comparative advantage perspective (Journal of Urban Planning and Development 144 (2)) pp 05018006

[9] Voronin A A Lepeshkina L A Krutova O V Klevtsova M A Yeprintsev S A 2019 Assessment of Environmental-Landscape Condition of Territories Botanical Gardens of Forest-Steppe Zone (IOP Conference Series: Earth and Environmental Science 224) pp 012053.

[10] Seddighi M J Ayali H 2008 Environmental management techniques in sustainable landscape

(Proceedings of the 2008 International Conference on Information and Knowledge Engineering IKE) pp 463-469
[11] Mazhitova G Z Janaleyeva K M Berdenov Z G Doskenova B B Atasoy E 2018 Assessment of the sustainability of landscapes of the North-Kazakhstan region to agricultural impact (News of the National Academy of Sciences of the Republic of Kazakhstan, Series of Geology and Technical Sciences 3 (429)) pp 90-95
[12] Topchiy D Kochurina E 2018 Environmental situation in construction, reconstruction and re-profiling of facilities in high-density urban development (MATEC Web of Conferences 193) pp 05012
[13] Li W Saphores J D M Gillespie T W 2015 A comparison of the economic benefits of urban green spaces estimated with NDVI and with high-resolution land cover data (Landscape and Urban Planning 133) pp105-117.

IOP Conf. Series: Materials Science and Engineering **582** (2019) 012013 doi:10.1088/1757-899X/582/1/012013

Style Transfer for CNC Machine Input Data Preprocessing

E I Nikolaev[1], V V Zaharov[2], N I Zaharova[1]

[1]North-Caucasus Federal University, 2, Kulakov Ave., Stavropol, 355029, Russia
[2]Stavropol State Agrarian University, 12, Zootekhnichesky lane, Stavropol, 355017, Russia

E-mail: notdeveloper@gmail.com

Abstract. Advances in deep neural networks have led to impressive results in recent years. The new technologies such as cross-domain adaptation, reinforcement learning and generative adversarial networks have shown a real promise for industrial and real-life applications. In this paper, the results of the experimental research on designing, training and implementation of the preprocessing algorithm for the computer numerical control machine input were presented. The algorithm of neural network transfer of artistic style has demonstrated wide possibilities in the field of generating graphic content. This paper demonstrates the possibility of using a generating neural network for the synthesis of stylized images that can be used as input images for a computer numerical control machine. Thus, the proposed algorithm is pre-processing the input image. The design feature of the laser engraver does not allow styling using an arbitrary style image, so dotted or linearized binary images are used as a style. The proposed preprocessing algorithm allows synthesizing binary images reproduced by a laser engraver. At the same time, image generation is performed in one forward pass of the generating neural network.

1. Introduction

Transferring the style from one image to another image is an interesting yet difficult problem. There have been many efforts to develop efficient methods for automatic style transfer [1-5]. Recently, Gatys et al. proposed a seminal work [6]: It transfers the style of artistic images to other images using Convolutional Neural Networks (CNN). Several follow-up works improved upon this innovative approaches [7-10]. Despite the fact that this work has drawn lots of attention, one of the main disadvantage of style representation: the practical use of this technology is negligible.

In this paper, a novel application of neural style transfer by casting it as a preprocessing algorithm for computer numerical control (CNC) machines is proposed. There are different types of CNC machines: it's possible to build a classification based on the principle of control, the working tool, the type of material being processed. In the framework of this work, a CNC laser engraver is considered, which performs processing of wood products by burning material with a laser beam. The task of adapting the input image of photographic quality to the capabilities of the CNC machine is being solved. Modern CNC machine control systems are based on the G-code command system: it is a specialized language for industrial machines that allow formulating commands for controlling the working tool (for example, movement and power of a laser module). With this approach, the process of obtaining the resulting image I_t from the original bitmap image I_c takes a long time. In the case of using vector images I_c, the speed of creating a product increases significantly. To address this problem the preprocessing algorithm based on neural style transfer method is proposed. The input bitmap image I_c is preprocessed and intermediate image I_m is generated. The I_m will be used as an input image for the physical laser engraver. The proposed method of generation $I_m = s(I_c, I_s)$ is based on "simplify" the initial image I_c. The transformation $s = s(\cdot, I_s)$ is a stylization algorithm with a style I_s. Among graphical attribute transfer algorithms, the highest efficiency is demonstrated by the neural network art style transfer methods [6]. But in the framework of this study the style image I_s effects on the efficiency of the fabrication process of the image I_t by CNC machine. As a result, the target image I_t is synthesized by a

CNC machine using an intermediate image I_m: $I_t = G(I_m)$, where $G(\cdot)$ is the transformation of a digital image into a physical product using the CNC machine working tool (laser module).

2. Related works

Gatys et al. observe that deep convolutional neural network is capable of extracting semantic image content from an arbitrary photograph and some appearance information from the well-known artwork [6]. Gatys et al. themselves [11] propose several slight modifications to improve their previous algorithm [6]. Although the original stylization method based on online image optimization is able to yield impressive stylized images, there are some limitations. The most concerned limitation is the efficiency issue. Another category of the stylization methods addresses the speed and computational cost issue by exploiting image generation based on offline optimization to reconstruct the stylized result, i.e., a feed-forward network is optimized over a large set of images for one or more style images. Such methods are proposed by Johnson et al. [12] and Ulyanov et al. [13] respectively. These two methods share a similar idea, which is to pre-train a feed-forward style-specific network and produce a stylized result with a single forward pass at the testing stage. They only differ in the network architecture, for which Johnson et al.'s design roughly follows the network proposed by Radford et al. [14] but with residual blocks as well as fractionally strided convolutions, and Ulyanov et al. use a multi-scale architecture as the generator network. The fast neural network transfer strategy is the most suitable method for the task addressed in this research.

In the framework of this experimental research, the problem of the intelligent control system design is also addressed. The cyber-physical systems, intelligent manufacturing systems and their role in modern industry are described in [15]. The structure of the modern intelligent manufacturing systems and the terminology of the cyber-physical systems are described in [16].

3. Input image preprocessing model

3.1. Overview

The scheme of the CNC laser engraver is depicted in Fig. 1a. A typical mode of operation of the machine is the sequential burning of points in the material. It's possible to achieve high efficiency of the fabrication process by using a vector image as an input, but in case of using a bitmap image, the time of the burning process increases significantly. The main idea of this experimental study is to create a model for generating an intermediate image I_m for the initial image I_c to obtain a high efficiency of the fabrication process and preserve a similarity of the I_c and I_t. The model takes bitmap image I_c as input and produces plausible image I_m via deep generative model. The I_m is used as an input for the physical CNC machine tool. The construction of this CNC machine, in general, contains two axes X and Y and the laser module which moves along them. The input image is shown in Fig. 1b. CNC laser engraver generates the picture by burning material by the laser beam. Therefore, it synthesizes a binary bitmap, but sometimes it's possible to implement a grayscale bitmap drawing by controlling laser module power. In this research, the approach with binary input images is considered. There are two different modes of the laser engraver functioning. The first method is to control the displacement and power of the laser module using a binary input image constructed from randomly located points. Such type of input is depicted in Fig. 1c and Fig. 1f. The second one is based on using a binary image constructed from horizontal lines (Fig. 1d, Fig. 1g). This mode provides high speed of the laser engraver, as the CNC control system manipulates with line segments in the manufacturing process.

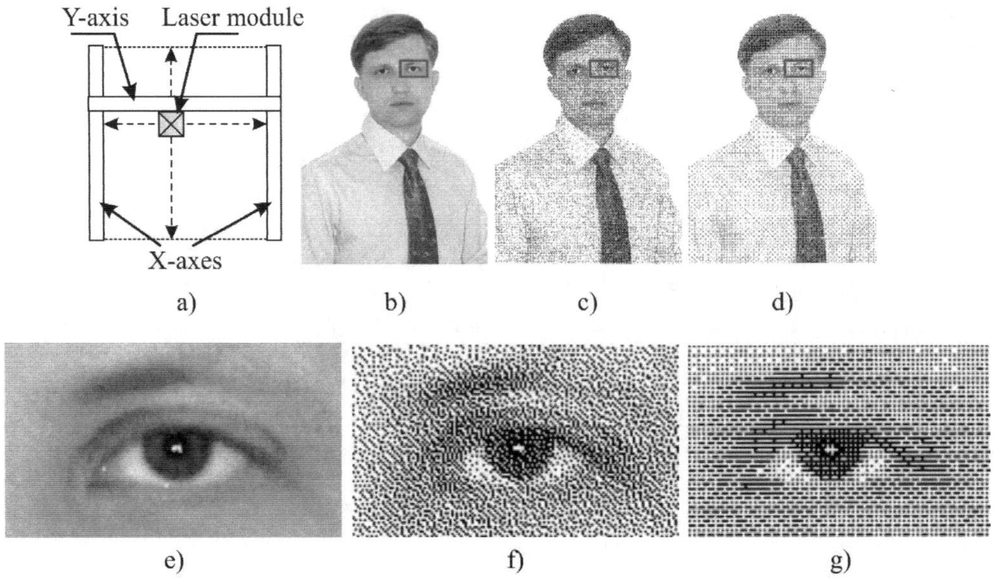

Figure 1. CNC machine basics: a) laser engraver scheme; b) input image; c) dotted image; d) linearized image; e) zoomed input fragment; f) zoomed dotted fragment; g) zoomed linearized fragment.

3.2. Online Neural Style Transfer

The original neural style transfer algorithm [6] tries to seek a stylized image I_t that minimizes the objective:

$$I_t = \arg\min_I \mathcal{L}_{total}; \qquad (1)$$
$$\mathcal{L}_{total} = \alpha \cdot \mathcal{L}_c(I, I_c) + \beta \cdot \mathcal{L}_s(I, I_s);$$

where $\mathcal{L}_c(I, I_c)$ compares the content representation of a given content image I_c to that of the stylized image I; $\mathcal{L}_s(I, I_s)$ compares the Gram-based style representation derived from a style image I_s to that of the stylized image I; α and β are used to balance the content component and style component in the stylized result.

The content loss $\mathcal{L}_c$, defined in a form (2), represents the squared Euclidean distance between the feature representations $\mathcal{F}^l$ of the content image I_c in layer l and that of the stylized image I.

$$\mathcal{L}_c(I, I_c) = \sum_{l \in \{l_c\}} \left\| \mathcal{F}^l(I_c) - \mathcal{F}^l(I) \right\|, \qquad (2)$$

where $\{l_c\}$ denotes the set of layers which are used for the content style representation. The example of the initial image and the subset of activation maps from VGG19 [17] are depicted in Fig. 2.

The style loss $\mathcal{L}_s$ exploits Gram-based visual texture modelling technique to represent the style. Therefore, the style loss is defined by the squared Euclidean distance between the Gram-based style representations of I_s and I (3).

$$\mathcal{L}_s(I, I_s) = \sum_{l \in \{l_s\}} \left\| \mathcal{G}(\mathcal{F}^l(I_s)) - \mathcal{G}(\mathcal{F}^l(I)) \right\|, \qquad (3)$$

where $\mathcal{G}$ is the Gram matrix to encode the second order statistics of the set of filters; $\{l_s\}$ represents the set of layers for calculating the style loss.

The Gram matrix $\mathcal{G}$ is the inner product between the vectorised feature maps i and j in layer l:

$$\mathcal{G}_{ij}^l = \sum_k \mathcal{F}_{ik}^l \cdot \mathcal{F}_{kj}^l, \qquad (4)$$

where $\mathcal{F}_{ik}^l$ is the activation of the ith filter at position k in layer l.

Figure 2. Neural network image representation: initial image I_c and a subset of the activation maps.

The choice of $\{l_c\}$ and $\{l_s\}$ empirically follows the principle that the usage of lower layer tends to retain low-level features (e.g., colours, simple strokes), while the usage of a higher layer generally preserves more high-level semantic content information. $\mathcal{L}_s$ is usually computed with lower layers and $\mathcal{L}_c$ is computed with higher layers. In framework of the current experimental research are used: $\{l_s\} = \{relu1\_2, relu2\_2, relu3\_2, relu4\_2, relu5\_2\}$ and $\{l_c\} = \{relu4\_2\}$, and VGG19 is used as a feature extractor network. To control a stylization quality the parameters α and β (1) are used. The results of stylization with different stylization rate (α/β) are shown in Fig. 3. The content image I_c is depicted in Fig. 2. In the considered algorithm in the formulation [6] there are a large number of parameters and hyperparameters, which can be tuned to improve the quality of stylization. The crucial problem arising in the process of image stylization according to the algorithm [6] is a performance. Optimization of the objective function (1) is a long iterative process based on gradient-based methods. In order to increase the efficiency of the neural style transfer as a preprocessing stage for the CNC machine input, in this study, the generative deep model is trained.

3.3. Offline Neural Style Transfer

In the framework of this research, the modification of the networks proposed in [12, 13] is used. The architecture of the generative neural network is depicted in Fig. 4a. It contains a sequence of convolutional layers $conv\ n \times n\ [in, out, num]$, where $n \times n$ is a filter size, in and out are the input and output image sizes, and num is the number of filters; residual blocks (Fig 4b); transpose convolution layers. Convolution layers contain instance normalization blocks [18].

The generative model is trained by using COCO [19] dataset, which contains 118,287 samples. The crucial task is that it is important to select the style image I_s which provides intermediate image $I_m = s(I_c, I_s)$ with the most suitable structure. In the framework of this experimental research the I_m is not just a stylized image, but also a map according which CNC machine tool will be moved.

The training process for the deep model depicted in Fig 4a is similar to the one in case of using online neural style transfer procedure, but rather than using per-pixel loss functions depending only on low-level pixel information, we train generative networks using perceptual loss functions that depend on high-level features from a pretrained loss network (VGG19). During training, perceptual losses measure image similarities more robustly than per-pixel losses, and at test-time, the transformation networks run in real-time. The scheme of the training pipeline is depicted in Fig. 4c. During online training, the L-BFGS-B method is used, but for generative model optimization in offline mode, Adam optimizer is used.

IOP Conf. Series: Materials Science and Engineering **582** (2019) 012013 doi:10.1088/1757-899X/582/1/012013

Figure 3. Neural Style Transfer with different stylization rate (α/β) values: a) style image I_s; b) $\alpha/\beta = 10^{-2}$; c) $\alpha/\beta = 10^{-4}$; d) $\alpha/\beta = 10^{-7}$.

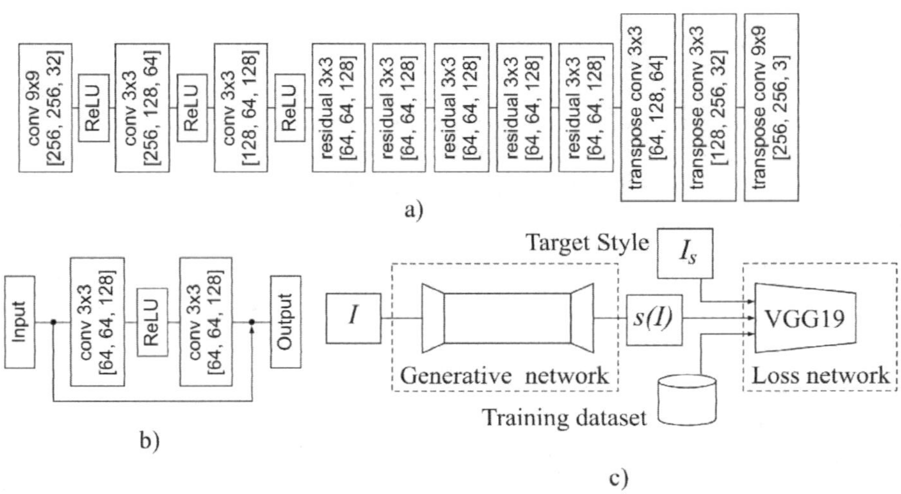

Figure 4. The Generative Network Architecture: a) neural network graph; b) residual block graph.

4. Experiments

The experiment with three different target styles was performed: a dotted style (Fig. 5a), linearized style (Fig. 5b) and vector style (Fig. 5b). The vector style aims to vectorize the input image.

Figure 5. Target styles for training generative models.

The result images are depicted in Fig. 6. The proposed algorithm of the photographic bitmap image preprocessing produces the binary image which is ready for drawing by a laser engraver. The quality of the synthesized image and the efficiency of the fabrication process depend on the initial image structure and style image structure.

Figure 6. Results of the preprocessing algorithm.

In Fig. 6 the vectorized style is shown for comparison purpose. The high degree of similarity between the synthesized image obtained using a vectorized style and the original image demonstrates the prospects of using a complex style for the pre-processing task. But the complexity of the reproduction of the resulting image by a laser engraver requires additional transformations, that is, lead to the need to re-apply the preprocessing algorithm.

5. Conclusion and future work

The presented experimental results can be applied not only to control the laser engraver, but also for any 2D CNC machine. The proposed preprocessing algorithm can be improved by reformulating the problem: the problem to be solved is related to the domain adaptation problem class. Solving the problem of domain adaptation can significantly improve the quality of preprocessing and, accordingly, increase the efficiency of the laser engraver. The neural network art style transfer used in this study is a special case of domain adaptation.

Another area of research is the development the control system for generating commands for physical equipment based on reinforcement learning algorithms. A similar approach is used in [20]. The use of technologies of deep generating neural networks in conjunction with reinforcement learning methods will allow you to synthesize images using CNC machines in a human-like manner.

References

[1] Hertzmann A, Jacobs C E, Oliver N, Curless B and Salesin D H 2001 Image analogies. *In Proc. of the 28th Ann. Conf. on Computer Graphics and Interactive Tech.* (ACM) pp 327-340.
[2] Efros A A, Freeman W T 2001 Image quilting for texture synthesis and transfer. *Proc. of the 28th Ann. Conf. on Computer Graphics and Interactive Tech.* (ACM) pp 341-346.
[3] Efros A A, Leung T K 1999 Texture synthesis by non-parametric sampling. *Proc. of the Int. Conf. on Computer Vision* **2** pp 1033-1050.
[4] Shih Y, Paris S, Barnes C, Freeman W T and Durand F 2014 Style transfer for headshot portraits.

ACM Transactions on Graphics, **33**(4):148.

[5] Kwatra V, Essa I, Bobick A and Kwatra N 2005 Texture optimization for example-based synthesis. *ACM Transactions on Graphics*, **24**(3):795-802.

[6] Gatys L A, Ecker A S and Bethge M 2015 A neural algorithm of artistic style *arXiv preprint* arXiv:1508.06576.

[7] Li N, Wang J and Hou L 2017 Demystifying neural style transfer *Proc. of the Twenty-Sixth Int. Joint Conf. on Artificial Intelligence IJCAI-17* pp 2230-2236.

[8] Li S, Xu X, Nie L and Chua T-S 2017 Laplacian-steered neural style transfer *Proc. of the 2017 ACM on Multimedia Conf.* (ACM) pp 1716-1724.

[9] Ruder M, Dosovitskiy A and Brox T 2016 Artistic style transfer for videos. *In GCPR.*

[10] Ledig C, Theis L, Huszar F and Caballero J 2016 Photo-realistic single image super-resolution using a generative adversarial network *arXiv preprint* arXiv:1609.04802.

[11] Gatys L A, Ecker A S, Bethge M, Hertzmann A, Shechtman E 2017 Controlling perceptual factors in neural style transfer *Proc. of the IEEE Conf. on Computer Vision and Pattern Recognition*, pp 3985-3993.

[12] Johnson J, Alahi A and Fei-Fei L 2016 Perceptual losses for real-time style transfer and super-resolution *In European Conference on Computer Vision*, pp 694-711.

[13] Ulyanov D, Lebedev V, Vedaldi A and Lempitsky V 2016 exture networks: Feed-forward synthesis of textures and stylized images *Int. Conf. on Machine Learning*, pp 1349-1357.

[14] Radford A, Metz L and Chintala S 2015 Unsupervised representation learning with deep convolutional generative adversarial networks *arXiv preprint* arXiv:1511.06434.

[15] Chen J, Yang J, Zhou H and Xiang H 2015 CPS Modeling of CNC Machine Tool Work Processes Using an Instruction-Domain Based Approach Engineering (Beijing, China: Academy of Engineering) **1**(2) pp 247-260.

[16] Lee J, Bagheri B, Kao H A 2015 A Cyber-Physical Systems architecture for Industry 4.0-based manufacturing systems *Manufacturing Letters* **3**:18-23.

[17] Simonyan K, Zisserman A 2014 Very deep convolutional networks for large-scale image recognition. arXiv:1409.1556.

[18] X. Huang and S. Belongie. Arbitrary style transfer in real-time with adaptive instance normalization. Proceedings of the IEEE International Conference on Computer Vision, 2017, pp. 1501-1510.

[19] Lin T, Maire M, Belongie S and Bourdev L D 2014 Microsoft COCO: Common Objects in Context. arXiv preprint arXiv:1405.0312.

[20] Ganin Y, Kulkarni T, Babuschkin I, Eslami S M and Vinyals O 2018 Synthesizing Programs for Images using Reinforced Adversarial Learning *Proc. of the 35th Int. Conf. on Machine Learning (ICML)* ed J Dy, A Krause (PMLR) **80** pp 1652-1661.

IOP Conf. Series: Materials Science and Engineering **582** (2019) 012014 doi:10.1088/1757-899X/582/1/012014

Mathematical Modeling and Information Technologies in the Management of Tractor Maintenance Operations

O V Myalo [a], A P Solomkin [b], S P Prokopov [c], A S Soyunov [d]

Federal State Budgetary Educational Institution of Higher Education "Omsk State Agrarian University named after P.A. Stolypin, 644008, Omsk, ul. Institutskaya Square, 2

E-mail: [a] ov.myalo@omgau.org, [b] a.solomkin@ibox.ru, [c] sp.prokopov@omgau.org, [d] as.soynov@omgau.org

Abstract. Obtaining the greatest amount of crop with the lowest labor and material costs in agriculture is possible by reducing the operating costs of the equipment used. Reducing the machine and tractor fleet cost is possible only if high performance machines are provided with a sufficient level of reliability of the equipment used. In order to ensure trouble-free machines operation during the stressful agricultural period, it is necessary to determine the level of technical equipment maintenance sufficient to prevent sudden failures and to perform the necessary maintenance and repair operations for the units outside the stressful agricultural period. The article presents an algorithm for calculating the incorrect determination probability of the failure-free tractor operation. Using the probabilistic characteristics of the errors measurement distribution according to the normal law and the tractor reliability indicators according to the Weibull law, it is possible to determine the risks associated with the tractor operation during busy agricultural periods. The obtained mathematical models were used while computer program algorithm writing for planning the maintenance and repair of the agro-industrial complex enterprises fleet.

1. Introduction

In tense periods of the main technological operations implementation at the agro-industrial complex enterprises, the need for stopping the machine-tractor units for maintenance and repair, as part of a planned-preventive system, arises from the timeliness and performance quality, which directly depends on the agricultural enterprise yield and profitability. At the same time, the maintenance and repair work volume increases proportional to the volume of technological operations performed, that is, a contradiction arises: if maintenance is not carried out in time, especially complex and labor-intensive TO-2 and TO-3, the failure rate increases and the tractor fleet is forced to stand idle for repairs and the crop is lost, and if carry out maintenance on time, the number of failures decreases, but maintenance downtime increases, which also leads to delays in the execution of operations and yield losses. The optimal solution here is difficult to find and the mechanic tractor operating decides in favor of carrying out field work to the detriment of maintenance work, which leads to premature wear and equipment failure. To eliminate this contradiction, the machine-tractor fleet maintenance and repair system was developed and introduced at the enterprises of the agro-industrial complex on the basis of scheduling. The main system feature is that complex maintenance services of TO-2 and TO-3, which, according to the principle of a preventive system, fell out during busy agricultural periods, are carried out before the onset of a busy period when tractors are idle [1, 2]. It was assumed that, by conducting the upcoming TO-2 and TO-3 using diagnostic tools, not only maintenance downtime, but also to eliminate failures in the period when it is necessary to achieve maximum productivity during agricultural work, will be eliminated. At the same time, the effect E taken from conducting TO-2 and TO-3 (TO-2 is also performed in the amount of TO-3) before agricultural work, which is determined by the formula:

$$E = C_1 + C_2 - C_3, \tag{1}$$

where C and C are the costs associated with the equipment downtime reduction for maintenance and the failures elimination in the upcoming stressful field work period;

C_3 - additional costs associated with an increase in the work volume on the tractors maintenance and diagnosis before the stressful period onset.

2. Objects and methods

The main research methods were used methods of mathematical modeling and statistical processing of the results. On the basis of the dependencies obtained, a program for determining the tractor resource based on the Excel Analysis Package Package was written.

Based on the practical application of the criterion of expediency, it does not take into account a number of quite important factors that reduce the effectiveness of this progressive form of maintenance of machines. When applying a form of maintenance using the scheduling system, it is not known whether the machine will work, carrying out the forthcoming amount of field work without causing failures. Will there be "bogus failures" during maintenance? The appearance of different failures types in this system depends on the accuracy of measuring diagnostic parameters and determining the probability of failure-free tractor components operation, which is not taken into account in equation (1).

3. Research results

Let x be the true parameter value, and y the measuring result the same parameter with a diagnostic tool to take into account the influence of the probabilistic processes characteristics of technical parameters maintenance and diagnostics on the expediency criterion value.

In this case: f (x) is the distribution density of the tractor unit status parameter, for example, the gap in the conjugation;

$\tau = (y\text{-}x)$ - parameter measurement error;

$f(\tau)$ is the density of the measurement error distribution;

x and y are random, independent values.

As a result of measuring the value of x of the parameter $\boldsymbol{X}$, according to the indications of the diagnostic device, we read the measurement result of y.

The following situations are possible:

$S_{\bar{x}}$ - value $\bar{x}$ is within the tolerance of a parameter change (for example, the nominal and allowable gap in the interface);

S - value is outside the tolerance range;

S_y- measurement of y is within the limits of the tolerance field;

S - measurement result is outside the tolerance range.

An event with the x value of the parameter X and the measurement result y pairwise compatible. Then the following complex events are possible:

S_{xy} is the x value and the measurement result y is within the tolerance range (gap in the conjugation);

$S_{\bar{x}}$ - value and measurement result $\bar{x}$ are outside the tolerance range.

After processing the data we can conclude that in both cases the correct conclusion about the node state is made.

S_x - the value of x is within the tolerance range, and the measurement result is outside it;

S_y - the value is outside the tolerance limits, and for the measurement result - within the limits.

These events constitute a complete system and its total probability is equal to one:

$$P(xy)+P(S\bar{x})+P(Sx)+P(S\bar{x}y)=1 \qquad (2)$$

The probabilities $P(xy)$ and $P(S\bar{x})$ are the probabilities of correctly determining the node state being diagnosed, and $P(Sx)$ and $P(S\bar{x}y)$ are incorrect.

For the tractor owner using the machine, it is important to know the risks from undetected failures $P(Sx)$ and $P(S\bar{x}y)$ To clarify them, for example, $P(Sx)$, consider the event S. Let the conditional probability $P(S/\tau)$ be known - the occurrence of the event S provided that the event τ has already occurred (measurement error). Then, according to the formula of total probability, we can write down [3]:

$$P(x)=\int P\{Sx/\tau\}f(\tau)d\tau, \qquad (3)$$

where $f(\tau)$ is the measurement error density.

The conditional probability *P{Sx/τ}* must be determined at a fixed value of τ (the measurement error has already occurred).

Depending on the instrumental determination outcomes of the technical object state, τ may be greater or equal to zero $\tau \geq 0$ or $\tau \leq 0$.

Consider cases for technical condition parameters, the value of which increases during operation of a component of a car, for example: crankcase consumption when the engine is worn, the gaps in the interface (case I) increase (Fig. 1), or the parameter values decrease, for example, fuel injection pressure at wear of injector nozzles; pressure generated by the fuel pump (case II) (Fig. 1).

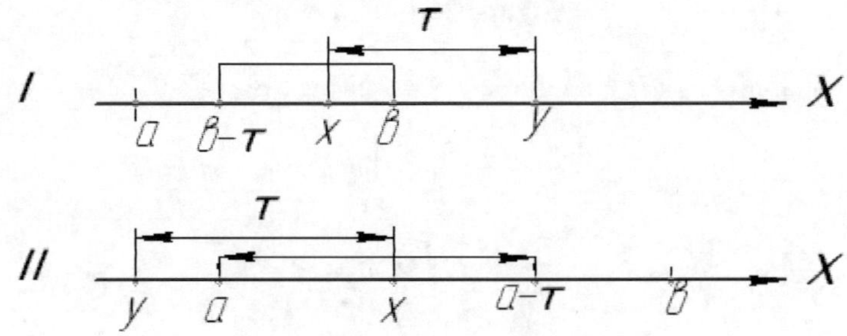

Figure 1. Sx situation

where a is the nominal value of the parameter; $в$ - value of the parameter allowed without repair; $aв$ - field of tolerance; x is the true value of the state parameter $\boldsymbol{X}$ of the tractor unit or aggregate; y is the measured value of this parameter using a diagnostic device; τ is the magnitude of the measurement error.

Thus, for case ***I*** with $\tau \geq 0$ (Fig. 1. I) with fixed τ:

$$P\{Sx\bar{y}/\tau\}_{\tau\geq 0} = \int_{в-\tau}^{в} f_1(x)dx \tag{4}$$

For case ***II*** with $\tau \leq 0$ (fig. 1. II)

$$P\{Sx\bar{y}/\tau\}_{\tau\leq 0} = \int_{a}^{a-\tau} f_1(x)dx \tag{5}$$

In the first case, the probability that the value of x will be within the tolerance field *(a, b)*, and the result of measuring y is outside it, corresponds to the probability that x is smaller or larger *(b-τ)*. In the second case, an identical picture is formed, only the upper limit of integration will be smaller *(a-τ)*. To go to the full probability *P (Sx)* (formula 3) and following [3], integrating with at > a (case I), we get:

$$P(Sx\bar{y}) = \int_{в-\tau}^{в} f_1(x)dx \times \int_{0}^{в} f_2(\tau)d\tau \,|_{\tau>0} \tag{6}$$

where $f_2(\tau)d\tau$ is the measurement error distribution function.

For case ***II***, *P(Sx)* will be:

$$P(Sx\bar{y}) = \int_{a}^{a-\tau} f_1(x)dx \times \int_{в-a}^{0} f_2(\tau)d\tau \,|_{\tau<0} \tag{7}$$

The situation when the measurement result is in the tolerance field *(a, b)* of the parameter change of the technical condition of the tractor component, and its true value is outside it, can be presented for cases (I, II) in the form (Fig. 2):

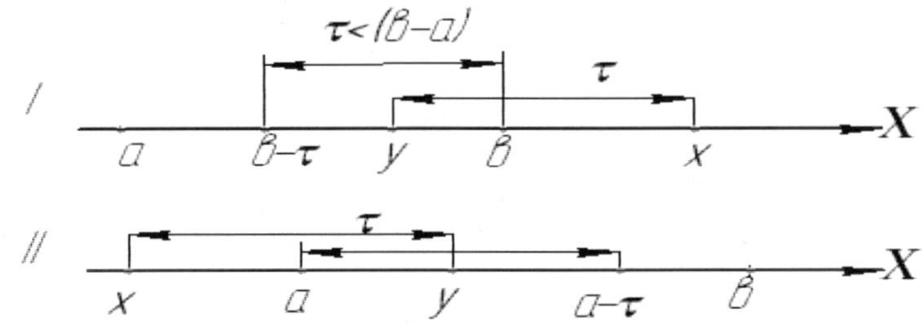

Figure 2. Situation *Sy*

The conditional probability *P(y/τ)* for *a* fixed τ is equal for case ***I*** (Fig. 2):

$$P\left(S\bar{x}y/\tau\right)_{\tau<0} = \int_{\text{в}-\tau}^{\text{в}} f_1(x)dx \tag{8}$$

For case ***II*** (fig. 2):

$$P\left(S\bar{x}y/\tau\right)_{\tau\geq 0} = \int_{a-\tau}^{a} f_1(x)dx \tag{9}$$

According to the formula of the total probability for the ***I*** case:

$$P\left(S\bar{x}y\right) = \int_{\text{в}-\tau}^{\text{в}} f_1(x)dx \times \int_{0}^{\text{в}-a} f_2(\tau)d\tau \tag{10}$$

For case ***II***:

$$P\left(S\bar{x}y\right) = \int_{a-\tau}^{a} f_1(x)dx \times \int_{0}^{\text{в}-a} f_2(\tau)d\tau \tag{11}$$

Taking into consideration all mentioned above, the probability of incorrectly determining the state of the diagnosed node for the first case is equal to:

$$P\left(Sx\bar{y}\right) + P\left(S\bar{x}y\right) = \int_{\text{в}-\tau}^{\text{в}} f_1(x)dx \times \int_{0}^{\text{в}} f_2(\tau)d\tau_{\tau>0} + \int_{\text{в}-\tau}^{\text{в}} f_1(x)dx \times \int_{-(\text{в}-a)}^{0} f_2(\tau)d\tau \tag{12}$$

For case ***II***:

$$P\left(Sx\bar{y}\right) + P\left(S\bar{x}y\right) = \int_{a}^{a-\tau} f_1(x)dx \times \int_{-(\text{в}-a)}^{0} f_2(\tau)d\tau + \int_{a-\tau}^{a} f_1(x)dx \times \int_{0}^{\text{в}-a} f_2(\tau)d\tau \tag{13}$$

In the future, the use of the dependencies we obtained is specified by the distribution laws of the measured parameter *f(x)* and measurement errors *f(τ)*. The most likely of these are the normal law and the Weibull distribution law. Normal law should be used to characterize measurement errors and parameters of the technical state of components and assemblies with $\text{V} = \frac{\sigma}{\text{m}} \leq 0{,}3$ (*V* - coefficient of variation) and the Weibull law is the characteristics of the parameters of the components and assemblies technical condition, usually associated with the wear of parts, where usually *v>0.3*. The densities of these distributions, respectively, will be:

$$f(\tau) = \frac{1}{\sigma_\tau \sqrt{2\pi}} e^{-\frac{(\tau_x - \bar{\tau})^2}{2\sigma_\tau^2}} \tag{14}$$

$$f(x) = \frac{\text{в}}{a}\left(\frac{x}{a}\right)^{\text{в}-1} e^{-\left(\frac{x}{a}\right)^{\text{в}}} \tag{15}$$

where σ_τ, is the standard deviation and mathematical expectation of measurement errors (or technical parameters σ_x and mx with $v \le 0.3$), and a are the Weibull law distribution parameters, x is the current value of the technical condition parameter of the tractor unit and unit, e - base of natural logarithm.

For the practical implementation of generalized dependencies (12, 13), we use the distribution law (14, 15) and introduce the concepts:

- the nominal value of the parameter X_n, corresponding to the state of the diagnostic parameter of the node of the new machine (in the tolerance field, this is "*a*" figure 1.2);

X_p is the permissible value of the parameter, at which failure recovery is not yet required (in the tolerance field it is "*b*");

X_n - its limiting value with fully worn parts, as a result of which a failure occurs;

δ_0 - is the minimum tolerance corresponding to the allowable residual resource from X_p to X_n.

$$P\left(Sx\bar{y}\right)+P\left(S\bar{x}y\right)=\frac{1}{\sigma_\tau\sqrt{2\pi}}\int_0^{(Xg-Xn)-X} e^{\frac{-\tau^2}{2\sigma_\tau^2}}d\tau\left[\exp-\left(\frac{X_p-X}{a}\right)^{в}-exp-\left(\frac{X_p-\tau-X}{a}\right)^{в}\right]+ \\ +\frac{1}{\sigma_\tau\sqrt{2\pi}}\times\int_0^{(X_д-X_n)-X} e-\frac{\tau^2}{2\sigma^2\tau}d\tau\left[\exp-\left(\frac{X_p-X}{a}\right)^{в}-\exp-\left(\frac{Xp-\tau-X}{a}\right)^{в}\right] \tag{16}$$

And for the narrowing parameters values of the tractor components and assemblies technical condition when worn:

$$P\left(Sx\bar{y}\right)+P\left(S\bar{x}y\right)=\frac{1}{\sigma_\tau\sqrt{2\pi}}\int_{-(X-X_n-\delta_0)}^{0} e^{-\frac{\tau^2}{2\sigma^2\tau}}d\tau\left[\exp-\left(\frac{X_p-\tau-X}{a}\right)^{в}-\exp-\left(\frac{Xp-X}{a}\right)^{в}\right]+ \\ +\frac{1}{\sigma_\tau\sqrt{2\pi}}\int_{-(X-X_l-\delta_0)}^{0} e^{-\frac{\tau^2}{2\sigma^2\tau}}d\tau\left[\exp-\left(\frac{Xp-X}{a}\right)^{в}-\exp\left(\frac{Xp-\tau-X}{a}\right)^{в}\right] \tag{17}$$

However, it should be determined that the risk of incorrectly determining the state of a node or aggregate with a probability *P(x)* is fraught with the fact that there is actually no failure, and according to the readings of diagnostic devices it is there, and there is a need for a more detailed study of the condition of the node or aggregate of the tractor and possibly disassembling it. This situation leads to additional costs of time and money in diagnosing, although this happens in an unstressed field work [4, 5]. Therefore, the first components of the dependencies in formulas (16) and (17) should be considered as the likelihood of additional costs when performing work during scheduling maintenance of tractors, and the second component as the likelihood of the risk of failure in the upcoming field work period, which usually does not taken into account.

To implement the obtained dependencies, it is necessary to know the distribution functions of the main tractor components and assemblies technical parameters – the key ones, and there are not so many of them that determine the performance of the tractor [6]. Values of permissible and limiting parameters are determined from the results of the experiment. The standard deviations of measurement errors are determined on the basis of the maximum measurement error $\varepsilon = 3\sigma\tau$.

The experimental data determine the of the variation coefficients value of the Weibull distribution, (*V*) by the coefficient (*V*) determine the differential

$f(x)=\frac{в}{a}\left(\frac{x}{a}\right)^{в-1}e^{-\left(\frac{x}{a}\right)^{в}}$ or integral $F(x)=1-e^{-\left(\frac{x}{a}\right)^{в}}$ functions, then implement the dependencies (16) and (17), setting the integration limits, based on specific conditions: degree of deterioration of the tested components and assemblies of the tractor (resources), duration of the upcoming field work period, and others [7].

In the practical implementation of the models obtained, it is possible to obtain the resource values of individual tractor units based on the specific operating conditions using mathematical processing of the Excel Analysis Package [8]. (Fig. 3).

	A	B	C	D	E	F	G	H	I
25	ВЫВОД ИТОГОВ								
26									
27	*Регрессионная статистика*								
28	Множественный R	0,905534508							
29	R-квадрат	0,819992745							
30	Нормированный R-к	0,619984684							
31	Стандартная ошибк	364,3629253							
32	Наблюдения	20							
33									
34	Дисперсионный анализ								
35		*df*	*SS*	*MS*	*F*	*Значимость F*			
36	Регрессия	10	5442906,461	544290,6461	4,099798	0,022505715			
37	Остаток	9	1194843,072	132760,3414					
38	Итого	19	6637749,533						
39									
40		*Коэффициенты*	*Стандартная ошибка*	*t-статистика*	*P-Значение*	*Нижние 95%*	*Верхние 95%*	*Нижние 95,0%*	*Верхние 95,0%*
41	Y-пересечение	1907,397948	8285,241279	0,230216343	0,823071	-16835,11996	20649,9159	-16835,12	20649,91585
42	X1	-2984,02571	40033,10404	-0,074538954	0,942212	-93545,19876	87577,1474	-93545,1988	87577,14735
43	X2	-151,137828	361,5780046	-0,417995082	0,685747	-969,0841008	666,808445	-969,084101	666,8084453
44	X3	-6,33703048	44,52652791	-0,142320337	0,889962	-107,0630345	94,3889736	-107,063035	94,38897357
45	X1^2	-24238,4469	14290,38349	-1,696136912	0,124095	-56565,54029	8088,64644	-56565,5403	8088,646441
46	X2^2	2,210097841	4,57652673	0,482920339	0,640682	-8,142724882	12,5629206	-8,14272488	12,56292056
47	X3^2	-0,01195106	0,012124364	-0,985706063	0,350034	-0,039378275	0,01547616	-0,03937827	0,015476157
48	X12	394,7218469	1633,126035	0,241697112	0,814432	-3299,66591	4089,1096	-3299,66591	4089,109604
49	X13	77,53501801	207,8271164	0,373074599	0,717728	-392,6025819	547,672618	-392,602582	547,6726179
50	X23	0,166983248	1,598502004	0,104462333	0,919093	-3,449079511	3,78304601	-3,44907951	3,783046007
51	X123	-1,00055025	7,707820809	-0,129809745	0,899572	-18,4368523	16,4357518	-18,4368523	16,4357518

Figure 3. Regression analysis results obtained using the Excel Analysis Package

The resulting mathematical models can be processed in publicly available programs in order to obtain an algorithm for planning a technical service system for a particular enterprise of the agro-industrial complex and managing the processes of maintenance and repair of the fleet with the lowest predicted costs and saving the unit life during the overhaul period [9, 10].

4. Conclusion

Based on the above, we can draw the following conclusions:

1. In order to ensure trouble-free machines operation during the stressful agricultural period, it is necessary to determine the level of technical maintenance of equipment sufficient to prevent sudden failures and to perform the necessary maintenance and repair operations for the units outside the stressful period of agricultural production.

2. To take into account the influence of the probabilistic characteristics of technical parameters maintenance and diagnostics processes on the expediency criterion value, mathematical dependencies of the tractor units and assemblies technical condition parameters were obtained for expanding parameters values as a result of wear (16), and dependencies for the narrowing values of tractor components technical parameters under wear (17).

3. Mathematical models are processed in publicly available programs. As a result, an algorithm for planning the system of technical service and managing the processes of maintenance and repair of agricultural enterprises fleet has been obtained.

References

[1] Theoretical substantiation of machine-tractor fleet technical maintenance system on the example of Omsk region agricultural enterprises / O V Myalo, V V Myalo, S P Prokopov, A P Solomkin, A S Soynov - IOP Conf. Series: Journal of Physics: Conf. Series 1059 (2018) 012005

[2] Ensuring Machine and Tractor Aggregates Operability. / G.V. Redreev - IOP Conference Series: Materials Science and Engineering, Volume 142, Number 1.

[3] Barzilovich, E. Yu. Maintenance models of complex systems / E. Yu. Barzilovich. - M.: High School, 1982. - 230 p.

[4] Formation of agricultural tractors fleet in the Krasnoyarsk Territory // N.I. Selivanov, V.V. Matyushev, N.I. Chepelev, I.A. Vasiliev // Achievements of Science and Technology in the Agricultural Sector - Moscow, 2017. № 9. Pp. 72 - 75.

[5] Technical and technological support of agricultural producers in the transition to an innovative economy / A.V. Kuchumov, A.V. Belokopytov // Achievements of science and technology of the agro-industrial complex - Moscow, 2017. V. 31. № 12. Pp. 78 - 81.

[6] Improvement of technical service in the agro-industrial complex based on the assessment and analysis of the technological level of repair enterprises / A.I. Anosova, M.K. Buraev // Achievements of science and technology of agrarian and industrial complex - Moscow, 2014. № 10. Pp. 65 - 68.

[7] APC Technique - quality service / N.M. Ivanov, A.E. Nemtsev, V.V. Korotkikh // Achievements of science and technology of the agro-industrial complex - Moscow, 2016. V. № 4. Pp. 81 - 82.

[8] Intellectual management of technological equipment of an enterprise in the conditions of digital production / Yu.G. Kabaldin, D.A. Shatagin, A.M. Kuzmishina // Proceedings of higher educational institutions. Mechanical Engineering, Moscow State Technical University Bauman (National Research University) - Moscow, 2018. T 7 (700) Pp. 30 - 41.

[9] Modeling the Technological Process for Harvesting of Agricultural Produce / S.D. Shepelev, V.D. Shepelev, Z.V. Almetova, N.P. Shepeleva, M.V.Cheskidov, // IOP Conference Series: Earth and Environmental Science115(1), (2018) 012053

[10] Basic robotecnical platform for implementation of accurate farming technologies / N.I. Tchernyshev, O.E. Sysoev , D.B. Solovev, Kiselyov, E.P. // Bulletin of Electrical Engineering and Informatics 7(4), (2018) P p. 522-528

IOP Conf. Series: Materials Science and Engineering **582** (2019) 012015 doi:10.1088/1757-899X/582/1/012015

Prospects of Using of Chemicals in Environmentally safe Disinfection of Liquid Waste from Livestock Enterprises

N V Limarenko [a], E V Krasnova, L A Pudeyan, O E Baryshnikova

Don state technical university, 1, Gagarin sq., Rostov-on-Don, 340000, Russia

E-mail: [a] limarenkodstu@yandex.ru

Abstract. The prospective types of impacts on liquid waste from livestock enterprises during their disinfection were considered. The analysis of chemical reagents, providing the maximum prolonged bactericidal effect, the minimum pathogenicity of the reaction by-products and the minimum cost of the reagent and its production technology was carried out. The kinetic models of disinfection used in the evaluation of the bactericidal productivity of chemical reagents were considered. The basic parameters of liquid waste from livestock enterprises were formulated to take into account the specificity and preservation of nutrient properties, which significantly improves the level of environmental management and environmental safety.

1. Introduction

The development of agriculture, in particular animal husbandry, in addition to addressing food security issues, is associated with the formation of a large amount of hazardous waste epidemiologically. The volume of epidemiologically hazardous fluids produced by the livestock production farm for 12.000 heads is about 250 m^3 / day.

The specificity of which is the high content of nitrogen N, phosphorus P_2O_5, potassium K_20 and other elements capable of exerting positive effects on the properties of soils [2, 8, 10], which makes rational their further utilization.

However, in addition to nutrients, this type of waste is also saturated with pathogenic organisms such as: bacteria of the group of intestinal sticks, enterococci, staphylococci, clostridia, salmonella and helminth eggs. Therefore, environmentally safe disposal is impossible without the suppression of the pathogenic component. One of the main operations of the utilization cycle of this type of waste is their disinfection.

Existing approaches to disinfection depending on the type of exposure can be divided into: chemical, physical, biological.

The main methods of chemical disinfection are: chlorination, ozonation, disinfection with silver ions, iodine, hydrogen peroxide, bromine.

The main methods of physical disinfection are: mechanical vibrations with frequencies from 20 Hz to 300 MHz, thermal effects (boiling, pasteurization), electromagnetic field: constant, variable and pulsed nature (ionizing, optical, ultraviolet, infrared, terahertz, microwave and radio wave spectra).

The main methods of biological disinfection are exposure to anaerobic and aerobic microorganisms.

According to the analysis of information sources [1-11], the most promising from the point of view of environmental safety is the use of methods of disinfection using complex effects. As part of this study, the most promising chemical reagents used for disinfecting liquid media of livestock enterprises are considered.

The main parameters ensuring an increase in the level of environmental safety demanded of reagents during disinfection are [1, 2, 6, 8]:

– steady prolonged bactericidal effect;

– the number and level of epidemiological danger of negative disinfection products formed during the reaction;

– speed reaction;
– the cost of the reagent and the cost of obtaining its working solution.

Accordingly, the task of choosing the most effective reagents for environmentally safe decontamination, taking into account the specifics of this type of waste, allowing to increase the level of environmental management during their disposal is relevant.

The purpose of the study is to select the most chemical reagent that ensures the effective and environmentally safe suppression of epidemiologically dangerous organisms when disposing of the liquid fraction of animal waste.

2. Materials and methods

An effective way to assess the bactericidal effect of a reagent is to use kinetic models of disinfection [3, 9, 11]. The most widely used kinetic model is the Chick-Watson model. The essence of mathematical models of bactericidal kinetics is to assess the inactivation effect of microorganisms under the action of a specific reagent. Let us consider in more detail the mathematical model of Chick-Watson bactericidal kinetics:

$$\ln\left(\frac{N}{N_0}\right) = -kCnt \quad (1)$$

where N/N_0 – is the ratio of the number of organisms at the moment of time t to the time zero, colony forming units (CFU);

C – concentration of chemical reagent, mg;
n – empirical coefficient / indicator (determined for each reagent);
k – mortality, CFU / min;
N_0 – the initial number of pathogens, CFU / ml;
N – is the number of surviving pathogens during the exposure, CFU / ml;
t – the duration of contact, min.

The adequacy of this model depends on the reliability of the basic law of the rate of inactivation, so using (1) it is possible to explain the change in the concentration of the reagent during exposure to the environment to be disinfected; also the pathogens to be studied must be of the same strain.

Also an effective way to assess the bactericidal performance of a reagent is a direct count of the CFU of pathogen units grown on nutrient ogare after exposure [1-6]. The direct kinetics of the residual CFU number is described by the following relationship:

$$X_{CFU} = \frac{\alpha \cdot V_k}{V}, \quad (2)$$

where a – is the total number of CFU in the sample, units;

$V_к$ – control volume, mm^3;
V – is the actual sample volume, mm^3.

3. Literature review

Let us consider in more detail the main principles of disinfection by chemical impacts. The essence of chemical decontamination methods is the use of elements with the properties of antibacterial agents of oxidative and reducing types [1-5, 7-10]. The oxidative group includes chlorine containing reagents: chlorine, chloramine, chlorine dioxide, chloride, as well as bromides, bromine atoms, ozone and hydrogen peroxide. Reducing agents include: formaldehydes, isothiazalones, isocyanates, quaternary ammonium compounds, and also chlorinated phenols.

In general, the mechanism of chemical methods of disinfection is reduced to the effects either on the cell wall of pathogen cells or on enzyme systems that support the metabolism in them. Also, under certain conditions, the combined effects of both mechanisms are possible [1, 2, 8-11].

The main factors affecting the performance of chemical methods are: the nature of the antibacterial reagent; reagent concentration; contact time between reagent and pathogen; temperature; type and concentration of pathogens; pH level of the disinfected medium [8-11].

According to the analysis of information sources [2, 8-11], the most promising methods for chemical disinfection of liquid media are the use of silver ions, iodine, hydrogen peroxide, bromine, chlorine-containing and ozone containing reagents.

The essence of bactericidal disinfection with silver (*Ag*) lies in its ability to adsorb on the surface of bacteria, which leads to the irreversible formation of sulfhydryl complexes on the membranes and disruption of the pathogen cell surface dehydration processes.

The advantages of this type of effects are: the presence of a stable prolonged bactericidal effect, the absence of negative by-products of the reaction.

The disadvantages of this type of exposure are: bacteriostatic action against certain types of pathogens; the need for an electrolyzer to increase the efficiency of entering into the environment direct dependence of the bactericidal effect on temperature; not effective against spore-forming bacteria; a strong dependence of the bactericidal effect on turbidity and the hardness of the medium; high specific cost of the reagent.

Based on the analysis performed, it can be concluded that this reagent is of little use for disinfecting liquid media of livestock enterprises due to its high cost and technological limitations: a long exposure time, the presence of a bacteriostatic effect (ie, the absence of complete lysis of pathogens).

The advantages of this type of effects are: the presence of a stable prolonged bactericidal effect, the absence of negative by-products of the reaction.

The disadvantages of this type of exposure are: bacteriostatic action against certain types of pathogens; the need for an electrolyzer to increase the efficiency of entering into the environment direct dependence of the bactericidal effect on temperature; not effective against spore-forming bacteria; a strong dependence of the bactericidal effect on turbidity and the hardness of the medium; high specific cost of the reagent.

Based on the analysis performed, it can be concluded that this reagent is of little use for disinfecting liquid media of livestock enterprises due to its high cost and technological limitations: a long exposure time, the presence of a bacteriostatic effect (ie, the absence of complete lysis of pathogens).

The essence of bactericidal disinfection with iodine (*I*) is the impact of hydrated cationic iodine particles with the nucleotide base of pathogens, namely the destruction of vital amino acids in proteins.

The advantages of this type of effects are: higher bactericidal efficacy at an unstable pH temperature compared with chlorine (about 11 ... 13%); effectiveness against viruses, due to conformational changes in their protein structure; a weaker dependence of the bactericidal effect on the turbidity and hardness of the medium; the presence of a prolonged bactericidal effect.

The disadvantages of this type of effects are: poor efficacy against protozoan parasites; contact duration from 3 min. (usually 20...35 min.), and at a temperature below – 5°C, the duration should be doubled; prior mechanical cleaning is desirable to increase efficiency; the difficulty of determining the exact initial concentration; change in organoleptic properties (to eliminate it it is necessary to use neutralizing agents additionally: sodium thiosulfate, ascorbic acid, etc., which significantly increases the cost); high unit cost; the by-products of the overdose reaction are cumulative in nature and may adversely affect the functioning of biosystems.

On the basis of the analysis performed, it can be concluded that this reagent is of little use for disinfecting liquid media of livestock enterprises due to the high cost and technological limitations: long exposure time, preliminary mechanical cleaning, as well as changes in organoleptic properties.

The essence of bactericidal disinfection with hydrogen peroxide (H_2O_2) lies in the high oxidative ability of the free radicals present.

The advantages of this type of effects are: prolonged bactericidal effect; preventing the formation of pathogen biofilms; increased bactericidal effect in combination with other reagents, for example, ozone (O_3).

The disadvantages of this type of effects are: the difficulty of preparing concentrated solutions; high specific cost of the reagent; for effective bactericidal action, high concentrations are necessary; long contact period (about 35 minutes).

Based on the analysis performed, it can be concluded that the use of this type of reagent for disinfecting liquid waste of livestock enterprises is ineffective due to its high cost and prolonged exposure.

The essence of bactericidal disinfection with bromine (Br_2) lies in its oxidative capacity aimed at pathogen mass transfer processes and the protein component in the case of viruses.

The advantages of this type of effects are: the possibility of using in three aggregative states; effective against viral pathogens; effective at pH > 9; steady prolonged bactericidal effect.

The disadvantages of this type of effects are: reactivity with ammonia; high price; decomposes under the action of ultraviolet radiation; does not have a stabilizer; to increase the efficiency of oxidation, it is necessary to use reagents with bromide ions (for example, active chlorine); the complexity of the preparation of the working solution (it is necessary to maintain the temperature of 50...60°C).

Based on the analysis performed, it can be concluded that this reagent in its pure form is of little use for disinfection of liquid media of livestock enterprises due to the high cost of the reagent and the process of obtaining a working solution, however, its use in combination with active chlorine can be quite promising, in that case if suppression of viral forms of pathogens is necessary.

The essence of bactericidal disinfection with chlorine containing reagents is the interaction of hypochlorous acid and hypochlorite ions with substances that are part of the protoplasm of pathogen cells and the implementation of oxidative processes leading to their lysis. Depending on the volume, the pH of the medium used: chlorine (Cl), bleach ($2Cl_2 + 2Ca\ (OH)_2 = Ca\ (OCl)_2 + CaCl_2 + 2H_2O$), sodium hypochlorite ($NaOCl$), chlorine dioxide ($ClO_2$), chloramine ($NH_2Cl$) is common, for which is the presence of active chlorine, which has a high oxidizing ability.

The advantages of this type of impact are: the destruction of organic compounds in the form of phenols, which are sources of changes in organoleptic properties; removal of iron and magnesium by their rapid oxidation and precipitation of oxides; the presence of a stable and prolonged bactericidal effect.

The disadvantages of this type of exposure are: the formation of disinfection by-products (trihalomethanes, bromates, chlorates and chlorites); possibility of toxicity in case of leakage; the dependence of the effect on the turbidity and rigidity of the medium to be disinfected; ineffectiveness against cysts (lamblia, *cryptosporidium*).

Based on the analysis, it can be concluded that the use of this type of reagent for disinfecting liquid waste from livestock enterprises is quite effective, since it contributes to the coagulation, sedimentation, filtration, deodorization processes, but the formation of negative by-products significantly reduces the level of environmental safety. Effective use of this type of reagent for solving the problem is possible, in conjunction with the impact, allowing to minimize its initial concentration.

The essence of bactericidal disinfection of ozone (O_3) containing reagents lies in its high oxidative capacity, which ensures the penetration of bacterial cells into the cytoplasm and their subsequent destruction, as well as interaction with mineral, organic substances.

The advantages of this type of effects are: efficacy against Giardia and *cryptosporidium*; the ability to change the turbidity of the liquid being disinfected, a high rate of decomposition and conversion to oxygen compared to chlorine; the effect of decanogenization.

The disadvantages of this type of exposure are: the formation of by-products, including: aldehydes, ketones, organic acids, bromine containing trihalomethanes (including bromoform), bromates (in the presence of bromides), peroxides, bromoacetic acid; requires the use of biologically active filters for the subsequent removal of the resulting negative by-products; does not provide long prolonged bactericidal effect; rapid decomposition into oxygen of the residual reagent is possible only at a temperature of 60...350°C; the impossibility of storage and, respectively, immediate use; high price.

Based on the analysis performed, it can be concluded that the use of ozonation causes a high degree of saturation of the disinfected liquid with an ozone-air mixture, which increases its oxidative capacity, which makes it corrosively active. At the same time, corrosive activity increases with increasing temperature, which is a serious technological limitation, with a rather high specific cost of the reagent. Accordingly, the use of ozone in its pure form for the disinfection of liquid waste of livestock enterprises is ineffective due to the above disadvantages and the lack of a sustainable, long-term, prolonged bactericidal effect.

4. Results and Discussion

Based on the analysis of information sources and formulated requirements for chemical reagents, using the method of coding and categorization, we consider the effect of parameters on reagents in coded variables (Table 1), according to the following scale: "+1"– maximum; "0" – is neutral; "-1" – the minimum.

Table 1. Maximum / Minimum Effect Levels of coded variables on reagent

№	Parameter / Reagent	*Cl*	O_3	*Ag*	Br_2	H_2O_2	*I*
1	Prolonged bactericidal effect	1	-1	1	1	1	0.4
2	Pathogenicity of by-products formed by the reaction	0.3	0.2	-1	1	1	1
3	The cost of the reagent and the cost of its receipt	-1	0	1	1	1	1

In Figure 1 shows a graphical interpretation of parameters characterizing reagents. According to the analysis of information sources, the most promising reagents in terms of their use in complex exposures in the disinfection of liquid waste from livestock enterprises are chlorine-containing. Chlorine-containing reagents provide the most stable, prolonged bactericidal effect, the by-products formed during the reaction do not represent a high epidemiological hazard, and due to the additional physical effects, their volume can be minimized by reducing the initial concentration.

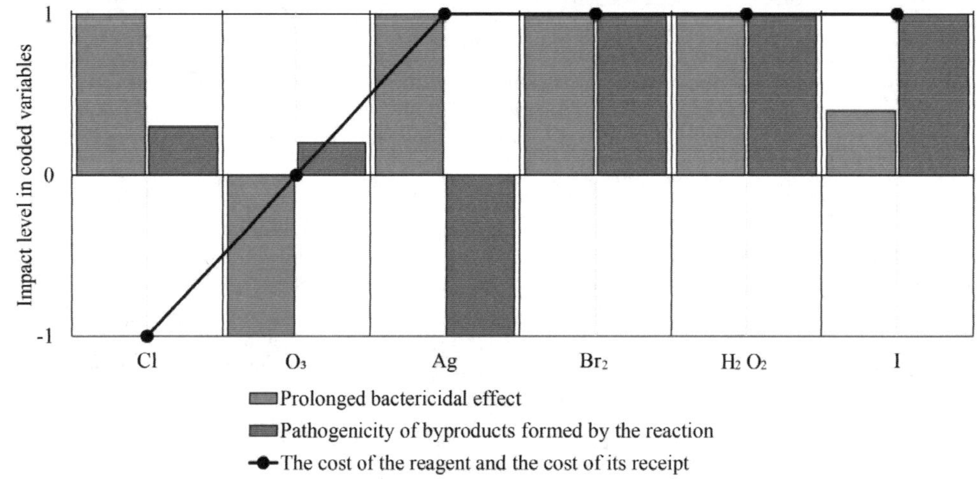

Figure 1. Comparison of the effectiveness of the use of chemical reagents in the disinfection of liquid waste of livestock enterprises

The cost and technological especially the preparation of the reagent and the working solution are minimal. Based on this, it can be concluded that chlorine-containing reagents are the most promising in the application of complex effects, since they provide an increase in environmental safety during the disinfection of liquid waste from livestock enterprises.

5. Conclusion

It has been established that from the point of view of environmental safety when disposing of liquid waste from livestock enterprises, the following requirements must be followed: the presence of a sustained, prolonged bactericidal effect; minimum quantity and level of pathogenicity of reaction by-products; minimum cost of reagent and technology and its production. It was proved that the most effective increase in the environmental safety of the operation of disinfection of liquid waste from livestock enterprises is possible with the use of complex physico-chemical effects. The most promising reagents in the framework of the set conditions are chlorine containing, as they ensure maximum compliance with the selected parameters and, as a result, increasing of the economic effect and the level of environmental management.

References

[1] Bichai F, Polo-López M I and Fernández Ibañez P 2012 *Water Research* **46(18)** 6040-6050.

[2] Bondarenko A M, Lipkovich E I and Lipkovich I E 2017 *Journal of Industrial Pollution Control* **1(33)** 1163-1170.
[3] Briukhanov A, Subbotin I, Uvarov R and Vasilev E 2017 *Agronomy Research* **15(3)** 658-663.
[4] Carrey D M 2011 *Water Recycling and Water Management* 1-280.
[5] Chen K C and Wang Y H. 2012 *Environmental Technology* **33 (4)** 487-495.
[6] Dong S, Lu J and Plewa M J 2016 *Environmental Science and Technology* **50(21)** 11752-11759.
[7] Guan C, Yang J and Shan J. 2014 *Nongye Gongcheng Xuebao. Transactions of the Chinese Society of Agricultural Engineering* **30(23)** 253-259.
[8] Kim M and Lee H. 2014 *Water Science and Technology* **70(12)** 1961-1968.
[9] Shchegolkov A V, Trufanov B S , Hmyrov V D, Kudenko V B, Guryanova Y V and Guryanov D V 2017 *Nano Hybrids and Composites* **13** 130-135.
[10] Moskvicheva E V, Moskvicheva A V, Doskina E P, Sidyakin P A, Shchitov D V, Lykova E Y and Fesenko L N *Key Engineering Materials* **736** 187-190.
[11] Uvarov R, Briukhanov A, Subbotin I and Shalavina E 2017 Agronomy Research **3(15)** 915-920.

IOP Conf. Series: Materials Science and Engineering **582** (2019) 012016 doi:10.1088/1757-899X/582/1/012016

Usage of Bioorganic Nano-fertilizer Nagro as Opportunity of Ecologization for Winter Rye Development (Secale Cereale)

L G Pinchuk [a], A V Piyanyh[b], I A Sergeeva[c], O B Konstantinova[d]

ESBEI of Higher Education Kemerovo state agricultural institute, Markotseva str. 5, Kemerovo, 650056, Russian Federation

E-mail: [a]Ludmilapinchuk@mail.ru,[b]anton.pyanih2014@yandex.ru, [c]ira_sergeeva@mail.ru, [d] olykk@mail.ru

Abstract. On the basis of research of winter rye productivity dependence of Vlada and Tetra short cultivar from exposure of seed treatment and sowing with bioorganic nano-fertilizer Nagro in Kuznetskaya forest steppe of south-east in West Siberia (2015 – 2018 years), there was found out increasing of yield activity respectively to control (sowings without fertilizer) for Vlada cultivar – 15-28% and Tetra short – 4-13%. Number of falling increased for 5-23%. Wasn't spotted correlation between hydrothermal conditions of vegetation and influence of biologics on studied productivity index.

1. Introduction

Intensification of agricultural development must be realized with allowance to ecological implication and challenges, and it must be directed to not only increase productivity of agricultural plants, but to optimize biospheric process.

Organic (ecologic) agriculture - responsible minimization for usage of synthetic fertilizer, pesticides and plant growth regulators. One of the way to increase amount and quality of plant production is production development of organic product in the Russian Federation [1]. Annually world market of management organic form production is growing by a mean of 15%. In spite of higher prices, more people want to buy organic products [2].

"All advanced technologies of plant science will be directed to ecologizate production and increase non-wastefulness" [3]. For this necessary to use agro-technologies to decrease man-made burden on agrobiocenosis. One of promising trends of agricultural products production ecologization is usage of bioorganic, which have complex-positively effect on crop productivity and their environment [4, 5].

Modern market of biologicals gives opportunity to choose most effective, based on influence on quality and quantity productiveness of agricultural crop, especially during cultivation of bread crops.

Goal research – on the basis of dependence research of formatting yield and seed quality from seed treatment and sowing exposure, optimization of bioorganic fertilizer Nagro usage in Kuznetskaya forest steppe.

Bio-fertilizer Nagro is a next generation of fertilizer, which determined as nano-fertilizer and characterized by high efficiency, prolonged and selective ability on microbiological activity of soil [6, 7]. Also used by this fertilizer active materials of mineral fertilizer [8] has influence on germination of vegetative plant mass and their productivity [9, 10, 11].

2. Terms, materials and methods of research

Experimental part of research was conducted in 2015 – 2018 years in natural climatic conditions of Kuznetskaya forest steppe (north forest steppe zone of West Siberia's south-east foothill, Kemerovo region) [12]. Target of research – cultivars of winter rye (Vlada and Tetra short). Experimental design included six options: I option – seed treatment by bio-energetic, treatment after vegetation: universal + bio-energetic during tillering period at spring. II option – seed treatment with bio-energetic, first treatment – after vegetation – universal + bio-energetic during tillering period at spring; second treatment – universal during paniculation period; III option – seed treatment with bio-energetic, first treatment after vegetation – bio-energetic during tillering period at spring, second treatment – universal during paniculation period; IV option – treatment after vegetation – universal + bio-energetic during tillering period at spring; V option – first treatment after vegetation – universal + bio-energetic during tillering period at spring, second treatment – universal during paniculation period; VI option – first treatment after vegetation – bio-energetic during tillering period at spring, second treatment – universal during paniculation period. Control group – sowings without bio-fertilizer. Pre-seeding treatment was conducted with bio-energetic Nagro with norm 1 l/ton, universal Nagro with norm 1l/ha and bio-energetic – 0.2 l/ha. Volume of spray material 250 l/ha. Replication – threefold. Plot allocation – randomized. Size - 100 m^2. Precursor – clean steam.

Bio-fertilizer Nagro contain vitamins, amino-acids, phytohormones, bio-solvents, nano-silicium, bio-calcium, anti-oxidants, adapt-genetic materials, metabolites, nitrogen-fixer, anti-pheromone – thoribon, bacterial spores (Pseudomonas aureofaciens), express fungicidal activity, increase progression of plants rhizosphere's microbiota [13]. For treatment of seeds and sowings were used two forms of fertilizer: bio-energetic and universal bio-fertilizer.

Cultivation technology is realized in accordance with zonal recommendation [14]. Seed treatment was conducted with spraying device MAROLEX MINI 1000 7 days before planting. Sowings were labored by shoulder sprayer ECHO DM-4610. Harvesting was realizing in full ripe stage. Yield activity considered by continuous method with furthering translation on standard moisture regain (14%) and 100% purity. Falling number was evaluated in accordance with GOST methodic 27676-88 [15]. For mathematical treatment were used elements of analysis of variance[16].

Soil – black earth leached, average-deep, medium-humic, granulometric texture – hard-loamy [17]. Prosperity by labile forms of phosphorus and potassium (according to Chirikov) corresponding to 140 and 154 mg/kg of soil (GOST 26204-91) [18]. pH – faintly-acid (salt extract 6.0) (GOST 26483-85) [19], mass fraction of mould 8,9 % (GOST 26213-91) [20].

Hydrothermal conditions of plants vegetation during summer period (may – august) in terms of warm and moisture, estimated by value of hydrothermal index – HTI [21] characterized as semi-arid in 2015 and 2016 (HTI = 0.69 and 0.83 correspondingly). 2017 and 2018 years – closer to optimum needs of rye plants (HTI = 1.02 and 1.04 correspondingly) (Table 1).

Table 1 -Hydrothermal index, HTI Krasnoye, Kemerovo region

3. Monyh	4. Long time average annual 5.	6. Year			
		7. 2015	8. 2016	9. 2017	10. 2018
11. May	12. 0.94	13. 1.19	14. 0.89	15. 0.82	16. 2.00
17. June	18. 1.00	19. 0.41	20. 0.49	21. 0.76	22. 0.70
23. July	24. 1.37	25. 0.38	26. 1.40	27. 1.42	28. 1.75
29. August	30. 1.07	31. 0.78	32. 0.53	33. 1.07	34. 0.45
35. May - August	36. 1.09	37. 0.69	38. 0.83	39. 1.02	40. 1.04

3. Results and discussion

Winter rye Vlada cultivar yield activity has changed between 2.70 and 3.00 ton/ha, making 2.34 ton/ha in control, at a time when were used biologics Nagro in every experiment options for four years of research (Figure 1). In all experiment options was found out increasing of yield activity respectively to control group for 0.36 – 0.66 ton/ha which is 15 – 28 %.

Tetra short cultivar also was responding to usage of biologics Nagro, but its reaction for yield activity was weaker in comparison with Vlada cultivar. And its increasing respective to control sowing (2.61 ton/ha) varied in all experiment options from 0.11 to 0.33 ton/ha (2.72 – 2.94 ton/ha). In this way excess was 4 – 13%.

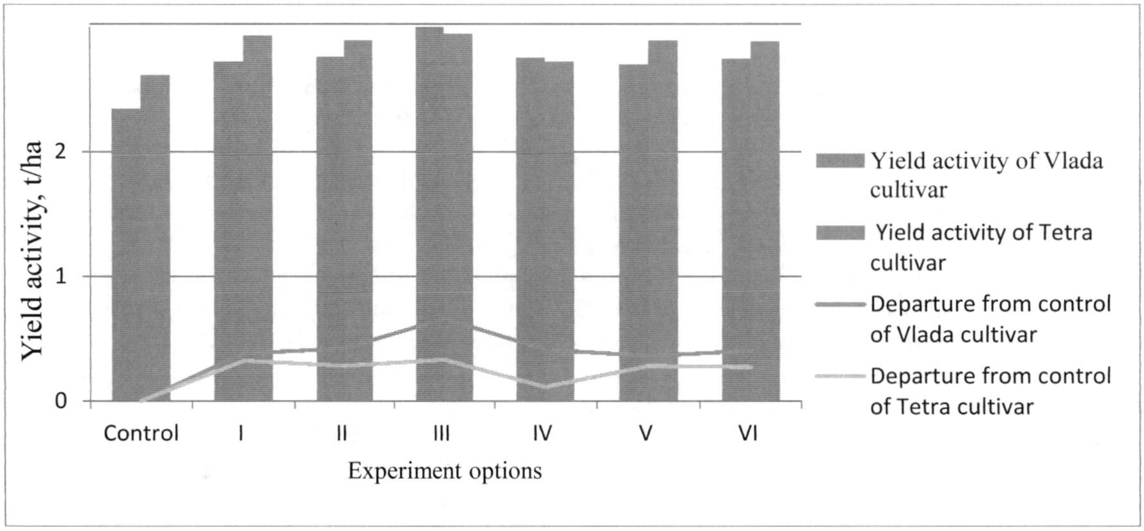

Figure 1 – Yield activity of winter rye and its departure from control weigh (ton/ha), 2015 – 2018 years.(Vlada cultivar HCP_{05} = 0.36; Tetra short cultivar HCP_{05} = 0.27).

Vlada cultivar has more effective sowing in third option – seed processing with bio-energetic, first processing after vegetation – bio-energetic during tillering period at spring, second treatment – universal during paniculation period. In Tetra short cultivar in third option – seed treatment with bio-energetic, first treatment after vegetation – bio-energetic during tillering period at spring, second treatment with bio-fertilizer universal during paniculation period, and first option – seed treatment with bio-energetic, treatment after vegetation: bio-fertilizer universal + bio-energetic during tillering period at spring. Herewith Vlada cultivar yield activity reach 3.00 ton/ha, exceeded control for 0.66 ton/ha (28%) then Tetra short cultivar correspondingly 2.94 and 2.93 ton/ha, 0.33 and 0.32 (13 and 12%). But it must be noted that Tetra short cultivar was less stable to response almost in all treatment options except fourth, providing formation of near to yield activity level.

Strong dependence of yield activity on year of research was spotted in both cultivars (V = 35 – 51 %), which is evident of strong influence of weather conditions on formation of yield activity. Both cultivar gave high yield activity in control and in test sowing in all options in 2016 year. (3.36 – 3.94 ton/ha), lowest in 2018 year (1.69 – 2.26 ton/ha). In comparison with other years, conditions have negative influence on yield activity, may be lower temperature (17.2°C) in July with intense rainfall (90.3 mm), severe drought and cold weather of August - seed-filling period. Also toward other years of research was early formation of snow cover (second decade and up to first decade was 9 mm). This can lead to strong decrease of temperature for tillering node and lose of productivity of further yield.

Therefore, usage of bio-fertilizer Nagro during cultivation of winter rye in all studied options has provided increase of yield activity for both cultivars for 4 – 28% with bigger response of Vlada cultivar. With strong dependence of yield activity from condition of research year (V = 35 – 51 %), synergies between nature heat supply infrastructure and water availability for plants and activity of biologics without finding.

In nature of response on treatment with bio-fertilizer Nagro to this order of falling for winter rye cultivars, Vlada and Tetra short were close (Figure 2). Both cultivars have positive influence of biologics in sowings from second to fifth, with increasing of control on 5 – 23 %.

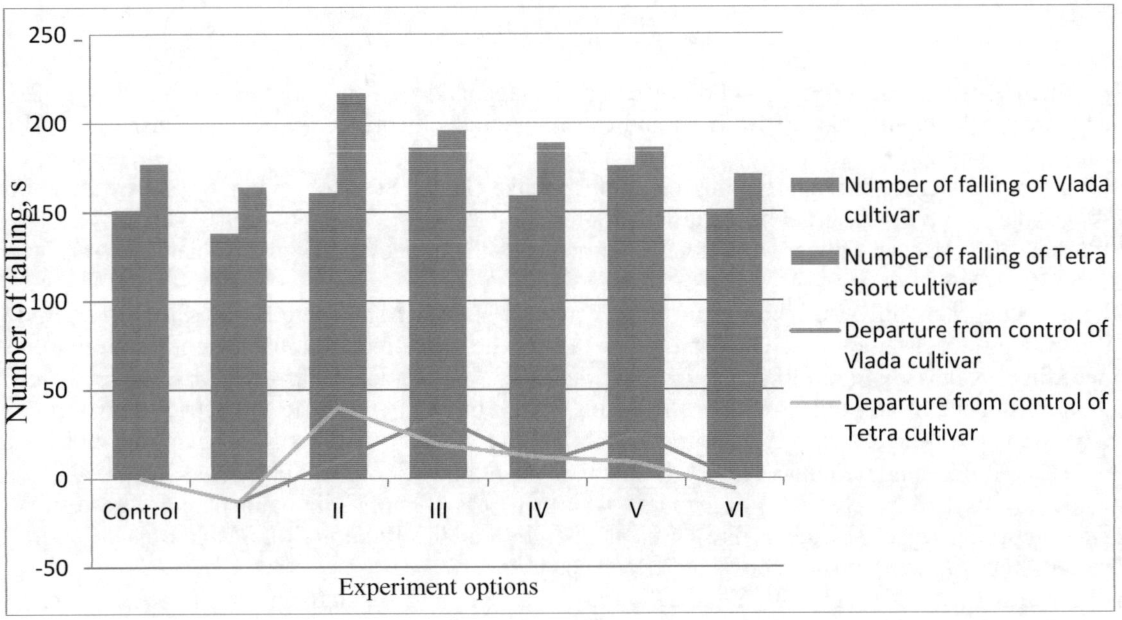

Figure 2 – number of falling of winter rye contrasted with usage of Nagro and its departure from control sowing, 2015 – 2018 years. (Vlada cultivar HCP_{05} = 15; tetra short cultivar HCP_{05} = 11).

Maximum increasing of number of falling, for Vlada cultivar, spotted in sowings that was treated with bio-fertilizer Nagro with options three, and number of falling for four years was 186 s in average, which is correspond to second-class seed in accordance with GOST R 53049-2008 [22]variety in years: from 112 to 254 s. Tetra short has biggest number of falling – 217 s in sowings, which were growled with second option, which is corresponding to first-class seed quality (variety in years: from 150 to 292 s). Also was noted strong variety of number of falling in years (V = 22 – 66 %) and in options (V = 64 – 68 %).

4. Conclusion

Usage of bioorganic nano-fertilizer Nagro during cultivation of winter rye of Vlada and Tetra short cultivars helped to increase yield activity for 4 – 28 %. in edaphoclimatic conditions of Kuznetsklaya forest steppe in south-east of West Siberia. With bigger respond of Vlada cultivars, most effective was sowing with third option. Yield activity reach 3.00 ton/ha, in years 2.26 – 3.93 ton/ha. Tetra short cultivars' yield activity reach 2.94 ton/ha, in years 2.13 – 3.94 ton/ha.

Biggest increase of number of falling for Vlada cultivar was spotted in sowings with third options (186 s – second-class seed, in years from 112 to 254 s) and for Tetra short cultivars – second option (217 s – first-class seed, in years from 150 to 292 s). Yield activity (V = 35 – 51 %), number of falling (V = 22 – 66 %). Between nature of moisture mode and influence of bio-fertilizer on yield activity and number of falling, and also between productivity index wasn't spotted big correlation.

References

[1] Chekmarev P.A., Glinushkin A.P., Startsev V.I. Production of organical products - competitive position of the Russian Federation APRK // Achievements of science and techniques APK. 2018. T. 32. № 3. P. 5-6.

[2] Bacilus methylotrophicus Strain NKG-1, Isolated from Changbai Mountain, China, Has Potencial Applications as a Biofertilizer or Biocontrol Agent Автор:: Ge, Beibei; Liu, Binghua; Thinn Thinn Nwet; с соавторами. PLOS ONE Том: 11 Выпуск: 11 Номерстатьи: e0166079 Опубликовано: NOV 10 2016.

[3] Methodological approach to forecasting effort for science and technology development of crop research field / E.V. Rudiy. S.V. Riumkin, M.S. Petukhova and other. // Achievements of science and techniques APK 2017. T. 31. № 10. P. 8-17.

[4] Influence of mineral fertilizer and biological active agents on yield activity of spring wheat // P.A. Chekmarev, S.V. Obushenko, V.B. Trots and other // Achievements of science and techniques APK 2018. T. 32. № 8. P. 28–31.
[5] Sturikova, Helena. Zinc, zinc nanoparticles and plants / Helena Sturikova, Olga Krystofova, Dalibor Huska, Adam Vojtech // Journal of Hasardous Materials. – 2018. – Vol. 349. – P. 101-110.
[6] Rahale C.S. Nano-Fertilizers to Increase Nutrient Use Efficiency. EverymansScience, 2017. VOL 52. BOOK 2. S. 91-94.
[7] The influence of Universal Bioorganic Nano Fertilizer NAGRO on Spring Barley Crop Productivity / V. Spruogis, A. Dautarte, A. Gavenauskas et al. // The sixth international Scientific Conference Proceedings. Kaunasdistrict. Lithuania: Akademija, 2013. Vol. 6.Book 3. Pp. 441–445.
[8] The influence of Organik Fertilizer "Bioplant Flora" on Winter Wheat Biometric indicators and Productivity / A. Gavenauskas,V. Spruogis, A. Dautarte et al. // The sixth international Scientific Conference Proceedings. Kaunas district. Lithuania: Akademija, 2013. Vol. 6. Book 2. Pp. 89–93.
[9] Adamovskaya M.N. Efficiency of recourse-saving process of main soil treatment and nano-fertilizer Nagro on spring wheat sowing in Pre-Urals steppe zone of the Republic of Bashkortostan: author's abstract PhD in agricultural science, Ufa, 2015. 17p.
[10] Pinchuk L.G., Piyanykh A.B. Yield activity, root system and vegetative mass of winter rye plants affected by usage of bioorganic nano-fertilizer "NAGRO" in Kuznetskaya forest steppe // Modern tendency of agricultural enterprise in world economy: materials of XV international research and development conference (Kemerovo, December 6-7, 2016 year) Kemerovo, 2016 (As part of Kuzbass agrofood forum). P. 136-143.
[11] Pinchuk L.G., Belous I.O. Alankina D.N. Evaluation of bioorganic fertilizer Nagro influence on yield activity of wheat in Kuznetskaya forest steppe // Modern tendency of agricultural enterprise in world economy: materials of XVI international research and development conference (Kemerovo, November 9-10, 2017 year) Kemerovo, 2016 (As part of Kuzbass agrofood forum). P. 370-375.
[12] Trofimov S.S., Kovalev R.V. Soil ecology and soil resources of Kemerovo region. Novosibirsk: Science. Siberian department 1975. 300 p.
[13] Baymukhanov E.N., Viurkov V.V. Efficiency of usage if biological "NAGRO" on yield activity and quality of soft and durum spring wheat in dry steppe zone pf Pre-Urals // Scientific-research magazine of West-Kazakhstan agricultural technical university n.a. Zhangir khana 2016. №2 (43). P. 3-6.
[14] Samarov V.M., N.N. Anokhina, O.V. Basis of agriculture and crop research of West Siberia: study guide for specialty of agronomy and mechanization of agriculture / under general editorship of V.M. Samarov. Kemerovo: Kuzbassvuzizdat 2003. 378 p.
[15] GOST 30498-97. Cereal crops. Definition of falling number. M.: IPK Standards publishing, 1997. 16 p.
[16] Dospekhov B.A. Methodic of field experiment. M.: ID Alliance , 2011. 352 p.
[17] Pinchuk L.G., Piyanykh A.B. Quality of winter rye seed as part of usage of bioorganic fertilizer Nagro // Achievements of science and techniques APK 2018. T. № 2. P. 52-54.
[18] GOST 26204-91 Soils. Definition of labile phosphorus and potassium by Chirikov's method in modification CIANO M.: Standards publishing 1992. 8 p.
[19] GOST 26490-85 Soils. Definition pH salt extract, exchangeable acidity, exchangeable cation, nitrate content, exchangeable ammonium and mobile sulfur by methods of CINAO M.: Standards publishing, 1985. 6 p.
[20] GOST 26213 – 91. Soils. Methods of definition for organic matter. M.: Standards publishing, 1992. 8 p.
[21] Selyanikov G.T. Specialization of agricultural districts on climatic indicator Crop research of USSR. 1933. t. 1. p. 1 – 15.
[22] GOST P 53049-2008. Rye. Technical regulations. M.: Standatdinform, 2009. 8 p.

IOP Conf. Series: Materials Science and Engineering **582** (2019) 012017 doi:10.1088/1757-899X/582/1/012017

Usage of Gray Cast Iron for Hardening of Agricultural Machines' Soil-Tilling Implement

O V Sankina [1, a], A S Sankin [2, b], A A Logov [1]

[1] Kemerovo State Agricultural Institute, 650053, Kemerovo, Markovtseva str, 5, Russian Federation

[2] Kemerovo Scientific-Research Institute of Agriculture Branch of Federal State Budgetery Establishment of Science of Siberian Federal Scientific Centre of Biotechnologies of Russian Academy of Science, 650510, Kemerovo region, Kemerovo district, Novostroika town, Tsentralnaya str., 47

E-mail: [a]olga_vk_06@mail.ru, [b]kemniish@mail.ru

Abstract. Research of gray cast iron was established to study physical-mechanical properties in microstructure while apply a layer on steel basis of work tool's soil-tilling. Gray cast iron went through preliminary heat treatment, where gray cast iron hardening with temperature of 750...1050 °C with following water quenching. Research program including: comparison analysis of reference specimen and specimens which were went through heat treatment. Researching material was applied on work tool's soil-tilling with two methods: arc surfacing and elcetrosparck surfacing. Microstructure of welding layer after arc welding and electrospark welding was studied. Therefore was found out that elecrosparck method gave opportunity to obtain wear resistant layer for a few run, also structure and basis of gray cast iron doesn't change. Coarse graining doesn't spot in pad weld, which is beneficial for properties. Measurement of welding layer's hardness is bigger than base metal and there is no softening of basis. Field tests of hardened products which have shown that to use effectively gray cast iron with heat treatment for hardening, because quality and coating thickness is meeting the requirements of abrasive resistance and there are no spotted significant changes of geometrical dimension for soil-tilling implement.

1. Introduction

At the present time any manufactory requires availability of more complex equipment which allows obtaining competitive products. Whereby for development of necessary equipment is using different materials – from common and easy of approach to rare and expensive. This is due to the specific conditions of equipment and tool's exploitation.

In relation to expensiveness of wear-resistant substance, the bases of parts are made of cheap material, but hardening only necessity parts. Usage of this method is exacerbated by the fact that obtaining of wear-resistant layer is fraught with difficulties of applying, if to be exact – there is changes of material structure and (or) its basis, also changing their physical-mechanical properties. This in turn restricts the wide application of given composition because of complexity of technological process and high cost.

Usually wear-resistant substances are using as layers of different thickness, which applying on part surface with different methods. Some parts requires additional machine processing after layer was applied, because it is difficult to achieve uniform thickness.

Parts are abraded to different types of wearing and load during exploitation. More often pars are drift out of geometrical dimension because of abrasive wear.

Operative parts of agricultural machines have more intensity abrade weariness in agricultural production. Operative parts of soil-tiling machines are developing from intermediate-carbon structural steel, which is sensitive to wear abrasion. To improve wear properties of operative parts, there is using of highly-alloyed hard metals, and as for alloying components using chrome, nickel, wolframium and other elements which improve self-cost of developing tool.

Deposit welding can be conducted by electrodes, solid and cored wire, cast-bar, mixture of different powders. At the present time the assortment of materials for welding are very big: sormite, cast tungsten carbide, stalinite, heat-resistant steel, cast iron and others.

During welding of operative parts with abovementioned coats there are following problems: lack of zone of adhesion of welding layer with basic material, overheating of basis, embrittlement of zone of adhesion and basis [1,2], which can lead to generation of micro-cracking with following destruction.

Transition zone (zone of adhesion) is basis which determines quality of weld joining. During analyze of joins obtained with sormite there was no transition section (Figure 1). Moreover there are graphitization zones if transition section after welding with electrodes. Such zones are completed cracks and they will lead to breakaway of layer even with minimum loads. Lack of transition zone and following decrease of properties of welding joins are related to degree of weld metal alloying [1, 2].

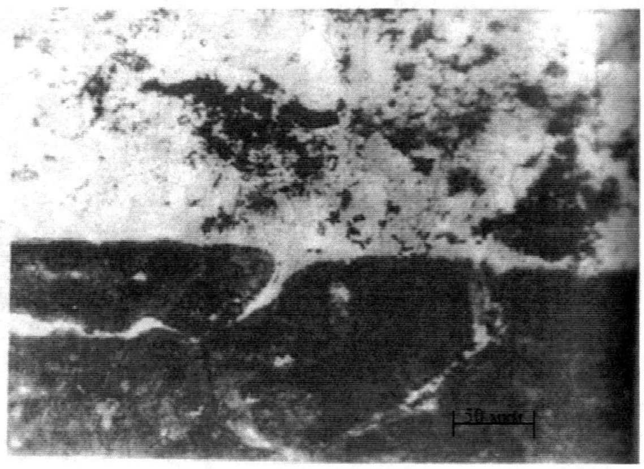

Fig. 1. Microstructure of composition steel 65G – sormite 1 using arc welding (x260)

Arc welding give an opportunity to obtain needed layer of welding metal, but such method leads to strong heating of basic metal in fusion zone and generation of coarse structure [2].

Goal of research – studying of physical-chemical properties and microstructure of cast iron which is welding on steel basis of soil-tilling operative parts and subjected to heat treatment, and to study abrasive resistance of current layer.

2. Equipment and devices used in studies

Heat treatment is more available method of changes of physical-mechanical properties and microstructure of electrodes from cast iron for welding. Heat treatment of gray cast iron consists of hardening in range of temperature 700 – 1050°C, heating during 30 minutes with following water quenching and dispatching with 300°C. Also was conducted cycle annealing (20 cycles) with temperature of 950°C with heating during 1 hour.

Obtained cast iron after heat treatment has improved durability 200…250 MPa and perlite-ledeburite structure without graphitizing, in comparison with nontreated reference specimen ($\sigma_в$=125…130 MPa). After dispatching with 300°C hardness of cast iron improves up to HRC 50…55 without generation of following hardening cracks. That is why gray cast iron is the new material for welding of soil-tilling implement.

Wear resistant layer on razor edge of soil-tilling implement for agricultural machines was applied by arc welding and electrosparck welding with different electrodes [2], subjected to heat treatment.

With the help of arc welding can be obtained layer with almost every thickness, but electrospark welding give opportunity to obtain needed thickness for a few run [2].

2.1 Results and discussion

Electrodes from formula of gray cast iron (in %): 3,8-4,2 C, 0,75 Si, 0,24 Mn, 0,08 S, 0,08 P, other – Fe, were subjected to heat treatment and had following mechanical properties (Table 1). Electrodes for arc welding were developed with trapezoidal shape, 7 mm section and 200 mm length, for electrospark welding – diameter 25 mm and length 150 mm.

Table 1 – Influence of heating temperature for hardening on mechanical properties of gray cast iron.

Heat treatment	$\sigma_{в}$, MPa	δ,%	Ψ,%
Without heating	125.0	2.0	1
Hardening 700°C, water quenching	200	3.1	1
Hardening 750°C, water quenching	202	1.7	1
Hardening 800°C, water quenching	214	1	1
Hardening 850°C, water quenching	232	1	1
Hardening 900°C, water quenching	241	1	1
Hardening 1000°C, water quenching	240	0.5	1
Hardening 1050°C, water quenching	251	0.5	1
Annealing 900°C, heating time 1 hours, 20 cycles	235	0.45	1

Initial heat treatment of developed electrodes before welding them on work surface gave opportunity to obtain perlite-ledeburite structure without graphitizing (Figure 2). Current structure also give opportunity to obtain more quality wear resistant layer which will work in abrasive environment.

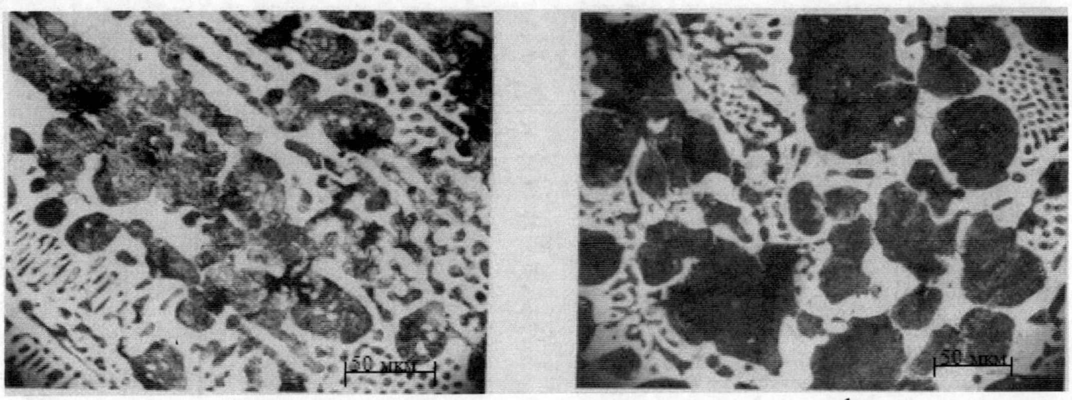

a b

Fig. 2. Microstructure of gray cast iron after heat treatment: a – hardening 900°C, water quenching; b – hardening 1050°C, water quenching (x300)

Applying of abrasion resistant coating was realized by arc welding in environment of protection gas and without protection.

Structure of welding layer after arc welding represented on figure 3.

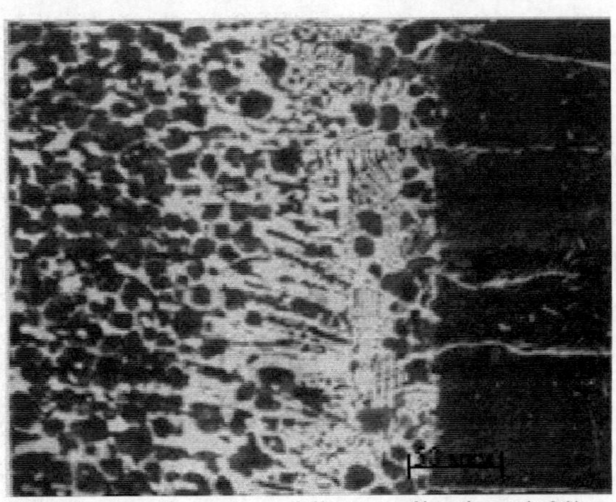

Fig. 3. Microstructure of welding layer after arc welding application. (x80)

Arc welding generate composition of wear resistance layer – basic material, herewith thickness of welding layer is enough for work in abrasive environment. But current method leads to strong overheating and increasing of grain effect (Fig. 3).

After conduction of literary analysis and established researches to obtain wear resistance layer on surface [1 – 10] there was trial of electrospark welding.

Current method gives good bond between layer with part, but not provide enough thickness for working of soil-tilling implement in abrasive environment which is soil itself. Also there is a problem related to permanent oxidation of electrode, during process of electrospark welding. It is requiring permanent stripping of defective layer [2]. Authors are offered way to delete defective layer from electrode's surface [2], that is why welding of heat treatment gray cast iron was realized with usage of current method.

Method of electrospark welding of wear resistance layer are not disturbing structure of basic material, because welding material is quickly cooling and necessary wear resistance are not possible to reach. There was decided to applicate layer for a few run, to solve this problem.

After realization of welding there was study of microstructure and evaluation of features of welding part on prepared samples.

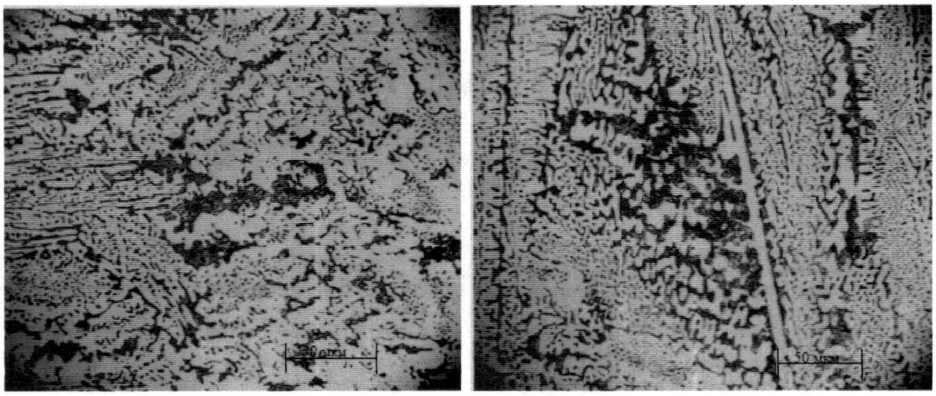

a b

Fig. 4. Microstructure of welding gray cast iron by electrospark welding with heat treatment: a – hardening 900°C, water quenching; b – 900°C, hardening, heating time 1 hour, 20 cycles. (x300)

There is lack of graphite insertion in structure of welding metal, which allow to infer of keeping structure of components, which were obtained after heat treatment and electrosparking welding. During the study there is also wasn't spotted increase of grain effect which is beneficial for features (Fig. 4).

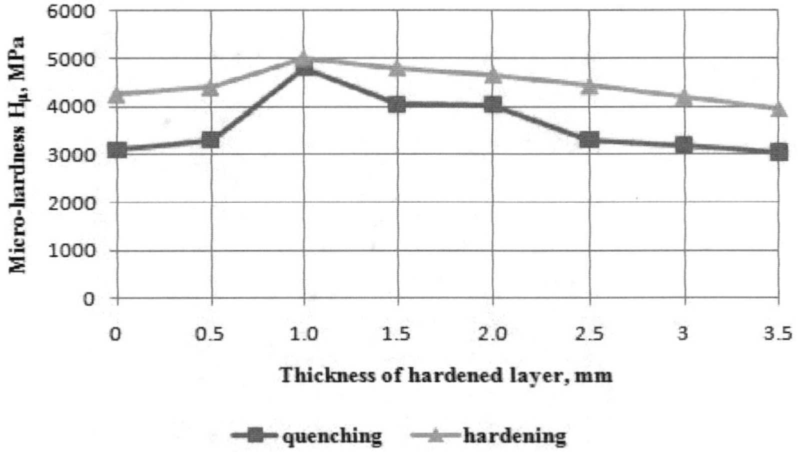

Fig. 5. Influence of electrode's heat treatment on changes of hardness of hardened layer in sample section

Measurement of hardened layer's micro-hardness (Fig. 5) shows that surface layer, which is applied by electrodes of gray cast iron and subjected to initial heat treatment, has improved hardness than basic material and it gives an opportunity to infer of higher wear resistance.

Hardened products were subjected to field tests on farms of Kemerovo region, the Russian Federation. Tests showed efficiency usage of hardening with given material for electrospark welding, because thickness and quality of welding layer is meet the necessary requirements of wear resistance. After soil-tilling implements work during one season there was observed insignificant changes of geometrical dimension, which can be recovered by all-known methods.

3. Conclusions

Usage gray cast iron as welding material, subjected to initial heat treatment with following application on surface of pars with electrospark welding, and working in environmental of abrasive wear, gives an opportunity to increase operational life and decrease purchase costs for new parts.

References

[1] Kurbanova, M.G.; Chernysh, A.P.; Sankina, O.V.; Buziyan, O.N., Increasing of wear resistance of functional surface of operative parts and mechanisms of grinding and blending of biogas unit. Achievements of science and technics APK 2013. №5. P.: 79 – 80.

[2] Chibryakov, M.V.; Sankina, O.V., Arc welding and electrospark welding of parts with unalloyed white cast iron. News of international agrarian education academy 2018. Bulletin №39 (2018). P.: 53 – 57.

[3] Shibe Vineet, Chawla Vikas, Characterization of Fe-C-Cr Based Hardfacing Alloys, Transactions of the Indian Institute Of Metals. Volume: 71. Issue: 9. P.: 2211-2220

[4] Martinazzi, Douglas; Lemos, Guilherme V. B.; Landell Renan M.; et al., Preliminary study on effect of rod geometry in FHHP between FE55006 nodular cast iron and SAE 8620 steel, PERIODICO TCHE QUIMICA. Volume: 16. Issue: 31. Published: Jan 2019. P.: 642-649.

[5] Carcel-Carrasco Francisco-Javier; Pascual-Guillamon Manuel; Salas-Vicente Fidel; et al., Influence of Heat Treatment in the Microstructure of a Joint of Nodular Graphite Cast Iron when Using the Tungsten Inert Gas Welding Process with Perlitic Grey Cast Iron Rods as Filler Material , Metals. Volume: 9. Issue: 1. Article Number: 48. Published: Jan 2019.

[6] Reisgen U.; Sharma R.; Wieland S.; et al., Improving the corrosion and wear resistance of grey cast iron components by surface welding with duplex stainless steel using regulated gas metal arc welding: Influence of dilution on corrosion properties, MATERIALWISSENSCHAFT UND WERKSTOFFTECHNIK. Volume: 49. Issue: 12. Published: Dec 2018. P.: 1520-1537.

[7] Winiczenko Radoslaw; Kaczorowski Mieczyslaw; Skibicki Andrzej, The microstructures, mechanical properties, and temperature distributions in nodular cast iron friction-welded joint, Journal of the Brazilian society of mechanical sciences and engineering. Volume: 40. Issue: 7. Article Number: UNSP 347. Published: Jul 2018.

[8] Saraev Y.; Gladkovsky S.; Lepikhin S.; Kamantsev I.; Lunev A.; Perovskaya M., Investigation of the Influence of Energy Parameter of the Covered-Electrode Welding on the Impact Strength Characteristics and Cracking Resistance of the Welded Joints Obtained, Obrabotka metallov-metal working and material science. Volume: 20. Issue: 2. P.: 100-115

[9] Chamim M.; Triyono; Diharjo Kuncoro, Effect of Electrode and Weld Current on The Physical and Mechanical Properties of Cast Iron Welding, International Conference on Engineering, Science and Nanotechnology (ICESNANO) Location: Solo, INDONESIA Date: Aug 03-05, 2016. Book Series: AIP Conference Proceedings. Volume: 1788. Published: 2017.

[10] Ambroza P.; Bockus S.; Kavaliauskiene L., Formation of build up layers microstructure by arc automatic overlay welding using secondary raw material powders, Archives Of Metallurgy And Materials. Volume: 58. Issue: 2. P.: 549-553. Published: 2013.

[11] Gabets D.A.; Markov A.M. Study of the influence of alloying elements on the structure and properties of gray cast iron operating under conditions of shock-friction wear. *Obrabotka metallov (tekhnologiya, oborudovanie, instrumenty), Metal Working and Material Science*, 2019, vol. 21, no. 1, P.: 70–81.

IOP Conf. Series: Materials Science and Engineering **582** (2019) 012018 doi:10.1088/1757-899X/582/1/012018

Simulation of Transcendental Deformation Mode of Interchamber Pillars and Ore Deposit in the Course of Development of Bauxite Resources at Great Depth on the Example of Mine «Sevuralboksitruda»

N I Kosukhin[1, a], D V Sidorov[1, b], V Yu Timofeev[2, c]

[1] Saint-Petersburg Mining University, 199106, St. Petersburg, V.O. 21st line, house 2, Russia
[2] Tomsk Polytechnic University, 634050, Tomsk, Lenina prospect 30, Russia

E-mail: [a]nikolay.kosuhin@uralkali.com, [b]sidorov-post@yandex.ru,

[c]tvy@tpu.ru (corresponding author)

Abstract. The publication describes the problem of transcendental deformation mode of interchamber pillar and ore deposit associated with the formation of hazardous zones in the rock mass in process of development of rockburst-hazardous rock masses at the great depth. Geomechanical software PRESS 3D URAL is used to assess the impact of the range of geological and technological factors on the mode of interchamber pillars deformation and ore deposit. The software allows to solve spatial problems in the elastic-plastic formulation. The results of step-by-step modeling of different mining and technical situations with a test of the bearing capacity of interchamber pillars showed that some of them can't withstand the mountain pressure and go to the transcendental mode of deformation with subsequent destruction. Identification of the destroyed pillars allows to adjust the boundary conditions of loading the ore mass. At the same time the neglect of transcendental mode of deformation of interchamber pillars can lead to an undervaluation of the stress level in the ore mass by 1.5 or more times. This can lead to an error in predicting the parameters of zones dangerous due to rock bumps.

Introduction

The problem of determining of transcendental deformation mode of interchamber pillars and ore deposit especially actual in the process of deposits development of the North Ural bauxite basin. The reason is due to the complexity of the mountain pressure control in the process of application chamber-and-pillar system of mining of ore deposits in which rock bumps are possible. The main feature of the chamber-and-pillar development system is the maintenance of room space by the ore pillars, which prevent the collapse. Safety of working conditions and accident-free operation is ensured by the stability of the roof exposures of room spaces and interchamber pillars, which depends on the correctness of the definition of their parameters (size of the roof span and size of the pillar). If the size of interchamber pillars chosen incorrectly the consequences of their destruction will be catastrophic and can lead to injury of workers, disruption of the production process, etc [1-5].

Experience in the development of deposits of the North Ural bauxite basin at great depths shows that the boundary part of the ore deposit can be in conditions of elastic or transcendental deformation. For example, in the case of presence in boundary part of the ore deposit strong and stiff varieties of bauxite (motley bauxite) it's characterized by elastic deformation. Transcendental deformation of the boundary part of the ore deposit is characterized by the presence of less strong varieties of bauxite. Early determination of the deformation mode of the boundary part of the ore deposit allows to determine stress state and, as a result, to identify areas of potential danger in which a rock burst can occur [6-12].

Thus, the study of deformation mode of interchamber pillars and ore deposit with respect of the influence of a wide range of geological, mining and technical factors is an relevant task for the theory and practice of the development rock-bump hazardous ore deposits.

Research objective

The purpose of research is to determine the regularities of changes stress-strain state in the ore deposit under transcendental loading of interchamber pillars in the process of underground development of ore deposits of complex geological structure to improve the efficiency of mines operation at great depth.

Methods

The regularities of formation zone of maximum stress state in the boundary part of the ore deposit at it's different thickness are shown in Figure 1a. Figure 1b shows a refined epure of stresses acting in the interchamber pillar of variable thickness.

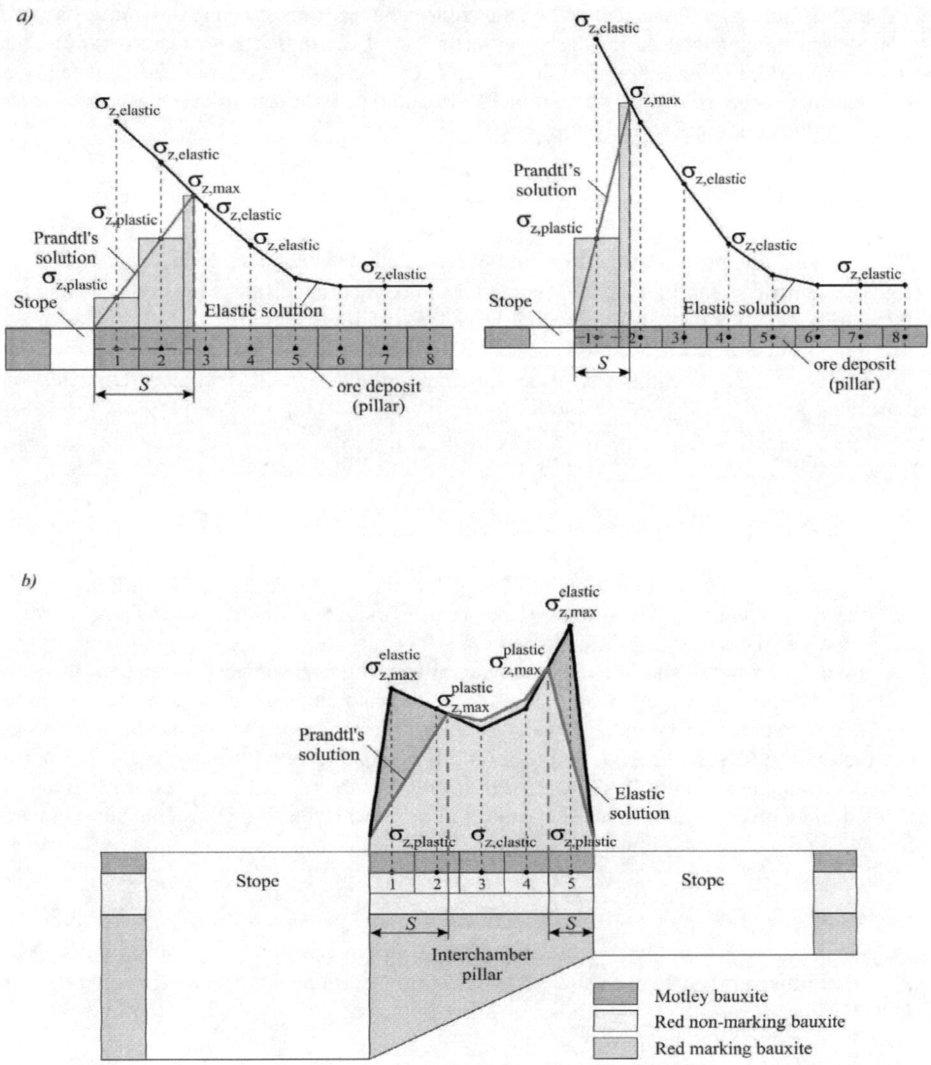

Figure 1. Regularities of limit state zone formation in ore deposits (*a*) and interchamber pillar (*b*) with variable thickness

The account of transcendental deformation mode of interchamber pillars in mined-out space is carried out according to the following scheme.

At the first stage, stresses in ore elements are calculated by taking into account their elastic deformation with using the modified method of boundary integral equations. This method is the basis of the software product PRESS 3D URAL, that allows to solve spatial problems of elasticity theory taking into account the natural and man-made flexibility of ore deposits and pillars [2, 10, 11]. The following mining-and-geological and mine technical factors are taken into account:

– mined-out spaces and construction units, which can be arbitrary in size and configuration;
– arbitrary location of construction units relative to boundaries of mined-out spaces;
– various physical and mechanical properties of enclosing rocks and ores;
– the changing thickness of the ore deposits;
– parameters of tectonic disturbances;
– the thickness of relieve slot.

At the second stage, the calculated elements are checked for resistance to allowable stresses, identify the destroyed elements. In the destroyed elements of the initial base the values of the elastic modulus are replaced by deformation modules. After that the estimation of the bearing capacity of the calculated elements is carried out. The transition of the calculated element to the transcendental state is determined from the condition:

$$\sigma_{z,total} \geq 1.4 \cdot \sigma_{comp} \cdot \frac{d}{m}, \tag{1}$$

where $\sigma_{z,total}$ – total compressive stresses acting in the calculated elements, MPa;
σ_{comp} – tensile strength of the ore sample for uniaxial compression, MPa;
d – linear size of the smallest side of the calculated element, m.

The value of the total compressive stress, acting in the design element, is determined from the following expression:

$$\sigma_{z,total} = \sigma_z(\xi, \eta) \cdot \gamma H + \gamma H, \tag{2}$$

where $\sigma_z(\xi, \eta)$ – additional stresses obtained in the calculated elements as a result of numerical simulation;
γH – weight of rocks column from the design element to the earth's surface, MPa.

At the final stage, the stress state of the ore deposit is estimated, taking into account the transcendental deformation mode of ore elements.

Assessment of stress state of interchamber pillars and ore deposits is carried out using the modified method of boundary integral equations, but in the calculated elements in which the condition (1) is satisfied, the values of the initial elastic modulus E_i are required to replace with the values of the reduced elastic modulus E_{ir}, having used the graphic dependence given in figure 2. Analysis of the graphical dependence (figure 2) shows that the value of the reduced modulus of elasticity E_{ir}, of the calculated element, depending on its shape coefficient, can be determined from the implementation of the following approximating linear dependence:

$$\frac{E_{ir}}{E_i} = a \cdot \frac{d}{m} + b, \tag{3}$$

where E_{ir} – the reduced modulus of elasticity (Young) of the calculated element depending on its shape coefficient, MPa;
d – the width of interchamber pillar, m;
m – the height of interchamber pillar, m.

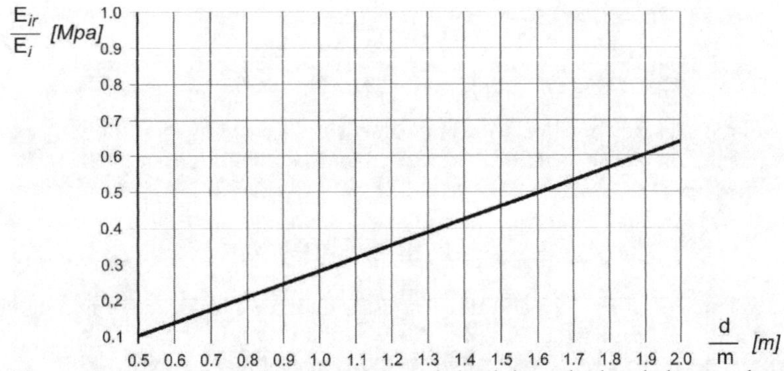

Figure 2. Determination of the reduced modulus of elasticity calculated element depending on its shape coefficient

The approximation parameters a and b, included in expression (3), are determined from the following two conditions: in case of uniaxial compression $\frac{\mathrm{d}}{\mathrm{m}} = 0.5$, we accept $\frac{E_{ir}}{E_i} = 0.1$ in case of volume compression $\frac{\mathrm{d}}{\mathrm{m}} = 3$ we accept $\frac{E_{ir}}{E_i} = 1$.

Taking into account the above, we obtain the following system of linear equations:

$$\begin{cases} 0.1 = 0.5 \cdot a + b \\ 1.0 = 3.0 \cdot a + b \end{cases} \tag{4}$$

Solving a system of linear equations (4) with respect to the approximation parameters a and b, we obtain: $a = 0.36$, $b = -0.08$. Substituting the approximation parameters a and b in the expression (3), we obtain the following expression to determine the value of the reduced elastic modulus E_{ir} for calculated element, depending on it's shape factor:

$$E_{ir} = \left(0.36 \cdot \frac{\mathrm{d}}{\mathrm{m}} - 0.08\right) \cdot E_i \tag{5}$$

In the case of non-uniformity structure of the calculated element, we can use the scheme shown in figure 3.

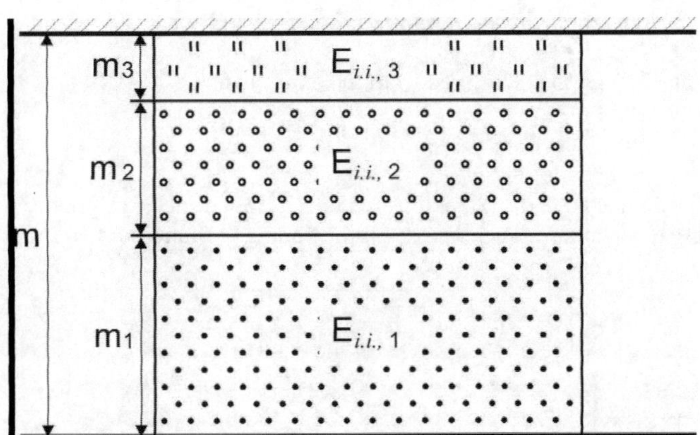

Figure 3. Scheme to determine the modulus of elasticity of the calculated element with its non-uniformity structure

The value of the elastic modulus in this case can be determined from the following expression:

$$E_i = \frac{\text{m}}{\sum_{i=1}^{n}\left(\frac{m_i}{E_{ii}}\right)}, \tag{6}$$

Control of the accuracy of calculations by value is carried out by the following expression. The indicator represents the ratio of the sum of the required loads in the mining-induced zone to the sum of the loads.

$$\delta = |\textstyle\sum_{i=1}^{n}(P_i \cdot \Omega_i) / \sum_{i=1}^{m}(Q_i \cdot S_i|, \tag{7}$$

where P_i – the calculated value of the additional stresses due to the influence of mining unit;
Ω_i– calculated element area, m^2;
Q_i – the load, which is removed from the soil of excavation ($\gamma \cdot H$);
S_i – the area of the element, located in the mined-out space, m^2.

The breakdown of the mining-induced zone into elements is considered satisfactory if the value deviates from the unit by no more than 10-15 %. For the numerical calculation of stresses in the ore deposit and pillars it is necessary to enter the initial information, characterizing the geological and technological conditions. At the first stage, the limits of integration are set to take into account the geometric dimensions of the boundary elements, belonging to both the rock mass and the developed space.

At the second stage, each boundary element of the ore deposit is assigned a Poisson's ratio and the modulus of elasticity of the enclosing rocks and ores, thickness of the design element and the thickness of relieve slot in the design element. In the elements, belonging to the developed space, boundary conditions are formed through the pressure angles, which allows to take into account the size of the developed space, the depth of stopping, incidence angle and thickness of ore deposits.

To clarify the mining-induced zone epure in the boundary part of ore deposit and pillars, taking into account the zone of the maximum stress state, the following provisions are used to calculate the parameters of the zone of the maximum stress state.

The width of mining-induced zone in the boundary part of ore deposit (pillar) is determined from the expression, developed by the scientists of the Institute VNIMI:

$$X_1 = \left(\frac{h}{2\tau_s} \cdot |K_{int}|\right)^{\frac{2}{3}} \cdot f(p), \tag{8}$$

where K_{int}– stress intensity factor at the edge of the pillar (ore deposit), МПа$\sqrt{\text{м}}$;
h – thickness of ore in the boundary part of ore deposit (pillar), м.

The stress intensity factor is determined from the expression:

$$K_{int} = \sqrt{2\pi R \cdot (\sigma_{z,p} - 1) \cdot \gamma H}, \tag{9}$$

where R – distance from the edge of the pillar (ore deposit) to the center of gravity of the design element, m. The expression for finding the function f(p) looks like:

$$f(p) = 0.63 \cdot (\left(\sqrt[3]{\sqrt{1+p}+1} - \sqrt[3]{\sqrt{1+p}-1}\right)^2, \tag{10}$$

where $p = 0.57 \cdot \left(\frac{1.4 \cdot \sigma_{comp}}{K_{int}}\right)^2 \cdot \frac{h}{2}$

The dependence for determining the value of the normal stress in the mining-induced zone maximum from the side of in mined-out space is as follows:

$$\sigma_{z,max} = -1.4\sigma_{comp}\frac{2X_1}{m}, \tag{11}$$

The proposed method of stress calculation in the mining-induced zone of the ore deposit allowed to adjust the software PRESS 3D URAL. This made it possible to determine stresses in the mining-induced zone, stress intensity factors, the size of maximum stress state zone in the boundary part of the ore deposit for any spatial configuration of the stopping zone, the ratio of the deformation characteristics of the host rocks and ores, parameters of relieve slot.

Results

At the first stage, as a result of numerical implementation, the distribution of stress intensity of the total compressive stresses in the ore deposit is obtained, taking into account only the elastic mode of deformation of the interchamber pillars and ore deposit (figure 4).

At the second stage, the bearing capacity of the calculated elements of the ore deposit was checked, as a result of which a calculated scheme of the site with destroyed elements was obtained (figure 5).

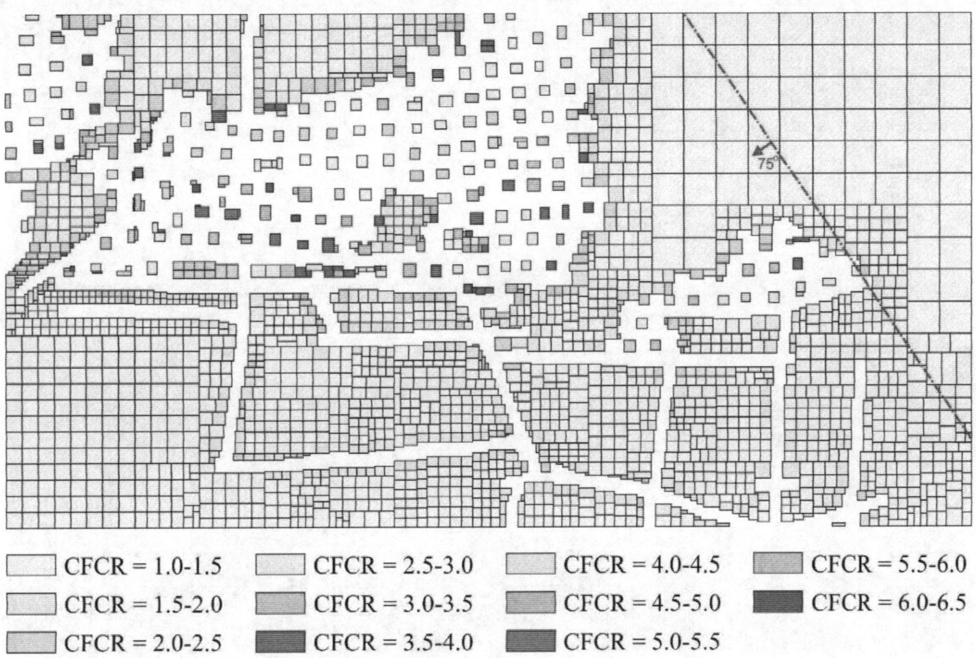

Figure 4. Assessment of the stress state of the ore Deposit by the modified method of boundary integral equations (CFCS – concentration factor of complete stress)

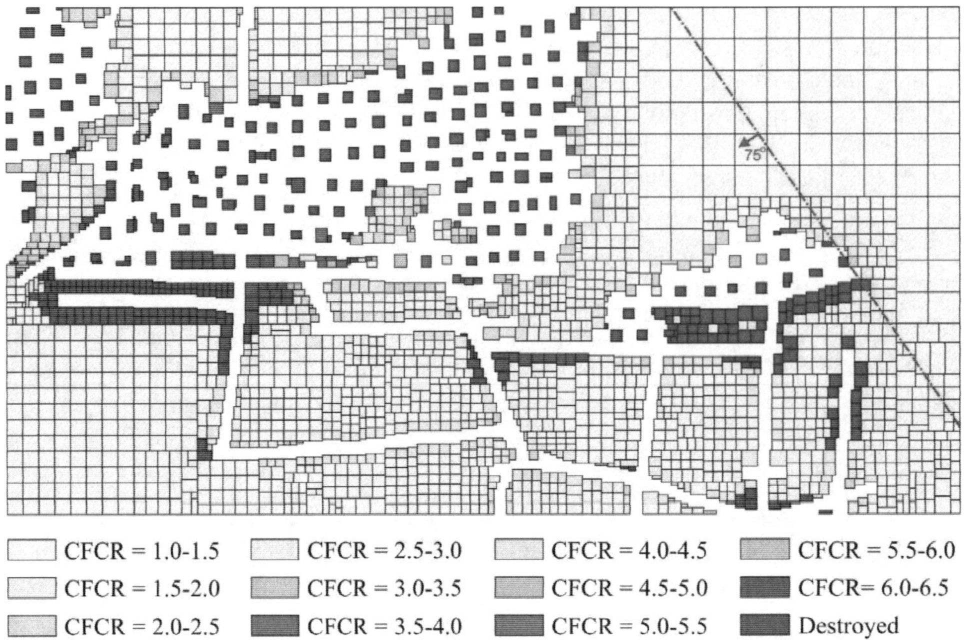

Figure 5. Determination of destroyed ore elements

At the final stage, as a result of adjusting the initial data (elastic modulus values), the following picture of the distribution of the coefficients of concentration of total compressive stresses in the ore deposit is obtained (figure 6).

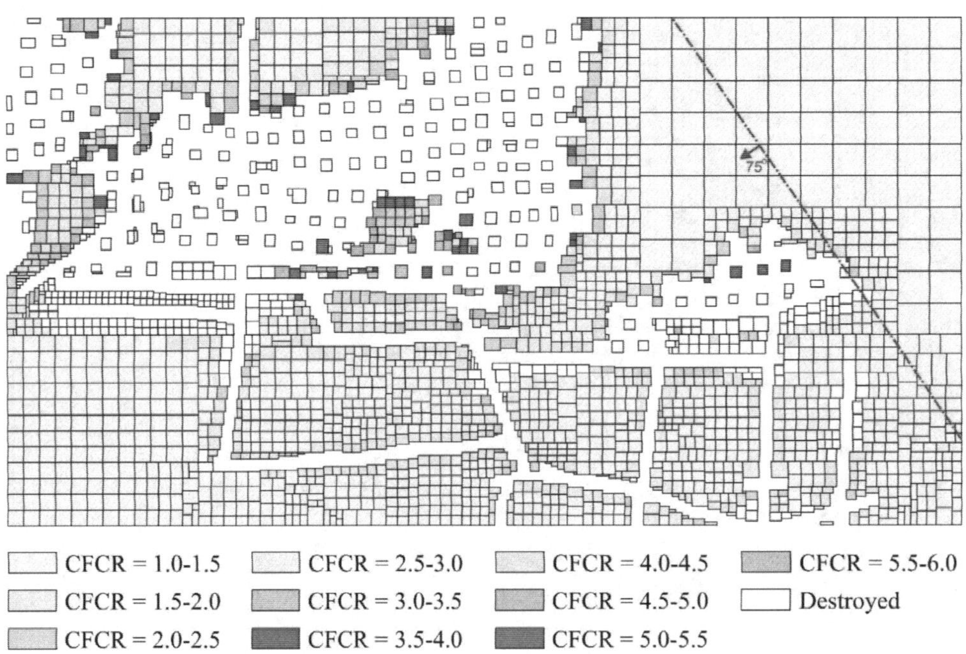

Figure 6. Results of the numerical evaluation of the stress state of the ore deposit, taking into account the destroyed ore elements

Figure 7 shows the detail of changing in the stress state of the untreated area of non-industrial mineralization left in the developed space. Analysis of the results shows that the destruction of inter-chamber pillars in the developed space leads to an increase in the stress concentration by 50 % or more.

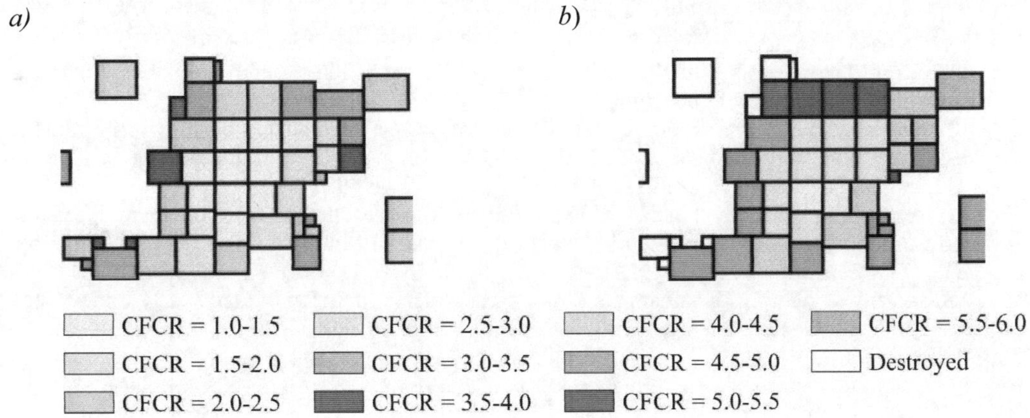

Figure 7. Changes in the stress state of the non-industrial mineralization site: a – without taking into account the destruction of the design elements; b – with the destruction of the design elements

Similar qualitative regularities of stress-strain state of interchamber pillars were identified as a result of widespread mining experimental studies, carried out on ores of Dzhezkazgan and Mirgalimsay, which is developing a copper-nickel deposits, using chamber-and-pillar system development. The reliability of the developed modified method of integral equations was confirmed by comparing the results of the numerical experiment with the results of mining and experimental work (mine "Kalinskaya" OJSC "SUBR"), as well as with experimental data of laboratory studies of the stress-strain state of interchamber pillars, using equivalent materials in the certified laboratory of rock modeling of the Scientific research center of geomechanics and problems of mining production in St. Petersburg mining University.

Conclusion

The problem of formation of hazardous zones in the ore massif during the development of rock-bump hazardous ore deposits at great depths is associated with the transcendental deformation of interchamber pillars in the developed space. To determine the transcendent mode of interchamber pillars deformation and the ore deposit with the influence of a wide range of geological and technological factors can be used geomechanical software PRESS 3D URAL. The results testing of the bearing capacity of interchamber pillars showed that some of them do not withstand the mountain pressure and pass into the transcendental mode of deformation with subsequent destruction, which requires adjustment of the boundary conditions of loading of the ore massif. The neglect of the transcendent deformation mode of interchamber pillars can lead to error in predicting the parameters of zones dangerous due to rock bumps.At the same time the neglect of transcendental mode of deformation of interchamber pillars can lead to an undervaluation of the stress level in the ore mass.

References

[1] Galchenko Yu.P., Eremenko V.A., Myaskov A.V., Kosyreva M.A. Solution of geoecological problems in underground mining of deep iron ore deposits // Eurasian Mining. – № 1, 2018. – pp. 35-40.

[2] Kosukhin N.I, Sidorov D.V., Beloglazov I.I, Timofeev V.Yu. Assessment of stress-strain and shock bump hazard of rock mass in the zones of high-amplitude tectonic dislocations // IOP Conf. Series: Earth and Environmental Science, 224, 2019, – pp. 1-5, DOI: 10.1088/1755-1315/224/1/012014.

[3] Ponomarenko T., Sergeev I. Quantitative methods for assessing levels of vertical integration as a basis for determining the economic and organizational sustainability of an industrial corporation // Indian Journal of Science and Technology, 2016, 9 (20).
[4] Ikonnikov D. A., Kovshov S. V. The Structure of Reclaiming Warehouse of Minerals at Open-Cut Mines with the Use Combined Transport //IOP Conference Series: Materials Science and Engineering. – IOP Publishing, 2017. – T. 221. – №. 1. – C. 012014.
[5] V. S., Serova A. A., Timofeev V. Y. Environmental-Toxicological Characteristics of Waters and Their Sources at Magnitogorsk With the Its Iron and Steel Industry //IOP Conference Series: Materials Science and Engineering. – IOP Publishing, 2016. – T. 142. – №. 1. – C. 012117.
[6] Eremenko V.A. Kontrol` geomexanicheskogo sostoyaniya geologicheskoj sredy` pri otrabotke Sheregeshevskogo mestorozhdeniya / V.A.Eremenko, A.A.Eremenko, V.N.Filippov i dr. // Gorny`j informacionno-analiticheskij byulleten. – 2007. – № 12. – C. 155–169.
[7] Gospodarikov A.P., Vykhodtsev Y.N., Zatsepin M.A., Mathematical modeling of seismic explosion waves impact on rock mass with a working / Journal of Mining Institute, № 226, 2017. pp. 405 - 411.
[8] Rasskazov I.Yu. Geodinamicheskoe sostoyanie massiva porod Nikolaevskogo polimetallicheskogo mestorozhdeniya i osobennosti proyavleniya udaroopasnosti pri ego osvoenii / I.Yu. Rasskazov, B.G. Saksin, V.I. Usikov, M.I. Potapchuk // Gorny`j zhurnal. – 2016. – № 12. – S. 13-19.
[9] Rasskazov I.Yu. Geomexanicheskaya ocenka uslovij razrabotki glubokix gorizontov polimetallicheskogo mestorozhdeniya «Yuzhnoe» / I.Yu. Rasskazov, G.A. Kursakin, M.I. Potapchuk // Fiziko-texnicheskie problemy` razrabotki polezny`x iskopaemy`x. – 2012. – № 5. – S. 125–134.
[10] Sidorov D.V. Prognozirovanie udaroopasnosti tektonicheski narushennogo rudnogo massiva na glubokix gorizontax Nikolaevskogo polimetallicheskogo mestorozhdeniya / D.V. Sidorov, M.I. Potapchuk, A.V. Sidlyar // Zapiski Gornogo instituta. – 2018. – T. 234. – S. 604-611. DOI: 10.31897/PMI.2018.6.604.
[11] Sidorov D.V., Ponomarenko T.V. The development of a software suite for predicting rock bursts within the framework of a system for ensuring geodinamic safety of mining operations // 17th International Multidiscip-linary Scientific GeoConference SGEM 2017, www.sgem.org, SGEM 2017 Conference Proceedings, ISBN 978-619-7408-02-7 / ISSN 1314-2704, 29 June - 5 July, 2017, Vol. 17, Issue 22, 633-638 pp, DOI: 10.5593/sgem2017/22/S09.079.
[12] Zhang, Z., Shimada, H., Sasaoka, T., Kai, W. Study on the overlying strata movements and stability control of the retained goaf-side gateroad // Memoirs of the Faculty of Engineering, Kyushu University. – Vol. 76, Issue 2, 2016. – pp. 1-17.

IOP Conf. Series: Materials Science and Engineering **582** (2019) 012019 doi:10.1088/1757-899X/582/1/012019

Concepts of Calculation Technological Equipment Efficiency Criteria

I G Abramova [a], D A Abramov, A A Ayzatulin

Samara University, Russia, Samara, 446086, c. Samara, Moscow highway, 34, Department of engine technology

E-mail: [a] abi_ssau@inbox.ru

Abstract. Various criterions can characterize the work of mechanic processing equipment. We know the "Compression of working time factor" and the «Overall Equipment Effectiveness» (OEE). The last criterion became known in the 80s-90s of the last century. Was performed the analysis of the efficiency index of OEE equipment. The first criterion was known in the 40s of the last century. They use different concepts. They are reflected in the article, the article also presents the calculation of indicators. An important indicator of OEE contains three components. The meaning of each of them reflects historically applicable indicators. The calculation of the coefficient "OEE" is disclosed through the indicators used in Russian practice. The element base of the Russian indicators is the opposite of the elemental base of the "OEE" indicator and together they form a single whole. The measurability of the indicators makes it possible to make a prediction about the efficiency of the equipment.

1. Introduction

Improving labor productivity by reducing the loss of time is an urgent problem in enterprises in the manufacture of products all the time. In the period of a planned economy in our country, methods of accounting and analysis of working time were widely used and covered in scientific and educational literature.

The ability to account for cost and time wasted in automatic mode appeared with the use of CNC machines and advanced information technologies in the production. At the end of the last century, the level of equipment of foreign industries was higher than the level of equipment of the machine park in our country. The methodology for assessing the effectiveness of equipment based on the use of digital data transmission was not born in our country. Foreign authors have proposed the indicator "OEE" to assess the effectiveness of equipment based on their goals, concepts, terminology.

However, it should be noted that long before the birth of the OEE indicator, Russian researchers presented similar indicators to readers. They reflected the savings in equipment operation time. It would not be fair to forget the history of the development of equipment efficiency indicators, born in our country. This article analyzes the methods and discloses methods for calculating the OEE indicator in order to establish links between the indicators for evaluating the performance of equipment used abroad and in our country. Author was trace and compare the calculation of indicators by methods that appeared at different times.

2. The concept and definition of the indicator "OEE" and its elements

Currently, the *OEE* (*Overall Equipment Effectiveness*) indicator has gained fame and is one of the "generally accepted key indicators of the overall efficiency of technological equipment" worldwide [1, 2]. The term itself was proposed in the late sixties of the last century by Seiychi Nakajima) [3], but began to be used outside of Japan only in the late eighties, gained fame at the turn of the century, especially in connection with the publication of the book by the American researcher Hansen, Robert C [4].

General parameter of effectiveness (*OEE*) determined by 3 following items:

- *A-Availability*: (percentage of scheduled time that the operation is available to operate.

IOP Conf. Series: Materials Science and Engineering **582** (2019) 012019 doi:10.1088/1757-899X/582/1/012019

- *P-Performance*: (speed at which the Work Center runs as a percentage of its designed speed).
- *Q-Quality*: (Good Units produced as a percentage of the Total Units Started).

$$\text{OEE} = \text{A} \cdot \text{P} \cdot \text{Q} \tag{1}$$

Each parameter can be represented by coefficients: For instance if A = 0,87; P = 0,50; Q = 0,95, than OEE = 0,87 · 0,50 ·0,95 = 0,42

«Availability» (A) is calculated [1, 2, 3, 4, 5, 6]:

$$A = \frac{\text{OT}}{\text{PPT}} = \frac{\text{PPT}-\text{DTL}}{\text{PPT}} \tag{2}$$

Where the designation is using:

OT - Operating Time (reducing stand by time),

DTL - Down Time Loss - all stand by or non-operational time by all reasons, loss at stops, including any unscheduled stops, such as: equipment breakdowns and failures, stops due to a shortage of raw materials or lack of storage space (transition time is also included in the OEE analysis, as it is one of the forms idle time, although the transition time cannot be eliminated, in most cases it can be reduced);.

PPT - Planned Production Time.

These elements of the formula (2) and other elements of time we present in Figure 1:

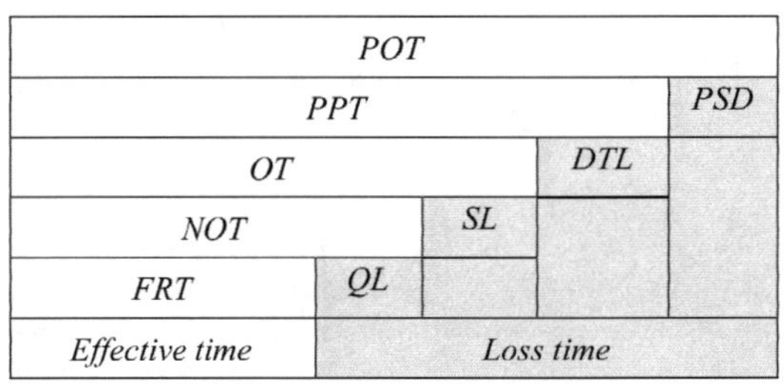

Figure 1 *- The composition of the elements of time in the calculation of the coefficient "OEE"*

POT - Plant Operating Time.

PSD - Planned Shut Down - this time to be excluded from efficiency analysis, since production at this moment is impossible.

SL -Speed Loss;

NOT -Net Operating Time,

QL - Quality Loss

FPT - Fully Productive Time.

The coefficient "Performance" is calculated according to the following dependencies:

$$\text{P} = \text{ICT} / (\text{OT}/\text{TP}) \tag{3}$$

or

$$\text{P} = (\text{TP}/\text{OT}) / \text{IRR} \tag{4}$$

Further accepted:

ICT – Ideal Cycle Time – theoretical, minimum time required for unit production;

IRR – Ideal Run Rate – theoretically, the maximum amount of production per unit of time is the inverse *ICT*;

OT – Operating time.

TP – Total Pieces – the actual number of units released during the operating time *OT*.

The third element in the criterion OEE is the element that defines the "quality level" (Q). The quality criterion takes into account the loss in quality, i.e. production, which does not meet the standards [1, 5]:

$$Q = GP / TP \tag{5}$$

The designation is accepted:

GP – Good Pieces – the actual number of units of standards products released during the operating time *OT*

3. The concept of time accounting for the rationing of manufacturing parts process

The concept of labor intensity or item-calculated time ($t_{\text{item-calc}}$) is used in domestic practice. Elements of time and the methods by which these elements can be calculated are marked in Figure 2.

t_{item} or t_{table} - time of item manufacturing;

t_{op} - operational time; $t_{op} = t_{base} + t_{\text{add}}$, t_{base} - base time, t_{add} - additional time.

t_{break} - break time; $t_{\text{break}} = t_{\text{maintenance}} + t_{\text{op.m}} + t_{\text{wp}}$.

$t_{\text{maintenance}}$ - maintenance time;

$t_{\text{op.m}}$ - time for operational management;

t_{wp} - operator's private time.

$t_{\text{p.e}}$ - preparation and ending time

The method of "Motion study" ("MS") [7] is based on the recording of activities in the workplace. The special form ("Workshop report for the month on motion study of the working day on the workplace") noted the actual work of each machine, taking into account two categories of time: productive work of the machine and time of organizational and technical losses. The time of loss revealed after processing the results of the report, the classification of the elements of time. After this procedure, the calculation of the ratio of compression and an increase in labor productivity.

It is especially worth emphasizing that this method was used in the late 30s and early 40s of the last century. The list of references intentionally contains the work of Soviet scientists [7], dated before the beginning of the Great Patriotic War. Such publications testify to the fact of the steady development of the domestic industry in those early years.

The ratio of time compression [8, 9] is shown below:

$$C_{\text{compr}} = \frac{T_{shift} - (t_{p.e} + t_{op} + t_{maintenance} + t_{op.m} + t_{wp})}{T_{shift}} \tag{6}$$

where C_{compr} - the ratio of time compression;

T_{shift} - observation time or time of one shift;

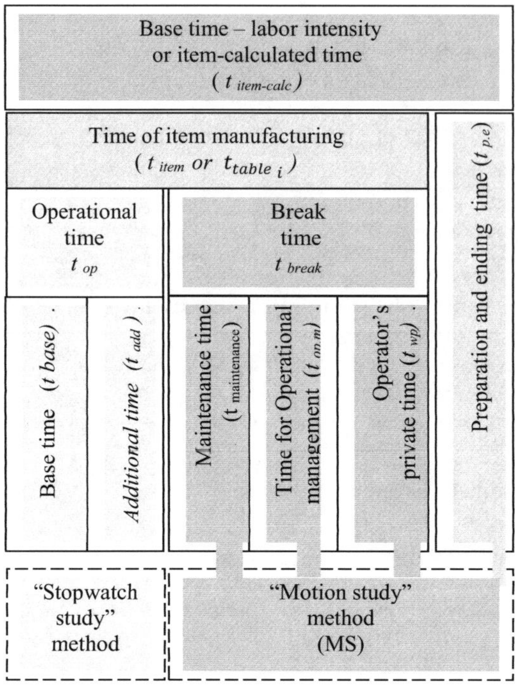

Figure 2. *The scheme of composition of elements of Base time – labor intensity or item-calculated time ($t_{item-calc}$) and methods that help to identify them.*

Time of effective work of the equipment (t_{eff}):

$$t_{eff} = t_{p.e} + t_{op} + t_{maintenance} + t_{op.m} + t_{pe} \tag{7}$$

This coefficient helps to determine proportional workforce productivity.

$$C_{\uparrow WP} = \frac{C_{compr}}{1-C_{compr}} \tag{8}$$

or

$$C_{\uparrow WP} = \frac{t_{loss}}{T_{shift} - t_{loss}} \tag{9}$$

where t_{loss} - down time loss, $t_{loss} = Tshift - t_{eff}$

4. Record of the indicator of "OEE" in terms that are used in the domestic literature

We can write the elements of the indicator "OEE" using indicators adopted in the domestic literature with the following notation: $OT = t_{eff}$; $DTL = t_{loss}$; $PPT = T_{shift}$.

Formula (2) can be recorded in accordance with the parameters commonly used in Russian manufacturing practice:

$$A = \frac{T_{shift} - t_{loss}}{T_{shift}} \tag{10}$$

We show the relationship indicators:

$$A = \frac{T_{shift} - t_{loss}}{T_{shift}} = 1 - \frac{t_{loss}}{T_{shift}} \tag{11}$$

$$C_{compr} = \frac{Tshift-(tp.e+t\ opr + tmaintenance+ twp)}{Tshift} = \frac{T_{shift}-t_{eff}}{T_{shift}} = \frac{t_{loss}}{T_{shift}} \quad (12)$$

$$A = 1 - C_{compr} \quad (13)$$

$$A + C_{compr} = 1 \quad (14)$$

From formula (14) it follows that the indicator of "Availability" and "Ratio of time compression " are the exact opposite in meaning and at the same time constitute a single whole (see Fig. 3).

When considering the second element of the indicator "OEE" - "performance" (P), we can note the following. The concept of the ideal cycle time, the ideal rate of production for the domestic reader is not quite familiar. Since the words "work efficiency", "productivity", "ideal cycle time" can be correlated with the direct process of equipment operation during part processing, then the time elements "OEE" should be correlated with the norm elements of piece (calculation) time. Using the notation adopted in domestic practice in the manufacture of one part of the i-th name on one workplace of a simple type, you can write: $ICT = t_{base}$; $OT = t_{itemi}$ or $t_{item-calc\ i}$;

Then the coefficient of "performance" will take the following form:

$$P_1 = \frac{ICT}{OT/TP} = \frac{t_{base}}{t_{itemi}/_1} = \frac{t_{base}}{t_{itemi}} < 1 \quad (15)$$

In total, while for production (N_{reali}) of items (i) of work benches :

$$P_N = ICT / (OT/TP) = \frac{\sum t_{item-calc\ i}}{\sum t_{basei} / \sum N_{real\ (i)}} = \frac{\sum N_{real\ (i)} \cdot t_{item-calc(i)}}{\sum t_{base\ (i)}} \quad (16)$$

$$PN = (TP/OT) / IRR = \frac{N_{real(i)} / \sum t_{base(i)}}{\sum t_{item-calc(i)}} = \frac{\sum N_{real(i)} \cdot t_{item-calc(i)}}{\sum t_{base(i)}} \quad (17)$$

or

$$P = ICT/(OT/TP)=(TP/OT)/IRR= \frac{\sum N_{real(i)} \cdot t_{item-calc(i)}}{\sum t_{base(i)}} \quad (18)$$

The third parameter of OEE defined a "quality level" (Q), by the following formula [1, 2]:

$$Q = \frac{N_{produced\ (i)} - N_{non-confomirng\ (i)}}{N_{produced(i)}} \quad (19)$$

where $N_{produced\ (i)}$ is a total production by defined time cycle (i),
$N_{non-confomirng\ (i)}$ – is a number of defective or non-conforming units by the same period.

In Russia a similar parameter "output level" is applied together with "level of non-conformities" (*q*):

$$q = \frac{N_{non-confomirng\ (i)}}{N_{produced\ (i)}} \quad (20)$$

$$Q = \frac{N_{produced\ (i)} - N_{non-confomirng\ (i)}}{N_{produced(i)}} = 1 - q \quad (21)$$

Or

$$Q + q = 1 \quad (22)$$

On Figure 4 correlations of those parameters drawn on pie chart:

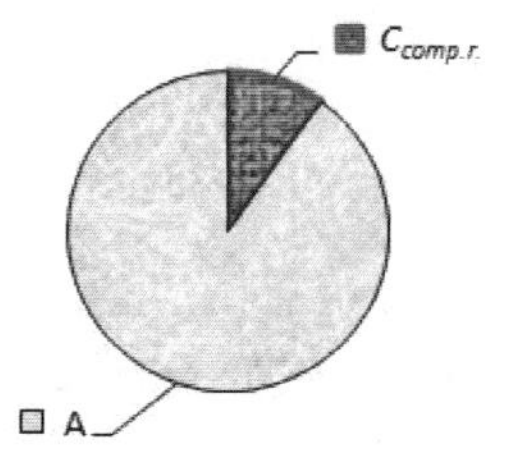

Figure 3. *Schematic determination of OEE's parameter of Availability (A) and time Compression rate Coefficient (C_{compr})*

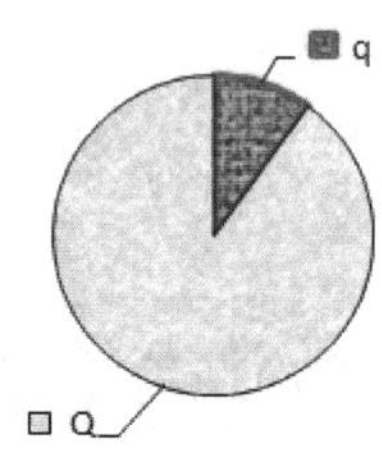

Figure 4. *Correlation of level of non-conformities (q), applied in Russia and quality level (Q), according to OEE.*

We insert into the formula (1) the expression (13), (18), (21).
The parameter *OEE* will be the form (23):

$$OEE = (1 - C_{compr}) \cdot \left(\frac{\sum N_{real\ (i)} \cdot t_{item.calc(i)}}{\sum t_{base\ (i)}}\right) \cdot (1 - q) \tag{23}$$

5. Conceptual approaches to the prediction of equipment effectiveness criteria

The parametrical analysis shows that applied in the past "Motion study" method is based on data collected by the direct verification of production activities. For instance, there is no additional explanation or training required for operators to count non-conforming products or standby time Application of OEE parameter is based on the determination of the loss and cost deduction methods.

Determination of loss and cost deduction leads us to the next question, how to calculate a time loss. Calculation of "Availability" parameter does not give us methods of loss time accountability.

Stand by time determination is also missing. We should understand that OEE methods are commonly applied with the support of IT and digitalization of the production process. By the integrated production statistical technics, those parameters are automatically calculated. However not all manufacturers could afford those It/automatic calculation software, therefore additional loss parameters might be added in the future. According to "Motion study" methods structure is clearly defined. Same time OEE does not give us an output for labor intensity/manpower calculation.

It was shown that OEE and some commonly used parameters applied in Russian manufacturing practice are assimilated each other, are uniting same core parameters and can be applied together for determination of the full picture of manufacturing process management.

It was shown that by applying parameters together (*OEE, Q, q,* C_{compr}) production activities can be assessed from a different angle by focusing the same target of continues effectiveness improvement of the production process.

When the value of the performance index of equipment is projected and rationing, it is necessary to have a clear classification of the elements of productive time and time loss.

6. Conclusion

The recording of the "OEE" coefficient, taking into account the notation used in the Russian literature, reveals indicators characterizing the concentrated cause of the loss of time. A special place is given to the method of "Motion study".

The purpose of the method of "Motion study" – is the rationing of labor and fixing time loss.

The purpose of the OEE indicator is not the rationing of loss time, but accounting equipment loss time.

The article shows the calculation of each parameter of the element base of the generally accepted in the world indicator of Overall Equipment Effectiveness ("OEE") in accordance with well-established concepts and definitions in domestic practice. This will allow for adequate adjustment and evaluation of equipment performance.

References

[1] OEE Primer, Calculating OEE, http://www.oee.com/calculating-oee.html
[2] Thomas R. Cutler. Improving overall equipment effectiveness with lean and value-stream mapping. 10/26/2010.http://www.qualitydigest.com/inside/twitter-ed/improving-overall-equipment-effectiveness-lean-and-value-stream-mapping.html#
[3] Shopin, A.G. OEE and downtime management: from theory to implementation in SIMATIC IT / A.G. Shopin, I.V. Zanin, "Automation in Industry", 2006, 09, p. 24 - 29
[4] Hansen, Robert C. Overall Equipment Effectiveness (Print-On-Demand Edition) Powerful Production/Maintenance tool for Increased Profits. 290 Pages, Softcover Published: January, 2001. ISBN: 9780831102180. http://new.industrialpress.com/overall-equipment-effectiveness-1.html
[5] Antonenko I., Kryukov I., Shestopalov P. Efficiency of using production equipment. http://www.cfin.ru/management/manufact/oee.shtml
[6] Optimal planning of Planned Preventative Maintenance (PPM) - a way to increase the efficiency of equipment used / G. Medvedeva, A. Musuridze, Yu. Tikhonova, A. Kryukov, A. Zaydullin // CAD and graphics, 2013 №1, p. 84 - 89.,
[7] Organization of engineering production, ed. by B.I. Katzenbogen. Approved All-Union Committee On Affairs Of High School under the Council of People's Commissars of the USSR approved as a textbook for engineering schools . Moscow, Leningrad, State United Scientific and Technical Publishing House of the People's Commissariat of Heavy Industry of the USSR, 1941. 648 pp. (Before the title auth. Abramovich, Dumler, Katzenbogen, Kornitsky and Heifetz).
[8] Organization and production planning at a machine-building enterprise. [Textbook for universities in the specialty "Economics and organization of the engineering industry"]. Ed. Prof. V.A. Letenko. M., "High School", 1972.
[9] Abramova, I.G. Basis of the organization production of the engineering enterprise [Electronic resource]: (lecture course and workshop): study guide / I. G. Abramova; Samara State aerospace university named by S.P. Korolev (national research institute) (SSAU). - electronic text data. - Samara: 2011. - 1 CD-ROM.

IOP Conf. Series: Materials Science and Engineering **582** (2019) 012020 doi:10.1088/1757-899X/582/1/012020

The Geokhod Bar Working Body With a Diameter of 1.8 m Characteristic Point Determination

Vladimir V Aksenov [1,a], Vladimir Yu Sadovets [2,b], Elena V Rezanova [2], Dmitriy A Pashkov [3,c]

[1]Research Center "Siberian NPO", 650000, Russia, Kemerovo, Sovetsky Avenue, 56.
[2]Kuzbass State Technical University named after T.F. Gorbachev, 650000, Russia, Kemerovo, Vesennyaya Str., 28.
[3]Siberian Branch of the Russian Academy of Sciences Federal Centre for Coal and Coal Chemistry Studies Coal Institute, 650610, Russia, Kemerovo, Leningradskiy Avenue, 10.

E-mail: [a] 55vva42@mail.ru, [b] vsadovec@yandex.ru, [c] pashkov.d.a@inbox.ru

Abstract. In the article the geokhod bar working body characteristic point with a graphical method is determined. The research relevance has been considered. The information about the geokhods' working bodies operating peculiarities is given. For the research objectives and tasks setting the geokhod bar working body characteristic point condition is defined. The 1.8 m diameter geokhod with a bar working body use conditions have been determined.On the ground of the research results the graph of changes of the geohod bar working body cutting force axial projection ($P_{O.CV}$) was constructed depending on the distance x by which the point is remote from the axis of rotation, as well as the geohod bar working body cutting force radial projection ($R_{IO.CV}$) dependence on the range from x to R_g..The geokhod bar working body characteristic point has been foundin a graphical manner, i.e. it is the interception point of graphs of changes in the cutting force axial projection ($P_{O.CV}$) in the range from 0 to x at a distance from the point rotation axis over a distance x and the cutting force radial projection ($R_{IO.CV}$) within the range from x to R_g at a distance from the point rotation axis over a distance x.

1. Introduction

One of the priority directions in the development of science, technology and aquipment in the Russian Federation is the need for effective use of space, including underground space [1]. However, the underground workings are laborious and expensive process in which the most urgent problems are increasing the speed of penetration, labor productivity and safety, reducing the cost of work [2,3,4].

A promising direction that addresses these issues is the application of geokhod technology for making holiow in the underground space where the basic element is geokhod [5,6,7].

Geokhods belong to a new class of tunneling equipment. A distinctive feature of geokhod is screwing of the hull into the rock array, which fulfills the function of the mobile support. In this way, pressure forces are created as a result of the introduction of propulsion elements into the contour array and the use of a normal reaction that occurs when the elements of geokhod interact with the rocks array. Early developments in this area led to the creation of the first samples of geokhods - screw-boring penetrator units (VPA) ELANG-3 and ELANG-4 (Fig. 1) [8,9,10]. At present, a prototype of geokhod with working body for destructing rocks of medium strength has been created (Figure 2) [11,12].

a) b)

Fig. 1. Experimental samples VPA ELANG-3 (a) and ELANG-4 (b)

Fig. 2. Experimental sample of geokhod with drum-like working body

At the current stage of development of the elements of geokhod technology, there are no variants with a variable number of cutting elements depending on the center of geokhod's rotation.

Therefore, the works aimed at developing technical solutions for the location of cutting elements relative to the center of rotation of geokhod for destructing rocks with a strength up to 1 of Protodyakonov hardness are topical.

Nowadays new technologies in the field of robot automation are being introduced at rapid pace. However the sphere of creating devices that can form cavities underground is less studied [1,2].

For this reason there the tasks appear connected with the development of new approaches, technologies, and machines that allowrobotically perform all processes for cavity formation underground [3,4].

The use of geokhod technology is one of the areas that allows an underground robot to form cavities underground.

Currently to develop the elements of geokhod technology it is necessary to develop constructive and technical decisions for the working bodies capable of forming cavities in the underground spaces in rocks with hardness up to 1 according to M.M. Protodiakonov scale [5-7].

Thus the work aimed at the substantiation of geokhods' working bodies parameters for rocks destruction with hardness of up to 1 is relevant.

2. The geokhod working body features

The geokhod original movement to the face involves the formation of complex surface geometry both of the face and of the working body. The surface of the mine at tunnel boring with the geokhod takes the form of several benched helicoid surfaces.

Point A (figure 3), located on the bar periphery, upon the given pitch distance of the external drive screw blade h_B, passing the track $2\pi R_g$ in one complete turn around the circumference, moves to the face at an angle β_1 to the plane perpendicular to the axis of geokhod rotation [8], wherein

$$\beta_1 = arctg\frac{h_B}{2\pi R_g}, \quad (1)$$

whereR_g is the geokhod head sectionradius.

Any point of the bar located at a distance of x (figure 1) from the geokhod axis of rotation moves to the face at an angle [8]:

$$\beta_x = arctg\frac{h_B}{2\pi x} \quad (2)$$

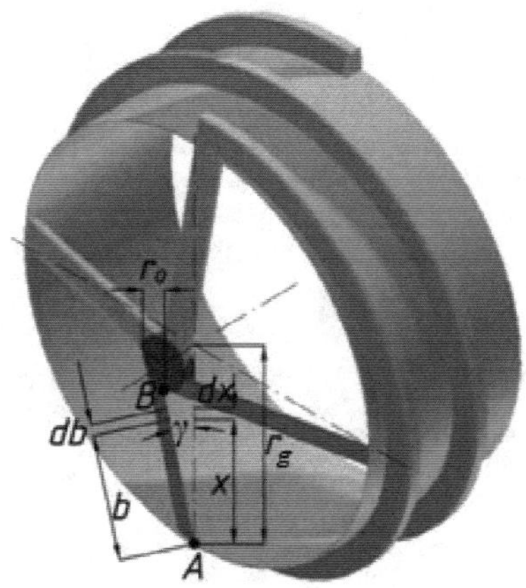

Fig.3 – The bar working body diagram

3. Methods of research

From the previously maderesearch, one of the characteristic points of the geokhod bar working body is the point at which the projection of the component ground resistance to cutting, depending on the width of the cut, on the axis of the geokhod rotation ($P_{0.CV}$) in the range from 0 to x, will be equal to the projection of the component ground resistance to cutting, depending on the width of the cut, on the plane perpendicular to the geokhod axis of rotation ($R_{IO.CV}$) within the range from x to R_g, i.e.

$$\int_0^x P_{O.CV}dx = \int_x^{R_g} R_{IO.CV}dx\,; \quad (3)$$

The projection of the component ground resistance to cutting, depending on the width of the cut, on the geokhod axis of rotation($P_{0.CV}$) in the range from 0 to x will be equal to [8-10]

$$P_{O.CV} = \frac{\varphi m_{cv} h_в^2 + h_в n P_{izn}}{2\pi n \cos\gamma}\cdot\left(\ln\frac{1}{tg\left|\frac{\beta_x}{2}\right|}\right) - \quad (4)$$

$$-\frac{h_в}{2\pi}\left(\varphi m_{cv}\frac{h_в}{n}\frac{ctg(\delta+\varphi_{TP})}{\cos\gamma} - \frac{ctg(\delta_1+\varphi_{TP})}{\cos\gamma}P_{izn}\right)\frac{1-\sin\beta_x}{\sin\beta_x}$$

where φ - coefficient that takes into account the influence of cutting angle;

m_{cv} - specific cutting force to overcome resistance of soil by the front at the cutting angle of 45^0, Pa;

h - depth of cutting, m;

P_{izn} - additional force of cutting which is to be exerted upon the bar section with a worn place or bluntness;

γ - angle of a bar slope to the plane perpendicular to the geokhod rotation axis, degree;

δ - angle of cutting, degree;

φ_{TP} - angle of friction, degree.

Projection of the component ground resistance to cutting depending on the width of the cut, on the plane perpendicular to the geokhod axis of rotation ($R_{IO.CV}$) within the range from x to R_g, will be equal to [8-10].

$$R_{IO.CV} = \frac{\varphi m_{cv} h_в^2 + h_в n P_{izn}}{2\pi n \cos\gamma} \cdot \frac{\sin\beta_x - \sin\beta_1}{\sin\beta_1 \sin\beta_x} + \\ + \frac{h_в}{2\pi}\left(\varphi m_{cv} \frac{h_в}{n} \frac{ctg(\delta+\varphi_{TP})}{\cos\gamma} - \frac{ctg(\delta_1+\varphi_{TP})}{\cos\gamma} P_{izn}\right) \cdot \ln\left(\frac{tg\left|\frac{\beta_x}{2}\right|}{tg\left|\frac{\beta_1}{2}\right|}\right) \quad (5)$$

One of the ways to solve equations is a graphical way. It is based on the construction of function graphs and determination of their intersection points. Thus, to find the characteristic point under condition (3) it is necessary to construct graphs of changes in the axial projection of the cutting force ($P_{0.CV}$) in the range from 0 to x at a distance from the point axis of rotation over adistance x according to formula (4) and the radial projection of the cutting force ($R_{IO.CV}$) in the range from x to R_g at a distance from the point axis of rotation over a distance x according to formula (5).

4. Results

On the basis of the mining conditions for underground workings by the geokhod with a bar working body presented in the table, there the cutting force axial projection ($P_{0.CV}$) in the range from 0 to x, where the variable x varies from 0 to R_g, as well as the cutting force radial projection ($R_{IO.CV}$) in the range from x to R_g, where the variable x varies from 0 to R_g in increments of 0.01 m were calculated according to formulas (4) and (5).

On the basis of calculations the graphs (figure 4) of changes in the axial projection of the cutting force ($P_{0.CV}$) in the range from 0 to x at a distance from the point rotation axis over a distance x and radial projection of the cutting force ($R_{IO.CV}$) within the range of x to R_g at a distance from the point rotation axis over a distance x.

There are the geohod bar working body cutting force projections values (H) on the Y-axis of the represented graphs of changes, and on the X-axis there is adistance x by which the point from the axis of rotation is remote.

Table – Mining conditions for underground workings by the geokhod with a bar working body.

Unit name	Designation	Unitofmeasurement	Unitvalue
The geokhod radius	R_g	m	0.9
The screw blade pitch distance	h_B	m	0.5
Coefficient taking into account the angle of cutting influence [11]	φ		0.74
Specific cutting force in the front of the slot at a cutting angle of 45^0[11]	m_{cv}	n/m^2	300000
Angle of slope of radial blade to the plane perpendicular to the geokhod rotation axis	γ	degree	0
Parameters characterizing the material resistance to elastic-plastic compression [11]	P_O P_{usl} h_{usl}	n/m n/m m	11300 61900 0.639
Angle of cutting	δ	degree	35
Angle of friction[12]	φ_{mp}	degree	31.4

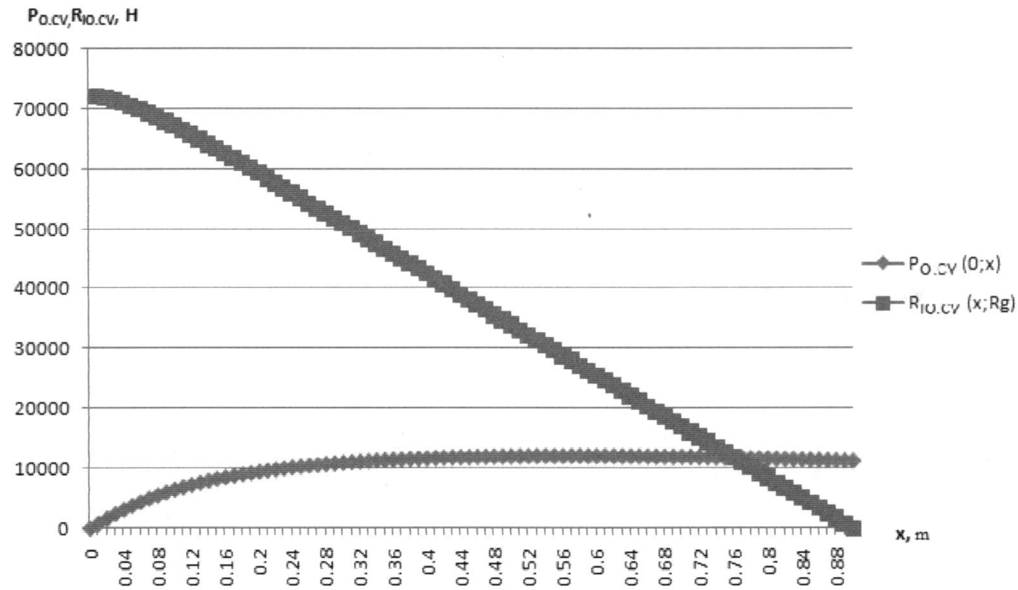

Fig.4 – Graphs of changes of the geohod bar working body cutting force projections depending on the distance *x* by which the point is remote from the axis of rotation.

5. Conclusion

The graphs in figure 2 show that the characteristic point for the geokhod bar working body with a radius of 0.9 m will be located at a distance of 0.76 m from the geokhodaxis of rotation.

With an increase in the cutting width by more than 0.57 m from the center of rotation the projection value of the cutting force on the geokhod axis of rotation begins decreasing. This is due to the change in the angle between the cutting force vector and the direction of the projection axis. Therefore it can be said that at a certain value of the cutting width from the geokhodaxis of rotation *x*, the geokhod working body will be self-sumping.

At a value of $x>0.1$ m the radial projection cutting force value is reducing in proportion to the distance from the geokhod axis of rotation.

References

[1] Nishi S., Seiki T. Planning and design of underground space use. // Mem. Sch. Eng. Nagoya Univ.- 1997. №1.

[2] Maidl, B. Hardrock Tunnel Boring Machines / B. Maidl, L. Schmid, W. Ritz, M. Herrenknecht. Berlin: Ernst& Sohn, 2008. 343 p.

[3] Wighman T. Think deeps – go underground // ENR: News Rec. 1998. №4.

[4] Aksenov V., Sadovets V., Pashkov D. The influence of parameters on the generatrix of the helicoid form guide of geokhod bar working body// E3S Web of Conferences The Second International Innovative Mining Symposium. 2017.

[5] Aksenov V.V., Efremenkov A.B., SadovetsV.Yu., Rezanova E.V. The geokhod structural portrait forming //Bulletin of the Kuzbass State Technical University. 2010. No. 01. p. 35-41.

[6] BeglyakovV.Yu. The geokhod working body with the face wasteinteraction surface parameters substantiation. Thesis of candidate of technical Sciences. - Yurga, 2012.-139p.

[7] Koperchuk A.V., Beglyakov V.Yu. The geokhodstarting device circuit design choice //Mining machinery and electromechanics. -2016. -№ 8 (126).-P. 15-18.

[8] Determination of the geokhod bar working body power parameters for rocks of little hardness breaking/ V.V. Aksenov, V.Yu. Sadovets, D.A. Pashkov // Bulletin of the Kuzbass State Technical University. - 2017. –No.3. –P. 116-126.

[9] Machinery for underground work / under Volkov D.P. general editorship – Moscow: Mechanical engineering, 1992. 187 p.

IOP Conf. Series: Materials Science and Engineering **582** (2019) 012020 doi:10.1088/1757-899X/582/1/012020

[10] Vetrov Yu.A., Bladinskiy V.L. Machinery for special underground work. – Kiev: Publishing office of Kiev University, 1980. 308 p.
[11] Vetrov Yu.A. Calculation of cutting forces and soil digging. – Kiev: Publishing office of Kiev University, 1985. 251 p.
[12] Zelenin A.M., Balovnev V.I., Kerov I.P. Machinery for underground work // Textbook for universities – Moscow: «Mechanical engineering», 1975. 424 p.

IOP Conf. Series: Materials Science and Engineering **582** (2019) 012021 doi:10.1088/1757-899X/582/1/012021

Informational - entropic Analysis of Dynamics of Road Safety Orderliness

Artur I Petrov[1,a], Sergey A Evtyukov[2], Victor I Kolesov[1], Daria A Petrova[3]

[1]Tyumen Industrial University, 72 Melnikaite str., Tyumen, 625000, Russian Federation
[2]Saint Petersburg State University of Architecture and Civil Engineering, 4 2nd Krasnoarmeyskaya str., Saint Petersburg, 190005, Russian Federation
[3]Ural Federal University named after the first President of Russia B. N. Yeltsin, 16a Chapaev str., Ekaterinburg, 620002, Russian Federation

E-mail: [a] ArtIgPetrov@yandex.ru

Abstract. The article considers issues of identification of all-Russian and regional (on the example of Kemerovo region) road safety provision (RSP) systems orderliness. Research presents methods of identification of road safety provision system orderliness. The main idea of this methodic is the calculation of relative entropy of formation process of final result of cause-effect chain «Population – The number of vehicles – The number of road accidents – The number of victims – The number of deaths in road accidents» performance. Statistics of road accident rate during 2004...2018 was analyzed for the calculation of relative entropy. The result of research concludes that the level or road safety provision system increases in Russia generally and Kemerovo region specifically. Meanwhile the development of this process in Kemerovo region lags relatively to all-Russian one.

1. Introduction

Every system is an aggregate of cooperating subsystems (elements of the system). It includes plenty of system connections that usually have nonlinear character. During the system functioning hundreds of different subsystems processes form and develop simultaneously. They make an impact on the final system result. It's extremely difficult to identify these processes, model them and assess their degree of influence on the system. Moreover, these actions often don't answer the question how to effectively manage the system. For understanding of changes trajectory of the controlled system in long-term (10 years and more), it is necessary to know the modifications tendency of system orderliness. Orderliness is a consolidating characteristic that describes the quality of system processes management.

2. Formulation of the problem

In the context of the current research *orderliness is a system property identifying the result of the performance of rules and forbiddances that structure the system and limit its changes.* It is a system of restrictions that decreases the chaos n the system and the probability of conflict manifestation.

Applied to the road safety sphere *orderliness is a «degree of rigidity» of transport system functioning management.* The higher the orderliness, the lower the chaos degree in system and vice versa. For assessing of orderliness of road safety provision system (RSP), we will use relative entropy *Hn* or negentropy *(1 – Hn)* [1].

Aforesaid allows to formulate two individual tasks of this research. Firstly, calculating relative entropy in the relation to processes of road safety formation for specific transport systems (for example, for the Russian Federation and Kemerovo region). Secondly, comparison of models *Hn = f(A),* built for road safety provision systems of the Kemerovo region and Russian Federation and making conclusions about current state or road safety provision in the Kemerovo region relatively to all-Russian result.

3. Methods of problem solving

Methods of identification of road safety relative entropy *Hn* were previously considered in [2, 3, 4, 5], but in this research they were modified and now consists of:

1. Separation of road accident rate formation process into blocks of the cause-effect chain (fig. 1) «Population *(P)* – The number of vehicles (N_{Vh}) – The number of road accidents (N_{RA}) – The number of victims (N_V) – The number of deaths (N_D)»

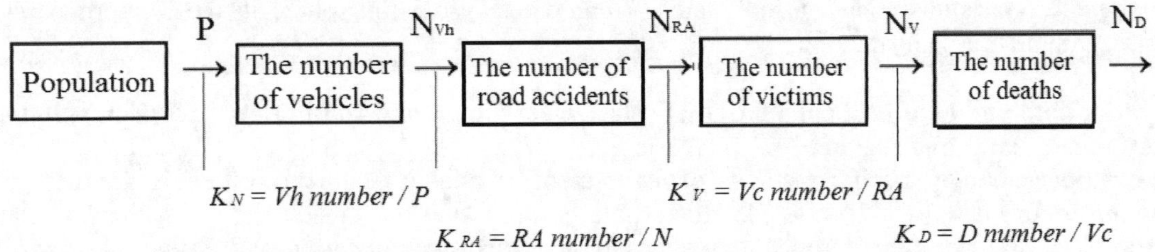

Figure 1. The cause-effect chain of road accident rate formation

2. Gathering of official statistics in the Russian Federation and Kemerovo region for each block of the cause-effect chain «Population *(P)* – The number of vehicles (N_{Vh}) – The number of road accidents (N_{RA}) – The number of victims (N_V) – The number of deaths (N_D)».
3. Calculation of transition coefficients K_i between blocks of the cause-effect chain of road accident rate formation (fig. 1). We will describe this process as 4 subprocesses that have specific coefficients K_N (transformation of number of population into the number of transport vehicles in transport fleet), K_{RA} (transformation of vehicle fleet into the number of road accidents), K_V (transformation of the number of into the number of road accidents victims), K_D (transformation of the number of road accidents victims into the number of lost in road accidents people).
4. Identifying of the positive of the contribution Q relatively to weights of appropriate elements of examined transformational process within the chain «Population – <...> – The number of deaths in road accident».

$$Q = Q_N + Q_{RA} + Q_V + Q_D = \ln(1/K_N) + \ln(1/K_{RA}) + \ln(K_V) + \ln(1/K_D) \tag{1}$$

The physical meaning of the positive of the contribution Q of different elements of the chain «Population – <...> – The number of deaths in road accident» into the final result of road accident rate is the measure of information amount or derivative of examined process entropy.

5. Identifying the structure of weight coefficients w_i for assessing the positive of the contribution Q of different elements of the chain «Population – <...> – The number of deaths in road accident». Availability of calculated values w_N, w_{RA}, w_V w_D of positive allows to solve the main problem of entropic analysis – assess the impact of different elements of the chain «Population – <...> – The number of deaths in road accident» on formation of final road accident rate. Above-stated researches were held for each year from period of 2004...2018.
6. Calculation of entropy H in road safety provision systems of the Kemerovo region and Russian Federation by classic C.E. Shannon's [6, 7] formula (2):

$$H = -\sum_{i=1}^{n} w_i \cdot \ln w_i \tag{2}$$

where n – system elements count (in our case $n = 4$);
w_i – weight coefficients, satisfying the normalization condition, $\sum_{i=1}^{n} w_i = 1$.

7. Calculation of relative entropy (3) of the road safety provision systems of the Russian Federation and Kemerovo region:

$$H_n = H/H_{max} = H/\ln(n) \quad (3)$$

8. Analysis of dynamics of relative entropy of road safety provision systems in the Russia and Kemerovo region during 2004…2018.
9. Comparison of *Hn* values of road safety provision systems of the Russia and Kemerovo region. Making of conclusions about current state of road safety provision sphere in the Kemerovo region relatively to all-Russian situation.

4. Initial data and results of calculation of relative entropy *Hn* of road safety provision systems in Russia and Kemerovo region

As mentioned above specific methods of assessment of human-technical systems orderliness were previously considered in [2, 3, 4, 5]. This article is presenting only results of estimation of relative entropy *Hn* of road safety provision systems in the Russian Federation and Kemerovo region in dynamics (during 2004…2018). Tables 1…2 show initial data [8] and results of calculation of relative entropy *Hn* of road safety provision systems in the Russian Federation and Kemerovo region accordingly.

Table 1. Initial data [8] that was used for analysis of road safety orderliness dynamics in the Russian Federation and results of calculation of relative entropy *Hn*

Year	Population, thousands of people	Vehicles fleet, thousands of units.	Automobilization, vehicles/1000 people	The number of road accidents in year, units	The number of victims, people	The number of deaths in road accidents, people	Value of relative entropy *Hn*
2004	144168.2	29025.3	201.3	208558	284865	34506	0.789
2005	143474.2	31087.6	216.7	223342	308821	33957	0.784
2006	142753.6	34968.2	245.0	229140	318086	32724	0.781
2007	142221.0	35885.3	252.3	233809	325514	33308	0.780
2008	142008.8	38696.0	272.5	218322	300819	29936	0.768
2009	141904.0	41206.4	290.4	203618	283143	27659	0.760
2010	142856.5	42062.7	294.4	199431	277202	26567	0.757
2011	142865.4	43325.3	303.3	199868	279801	27953	0.755
2012	143030.1	45471.1	317.9	203597	286609	27991	0.751
2013	143347.1	47881.8	334.0	204068	285462	27025	0.745
2014	143666.9	49540.4	344.8	199720	278748	26963	0.739
2015	146267.3	51591.9	352.7	184000	254311	23114	0.732
2016	146544.7	54014.3	368.6	173694	241448	20308	0.727
2017	146804.4	53500.0	364.4	169432	234462	19088	0.728
2018	146880.4	53047.0	361.2	168099	233067	18214	0.730

Table 2. Initial data [8] that was used for analysis of road safety orderliness dynamics in the Kemerovo region and results of calculation of relative entropy *Hn*

Year	Population, thousands of people	Vehicles fleet, thousands of units.	Automobilization, vehicles/1000 people	The number of road accidents in year, units	The number of victims, people	The number of deaths in road accidents, people	Value of relative entropy *Hn*
2004	2872.1	556.0	193.6	3601	5149	610	0.795
2005	2855.0	571.9	200.3	3614	5203	616	0.794
2006	2838.5	590.7	208.1	3612	5229	614	0.792
2007	2826.3	630.1	222.9	3608	5225	612	0.786
2008	2823.5	669.2	237.0	3601	5150	603	0.777
2009	2821.9	712.0	252.3	3413	4939	540	0.773
2010	2763.1	715.8	259.1	3406	4881	533	0.770
2011	2761.3	750.4	271.8	3497	5099	557	0.768
2012	2750.8	786.5	285.9	3913	5552	562	0.764
2013	2742.5	797.4	290.8	3710	5404	550	0.765
2014	2734.1	829.1	303.2	3527	5206	504	0.761
2015	2725.0	831.4	305.1	3232	4759	435	0.758
2016	2717.6	845.4	311.1	3054	4448	341	0.756
2017	2708.8	859.0	317.1	2952	4216	303	0.751
2018	2694.9	874.0	324.3	2780	4107	299	0.752

5. Estimation of dynamics of relative entropy *Hn* of road safety provision systems in the Russian Federation and Kemerovo region

Fig. 2 shows the trend of change of relative entropy or road safety provision systems in Russia and Kemerovo region in 2004…2018. Value of *Hn* decreases in both cases, therefore the degree of chaos declines in compared road safety provision systems and their orderliness increases. However in Kemerovo region this process proceeds slower than on average in Russia (fig. 2).

6. Discussion of results

Classical law of R. Smeed [9] and its later corrections [10] explain the decline of road accident rate with time by growth of motorization. M. Blinkin [11] believes that significant reasons of that phenomenon are «continuous self-education of nation» and progressive evolutional change of road safety paradigms. We will use idea about the connection between relative entropy *Hn* of road safety provision systems and motorization *A* to explain the dynamics of relative entropy of road safety provision systems (fig. 2).

IOP Conf. Series: Materials Science and Engineering **582** (2019) 012021 doi:10.1088/1757-899X/582/1/012021

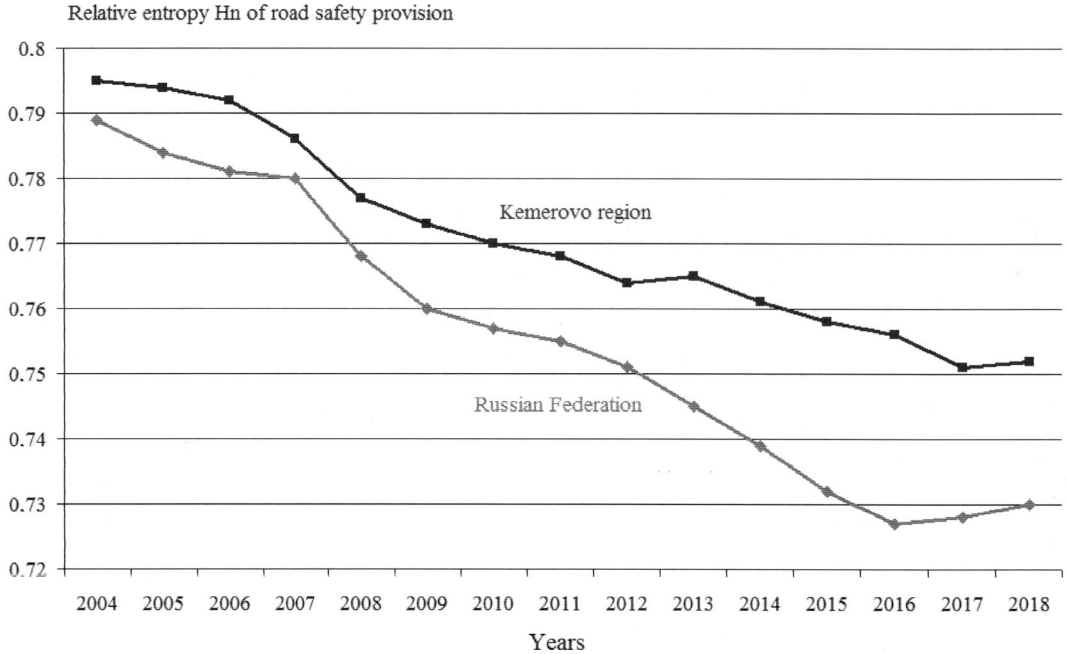

Figure 2. Dynamics of relative entropy *Hn* of road safety provision systems in the Russian Federation and Kemerovo region

6.1 Models Hn = f(A), built for road safety provision systems

Fig. 3 presents the model *Hn* = *f(A)* for the case of all-Russian road safety provision system. Fig. 4 shows the model *Hn* = *f(A)* for the case of road safety provision system of the Kemerovo region.

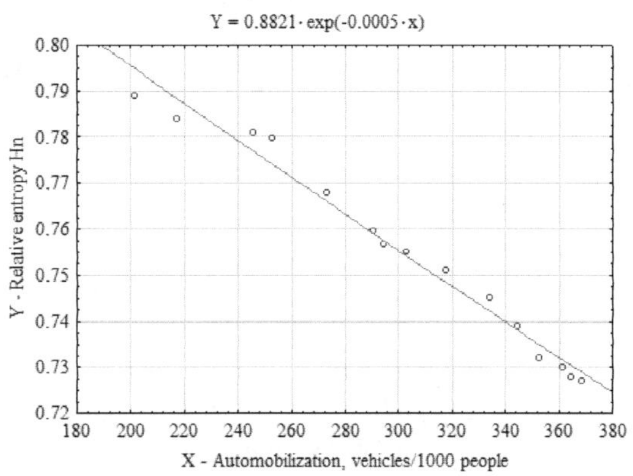

Figure 3. Model *Hn* = *f(A)* for the road safety provision system of the Russian Federation

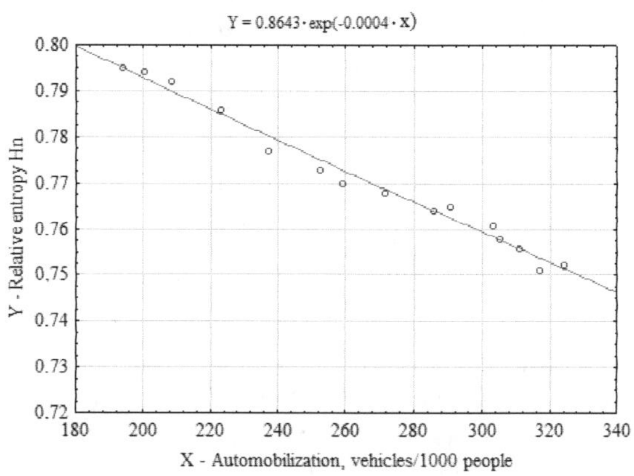

Figure 4. Model *Hn* = *f(A)* for the road safety provision system of the Kemerovo region

Parameter *b* of the model $Hn = a \cdot exp\,(b \cdot A)$ for the road safety provision systems in the Russia and Kemerovo region equals to 0.0005 and 0.0004 accordingly. It means that modification processes of road safety provision systems orderliness proceed to a lesser degree that in whole Russia.

From this it follows that road safety provision system of the Kemerovo region lags behind all-Russian analogs. It happens due to slow development of vehicle fleet and long transition to more progressive paradigms of road safety provision.

6.2 Identification of the model $Hn_{\,RSP\ of\ Kemerovo\ region} = f(Hn_{\,RSP\ of\ the\ Russian\ Federation})$

This model (fig. 5) was built to understand the correlation of trends of formation of relative entropy *Hn* of road safety provision systems.

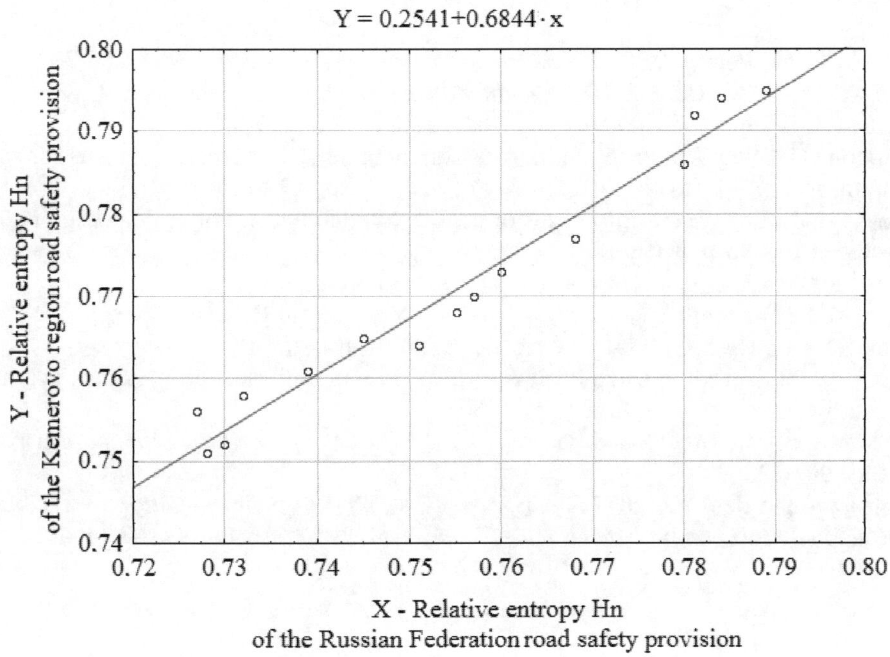

Figure 5. Model $Hn_{RSP\ of\ Kemerovo\ region} = f(Hn_{RSP\ of\ the\ Russian\ Federation})$

This statistical relation can be described by the model $Hn_{RSP\ Kem.\ reg.} = 0.25 + 0.68 \cdot Hn_{RSP\ RF.}$ It describes regional features of lag of road safety provision system of Kemerovo region relatively to all-Russian level with the high level of authenticity ($R^2 = 0.91$).

6.3 Problems of assessing the dynamics of road safety

Unfortunately, currently entropic analysis practically is not used in the analysis of the effectiveness of management decisions in the sphere of road safety assessment [12, 13]. People responsible for analytical work (for example, employees of Research Center of the State Road Safety Inspectorate of the Ministry of the Interior of Russia are used to assess progress using the method of simple comparison of actual first-level road safety indicators with the last year's ones [14].

That does not allow to make serious conclusions, excepting assessment of the simplest tendency. Usage of entropic analysis of road safety provision systems organization allows to estimate quality of professional activity of specialists who organize functioning of transport systems in country regions [13, 14].

Analysis of long-term dynamics of road safety provision systems organization in the Russian Federation and Kemerovo region, showed that the level of these systems organization is although relatively slowly, but raising. Entropy, as the measure of disorganization of these systems, is decreasing in both systems. This fact means that dynamics of the condition of road safety provision systems organization in the Russian Federation and Kemerovo region is positive and we should develop it [15].

7. Conclusion

Comparison of dynamics of relative entropy of road safety provision systems in the Kemerovo region and Russia shows that these processes proceed heterogeneously in different regions of country. The level of road safety provision system orderliness in Kemerovo region lags behind all-Russian state. It can be explained by slow development of motorization and the whole transport culture, determined by the evolution of road safety paradigms, of this region.

References

[1] Schrödinger E 1944 *The Physical Aspect of the Living Cell*
[2] Kolesov V and Petrov A 2017 *Transportation Research Procedia* **20** p. 305-310
[3] Petrov A 2018 *IOP Conf. Series: Earth and Environmental Science* **177** 012015

[4] Kolesov V and Petrov A 2018 *Transportation Research Procedia* **36** p. 286-294
[5] Petrov A 2019 *IOP Conf. Series: Earth and Environmental Science* **224** 012033
[6] Shannon C E 1948 *Bell System Technical Journal* **27** p. 379-423
[7] Shannon C E 1963 Works on the Information theory and Cybernetics (in Russian)
[8] State Inspection for Road Traffic Safety of the Ministry of Internal Affairs of Russia. Official website. Indicators of security status of traffic. Available from: https://www.gibdd.ru/stat/ [Accessed 10th April 2019]
[9] Smeed R J 1949 *Some Statistical Aspects of Road Safety Research*
[10] Kolesov V 2012 *Motor transport enterprise* **6** p. 54-55 (in Russian)
[11] Blinkin M and Reshetova E 2013 Traffic safety (in Russian)
[12] Petrov A and Kolesov V 2018 Problems of functioning of transport systems (Tyumen: Industrial University of Tyumen Press) 368-75
[13] Kolesov V I, Danilov O F and Petrov A I 2017 *IOP Conf. Series: Earth and Environmental Science* **90** 012059
[14] Kolesov V and Petrov A 2015 *Security concern and emergency situations* **3** 21-7
[15] Petrov A I 2016 *Economic and Social Changes-Facts Trends Forecast* **3** 154-72

IOP Conf. Series: Materials Science and Engineering **582** (2019) 012022 doi:10.1088/1757-899X/582/1/012022

Processes of Extraction of Non-Ferrous and Precious Metals from Alternative Sources of Raw Materials

Bashar Issa[a], T A Aleksandrova [b]

Saint-Petersburg Mining University, 199106, St. Petersburg, V.O. 21st line, house 2, Russia

E-mail: [a] basharissashtain@gmail.com, [b] alexandrova_tatyana@mail.ru

Abstract. The presence of precious metals in electronic waste such as gold, silver, platinum and palladium, as well as non-ferrous metals (copper, nickel, zinc, tin, etc.) make it attractive for recycling. In industry, both hydrometallurgical and pyrometallurgical methods are used to extract valuable metals from electronic waste. Universal method for processing of electronic scrap waste does not exist nowadays.

1. Introduction

The increase in demand for electronic equipment based on the development of technology. The service lifetime of existing electronic equipments is reduced, which leads to an increase in waste electronic equipment. The growth in electronic waste is growing and accelerating every year. At present, around 20 to 25 million tons of electronic waste are generated worldwide per year, where the great ratio of them from Europe, United States and Asia. Over the past 20 years, China and Latin America have become producers of a significant amount of electronic waste [1]. The mass of electronic scrap in Russia is about 1.5 million tons per year, and no more than 10% is recycled from them. The resulting waste, on the one hand, causes great harm to the environment, on the other hand, they represent the most valuable resources, surpassing natural sources in the content of useful components. All this creates prerequisites for the development in Russia and in the world of large-scale production of secondary metals [2].

Electronic waste is classified as hazardous materials, as there are components in household and industrial electrical apparatus such as batteries, capacitors, cathode ray tubes, etc. Electronic waste may consist of a large number of components of various sizes, shapes and chemical composition. Some of them contain hazardous metals, including mercury, lead, and cadmium.

The presence of precious metals in electronic waste such as gold, silver, platinum and palladium, as well as non-ferrous metals (copper, nickel, zinc, tin, etc.) make it attractive for recycling. In industry, both hydrometallurgical and pyrometallurgical methods are used to extract valuable metals from electronic waste. Universal method for processing of electronic scrap waste does not exist nowadays. The technologies used may have a number of advantages and disadvantages.

Electronic waste is a complex mixture of ferrous, non-ferrous, precious metals on plastic and ceramic bases. Printed circuit boards are one of the most important components of electronic equipment, on which most valuable metals are concentrated, but most of the toxic components are also concentrated in printed circuit boards.

Printed circuit boards are found in electrical and electronic devices (TVs, computers, mobile phones and laptops). As a rule, printed circuit boards consist of 40% of metals, 30% of plastics and 30% of ceramics [3]. To make them conductive, printed circuit boards are coated with metals (tin, silver or copper). There are two types of printed circuit boards that are used in mobile phones and personal computers. The first type of printed circuit boards is made of multilayered fiberglass coated with copper. Such printed circuit boards are used for small electronic equipment, such as mobile phones. The other type is made of monolayer fiberglass, cellulose paper or phenolic material, which is also coated with a layer of copper and is used for larger devices, such as computers and televisions. Polymers and industrial plastics are major components of printed circuit boards. These polymers contain polyethylene, polypropylene, epoxy resins and polyesters [1].

The printed circuit boards contain a large number of small components, so sampling for analyzing the composition of the printed circuit boards is difficult due to the heterogeneous and complex nature of the materials. As a rule, to analyze the composition of printed circuit boards, they are crushed to smaller sizes (less than 1-2 mm), and various methods are used to separate components, including electromagnetic separation, electrolytic dissolving, etc [4]. In [5] investigated the composition of waste printed circuit boards from mobile phones and personal computers. Pretreatment included crushing and then separation by magnetic and electrostatic methods. Chemical analysis was performed using dissolution in aqua regia. The analysis of the composition was performed using atomic emission spectroscopy. The results showed that the printed circuit boards of mobile phones had a higher (34,5%) copper content compared to personal computers boards (20%).

Electronic waste contains a wide range of elements, some of them are valuable, some of them are toxic or hazardous, and some of them are a combination of both. The valuable metals such as, gold, silver, platinum, palladium, copper, aluminum, nickel, tin, zinc, iron, etc. The toxic or dangerous metals, but the most interest are the following metals: mercury, beryllium, indium, lead, cadmium, arsenic, antimony, etc. The printed circuit boards also include: bromine, fluorine, chlorine, plastics, organic liquids, etc. When disposing of such electronic wastes without proper treatment, there is a high risk of environmental damage. The disposal of electronic waste underground has many disadvantages, including pollution of groundwater and soil, as well as the loss of a potential source of valuable metals.

Table 1 shows examples of the material composition of various equipment, as well as printed circuit boards from personal computers, cell phones and TVs. These figures are only indicative figures derived from specific samples, they may also vary within the same type of equipment, but the order of magnitude will be similar.

Table 1. The average content of components in electronic scrap

Component content%, (g/t)	Fe (%)	Al (%)	Cu (%)	Plastic (%)	Ag (g/t)	Au (g/t)	Pd (g/t)
TVs	28%	10%	10%	28%	280	20	10
Printed circuit boards	7%	5%	20%	23%	1000	250	110
Cell phones	5%	1%	13%	56%	1380	350	210
Audio equipment	23%	1%	21%	47%	150	10	4
Dvd players	62%	2%	5%	24%	115	15	4

For printed circuit boards, cell phones, precious metals make up more than 80% of the cost, for TV cards and DVD players, they make up just over 40% [6].

It should be noted that in the development of technologies for the processing of electronic scrap it is necessary to pay attention to the separation of toxic and hazardous components, as well as to minimize the loss of precious metals, which may adversely affect the economic performance of the process.

Most of the electronic waste is dumped, about 20% is burned [7]. In the last decade, many countries have developed legislation on electronic waste management. In some countries, laws have emerged that cannot be placed electronic waste underground or burn it in incinerators without isolating hazardous materials.

In addition, the export of electronic waste to underdeveloped countries is not allowed according to the international regulations. Underground disposal, burning in the air, and acid leaching adversely affect the environment, polluting drinking water and the atmosphere [8]. Therefore, the recycling of electronic waste is important in terms of minimizing pollution and resources management.

Recycling electronic waste to extract valuable components is important in terms of energy savings. Table 2 shows [7] six basic materials, the use or extraction of these materials from electronic

IOP Conf. Series: Materials Science and Engineering **582** (2019) 012022 doi:10.1088/1757-899X/582/1/012022

scrap provides significant energy savings compared to the energy conservation which is required for their primary production from raw materials.

Table 2. Energy savings during the recycling of some components compared to the energy consumption in their primary production [9, 10].

№	Component	Energy savings (%)
1	Aluminum	95
2	Copper	85
3	Iron and steel	74
4	Tin	65
5	Zinc	60
6	Plastic	>80

Moreover, the processing of electronic waste will reduce the burden on the mining industry in terms of ore mining for the primary production of metals. Consequently, this can help in solving problems with the extraction of metals, which exist in low concentrations in the ore of raw materials and their extraction is accompanied by a significant energy consumption. In fact, electronic waste is a rich source of precious and non-ferrous metals in comparison with crude ores. The amount of gold extracted from one ton of electronic waste in the processing of personal computers is greater than in the processing of 17 tons of gold ore. In some cases, recovery of precious and non-ferrous metals from electronic scrap is easier than from their ore [11].

2. Theoretical basis

Technological scheme of processing scrap of electronic industry can be represented as follows (Fig. 1). Raw materials coming from enterprises are sent for preliminary disassembly. At this stage, components containing precious metals are extracted from electronic computers and other electronic equipment. The enriched material is sent to smelt, as a result of that anodes are obtained, containing non-ferrous and precious metals.

Experience in the processing of secondary raw materials containing precious metals, by the largest national and international enterprises, allows us to conclude that the main steps can be: preparation of secondary raw materials (enrichment), burning and electrolysis.

When processing electronic waste, one of the main operations is burning. The main problem at the stage of pyrometallurgical processing is a large amount of plastics and other organic materials, their burning produces a significant amount of gases containing harmful components.

3. Experimental

From the above scheme, it can be seen that after the enrichment of the electronic scrap, the concentrate should be smelted to the anodes. As a result of smelting of the magnetic fraction, alloys with a high content of iron, nickel and cobalt can be obtained, and copper-zinc alloys can be obtained as a result of smelting of the non-magnetic fraction.

3.1. Smelting concentrate to anodes and the effect of tungsten on the extraction of precious metals

When smelting in an induction furnace of an iron-nickel concentrate, it was found that refractory tungsten, when oxidized, goes into slag. At the same time, analyzes of thin cross-sectional polished microscopic sections by Tescan TS 5130 MM microscope showed that tungsten collects gold (Fig. 2), which leads to its loss in accounting. During the smelting process, it was found that refractory tungsten deposited on the surface of a graphite crucible, carrying gold along with it.

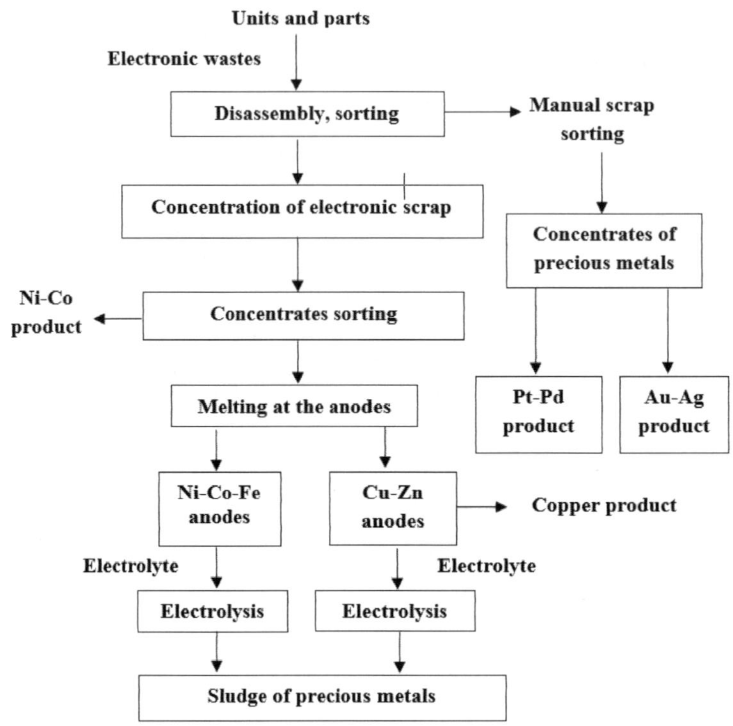

Fig. 1. Technological scheme for the processing of electronic waste

In Figure 2 it can be seen that platinum and palladium are evenly distributed, meanwhile some individual large areas with a high concentration of silver can be observed. Gold is evenly distributed in phases, but at the point of high concentration of tungsten, the gold content is maximum, and in this case tungsten is considered to be a collecting agent for gold.

This effect can be avoided if you get rid of tungsten by oxidizing it and converting it to slag. Then precious metals collected in tungsten will remain in the copper alloy.

The density of tungsten is almost close to gold – 19,3 g/cm^3, and the ionic radii differ twice, in this case it is possible to collect one metal into another. Tungsten at ordinary temperature is resistant to hydrochloric, sulfuric, nitric and hydrofluoric acids of all concentrations, as well as aqua regia. When heated to 80-100°C, the metal is resistant to hydrofluoric acid, very weakly interacting with hydrochloric and sulfuric acids. Slightly more noticeable, it interacts with nitric acid and aqua regia. In a mixture of hydrofluoric and nitric acids, as well as hydrogen peroxide, tungsten dissolves. This creates significant difficulties in the processing of electrolysis sludge, since non-oxidized tungsten is completely converted into the sludge, and its content in individual deliveries of electronic scrap exceeds the content of precious metals.

In the system of tungsten-iron these compounds are formed: W_6Fe_2, Fe_2W. The mutual solubility of the components in the solid state is insignificant; the interaction between tungsten and cobalt has a similar nature [12].

3.2. The effect of silicon on the process of electrochemical dissolution of anodes

On the basis of the St. Petersburg Mining University, smelting was carried out to anodes of electronic waste in the furnace "Tamman". 2 anodes from the magnetic fraction and 2 anodes from the nonmagnetic fraction were obtained.

The first anode was smelted from the magnetic fraction in a graphite crucible without any additives; during the smelting of the second anode, quartz sand (SiO_2) and potassium nitrate (KNO_3) were added.

A similar operation was performed in the smelting of the anodes from the non-magnetic fraction, followed by the production of anodes 3 and 4.

Studies of the obtained anodes were carried out by using a TESCAN VEGA 3 Sem scanning electron microscope and gave the following results on the composition of the anodes subjected to electrolytic dissolution in a laboratory setup (Table 3, 4).

Based on the data from Tables 3 and 4, it was concluded that silicon was recovered to its elementary state during smelting, or it is present in the anode as an oxide.

The anodes were subjected to electrochemical dissolution in a 1 mol/l solution of $NiSO_4 \cdot 7H_2O$ with the addition of H_2SO_4 with concentration 30 g/l. At a temperature of 25°C and a cathode current density of 230 A/m^2. We used an iron cathode and bath potential – 6 V.

As a second electrolyte, a 1 mol/l of $CuSO_4 \cdot 5H_2O$ was used with the addition of H_2SO_4 with concentration 30 g/l. The electrolyte temperature is 25°C, with copper cathode and cathode current density is 230 A/m^2. The bath potential – 6 V.

The result of this experiment was a decrease in the current density during dissolution of the obtained anodes, which was described in detail in previous works [12-16].

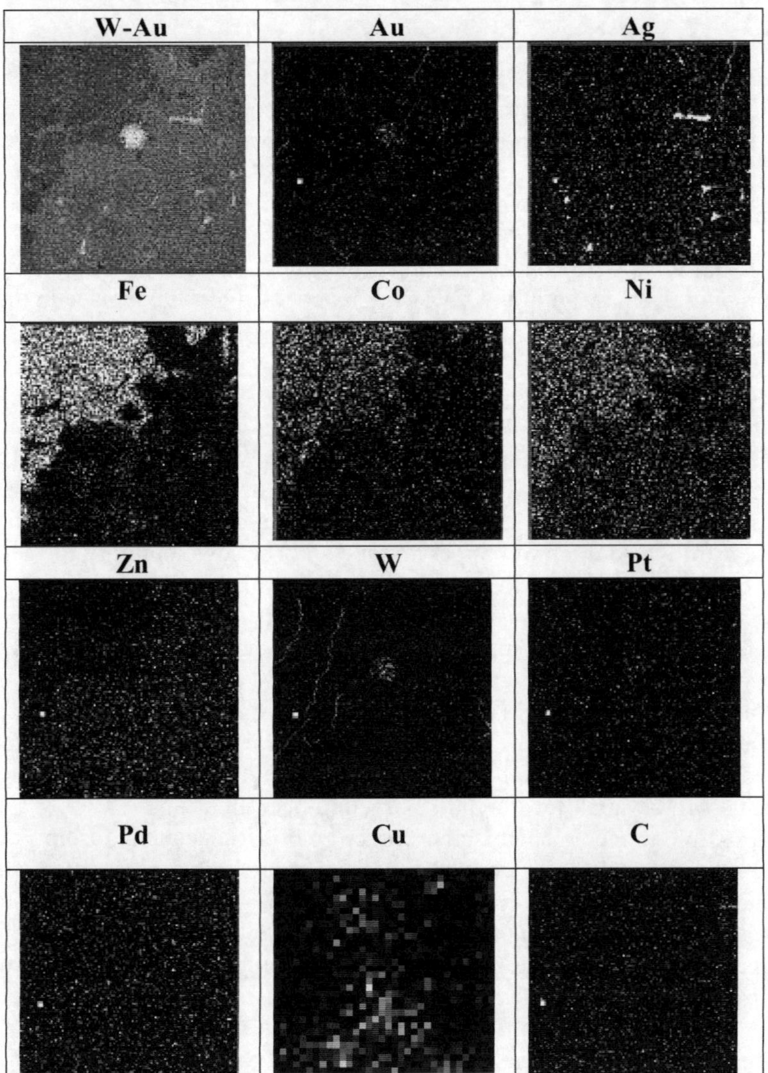

Figure 2. The results of smelting iron-nickel concentrate

IOP Conf. Series: Materials Science and Engineering **582** (2019) 012022 doi:10.1088/1757-899X/582/1/012022

Table 3. Compositions of Ni-Co-Fe anodes,%

Anode	Ni	Cu	Co	Zn	Fe	Si	Ag	Au	Pd	Pt	Total
1	49.99	6.36	2.14	0.06	40.40	-	0.18	0.82	0.05	–	100.0
2	42.30	23.67	3.38	0.03	21.51	5.70	2.48	0.54	0.39	–	100.0

Table 4. Compositions of Cu-Zn anodes,%

Anode	Ni	Cu	Co	Zn	Fe	Si	Ag	Au	Pd	Pt	Total
1	3.03	85.16	0.08	32.45	0.05	-	3.11	0.01	0.56	0.01	100.0
2	4.71	71.15	0.03	15.05	0.75	6.20	0.31	1.32	0.48	–	100.0

3.3. Recycling solutions

Electrolyte solutions obtained after the electrolysis process are enriched in nickel in the case of dissolving Ni-Co-Fe anodes in a nickel electrolyte and copper in dissolving Cu-Zn anodes in a copper electrolyte.

There is a need to process or purify such solutions from impurities or salt excess of non-ferrous metals.

And one of the ways is to converted copper and nickel salts into the hydroxide form, by the addition of calcium oxide, as a result of the following reactions which take place in the electrolyte:

$$NiSO_4 + H_2O + CaO = CaSO_{4(s.)} + Ni(OH)_{2\,(s.)} \quad (1)$$
$$CuSO_4 + H_2O + CaO = CaSO_{4(s.)} + Cu(OH)_{2\,(s.)} \quad (2)$$

After this treatment of the solutions, two kinds of cakes are obtained: nickel cake (up to 96% nickel in the cake) and copper cake, which can be recycled or sent to the consumer. The following method of processing used solutions is associated with the use of naphthenic acids for the subsequent extraction separation from a solution of copper and nickel. The low solubility of these acids in water, the possibility of their isolating them from waste of oil treatment and, as a result, the comparative cheapness determine the feasibility of using this reagent in the processing of metallurgical solutions. Naphthenic acids form complex compounds with a specific metal ion at a certain pH. This enables the selective isolation of metals from solutions.

In the future, it is possible to carry out the process of reextraction, obtaining a pure naphthenic acid and an aqueous solution of the salt of the corresponding cation. It is possible to use this method with significant concentrations of metal ions in solutions. But it may be that the concentrations of valuable components will be small for standard extraction operations.

If it is impossible to electrochemically extract gold or other metals from poor solutions, it is proposed to carry out an operation to convert these compounds into a frothy product for complete extraction. The method under study involves the transfer of metal compounds into a hydrophobic precipitate, followed by their isolation from the solution. Collecting agent, used as a precipitator, significantly increases the concentration in the resulting product, which will undergo separation. Equipment for this technology is standard laboratory or industrial equipment not required by special regulatory systems. High performance of this process allows its application in any industries. Such devices can be used to clean recycled industrial water. Such devices can be used to clean circulating industrial water. Foam apparatus – a turbo-aerator, constructional design of which provides the formation of an air (foam) layer with the lowest hydraulic resistance. The device is a vertical container with square cross section. The tank is divided into two zones: the upper chamber (expansion), designed to froth breakage and the lower (mixing) chamber. In a turbo aerator, the gas moves through a layer of foam that forms on the grid where the fluid is supplied, when the air is blown from below or when an air stream hits the surface of the liquid. This device is simple in design and quite effective. In this reactor, happens process of concentration.

Furthermore, in the presence of other cations in the solution, there is the prospect of selective extraction of metals from poor solutions adjusting just some of the process parameters such as temperature or pH value.

4. Conclusions

Based on the obtained data, it can be concluded that during the smelting of concentrates, the obtained anodes contain impurities that can affect the extraction of precious metals. Such impurities should be removed as much as possible at the smelting stage, before the hydrometallurgical processing stage. To these impurities include primarily tungsten, lead, tin.

Some impurities may have a beneficial effect on the electrochemical dissolution of the anodes. Silicon can be used as an impurity with which it is possible to regulate and control the process of electrochemical dissolution. The presence of silicon can reduce the dissolution potential of the anode, which leads to a reduction in power consumption, or allows you to increase the current density to accelerate the process of electrolytic dissolution, if necessary. Furthermore, the silicon additive allows to avoid the passivation process of the electrochemical dissolution of the alloy in sulfuric acid electrolyte.

Recycling solutions is an important ring in a technological chain. In addition, using the proposed options for processing solutions, it is possible not only to make the technology for processing electronic scrap closed loop, but also to be more profitable from an economic point of view, as well as more environmentally friendly. The proposed methods allow working not only with solutions containing a large amount of non-ferrous and precious metals, but also with poor solutions, which the metal content does not exceed 40 mg/$м^3$.

References

[1] Abdul Khaliq. Metal Extraction Processes for Electronic Waste and Existing Industrial Routes: A Review and Australian Perspective, Abdul Khaliq, Muhammad Akbar Rhamdhani, Geoffrey Brooks and Syed Masood, Resources 2014, 3, 152-179, doi:10.3390/resources3010152.

[2] Ivanova M. A. Study of the possibility of using international practices of processing of electronic scrap, master's thesis, Krasnoyarsk, 2016.

[3] Ogunniyi, I.O., Vermaak, M.K.G.; Groot, D.R. Chemical composition and liberation characterization of printed circuit board comminution fines for beneficiation investigations. Waste Manag. 2009, 29, 2140–2146.

[4] Luda, M.P. Recycling of printed circuit boards. Integr. Waste Manag. 2010, 2, 285–299.

[5] Yamane, L.H.; de Moraes, V.T.; Espinosa, D.C.; Tenório, J.A. Recycling of WEEE: Characterization of spent printed circuit boards from mobile phones and computers. Waste Manag. 2011, 31, 2553–2558.

[6] Hagelüken, C. Improving Metal Returns and Eco-Efficiency in Electronics Recycling – A Holistic Approach for Interface Optimization between Pre-Processing and Integrated Metals Smelting and Refining. In Proceedings of the IEEE International Symposium on Electronics and the Environment, Scottsdale, AZ, USA, 8–11 May 2006.

[7] Office of Solid Waste. Electronic Waste Management in the United States—Approach 1; U.S Environmental Protection Agency: Washington, DC, USA, 2008; p. 56.

[8] Guo, J.; Xu, Z. Recycling of non-metallic fractions from waste printed circuit boards: A review. J. Hazard. Mater. 2009, 168, 567–590.

[9] Cui, J.; Forssberg E. Mechanical recycling of waste electric and electronic equipment: A review. J. Hazard. Mater. 2003, 99, 243–263.

[10] Institute of Scrap Recycling Industries (ISRI). Scrap Recycling: Where Tomorrow Begins; ISRI: Washington, DC, USA, 2003; pp. 16–24.

[11] Rankin, J. Minerals, Metals and Sustainability—Meeting Future Materials Needs; Commonwealth Scientific and Industrial Research Organization (CSIRO): Melbourne, Australia, 2011.

[12] Alexandrova T.A., Gorlenkov D.V., Romanova N.A. Researching of influence of tungsten, silicon, and impurities oxidation on electrolytic dissolution of Cu-Zn and Fe-Ni-Co anodes / // Periódico Tchê Química (Southern Brazilian journal of chemistry) Vol. 14, N. 28, pp. 9-17, 2017, ISSN 2179-0302.

[13] Aleksandrova T. A., Telyakov N. M., Telyakov A. N., Gorlenkov D. V. Effect of Tungsten on Precious Metal Extraction During Processing of Radio-Electronic Scrap // «Springer», Metallurgist, July 2017, Vol. 61, Iss 3-4, pp. 188-192, ISSN: 0026-0894.

[14] D.V. Gorlenkov et al., "Selection of Complete Recovery of Precious Metals in the Processing of Copper-Nickel Alloys in Hydrometallurgical Way", Materials Science Forum, Vol. 927, pp. 190-194, 2018.

[15] Beloglazov I.I., Petrov P.A., Gorlenkov D.V. Development of an Algorithm for Control Metallurgical Processes of Fluidized Roasting Using an Adaptive Controller. Vol. 1059, IX International Scientific Practical Conference "Innovative Technologies in Engineering" 24-26 May 2018, Yurga, Russian Federation. Publishing IOP Conf. Series: Journal of Physics: Conf. Series 1059 (2018).

[16] Bazhin V. Yu., Aleksandrova T. A., Kotova E. L., Gorlenkov D. V., Suvorov, R. S. Metallurgists of the mining university and development of coinage. 245 years of history // Publ. St. Petersburg mining university (St. Petersburg), "Journal of mining Institute", Vol. 230, 2018, pp. 131-138.

IOP Conf. Series: Materials Science and Engineering **582** (2019) 012023 doi:10.1088/1757-899X/582/1/012023

Software for Reduction of Energy Capacity of Cargo Transportation by Road Transport

D Chainikov[1,a], N Relyan[2]

[1]Motor transport operation Department, Industrial University of Tyumen, 625039, Tyumen, Russian Federation

[2]Foreign Languages Department, Industrial University of Tyumen, 625039, Tyumen, Russian Federation

E-mail: [a] 4ainikov@mail.ru

Abstract. The article deals with the problem of improving the efficiency of vehicle use by reducing the consumption of fuel and energy resources of trucks during the carriage of goods. A solution is proposed, based on the development of a special software product that allows taking into account the vehicles` adaptability of different weight categories to the mass of the transported cargo. It is offered the software to calculate the objective fuel consumption norms on cargo transportation.

1. Introduction

On September 21, 2018, the Ministry of Transport of the Russian Federation issued a decree on Amendments to the Methodological Recommendations [1] of the "norms of fuel and lubricant consumption in road transport, put into effect by order of the Ministry of Transport of the Russian Federation from March 14, 2008 n am-23-r. It is known that the existing standards are complemented by new models of vehicles and correct the raising factor for the operation of vehicles in populated areas with different populations. The changes and criticism of the current system of fuel consumption rationing in these changes were not taken into account.

Improving the efficiency of vehicle use includes a number of activities. One of the most important and significant in terms of importance is the course of improving the economic efficiency of vehicles.

The basis of the economic efficiency of vehicles is the concept of fuel efficiency and operating fuel consumption. Fuel efficiency, as you know, involves the performance of a given work with minimal fuel consumption [2]. Operating fuel consumption can be divided into actual and regulatory. If the actual consumption is the amount of fuel that was used to perform a given job and can vary widely depending on the conditions in which the vehicle was operated, on its technical condition, and the honesty of the driver. The standard consumption is a fixed value, an established measure, determined for a particular vehicle when it is used under given conditions. The latter is based on the norm, which in practice is calculated according to the Methodological Recommendations using the base rate of fuel consumption.

The economic efficiency of vehicles in operation is achieved mainly due to the saving of fuel and energy resources. Savings are the benefits that come when carefully spending something. Savings, in fact, is a reduction of losses due to the use of resource-saving technologies, prudent management, high organization of labor. Consequently, to save fuel means to spend it economically, and economical operation of a vehicle is its use, which makes it possible to profit when using fuel economically.

In order to figure out how to increase economic efficiency, it is necessary to understand how, in practice, fuel economy is obtained, to identify the source of its formation. Based on the above concepts, we can conclude that in the operation of fuel economy is determined by the level of deviation of the actual operating fuel consumption from the standard. At the same time, savings are generated if the difference between the actual and standard consumption is negative, otherwise there is an overrun.

The discrepancy between the calculated and actual values leads to the fact that truck drivers have the opportunity to drain the excess fuel and sell them on the private market. Only a well-established rate of fuel consumption will avoid it. However, the exact norm can only be obtained as a result of taking into account the influence of complex and changeable factors. Obtaining an objective standard is possible only as a result of the development and use of special software.

2. Methodology

2.1 The choice of the research object

The article proposed a method to increase the efficiency of trucks transporting goods in the city and on public roads.

Currently, a reduction in actual fuel consumption will not lead to real savings [3,4]. The reason is the lack of an objective, scientifically based standard. Even the use of specialized equipment - instrumental means of monitoring actual fuel consumption does not allow achieving high fuel efficiency without an objectively established limit of fuel consumption. In addition, the use of instrumental control fuel consumption currently faces a number of problems. The high cost of equipment and the need to control and analyze data are the main ones.

Thus, in order to increase economic efficiency due to real fuel economy, it is necessary, first of all, to accurately and objectively determine the standard fuel consumption for vehicles of different brands and models. After that, one should achieve high fuel efficiency by reducing actual fuel consumption.

For the development of scientifically based fuel consumption rates, it is necessary to take into account both the design features of automobiles and the conditions in which they will be used. One of the most effective ways to develop fuel consumption rates is to use a space-time approach.

The basis of the space-time approach is that for vehicles used in variable and harsh conditions, the property of adaptability is introduced. Along with reliability, it was proposed to consider adaptability as one of the most important properties in shaping the quality and efficiency of vehicles [5]. If reliability is manifested in time, then adaptability is in space (operating conditions). The adaptability of a vehicle depends on its design features and on the conditions in which it is used and is a property to maintain at the nominal level the values of quality and efficiency indicators when the operating conditions deviate from standard ones.

2.2 Determination of the objective rate of fuel consumption of trucks for the carriage of goods

An indicator was obtained reflecting the influence of the mass of the transported cargo on the fuel consumption of vehicles with different curb weight.

$$K_p = \frac{q_{so} \cdot m_{cargo} \cdot (a + 10 \cdot \psi)}{H_s}, \tag{1}$$

where q_{so} – specific indicator fuel consumption of the engine, l / 100Nkm;
ψ – коэффициент сопротивления дороги в заданных условиях road resistance coefficient under specified conditions;
m_{cargo} – mass of transported cargo, kg;
a – vehicle acceleration, m / s^2;
H_s – base rate of fuel consumption, l / 100km.

Fig. 1 shows the change in the K_p coefficient calculated by the formula (1) from the mass of the transported cargo for two vehicles with different curb weights.

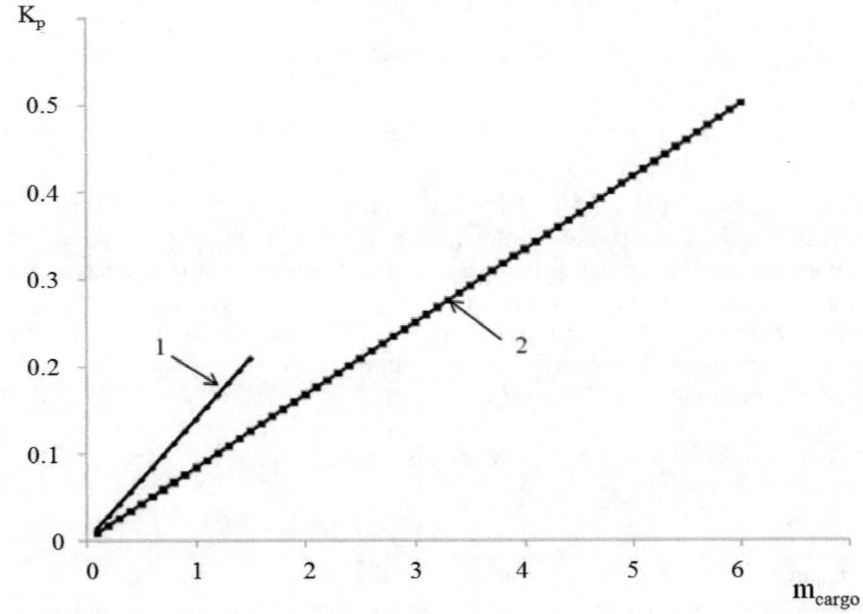

Figure 1. The dependence of the K_p coefficient on the mass of the transported cargo for on-board vehicles with a gasoline engine: 1 - GAZ-2705 q_n = 1500 kg.); 2 - ZIL-4331 (q_n = 6000 kg.)

Figure1 shows that the same change in the basic operating conditions (the mass of the load) in vehicles with different weight parameters leads to a different change in the coefficient K_p. Thus, the K_p coefficient characterizes the influence of the mass of the transported cargo on the fuel consumption of automobiles, and depends on their weight parameters.

The values of the coefficient K_p for suburban and urban (with frequent stops) modes of movement of on-board vehicles and vans with different types of engines are given in Table 1. The coefficient of road resistance ψ is assumed to be equal to ~ 0.018. The average density of gasoline is taken to be 0.74 kg / l, diesel fuel is 0.85 kg / l, and their lower calorific capacity is 44 and 42.5 MJ / kg, respectively.

Table 1. Coefficient K_p values

Type of engine	Coefficient K_p during country driving	Coefficient K_p in city traffic mode
gasoline	$\frac{18.8\cdot10^{-4}\cdot m_{cargo}}{H_s}$	$\frac{10.7\cdot10^{-3}\cdot m_{cargo}\cdot(a+0.18)}{H_s}$
diesel	$\frac{10.2\cdot10^{-4}\cdot m_{cargo}}{H_s}$	$\frac{6\cdot10^{-3}\cdot m_{cargo}\cdot(a+0.18)}{H_s}$

Due to the difference between the working cycles of diesel and gasoline engines, the values of the coefficient K_p for them are calculated separately. In the system of fuel consumption rationing, the features of different types of engines are taken into account and the value of fuel consumption for the performance of transport work was taken to be different. Currently, the norm is set for vehicles using diesel fuel - 1.3 l / 100tkm and for using gasoline - 2 l /100tkm.

The study of hypotheses about the form of the mathematical model of the dependence of fuel consumption on the mass of the load from the point of view of their applicability allowed us to identify a piecewise linear model of the form:

$$Q_f = Q_o + m\cdot\left(\sum_{i=0}^{n_1} S_{1i} + \sum_{i=0}^{n_2} S_{2i} + \sum_{i=n_2}^{n_3} S_{3i}\right), \quad \text{l/100 km} \qquad (2)$$

where m – unit of mass of the transported cargo (1 ton);

S_1, S_2, S_3 – values of sensitivity parameters in the corresponding interval of K_p values, l / 100 tkm;

n_1, n_2, n_3 – the number of full tons of cargo (trailer / semi-trailer) in the corresponding K_p value range;

The use of new patterns that characterize different adaptability of vehicles, in particular to different masses of cargo, allows them to operate with maximum efficiency. Therefore, taking into account the data obtained (Figure 1 and Table 1), it is necessary to determine the levels of adaptability to the mass of the cargo in terms of fuel consumption for vehicles with diesel and gasoline engines separately.

In accordance with the space-time concept [6,7], one of the quantitative indicators of adaptability is the coefficient of adaptability. It shows how many times the value of the indicator of quality and efficiency in these conditions differs from the base value. The use of this coefficient will allow to distribute all vehicles into groups depending on their level. The coefficient of adaptation of vehicles to the mass of the transported cargo will be determined by the ratio:

$$A = \frac{Q_o}{Q_f} \tag{3}$$

where Q_o – value of fuel consumption of an empty vehicle, l / 100km;
Q_f – the actual value of the fuel consumption of the loaded vehicle, calculated using the coefficient K_p, l / 100km.

In accordance with formula (2), vehicles are divided into three levels of adaptability: high, medium, low. Table2 and Table 3 show an example of the division of vehicles into groups, depending on the level of adaptability.

Table 2. The adaptability levels of vehicles with gasoline engines to the mass of transported cargo

Name of adaptability range	Curb weight, characteristic for a given range of adaptability, t.	Vehicles	Full weight class
High	6000 and more	URAL-377H	5 – 7
Medium	3500	ZIL-431410	3 – 4
Low	2000 and less	YerAz -762A	1 – 2

Table 3. The adaptability levels of vehicles with diesel engines to the mass of transported cargo

Name of adaptability range	Curb weight, characteristic for a given range of adaptability, t.	Vehicles	Full weight class
High	9000 and more	MAZ-6303	6 – 7
Medium	6500	ZIL-534330	4 – 5
Low	4000 and less	ZIL-5301BE	1 – 3

The values of the coefficient of adaptability to the mass of the transported cargo in terms of fuel consumption were obtained analytically. When calculating using ratios (1), (3), q_{so} values were used, which, depending on the type of engine installed on the vehicle, take values from $5.4 \cdot 10^{-3}$ to $11.6 \cdot 10^{-3}$ l/100 Nkm, as well as flow values fuel of vehicles in the empty state. The presented results were obtained for the case of transportation of 1 ton of cargo. The result of statistical analysis showed that the density function of the distribution of values of the coefficient of adaptability corresponds to the normal law (Fig. 2).

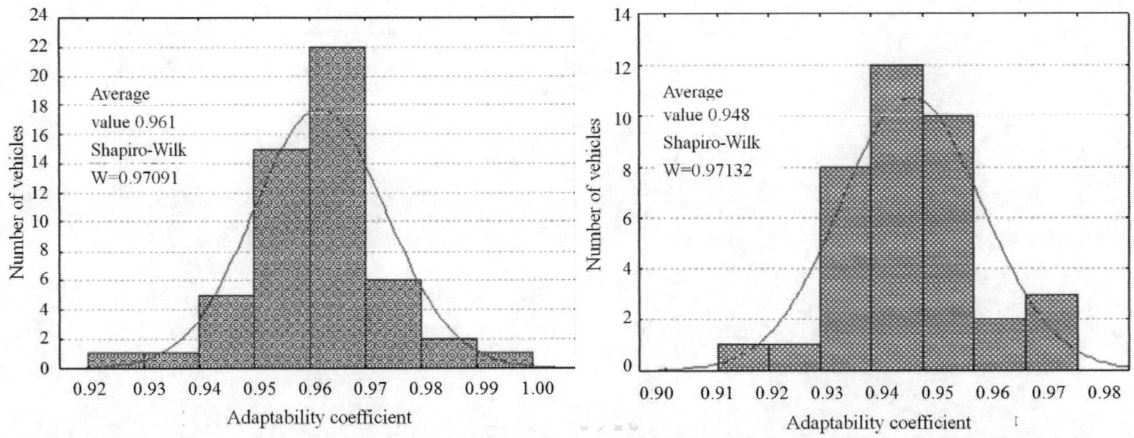

Figure 2. Histogram and distribution curve of the values of the adaptability coefficient for vehicles with a diesel (left) and gasoline engine (right)

To assign a particular vehicle to a particular range of adaptability, it is enough to determine what type of engine is installed in the vehicle and then find out what class the vehicle is in full weight.

Figure 2 shows the dependence of the adaptability coefficient A on the mass of the transported cargo for vehicles with petrol and diesel engines, which belong to different classes according to the total weight.

3. Results and analysis

To determine the amount of fuel consumption during a cargo transportation, an experiment was performed using the data of the on-board computer and the method of topping up the tank to full. The cargo mass was measured by floor scales MP 60 MDA-7 p / d-10 "Guliver". Figure 3 illustrates the results of the experiment for three vehicles: Volvo FM12, MAZ-437040, GAZ-3302.

Figure 3 shows that vehicles with different gross weights have different dependences of the adaptability coefficient on the mass of the load. That is, vehicles with excellent weight characteristics react differently by changing fuel consumption by the same weight of cargo. This means that they have different adaptability. In this regard, the size of the fuel consumption rate for the performance of transport work must be appointed differentially.

Based on the obtained S values, a table of the values of the fuel consumption rate for the performance of transport work was compiled. The work was differentiated depending on the adaptability levels of vehicles of different brands and models, for different severity intervals (Tables 4 - 4).

The use of objective differentiated standards will save fuel, primarily due to the elimination of the possible causes of its overspending: theft, poor technical condition of the vehicle, unskilled driving [8].

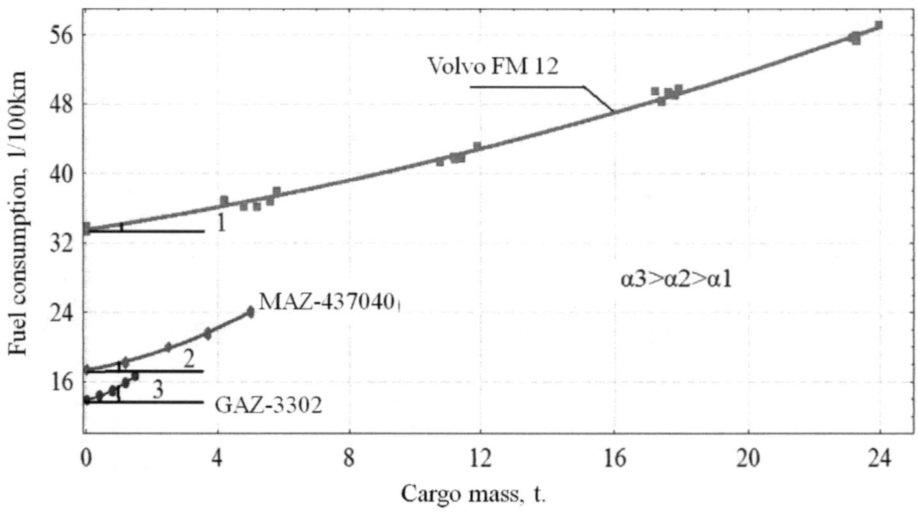

Figure 3. The dependence of fuel consumption of vehicles on the mass of the cargo

Table 4
Fuel consumption rate for transport work of vehicles with a diesel engine

Level of vehicle adaptability	The value of the fuel consumption rate for the transport of 1 ton of cargo for the severity intervals, l / 100 tkm		
	Temperate	Moderately severe	Severe
High	0.78	0.92	1
Medium	0.91	1.03	1.12
Low	1.02	1.14	1.23

Table 5
Fuel consumption rate for transport work of vehicles with a gasoline engine

Level of vehicle adaptability	The value of the fuel consumption rate for the transport of 1 ton of cargo for the severity intervals, l / 100 tkm		
	Temperate	Moderately severe	Severe
High	1.56	1.74	1.82
Medium	1.64	1.8	1.9
Low	1.73	1.88	1.97

For ease of use in any enterprise that uses automobile transport and rationing the fuel consumption of automobiles, a software implementation of this Methodology has been proposed. The program is created in C ++, has an intuitive interface and allows you to get the value of the fuel consumption of a particular vehicle to perform a given transport work (figure 4).

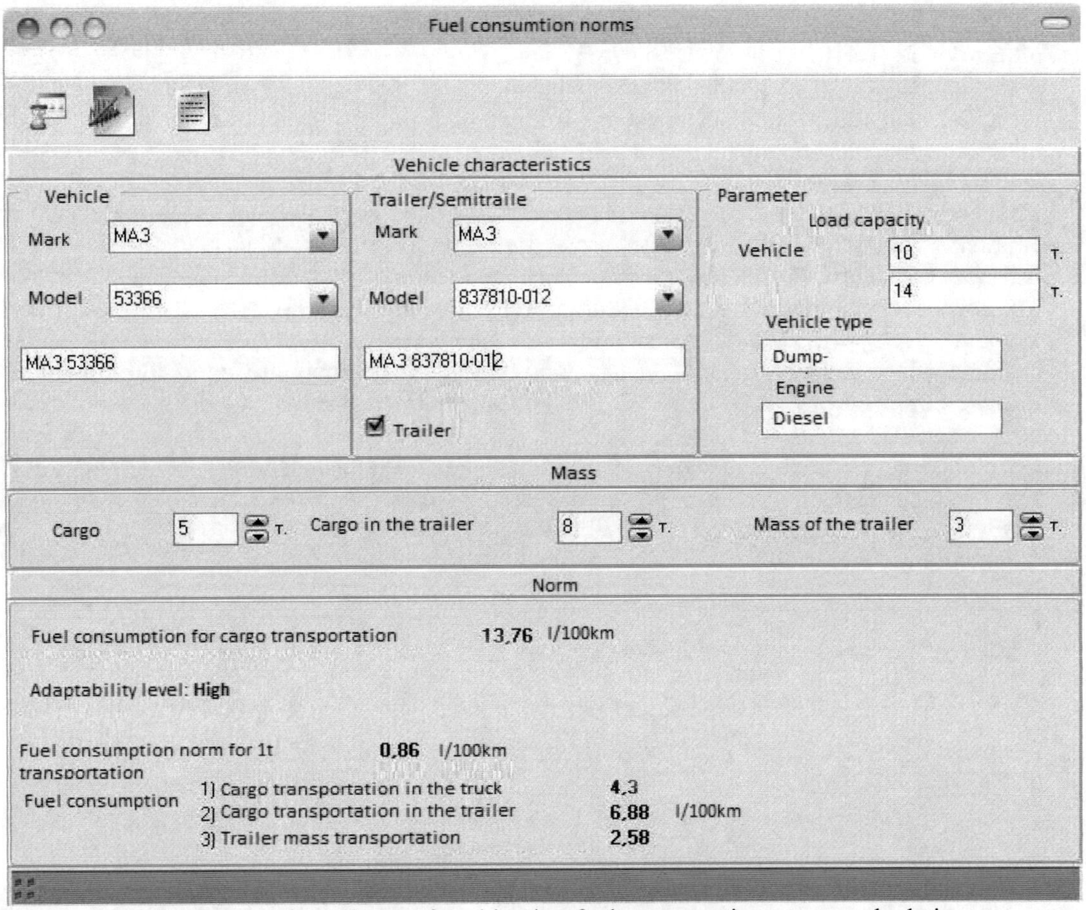

Figure 4. The software for objective fuel consumption norms calculation

3. Conclusion

Thus, we can conclude that for vehicles with gasoline and diesel engines belonging to different weight groups, the ranges of fitness are defined. At the same time, there is no need to make calculations for assigning a vehicle to one group or another. In practice, it will improve the efficiency of the use of vehicles mainly due to the use of the most suitable rolling stock for the transportation of a given mass of cargo, and as a result of the establishment of a differentiated rate of fuel consumption for transport work.

The economic effect obtained through the use of software created on the basis of objective rationing, taking into account the various levels of fitness of vehicles, in comparison with the current standards for the studied vehicles ranges from 3.7 to 8 rubles / 100 km per vehicle (prices in March 2019) when transporting one ton of cargo.

References

[1] Order of the Ministry of Transport of the Russian Federation of March 14, 2008 No. AM-23-p (as amended on 09.20.2018) "*On the implementation of the methodological recommendations" Consumption rates of fuels and lubricants for road transport* " Ministry of Transport, Road Department, JSC SaRIoMT - 2018. 89 p.

[2] Wang, R., Wu, Y., Ke, W., Zhang, S., Zhou, B., Hao, J. Can propulsion and fuel diversity for the bus fleet achieve the win-win strategy of energy conservation and environmental protection? (2015) *Applied Energy* - Volume **147**, pp. 92-103, doi: 10.1016/j.apenergy.2015.01.107.

[3] Bondarenko, E.V., Filippov, A.A. Evaluation of use of certain types of motor fuel for environmental safety criteria // *Autogas filling complex + alternative fuel* – Volime **52**, pp. 31-35, 2010.

[4] Ling-YunHE[abc]Lu-YiQIU[d1], Transport demand, harmful emissions, environment and health co-benefits in China // *Energy Policy*. Volume **97**, 2016, Pages 267-275.

https://doi.org/10.1016/j.enpol.2016.07.037

[5] Anisimov, I., Ivanov, A., Chikishev, E., Chainikov, D., Reznik, L. Assessment of gas cylinder vehicles adaptability for operation at low ambient temperature conditions (2014) *WIT Transactions on Ecology and the Environment*, **190** volume 1, pp. 685-695, doi: 10.2495/EQ140651.

[6] Chainikov D., Chikishev E., Anisimov I., Gavaev A. Influence of ambient temperature on the co2 emitted with exhaust gases of gasoline vehicles *Innovative Technologies in Engineering* VII International Scientific Practical Conference. Conference Proceedings. National Research Tomsk Polytechnic University. 2016. C. 12109.

[7] Ertman J.A., Ertman S.A., Anisimov I.A., Chainikov D.A., Chikishev Y.M. Complex characteristic estimation of vehicle adaptability for low temperature operation conditions // *Research Journal of Pharmaceutical, Biological and Chemical Sciences*. Volume **6**, Issue 1, 2015, Pages 1761-1770.

[8] Oprešnik, S.R., Seljak, T., Bizjan, F., Katrašnik, T. Exhaust emissions and fuel consumption of a triple-fuel spark-ignition engine powered passenger car // *Transportation Research Part D: Transport and Environment* - Volume **17**, Issue 3, 2012, pp. 221-227, doi: 10.1016/j.trd.2011.08.002.

IOP Conf. Series: Materials Science and Engineering **582** (2019) 012024 doi:10.1088/1757-899X/582/1/012024

The Assessment of the Level of Pollution of Slime Pits with Heavy Metals

O A Nemtsova [1,a], I I Romantsov[1], I I Avdeeva[1], P N Tkachenko[2], A V Nekhorosheva[3]

[1] National Research Tomsk Polytechnic University, 30, Lenin av., Tomsk, 634050, Russia

[2] Civil Defence Academy EMERCOM of Russia, Novogorsk district, Khimki, Moscow region, 141435, Russia

[3] Ugra State University, 60, Kievskaya st., Khanty-Mansiysk, 628007, Russia

E-mail: [a] nemtsova@tpu.ru

Abstract. In drilling and running oil and gas wells the great amount of toxic waste is formed and in the surface of the lithosphere and hydrosphere the great amount of different chemical elements and compounds access, they are needed to be isolated from the environment. The most useful way of isolation is their utilization in the slime pits on the wells territory. Heavy metals are dangerous waste in the drilling slime; they accumulate in soil and under some conditions they in form water-soluble parts and migrate in to the soil. The aim of this paper is to organize the chemical and analytical research of the amount of heavy metals in the drilling slime, and research the mechanism of their spreading on different depth and square in the slime pits typical for the oil site placed in KHANTY-Ugra.

Introduction

The intensiveness development of the mineral-material complex leads to the annual growth of the drilling volume first of all it is connected with searching and prospecting of the mineral resources in the production of liquid, gaseous and mature mineral resources in the bore hole use.

The borehole drilling influence negatively on the engineering and on all the components of the environment. The most dangerous and negative influence is subjected on the natural eco-system, where is the storage of the drilling waste; the imperfective aspect of the drilling technologies and utilization of the drilling sludge. The placement of the drilling objects (in the environment) containing toxic substances in the environment – is the main reason for the deterioration of the environment on the drilling territories.

On the territories on the West Siberia for a meter of the drilling work form approximately 0.2 – 0.6 m^3 of the drilling sludge. The less index is marked with the technologies with high recycling degree. For the 2000 meters` wells, the volume of the drilling sludge in the West Siberia is 500 m^3 in average [3].

Therefore, on the great volume of the drilling sludge storage on the territory of Russian Federation and their ecological danger to the environment, till now, there are no any effective technological resolutions on their utilization. That`s why it is essential to take the complex quantative mark on the negative influence of the drilling sludge on the environment and work new methods of utilization and prevention out and using drilling sludge.

The drilling sludge waste are the most dangerous for the objects of the environment, they keep and store on the territories of the bore-holes, as a rule, in the foundation pits-slime ponds, organized in the mineral or earth banks. As a rule, while designing 5 types of slime pits are foreseen with the general volume of 1000 m^3 for the drilling sludge, for collecting the drilling sewage; for collecting the fluide formation; for collecting household sewage; for the water sediment placed near the water scoop; for the horizontal outer flame [6].

The waste consists of the wide range of deposits of mineral and organic origin, presented by the materials and chemical reagents, used for the preparation and working chemical treatment for drilling solutions.

The drilling solutions, in their turn, are of water state and consist of 85 - 89 % bentonite clay – 10 – 11 %. The last 1 – 5 % may consist of different libricative, antiseptic, foam suppressor, anti-filtrate and hydrophobizing liquids [1].

If the hydro-isolation of the bottom and the walls of the slime pit poor, the pollution of the territories takes place. According to the bio-testing results, the drilling waste belongs to the IV class dangerous compounds [5]. The waste keeping in slime pits depend on the technological drilling features and the compound rock in the bottom in the area of slime pits the high concentration of chloride, oil products and heavy metals: Cd, Pb, Zn, Cr and others are mentioned.

The objects and the research methods

The test specimens of the drilling slime were chosen as the objects for researching from the typical field in KHANTY-Ugra. The test specimens were taken in winter.

The researching pit of the first site section of the following size: 90 m length and 50 m width. The depth of the slime pit was 350 sm. The power of the waste was 280 – 290 sm. The width of the ice in the pit was 70 sm. The depth of the testing level at the time of taking samples was between 80 – 150 sm from the one side and between 80 – 90 sm from the other side.

During the drilling the slime disposal to the pit occurred on the sludge setting tank at a distance of 3 – 6 meters of the edge diking of the pit and at 5 – 7 meters height from the bottom.

On the field the cluster`s pit consists of 3 sections. Production oil wells are drilled in groups of 4 wells. The distance between the wells in groups – 5 m, between wells` groups – 15 m. The collection of the drilling sludge occurs on the right part of the pit. The first point of dumping of the 1 group, the first well is at a distance of 38 m from the left edge of the clusters` pit (picture 1).

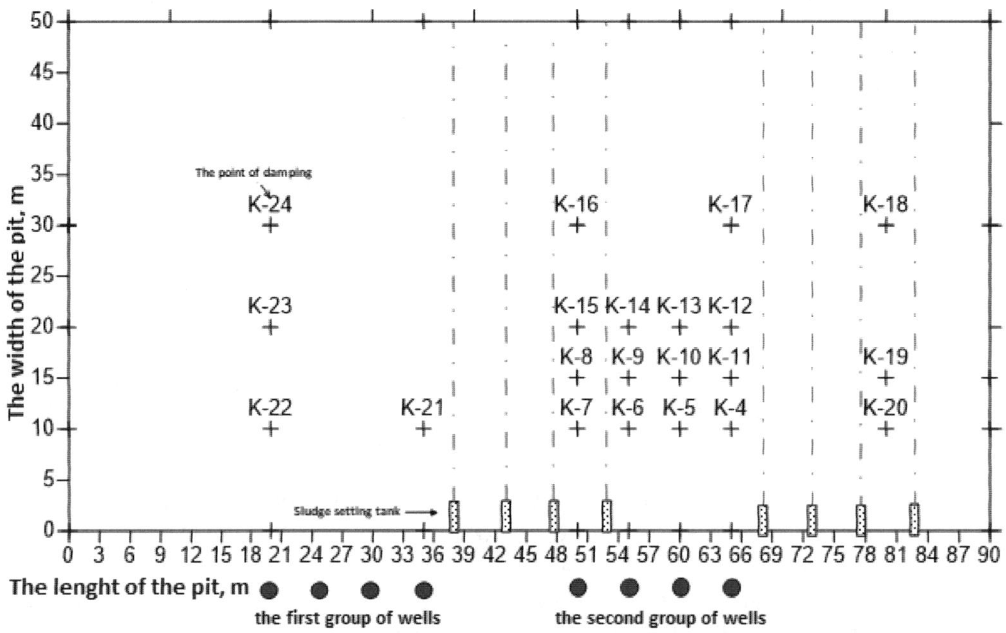

Picture 1. The scheme of screening taking sludge samples in the first section of the cluster`s pit № 12

The slime samples were taken by the self-constructed sampler. The construction of the sampler allowed taking samples of different opportunities forming them according to the depth, keeping structural features of the sludge, its different features (picture 2).

IOP Conf. Series: Materials Science and Engineering **582** (2019) 012024 doi:10.1088/1757-899X/582/1/012024

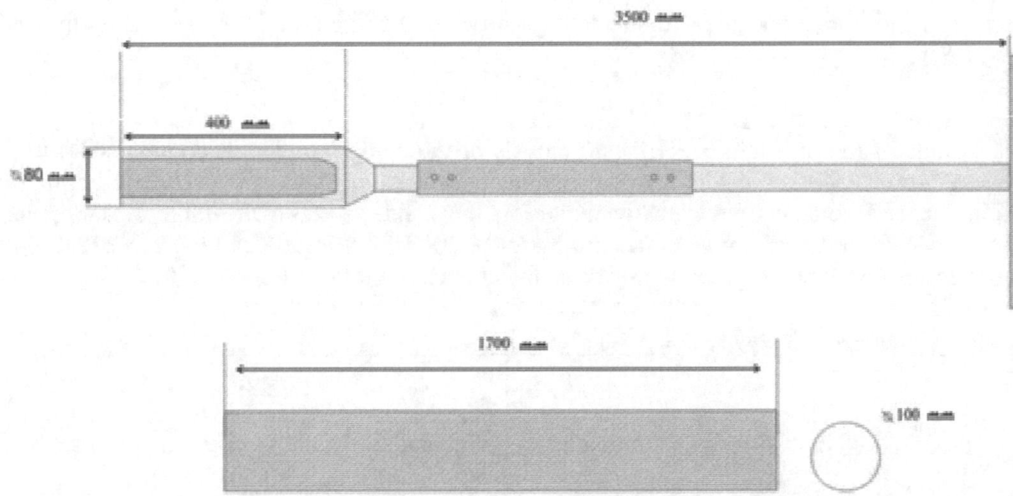

Picture 2. The sampler of the sludge

All in all, from the first pit were taken 152 sludge samples out of the 21st point. The general scheme of taking sludge samples at different depth of the pit is given on the picture 3.

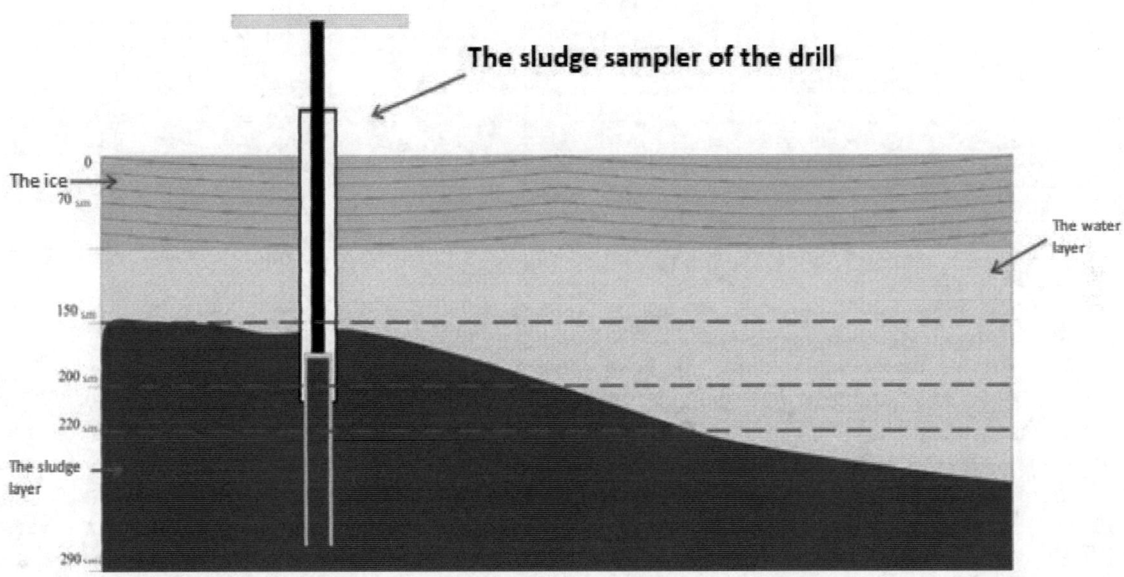

Picture 3. The scheme of taking sludge samples from different depth of the pit

The sludge sample was air-dried, freed from different parts, rubbed to power and winnowed it. Then it was weighed and mixed with bi-distilled water for 30 minutes, and then it was filtered.

The extraction of the mobile metal forms was done by the acetate-ammonium buffering solution with pH 4.8. From the extract of the solution the alloy quote was taken. For each part of the experiment the testing probe was prepared, which was lead through for each part of the experiment, the test probe was prepared and it had been done during the procedure of the analysis, it was taken into the account at the final element calculation in the probes. In each probe the scandium solution was added (1 mg/gr) for the calculation of the matrix influence and the drift of the instrument.

The determination of the volume of Ni, Cu, Pb, Zn, Mn was done by the atom-emission method of spectrometry with the inductance connected plasma on the Optima 2000 DV (Perkin Elmer Instruments, USA).

Results and discussion

The results of the quantative chemical analysis of the sludge probes in comparison with the positive concentrations (MPC – maximum-permissible concentration, APC – approximate-permissible concentration) general content of the heavy metals in the soils, given in the table 1. The general characteristics of MPC and APC are given according to the HS 2.1.7.2041-06 and HS 2.1.7.2042-06. In all probes the volume of the heavy metals prevailed the characteristic of MPC.

Table 1. Contents of heavy metals in sludge probes

The code of the point	Ni, mg/kg	Pb, mg/kg	Cu, mg/kg	Mn, g/kg	Zn, mg/kg
K-4	22.9	15.2	22.9	1.1	41.5
K-7	25.0	16.9	65.4	1.2	320.4
K-8	19.2	100.8	85.5	0.7	406.4
K-12	22.1	60.4	61.3	0.8	296.2
K-15	33.0	47.9	40.6	0.9	114.0
K-16	23.3	34.5	35.6	0.7	119.6
K-17	25.0	43.9	50.7	0.8	216.8
K-18	18.8	32.1	68.2	0.8	366.6
K-20	25.1	77.7	51.2	1.1	267.6
K-22	16.0	39.8	18.9	0.4	55.4
K-23	21.5	38.4	28.7	0.7	98.0
K-24	20.1	22.3	23.7	0.9	61.5
Average	22.9	50.5	49.7	0.9	219.9
MPC the mobile form, mg/kg	4.0	6.0	3.0	80 mg/kg	23.0
APC the general content	80.0	130.0	132.0		220.0

The volume of the Cu in the analyzed probes is being changed widely and give rise to oscillations on the pit`s square in 3.5 – 8 times. Maximum concentrations are mentioned on the right-side of the pit in the middle and upper layers. It can speak of the fact, that the main source of Cu and other elements are the mountain rock. The main sources of Cu are sulphide ores and cooper sandstones. 200 minerals consist of cooper. In waters the cooper contents consist of approximately from several milligrams or to few tens of milligrams to the litre. Cooper is a strong complex element and can occlude to the weighted substance by the absorption on the hydroxides` metal surface (Fe, Al, Mn), the ionic exchange with the clay minerals and the interaction with other high-molecular connections on the surface of the weighted particles [12].

The high concentration of manganese on the square of the pit is firstly explained by the regional specific feature of KHANTY-Ugra – the high excess abundance of manganese in the rock. But even with the high concentration of manganese, its maximum quantity is marked in the places of sludge dumping to the pit.

The main sources of manganese coming to the surface waters are the iron-manganese ores and some other minerals and many other sources (sewage from the on riched plants, metallurgical factories and other chemical plants). The great amount of manganese is formed as a result of disappearing of living beings the migrations abilities of manganese depends on many factors, such as the reduction-oxidation situation, pH, complex substances, the intensity of the absorption processes on the weighted particles.

The greatest Zn concentrations are mentioned in the middle and in the upper layer in the right side of the pit. Hypothetically, Zn transition is mostly the result of the natural agents – sedimentary rocks and mineral bed waters. Secondly, in some cases, in dominated sources is the managing man`s activity. The greatest Zn part is in the water – aqueous solution and its solubility grows with the growth of pH. The main place in its migration takes its complex compounds. The sorption processes are typical for Zn especially on clay particles, Fe, Al hydroxide, silicate minerals.

The greatest Pb concentration is in the upper and middle layer, of dumping sludge in the pit. The difference in the concentration in these layers is in the 14 – 20th times. In nature Pb is met in the endogenic and exogenic minerals. The dissolution of these minerals is one of the Pb sources in the surface waters. Buy now the main source is the anthropogenic sources. The Pb content in clean waters can be from the decimal part to the few milligrams in the litre. The Pb solubility depends on pH and is greatly dropped in the alkaline solutions. It is typical for Pb to have the high ability to form complex connection and the absorption based on the metal hydroxide and clay particles [12].

The Ni content is quiet enough on the square of the pit and on its depth. The difference in concentration is twice. Ni is met in nature mainly in the arsenic connections or sulphur connections. The nickel concentration in surface waters is from the milligram parts to 10 – 20 milligram in the litre. Ni contents are rich in waters with Ni rock. The main Ni source is the people`s activity (chemical plants, burning of the fuels). Among heavy metals, Ni is characterized as the metal with the middle complex its greatest part is migrated in the weighted state [12].

Conclusion

The result of the experiment is the following: the drilling sludge in the pit have marked non-heterogeneous allocation of the heavy metals. As a rule, maximum concentrations of the toxic elements in the pit are in the places of the sludge dumping from the deep horizons with gradual lowering of the concentrations the further one gets from these points. The left part of the pit the drilling consists of much less heavy metals. The deep layer of the drilling sludge in this part is less. Both effects are explained by the fact that the sludge extracted during the drilling the first group wells` conductors, firstly spreading uniformly in the pit and form the outcome layer in the left side of the pit. The conductors` sludge has low concentration of the heavy metals.

Taking account of the non-heterogeneous allocation of the pollution in the pit, it is recommended to use slightly polluted drilling sludge from the left part of the pit (mainly out of conductors` drilling) for the preparation of the disposal mixture with sand and peat. The remained part of the sludge is useful to reduce using the method of forming or mix non- polluted or more polluted layers before preparing the solid mixture. It is necessary to test the out coming content of the heavy metals for the ecological safety of putting sludge waste in the environment, soil.

References

[1] Saksonov, M.N., Balayan, A.E., Barkhatova, O.A., Kashina, N.F., Stom, A.D. Biotesting methods in assessment of toxicity of boring solutions and slimes at oil and gas production. International Multidisciplinary Scientific GeoConference Surveying Geology and Mining Ecology Management, SGEM. 2017. 17(14). Pp. 555-562.

[2] Soromotin, A.V. Ecological consequences of different stages of the development of oil and gas deposits in the taiga zone of the Tyumen' oblast. Contemporary Problems of Ecology. 2011. 4(6). Pp. 600-607.

[3] Solodovnikov, A.Yu. Results of ecological monitoring of natural environments of territory of new Surgutneftegaz OAO deposits. Neftyanoe khozyaystvo - Oil Industry. 2007. (2). Pp. 120-125.

[4] Solodovnikov, A.Yu. Transformation of tundra and north-taiga ecosystems at drilling sites construction. Neftyanoe khozyaystvo - Oil Industry. 2006. (2). Pp. 112-115.

[5] Klimova, A.A., Mishunina, A.S., Azarova, S.V., Fominykh, D.E., Yazikov, E.G. Determining the toxicity of drilling muds using the methods of biotesting, case study of Tomsk region the territory. Neftyanoe Khozyaystvo - Oil Industry. 2018. (4). Pp. 108-111.

[6] Pichugin, E.A., Shenfeld, B.E. To the issue of differences of drilling sludge and oil sludge. Ecology and Industry of Russia. 2017. 21(7). Pp. 14-19.

[7] Ruposov, V.L., Alexejenko, V.V., Mashovich, A.Y. On the technology of utilizing drilling sludge for the purpose of building material production. IOP Conference Series: Earth and Environmental Science. 2019. 229(1), 012024.

[8] Oreshkin, D.V., Chebotaev, A.N., Perfilov, V.A. Disposal of drilling sludge in the production of building materials. Procedia Engineering. 2015. 111. Pp. 607-611.

[9] Zhu, Z., Li, J., Xie, S., Yuan, G. Research and application on resource utilization technology of oily waste drilling fluid. Petroleum Science and Technology. 2011. 29(14). Pp. 1470-1481.

[10] Gorlenko, N.V., Timofeeva, S.S. Assessment of environmental damage from oil sludge to land resources in the Irkutsk region. IOP Conference Series: Earth and Environmental Science. 2019. 229(1), 012019.

[11] Ramirez, D., Kowalczyk, R.M., Collins, C.D. Characterisation of oil sludges from different sources before treatment: High-field nuclear magnetic resonance (NMR) in the determination of oil and water content. Journal of Petroleum Science and Engineering. 2019. Pp. 729-737.

[12] Linnik P.N., Nabivanets B.I. Forms of migration of metals in fresh superficial reservoirs. Leningrad: Hydrometeoizdat Publisher. 1986. 511 p.

[13] Pichugin E.A. Estimation of influence components included in the drilling sludge on soil KHANTY-Ugra. Bulletin PNRPU. Chemical technology and biotechnology. 2015. No. 2. Pp. 75 – 83.

[14] Soromotin A.V., Pislegin A.V. Heavy metals in bottom sediments of sludge pits in exploration wells in the Western Siberia. Geoecology. Engineering geology. Hydrogeology. Geocryology. 2015. No. 6. Pp. 514 – 520.

[15] Golybev E.V., Soromotin A.V. Structure and properties of drilling sludge of Western Siberia. World of science, culture, education. 2010. No. 6 (25). Pp. 319-320.

[16] Avdeeva I.I., Nekhorosheva A.V., Kirzhakov I.F., Akhmedzhanov R.R. Environmental hazard assessment for drilling slime pits in the open environment of KHANTY-Ugra. XXI century. Technosphere Safety. 2016. V. 1. No 4. Pp. 39 – 47.

[17] Solodovnikova A.U., Soromotin A.V. Experience of drilling waste utilization in Khanty-Mansi autonomous district – Yugra. Environmental protection in the oil and gas sector. 2015. No. 12. Pp. 44-48.

IOP Conf. Series: Materials Science and Engineering **582** (2019) 012025 doi:10.1088/1757-899X/582/1/012025

Relevance for Using Machines Appropriate to Strip Tillage

V V Myalo [a], E V Demshuk [b], D E Kuzmin [c], A S Soyunov [d], U K Sabiev [e]

Federal State Budgetary Educational Institution of Higher Education "Omsk State Agrarian University named after P.A. Stolypin 644008, Omsk, ul. Institutskaya Square,

E-mail: [a] vv.myalo@omgau.org, [b] ev.demchuk@omgau.org,
[c] de.kuzmin35.06.04@omgau.org, [d] as.soyunov@omgau.org, [c] uk.sabiev@omgau.org

Abstract. To increase the quantity and quality of any product type, it is necessary to upgrade machines, improve production processes, and minimize costs. The treated surface density, the amount of removed and transferred material should be minimal, but sufficient to create optimal conditions for the subsequent operations implementation of the technological chain, which requires the production processes adaptation to specific conditions. Therefore, the technology optimization and the machines working bodies improvement is a relevant area for research.

1. Introduction

Agricultural production process modernization and improvement is quite relevant taking into consideration the current increase in fuel prices and other production cost growth. Soil preparation is considered to be the most energy-intensive technological process for growing crops. The following block diagram represents tillage technology (Figure 1):

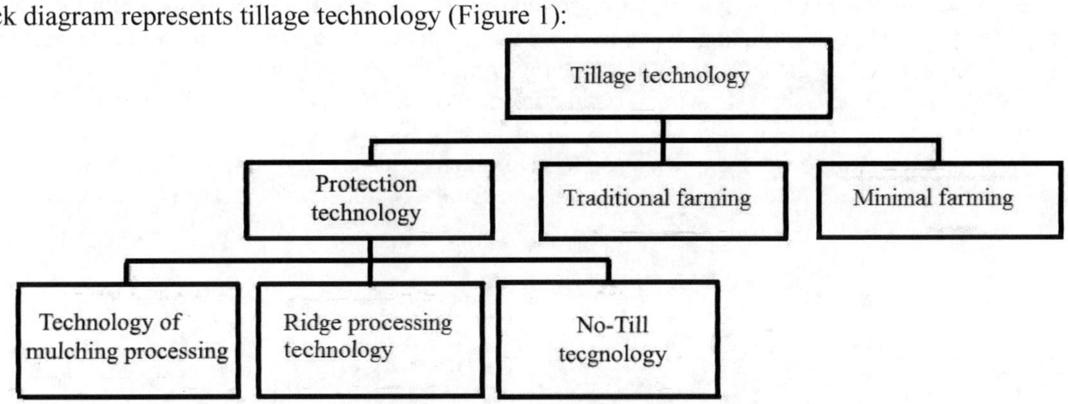

Figure 1 - Tillage technology

The tillage technologies are divided according to the soil impact degree:

• conservation technology - the technology by which soil conservation tillage is performed;

• traditional farming - where there is a maximum impact on the soil, since it is based on a plow;

• minimal farming - when there is minimal impact on the soil, and its preparation is carried out directly during sowing.

Processing method choice depends on the soil type, the ratio between crops in crop rotation, climatic conditions, infectious pressure of pathogens living in the soil, the prevailing form of organic fertilizer, wind and water erosion danger. The degree and depth of soil compaction propagation, the compaction depth and the area share of wheel tracks, the moisture content and carrying soil capacity, the quantity, distribution and crop residues properties, as well as the weed type and density should be taken into account.

Conservation technology can be divided into 3 tillage technologies (Figure 1):

• technology of the mulching processing, when plant residues accumulate on the field surface. The latter, in turn, protects the soil from wind and water erosion, and also contributes to the moisture

IOP Conf. Series: Materials Science and Engineering **582** (2019) 012025 doi:10.1088/1757-899X/582/1/012025

retention in the soil during hot summer days, when the soil warming in the upper layer can reach temperatures of 50°C and above.

• ridge processing technology allows to delay melt water flow, which carries out the fertile particle removal from the soil surface. Also ridges contribute to the rapid soil warming in areas with insufficient sunny day number.

• no-till technology - a technology that minimizes the impact on the soil while sowing or planting a crop. In the period between the change of predecessors, the soil is not processed.

According to the definition of the company Conservation Technologie Integration Center (CTIC), which is the center for the introduction of preservative technology, various groups of technologies classification for growing crops is presented in the following form.

2. Results and Discussion

Traditional farming is a plow technology, an intensive tillage technology. The minimum treatment - 15 ... 30% of the soil surface remains covered with crop residues, the fight against the erosion formation is carried out. Technology ridge tillage (Ridge-Till) - ridges are created on the field surface, crop residues remain between the ridges. The technology of mulch tillage (Mulch-Till) - the soil surface before sowing is treated with fully deep-rippers, cultivators or disc harrows. No-Till and Strip-Till: tillage methods where the soil remains untreated between the harvest and the sowing company. In the strip tillage technology process, narrow bands are processed in the autumn period, which allows you to simultaneously apply mineral fertilizers [1].

The band method is designed to reduce the level of tillage, but at the same time it contributes to maintaining the conditions necessary for plant growth. According to A. McGuire, the band tillage technology is promising, due to the fact that when cutting strips, the untreated surface will be about 70-80% of the field total area [2]. Strip-Till strip farming technology is suitable for many regions of Central Russia. The following can be distinguished from this soil-climatic zone features: a thin fertile soil layer, an abundance of weeds with a broad botanical composition, low fertility with a low content of organic matter, swimming soils, a short growing season compared with Western Europe. Scientists have noted a number of benefits from the technology Strip-Till use: reduced labor time and the technological operations number, strip loosening allows you to get higher yields when used properly [3, 4, 5].

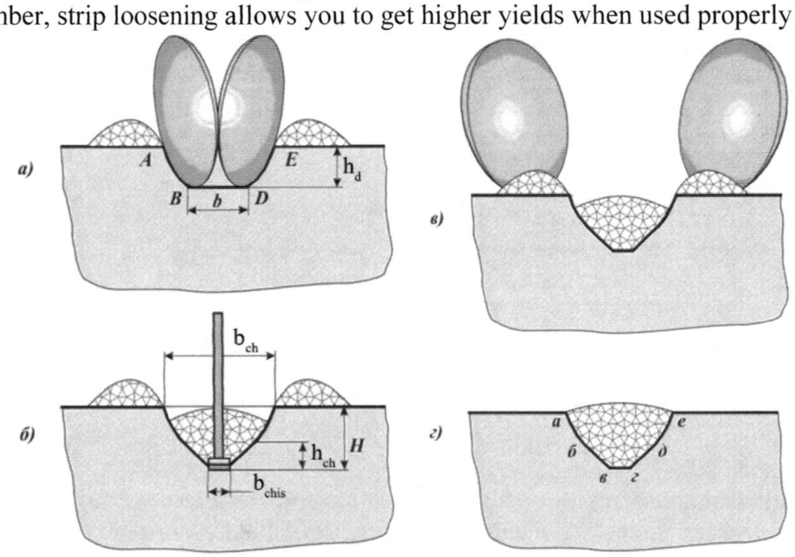

Figure 2. Strip-till strip tillage process flow diagram a - spherical disks loosening; b - chiselling; c - treated strip closing; g - treated strip horizontal section.

When processing the soil using the Strip-till technology, vertical volumetric loosening is reduced to three mandatory operations:

- soil loosening at a depth hd with its excavation and sowing strip formation of width b (Figure 2, a);
- open furrow chiselling by the working body with a chisel of width bchis to depth hch (Figure 2, b) provided that the strain width bch does not exceed the strip width b processed by the discs (bch ≤ b);
- processed strip closing by disk working bodies with subsequent rolling (Figure 2, c).

Work on this technology also has negative consequences in the form of soil layers mixing, abrasion of cleaners and closing rows organs microparticles and working surfaces. That contradicts the resource-saving technologies basics, obliging to carefully treat the soil, to preserve it [6].

In the Volgograd region territory, the strip tillage technology has been used since 2012 on sunflower and maize. It is southern low-power black soil climatic zone, with an average annual rainfall of 360...470 mm. The yield increase, when using this treatment, was about 19%, while the high rates of oil content remained [7]. The strip tillage allowed to increase the moisture absorption rate by 3...5 times, to improve the surface soil layers aggregates stability by 30%, to increase the soil organic matter by 0.1-0.15% per year.

In the Republic of Bashkortostan strip cultivation in the cultivation of maize and sunflower has been used since 2011. Geographically, the Republic of Bashkortostan is located in the Southern Urals and in the Ural Region, the climatic conditions within the republic are differentiated due to the presence of the Ural Mountains. Due to the saving technologies use in the farms of the SEC "Krasnaya Bashkiriya" Abzelilovsky, GUSP "Tavakan" Kugarchinsky, MTS "Zirgansky" Mezeuzovsky districts of the Republic of Bashkortostan, they experienced a significant increase in the soil fertility yield and reproduction [8, 9].

Saving agriculture technologies were introduced in the Belgorod region, where the climate is temperate continental with snowfalls and thaws in winter and hot, often with droughts and hot dry winds, in summer. The advantages of the Strip-Till tillage technology over the No-Till direct sowing technology were obtained: better plants root system development, earlier sowing start possibility [10].

In Germany, the Strip-Till tillage technology has been applied for the past 10 years. This technology advantages compared with the traditional treatment, are as follows: soil natural fertility is preserved; the possibility of growing row crops on slopes from 3 to 5 degrees, with the mandatory cultivation of intermediate crops; soil erosion is being fought; crop yield increases by 10...20%; reduced costs of mineral fertilizers, due to their precise introduction; reduced fuel and lubricant costs; reduced labor costs.

After market and relevant literature studying, we can conclude that initially machine development for the strip technology was carried out in a simpler way. The machines for ordinary tillage that existed at that time were refined and adapted to the strip technology. Initially, such machines were used only for row crops (corn, soybean, sunflower, etc.) [8]. The gun frames were borrowed, and the corresponding working bodies were already installed on them. Further modernization was aimed at accuracy improving of the soil topography copying, as well as characteristics improving of the technological process. At the same time, methods were developed for moving along individual working body depth. If we classify existing machines by this parameter, we can distinguish three types:

Type I: Cutting and loosening working bodies are rigidly fixed to the frame. The frame itself over the entire working width is recessed due to the adjustment of the support wheels. Clearing, combing and grinding working bodies, due to their design peculiarities, are able to repeat the soil relief. Car models produced by Unverferth, Carrotech, Schlagel, JohnDeere and BlueJet are widely used in agriculture.

Type II: A subframe is mounted on each machine section, on which individual working bodies are mounted. Due to this, this type gains advantages over the first type due to more accurate soil relief copying by each separate machine section. Necessary working bodies can also be installed and adjusted on this subframe (Figure 3). Manufacturers of this design machines: Strip-Cat, Orthman, B & H manufacturer, BighamBrothers, Hiniker, Carter, Yetter, Kongskilde, and Sunflower / Agco. This model is the most popular presented on the market.

Type III: Based on its own system, Kuhn / Krause has developed a machine that, like in type I, has a cutting tool fixed to the machine frame. The working body, which purpose is loosening (for the equipment of this company type, Krause is the rack) is attached to the machine frame with a parallelogram (like type II), and the grinding working body leads the tool in depth (Figure 4). Such a functional principle is a distinctive model feature of the third type. Accordingly, the grinding working bodies should be made of wear-resistant materials. The exclusion from the grinding working body section will directly lead to a change in the loosening working body depth, which will complicate the regulating process such processing.

Tools for band tillage, common in the Russian Federation: Orthman (1tRIPr), KRAUSE (Gladiator), as well as their domestic counterpart of DorAgroMash Orlik are quite expensive, which affects the final products price.

Figure 3. Orthman (1tRIPr)

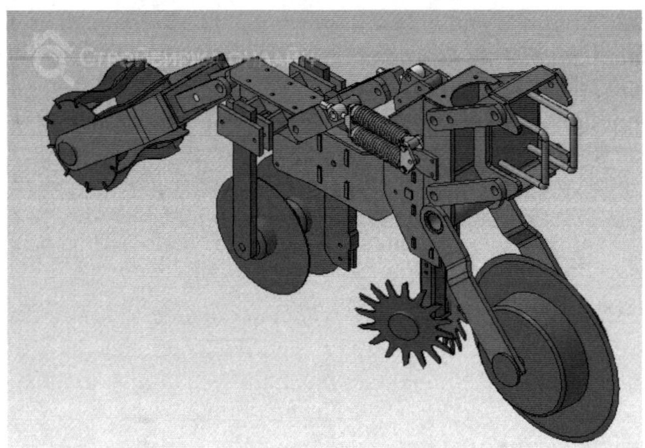

Figure 4. Instrument "Orlik" company DorAgroMash

The tools presented do not allow for the mineral fertilizer application, which draws another operation, respectively, and fuel and lubricants cost, time, and affects the production unit cost. From the domestic widespread development of technical tools for the sexual tillage we can distinguish the Orlik tool from DorAgroMash company. It is presented in several variations and has a lower cost compared to foreign counterparts.

Table 1. Production technology impact on direct technical costs (sunflower)

Technology	Fuel costs, %	Agricultural machinery amortization, %	Tractor amortization, %	Operator's salary, %
Chizel, mechanics	100	100	100	100
Chizel, plant protection products	88.6	103	94.2	83.2
Strip-till	64.5	75.5	78	63

Based on the data in Table 1, it can be concluded that, in terms of technical costs, Strip-till technology is more profitable and more promising than other tillage types. Technical costs are affected by machines number reducing when using Strip-till technology, as well as improving productivity in tillage.

IOP Conf. Series: Materials Science and Engineering **582** (2019) 012025 doi:10.1088/1757-899X/582/1/012025

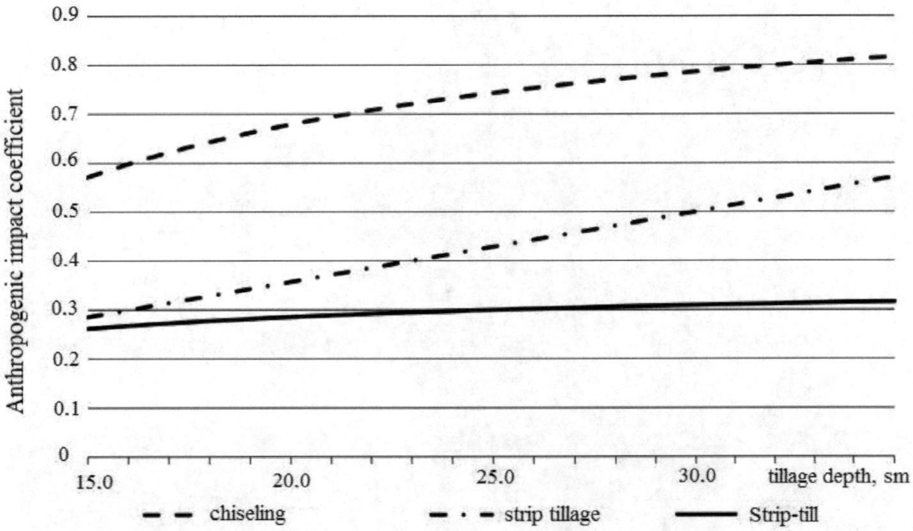

Figure 5. Anthropogenic impact on the soil, depending on the treatment type and depth

With regard to the territory of the Omsk region, which has 3 soil zones (Figure 6):

• forest-steppe zone occupies the largest part of the Omsk region territory (about 51%). Black soil-meadow, meadow-black soil and meadow soils, as well as alkaline soil complexes predominate in the forest-steppe zone soil cover. The average humus content varies within 4...5%. For agricultural needs in this zone, 3 744 000 hectares of land have been developed, of which 56% is arable land.

• The forest zone is the most limited by the agricultural land area (600 thousand hectares). This zone is lower than all in relation to the ocean level, and is also swamped, and as a result is over-humidified; groundwater occurs at a depth of 3 meters. The land fund is mostly represented by podzolic, marsh and meadow soils, which are mostly acidic, thin, with a low humus supply (up to 3%), phosphorus and nitrogen.

• The steppe zone covers only about 9% of the Omsk Region territory, but it is the most agriculturally developed. Treated land in individual farms reach up to 95%. In the land fund of the steppe zone, ordinary and southern black soil, as well as carbonate and solonets, predominate. The humus amount is within 4...9%.

Strip tillage in the rapeseed cultivation can be divided into two branches: a separate principle and a combined principle. A prominent representative of the first is the band tillage technology from AMAZONE. This principle can be represented as follows: band tillage with the possibility of mineral fertilizers, the direct seeds application is carried out by a separate operation. The second principle is adhered to by HORSCH, which allows for complex work, i.e. in one working cycle, the technician performs soil treatment, fertilizers application and seeds sowing. These principles advantages are presented in table 2.

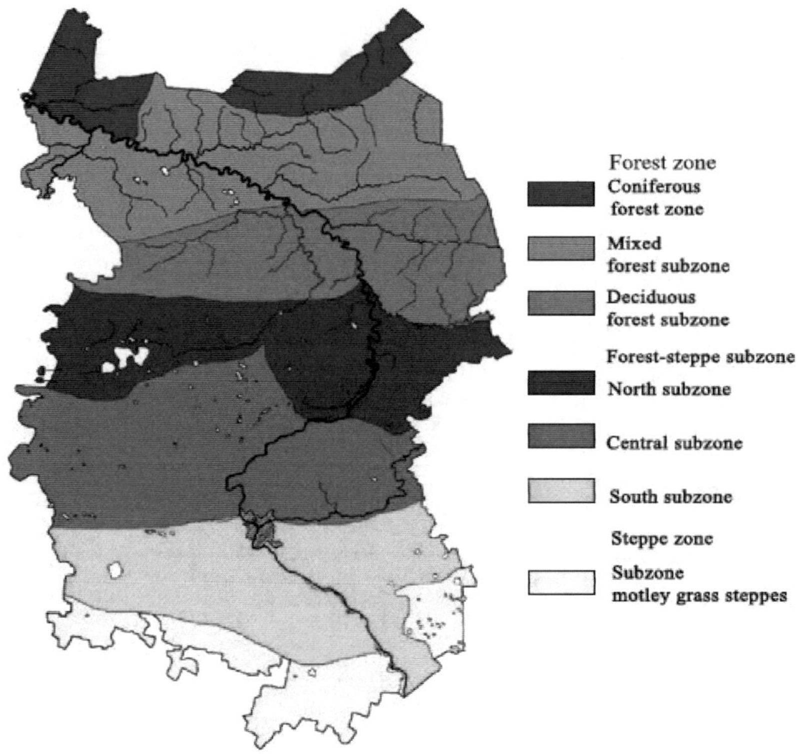

Figure 6. Omsk region climatic condition map

Table 2. The main advantages of band tillage separate and combined principles

Separate principle	Combined principle
Choosing the optimal time for making seeds	Labor saving
Sowing rate does not depend on tillage speed	Lower energy consumption
Tillage can be carried out around the clock, during daytime sowing	No need to use RTK-GPS steering system
Heavy soils (clay > 10%) can lower moisture before sowing	Weather conditions influence elimination between tillage and sowing

The band tillage use in the rapeseed cultivation in the soil and climatic conditions of the Omsk region territory can be divided into two branches: a separate principle and a combined principle.

3. Conclusion

In the Omsk Region, agrochemical soil condition monitoring is constantly carried out. Due to the humus content in the soil, acidity, the soil availability with phosphorus and potassium, pollution with heavy metals, as well as chemical elements, are taken into account. Thus, according to the regional agrochemical service data, practically 87% of arable land needs phosphorus-containing substances in combination with fertilizers additionally. The potassium content in the soils of the region is significantly higher than the phosphorus content in percentage terms. In the steppe zone and the southern forest-steppe, 98% of the land has a high and very high potassium supply. The minimum potassium content in the northern zone, and it is 87%. The annual humus loss as a result of agriculture reaches 0.4 t / ha. To date, there has been a decrease in the gross humus reserves in arable soils to 10...15% from the land development start, especially in the south of the region. Therefore, tillage with a strip method with the possibility of mineral fertilizer application is relevant.

The Strip-till technology use will reduce the fuel materials cost up to 35.5% compared to the classic chisel, and 24% relative to the minimum technology. Such a change in costs is due to the reduction in the number of technological operations of mechanical tillage.

IOP Conf. Series: Materials Science and Engineering **582** (2019) 012025 doi:10.1088/1757-899X/582/1/012025

According to the site "selhozportal.rf" the area of arable land in the Omsk region in 2018 is 286 thousand hectares. Tilled crops occupy about 40% of the area, no more than 20% of enterprises work on saving technologies. Such conservatism when choosing a cultivation technology is primarily related to the fact that the Omsk region is in the zone of "risky farming". The slightest mistake in technology application leads to significant losses during the harvest. Thanks to the rational band tillage use in tilled crops production can be achieved: increase yields, reduce fuel and lubricants and mineral fertilizers cost. To this end, it is necessary to develop tools that will allow strip tillage with simultaneous mineral fertilizers application, taking into account the agroclimatic features of the machine application area.

References

[1] Bauer B. Strip-Till, mit Streifen zum Erfolg / B. Bauer // DLG- Verlag, 2012

[2] Mc Guire A. High Residue Farming under Irrigation: Strip-Till / A. Guire // Washington State University Extension EM036E, 2011

[3] Dutzi Dr. S. Strip-Till-Drilltechnik der Zukunft. Erfahrungen und Ergebnisse aus Praxisversuchen / Dr. S. Dutzi // Saaten-Union Winterforum, Vortrag, 2012

[4] Bauer B. Platzierte DUngung- Starthilfe fur Ihren Raps, Gartersleben, Institut fur Pflanzengenetik und Kulturpflanzenforschung / B. Bauer // Top- Agrar Artikel, 2011

[5] Two-Cotton Sowing Soundry of Grain Crops with Different Mineral Fertilizer Level / E.V. Demchuk, U.K. Sabiev, V.V. Mylo, A.S.Soyunov, V.S. Koval, -. IOP Conference Series: IX International Scientific Practical Conference "Innovative Technologies in Engineering". - IX International Scientific Practical Conference "Innovative Technologies in Engineering", Published under licence by IOP Publishing Ltd ,Volume 1059, conference 1, http://iopscience.iop.org/article/10.1088/1742-6596/1059/1/012009

[6] Moisture and energy-saving technology of tillage and seeding in severe arid conditions / N.K. Mazitov, R.S. Rakhimov, Ya.P. Lobachevsky, R.L. Sakhapov, N.Kh. Galyautdinov, L.Z. Sharafiev // Technique and equipment for the village. 2013. №3. Pp. 2-6.

[7] Methodological approaches to the technologies and technical means choice in the grain production in Siberia / Yu.N. Blynsky, B.D. Dokin, O.V. Yolkin, N.M. Ivanov, V.L. Martynov // Siberian Bulletin of Agricultural Science. 2016. №2 (249). Pp. 105-109.

[8] The bioenergetik approach to evaluation of arable land fertility / M.G. Ishbulatov, R.A. Mindibaev, Kh.M. Safin, A.G. Baikov, I.R. Miftakhov, G. Baygildina, N.A. Zamanova, R.R. Khisamov, R.G. Yagafarov // Journal of Engineering and Applied Sciences. 2018. T. 13. № S10. Pp. 8353-8359.

[9] Basic Parameters Substantiation of the Cultivator Working Body for the Continuous Tillage in the System of Ecologically Safe Resource-Saving Agriculture / V V Myalo, O V Myalo, E V Demchuk, V V Mazyrov // IOP Conference Series: Earth and Environmental Science, Volume 224, Number 1 IOP Conf. Series: Earth and Environmental Science 224 (2019) 012023 https://iopscience.iop.org/article/10.1088/1755-1315/224/1/012023/pdf

[10] Trusov A.S. No-Till and Strip-Till technologies - main advantages (Experience of Grain Belogorye LLC) / A.S. Trusov // Science and technology achievement of agrarian and industrial complex, №12. - 2012

IOP Conf. Series: Materials Science and Engineering **582** (2019) 012026 doi:10.1088/1757-899X/582/1/012026

Optimization of the Work of Road Transport With the Help of Navigation Equipment on the Example of the Branch of "Mostostroy-11", JSC, TF "Mostootryad-36"

E Chikishev [1,a], A Fisher[1], V Alontsev[2]

[1] Motor transport operation, Industrial University of Tyumen, 625039, Tyumen, Russian Federation

[2] Department of sports improvement, Nosov Magnitogorsk State Technical University, 455000, Magnitogorsk, Russian Federation

E-mail: [a] chikishev_e@mail.ru

Abstract. The article deals with the control of fuel consumption in automotive engineering with the help of navigation equipment. As a result of the experiment on trucks VOLVO FMX 6x4 and Mercedes-Benz Actros 3346S it was found out that the actual fuel consumption is 2-16% lower than the standard. Thus, using the complex of navigation equipment in transport of the company allows not only to significantly reduce fuel costs, but also to optimize control over road flight performance, and also the regime of work and rest of the drivers. As a result of the economic calculation, it was established that the equipment costs are paid off in less than 1 month.

1. Introduction

The development of the organization of the transport process largely depends on the level of efficiency of the management activities of its management. At the same time, the costs of organizing transportation are formed from a number of different factors: the type of rolling stock, the price of fuels and lubricants, maintenance and repairing, staff salaries, etc. [1-6].

In order to be competitive in the market of services, motor transport enterprises need to optimize the cost of operating vehicles. One of the methods is the adjustment of fuel consumption rates. In our country there are several methods of accounting for such norms:

1. According to the order of the Ministry of Transport of Russia of March 14, 2008 No. AM-23-r (ed. No. IA-159-r of September 20, 2018) "On the implementation of the methodological recommendations "Consumption rates of fuels and lubricants for automobiles transport"".

2. With the help of electronic control of fuel consumption according to the order of the Ministry of Transport of the Russian Federation dated July 31, 2012 No. 285 "On approval of requirements for navigation means operating using navigation signals of the GLONASS or GLONASS / GPS systems and intended for the mandatory equipment of vehicles of category M, used for the commercial passenger traffic and category N used for the transport of dangerous goods".

In connection with the development of information technology, using of navigation equipment in transport to control fuel consumption (both online and offline) has an advantage over regulatory for several reasons [7-12].

In determining the actual consumption rates of fuels of a specific make and model of a car using order No. AM-23-r, many factors must be taken into account:

1. Natural and climatic conditions (ambient temperature, wind speed and direction, humidity, precipitation, etc.).

2. Road conditions (road category, type of roadway, slope and curvature of the road, height above the sea level, etc.).

3. Transport conditions (mass and type of cargo transported, use of a trailer or semi-trailer, speed conditions for cargo delivery, presence of traffic lights, traffic flow, etc.).

4. Technical characteristics of the vehicle (age, mileage, type of fuel used, etc.).

This effect is expressed in the form of appropriate surcharges to fuel consumption rates.

There is no need to take into account all the above factors if we use the navigation equipment, because the fuel consumption monitoring system shows the consumption in actual operating conditions.

But installing navigation equipment is a very expensive procedure. Accordingly, it is necessary to carry out research in the field of economic feasibility of using such equipment. Therefore, for motor transport enterprises engaged in the carriage of goods and passengers, automating the process of controlling fuel consumption is an urgent task, which makes it possible to optimize the cost of fuel.

An experiment in our work will be presented on the use of satellite monitoring of vehicles based on the Russian satellite navigation system GLONASS (Global Navigation Satellite System) for controlling moving objects.

The installation of such equipment allows in real-time do the monitoring of vehicle performance, including fuel consumption and the schedule of drivers. The research was carried out on the transport branch of the company JSC "Mostostroy-11" TF "Mostootryad-36" in Tyumen (Russia).

2. Methodology

For the effective operation of the navigation equipment the work of the vehicle fleet was automated with the help of the integrated transport information system "TIS-online". In fig. 1 is a screenshot of the main program window. This program keeps track of the movement of fuel, speed control, receipt and processing of applications from objects, electronic waybills with the possibility of automatic calculation of fuel consumption and driving time.

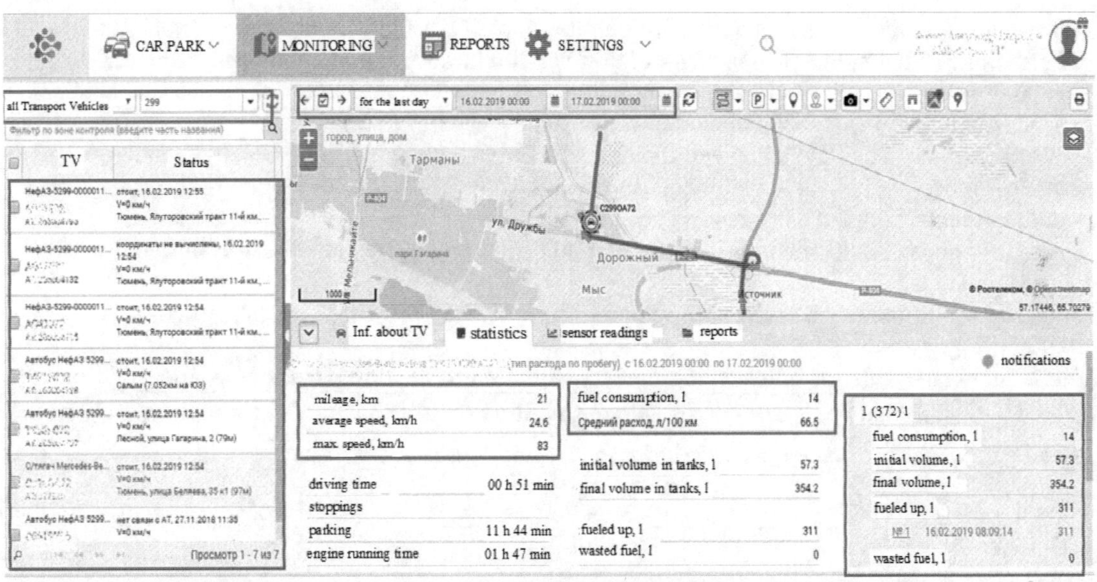

Figure 1 – Program "TIS-online"

The purpose of the research - the optimization of the auto enterprise by reducing the cost of fuel with the help of using the navigation equipment. The period of our research is from 08/01/18 to 12/27/18. There were four new cars of 2018, which were tested by the engine, components and assemblies within a month. Conducting an experiment on new cars minimizes errors in determining the actual fuel consumption caused by their technical condition.

There were two brands of cars on which the appropriate navigation equipment was installed - 2 VOLVO FMX 6x4 dump trucks (fuel consumption 52.4 l / 100 km) and 2 truck tractors Mercedes-Benz Actros 3346S (fuel consumption 65 l / 100 km). These vehicles delivered bulk goods (sand and crushed stone) along pendulum routes during the building of an automobile bridge in Tyumen.

In the experiment, a comparative analysis of the fuel consumption obtained according to the navigation equipment during operation and at the rate of the order of the Ministry of Transport No. AM-23-r.

3. The conduct of an experiment

As a result of the conducted research, according to the TIS-online software product, the actual values of fuel consumption for the four vehicles under study were established by months (August-December 2018 for VOLVO and September-December 2018 for Mercedes-Benz), their mileage was received, standard fuel consumption is presented according to the order of the Ministry of Transport of Russia No. AM-23-r, as well as the difference in fuel consumption between the actual and standard value. The error of measuring equipment according to the technical documentation is no more than 3%. This research presents the results taking into account the accuracy of the equipment.

Analysis of the operation of the first dump truck VOLVO FMX 6x4 is presented in Table.1.

Table 1. Summary table for mileage and fuel consumption of the dump truck VOLVO FMX 6x4 №1

Month	Distance, km	Standard fuel consumption (52.4 l + surcharge = 55 l per 100 km), l.	Fuel consumption for TIS-online, l	Difference, l
August	7815	4298	3955	343
September	9221	5071	4985	86
October	7947	4371	4255	116
November	6885	3787	3638	149
December	5483	3016	2783	233
Total	**37351**	**20543**	**19616**	**927**

The total surcharge to the fuel consumption rate for a VOLVO FMX 6x4 car is 2.6 l / 100 km.

The average actual value of fuel consumption per 100 km. for the first car was:

- in August - 50.6 l. (8% below the standard);
- in September - 54 l. (1.8% below the standard);
- in October - 53.5 l. (2.7% below the standard);
- in November - 52.8 l. (4% below the standard);
- in December - 50.7 l. (7.8% below the standard);

Thus, it has been established that for the first investigated car, for the observed period, the savings amounted to 927 liters.

Analysis of the operation of the second dump truck VOLVO FMX 6x4 is presented in Table. 2.

Table 2. Summary table for mileage and fuel consumption of the dump truck VOLVO FMX 6x4 №2

Month	Distance, km	Standard fuel consumption (52.4 l + surcharge = 55 l per 100 km), l.	Fuel consumption for TIS-online, l	Difference, l
August	7842	4313	3859	454
September	10087	5548	5325	223
October	8509	4680	4679	1
November	10418	5730	5208	522
December	5243	2884	2779	105
Total	**42099**	**23155**	**21850**	**1305**

The average actual value of fuel consumption per 100 km. for the second car was:

- in August - 50.4 l. (8.4% below the standard);
- in September - 49.4 l. (10.2% below the standard);
- in October - 50 l. (9.1% below the standard);
- in November - 49.9 l. (9.3% below the standard);
- in December - 53 l. (3.6% below the standard);

Thus, it has been established that for the first investigated car, for the observed period, the savings amounted to 1305 liters.

An analysis of the operation of the first Mercedes-Benz Actros 3346S truck tractor is presented in Table. 3.

Table 3. Summary table for mileage and fuel consumption Mercedes-Benz Actros 3346S №1

Month	Distance, km	Standard fuel consumption (65 l + surcharge = 70 l per 100 km), l.	Fuel consumption for TIS-online, l	Difference, l
August	1396	977	859	118
September	3227	2259	1884	375
October	3634	2544	2351	193
November	3442	2409	2087	322
December	**11699**	**8189**	**7181**	**1008**

The total premium to the rate of fuel consumption for a Mercedes-Benz Actros 3346S is 5 l / 100 km.

The average actual value of fuel consumption per 100 km. for the first truck tractor:

- in September - 61.5 l. (12% below the standard);
- in October - 58.4 l. (16.6% below the standard);
- in November - 64.7 l. (7.6% below the standard);
- in December - 60.6 l. (13.4% below the standard);

Thus, it was found that for the investigated car, for the observed period, the savings amounted to 1008 liters.

Analysis of the operation of the second Mercedes-Benz Actros 3346S truck tractor is presented in Table. 4.

Table 4. Summary table for mileage and fuel consumption Mercedes-Benz Actros 3346S №2

Month	Distance, km	Standard fuel consumption (65 l + surcharge = 70 l per 100 km), l.	Fuel consumption for TIS-online, l	Difference, l
August	2989	2092	1819	273
September	2564	1795	1633	162
October	1898	1329	1235	94
November	5371	3760	3207	553
December	**12822**	**8976**	**7894**	**1082**

The average actual value of fuel consumption per 100 km. for the second truck tractor:

- in September - 60.9 l. (13% below the standard);
- in October - 63.7 l. (9% below the standard);
- in November - 65.1 l. (7% below the standard);
- in December - 59.7 liters. (14.7% below the standard);

Thus, it was found that for the investigated car, for the observed period, the savings amounted to 1082 liters.

4. Conclusions

After analyzing the fuel consumption of the investigated cars, it can be noted that the actual fuel consumption according to the navigation equipment and the TIS-online software product in all cases turned out to be lower than the standard. This allowed the company to save on fuel costs.

On average, the actual fuel consumption is lower than the standard by 2-16%, depending on the climatic, road and transport conditions.

At the price for diesel fuel 46 rubles. per liter (in prices of January 2019), for the research period, the company saved on the purchase of fuel:

- on the first car VOLVO FMX 6x4 - 42642 rubles.
- on the second car VOLVO FMX 6x4 - 60030 rubles.
- on the first car Mercedes-Benz Actros 3346S - 46368 rubles.
- on the second car Mercedes-Benz Actros 3346S - 49772 rubles.

The cost of installing navigation equipment per vehicle is 40000 rubles. Monthly subscription fee - 700 rubles. Consequently, the payback of using the electronic control of the vehicle is less than a month.

Currently, the company TF "Mostootryad-36" on the entire wheeled vehicles (230 units) installed such navigation equipment, which allows significant savings on fuel costs.

Besides the actual fuel consumption, this equipment allows to fix and control:

- the timing of the trip and the deviations from the specified route of the car;
- places of refueling vehicles, drain or excessive fuel consumption;
- vehicle traffic safety (speed, sharp acceleration and braking, mode of work and rest of drivers, etc.);
- planning and accounting of the passed technical services and car repairs, accounting of installation and removal of tires, the rechargeable battery, etc.

Installation of navigation equipment allows you to automate control over the operation of the automobile fleet of enterprises, which allows you to plan and optimize costs, as well as improve traffic safety.

References

[1] Beziat, A., Koning, M., Toilier, F. Marginal congestion costs in the case of multi-class traffic: a macroscopic assessment for the Paris region // Transport Policy, volume 60, November 2017, pp. 87-98, doi: 10.1016/j.tranpol.2017.08.008.

[2] Courtonne, J.-Y., Longaretti, P.-Y., Dupré, D. Uncertainties of Domestic Road Freight Statistics: Insights for Regional Material Flow Studies // Journal of Industrial Ecology, volume 22, issue 5, 2018, pp. 1189-1201, doi: 10.1111/jiec.12651.

[3] Liimatainen, H., van Vliet, O., Aplyn, D. The potential of electric trucks – an international commodity-level analysis // Applied Energy, February 2019, pp. 804-814, doi: 10.1016/j.apenergy.2018.12.017.

[4] Merchan, A.L., Léonard, A., Limbourg, S., Mostert, M. Life cycle externalities versus external costs: the case of inland freight transport in Belgium // Transportation Research Part D: Transport and Environment, volume 67, February 2019, pp. 576-595, doi: 10.1016/j.trd.2019.01.017.

[5] Mokhirev, A., Gerasimova, M., Pozdnyakova, M. Finding the optimal route of wood transportation // IOP Conference Series: Earth and Environmental Science, volume 226, issue 1, 2019, pp. 012053, doi: 10.1088/1755-1315/226/1/012053.

[6] Wolff, M.G.D.C., Caldas, M.A.F. A Model for the Evaluation of Brazilian Road Transport: A Sustainable Perspective // Journal of Advanced Transportation, volume 2018, 2018, pp. 5274789, doi: 10.1155/2018/5274789.

[7] Jiménez, F., Naranjo, J.E., Anaya, J.J., García, F., Ponz, A., Armingol, J.M. Advanced Driver Assistance System for Road Environments to Improve Safety and Efficiency // Transportation Research Procedia, volume 14, 2016, pp. 2245-2254, doi: 10.1016/j.trpro.2016.05.240.

[8] Kubáňova, J., Kubasákova, I. The introduction and impact of tachographs on road freight transport // Transport Means - Proceedings of the International Conference, volume 2018-October, 2018, pp. 1085-1090.

[9] Laranjeiro, P.F., Merchán, D., Godoy, L.A., Giannotti, M., Yoshizaki, H.T.Y., Winkenbach, M., Cunha, C.B. Using GPS data to explore speed patterns and temporal fluctuations in urban logistics: The case of São Paulo, Brazil // Journal of Transport Geography, volume 76, April 2019, pp. 114-129, doi: 10.1016/j.jtrangeo.2019.03.003.

[10] Ranacher, P., Brunauer, R., Van der Spek, S.C., Reich, S. A model to estimate and interpret the energy-efficiency of movement patterns in urban road traffic // Computers, Environment and Urban Systems, volume 59, pp. 152-163, 2016, doi: 10.1016/j.compenvurbsys.2016.06.006.

[11] Simdiankin, A., Byshov, N., Uspensky, I. A method of vehicle positioning using a non-satellite navigation system // Transportation Research Procedia, volume 36, 2018, pp. 732-740, doi: 10.1016/j.trpro.2018.12.098.

[12] Wang, W., Uehara, M., Ozaki, H. Proposal of navigation system based on system optimal traffic assignment // Advances in Intelligent Systems and Computing, volume 772, pp. 184-195, 2019, doi: 10.1007/978-3-319-93659-8_16.

IOP Conf. Series: Materials Science and Engineering **582** (2019) 012027 doi:10.1088/1757-899X/582/1/012027

Substantiating the Parameters of a Two-drum Roller for Tillage

I R Rakhimov [1,a], **N A Telichkina** [2,b], **O S Batraeva** [2], **A A Shabunin** [2], **S M Krasnozhon** [2]

[1] OOO "Chelyabinsk Compressor Plant", Russia
[2] Institute of Agroecology, South Ural State Agrarian University, Russia

E-mail: [a] ildarr@bk.ru, [b] telichkina76@mail.ru

Abstract. The article reveals the research urgency. A device for an experimental two-drum roller, a flow-chart of its operation, a design model and an analytical relationship for substantiating its main design parameters are described. The dependence of the horizontal component of the roller on its operating width at different diameters of the inner drum is shown. It is proved that to reduce energy costs and meet the agrotechnical requirements for tillage, the diameters of the inner and the outer drums (D_{k2} and D_{k1}) should be 0.25 and 0.5 m, respectively, with the number of straps on the outer and inner drums being n_{s1} = 16 pcs and n_{s2} = 30 pcs. The results of experimental studying the two-drum roller made according to the substantiated parameters are given. The experimental studies proved the ability of the two-drum roller to provide the due structural composition of the seed layer, with soil lumps exceeding 50 mm being totally absent, a better evenness of the soil surface as compared to the one obtained with the traditional type of treatment, earlier and friendly sprouting of spring rape and the yield increase by 21.7%.

1. Introduction

Tillage for crop production being the most energy-intensive process, the main task for agricultural engineering is to find ways to reduce costs, with agrotechnical requirements for tillage being met. Most researchers see the solution to this problem in using combined units that can perform a full range of pre-sowing tillage activities in one pass, as well as modifying the working parts [1-5].

Rollers as integral parts of any tillage and sowing unit are used for the main and surface tillage for grinding large lumps, leveling and mulching the soil surface, removing weed roots on the soil surface, creating seedbeds during pre-sowing treatment and creating the contact of seeds with the soil when sown [6-8].

The outer diameter of all types of rollers depends on the size of the lumps occurring on the soil surface after the main or surface tillage. According to researches, the roller diameter is to be determined by pinching the soil lump between the roller working surface and the soil surface without soil shafts being formed in front of it.

Depending on the purpose, type and condition of the soil, the diameters of lumps, as well as the requirements for soil tillage, one- and two-drum rollers with different diameters are used, with rods of various types and sizes at different angles to the direction of movement [9,10,11].

Research urgency. Despite the wide range of existing designs of tillage rollers, there is a need to construct a roller to provide high-quality pre-sowing treatment necessary for crops, especially with small seeds.

Research purpose: the traction resistance determination of a two-drum roller and the substantiation of its design parameters.

IOP Conf. Series: Materials Science and Engineering **582** (2019) 012027 doi:10.1088/1757-899X/582/1/012027

2. Methods

The experimental roller consists of two drums: the inner and outer ones. The inner drum is placed in the inner space of the outer drum without any rigid kinematic connection with it. It is connected to the cultivator frame with a hinged-powered device equipped with a spring loader. The operating surfaces of both drums are formed by one-side sharpened straps rigidly fixed in the side discs. The outer drum rotates due to the adhesion of the straps with the soil due to acting traction force, with the rotation of the inner drum being ensured by the friction force occurring at the point of contacting [12].

The two-drum roller with the straps of the outer drum compacts the pre-seed layer and destroys large soil lumps. Soil lumps passed between the straps inside the roller are treated with the inner drum straps. The soil treated this way is thrown onto the field surface to form a loosened mulch cover. Thus, the necessary differentiation of the seed layer concerning its density and structural composition happens.

The parameters for the outer roller drum being chosen, its kinematic and design parameters can be substantiated.

When operating, the following forces act on the two-drum roller (Figure 1): G_1 is the gravity of the outer drum, kN; G_2 is the gravity of the inner drum, kN; P is the resistance force of the roller acting along the lead, kN; R is the soil resistance force occurring when the roller straps enter the soil, kN; R_x is the force of interacting the roller with the soil, kN; T is the interaction force between the rollers, kN; F is the friction force between the rollers, kN; F_0 is the compression force of the loader spring, kN.

The horizontal component of the roller traction resistance Px, kN, depends on its design parameters and physical-mechanical properties of the soil

$$P_x = \left(\frac{G_1}{\sin\alpha} - \frac{R_x - \delta \cdot b \cdot l_{outerst} \cdot n_{s0} \cdot \cos\alpha_0}{f \cdot \sin\alpha} + \frac{\pi \cdot D^2_{k1inner} \cdot b \cdot c \cdot \rho}{4 \cdot \sin\alpha} \right) \cdot \cos\alpha \, , \qquad (1)$$

where α is the angle of the lead inclination, degrees, δ is the soil shear resistance coefficient; b is the operating width, m; $l_{outer\ st.}$ is the thickness of the outer drum strap, m; n_{s0} is the number of straps contacting with the soil, pcs.; f is the coefficient of the soil friction with steel; $D_{k1\ inner}$ is the diameter along the inner surface of the outer drum straps, m; c is the roller filling coefficient; ρ is the specific gravity of the soil, kN/m^3.

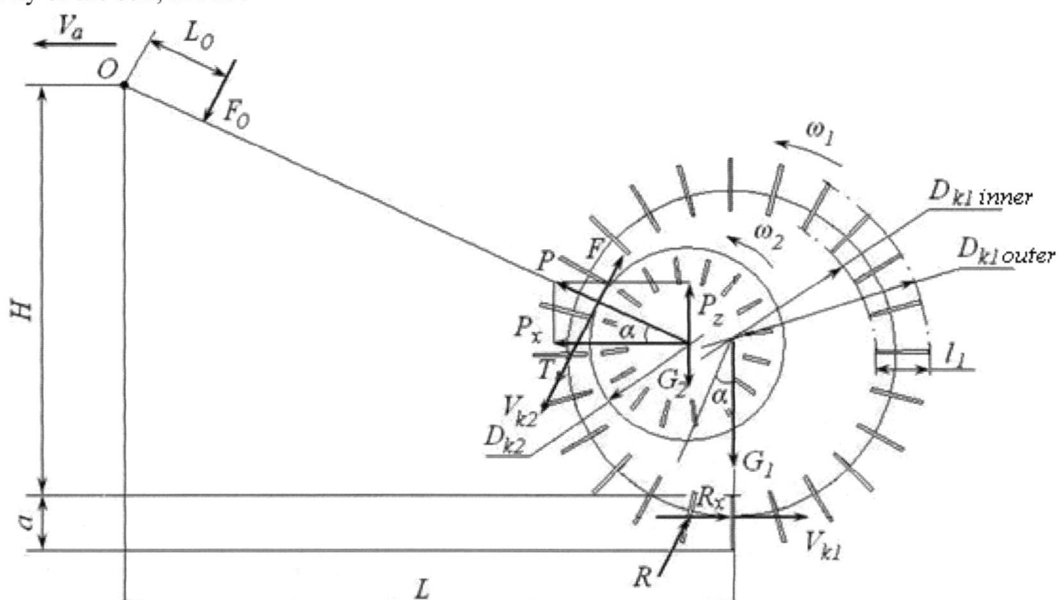

Figure 1 – The diagram for determining the forces acting on the two-drum roller

The horizontal component of the roller traction resistance is determined according to Formula (1) and its main design parameters are justified. The initial data for the calculations are given in Table 1.

Table 1 The initial data for the calculations

No.	Parameter	Measurement unit	Designation	Value
1	The roller depth	m	a	0.10
2	The angle of the lead inclination	degree	α	20.00
3	The specific gravity of the soil	kN/m^3	ρ	25.50
4	The soil shear resistance coefficient	кN/м^2	δ	10.00÷30.00
5	The roller filling coefficient		c	0.10
6	The outer drum rolling coefficient on the soil		η_{k1}	0.95
7	The inner drum slipping coefficient relatively to the outer drum		η_{k2}	0.90
8	The coefficient of the soil friction with steel		f	0.50÷0.80
9	The thickness of the straps of the outer and inner drums	m	$l_{outer\ st.}$, $l_{inner\ st.}$	$4 \cdot 10^{-3}$
10	The outer drum diameter	m	$D_{k1\ outer}$	2 D_{k2}
11	The inner drum diameter	m	D_{k2}	0.20÷0.35
12	The operating width	m	b	1.00÷3.00

3. Research results and their discussion

The results of the calculations are presented as graphs (Figure 2).

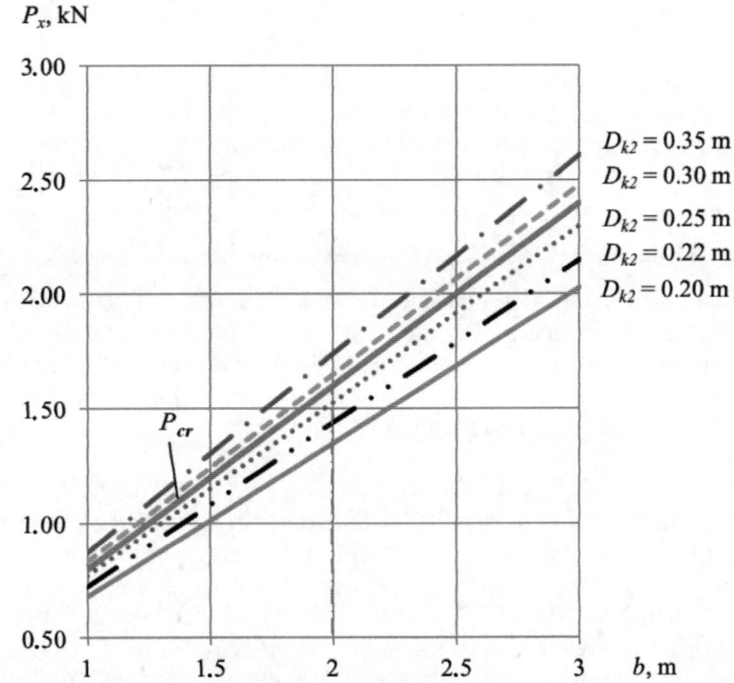

Figure 2 – The dependence of the horizontal component of the roller traction resistance P_x on its operating width b with various diameters of the inner drum D_{k2}

The specific resistance of a traditional roller with a non-smooth surface being k_{sp} = 0.6÷0.9 kN/m, its critical traction resistance P_{cr}, kN, (see Figure 2) can be defined as

$$P_{cr} = k_{sp} \cdot b \text{, кN,} \quad (2)$$

where b is the operating width, m.

To substantiate the diameter of the experimental two-drum roller, we accept the condition

$$P_x \le P_{cr}. \quad (3)$$

To fulfill this condition, the diameter of the inner drum is recommended to be D_{k2} = 0.25 m, with the diameter of the outer drum being $D_{k1\ outer} = 2D_{k2}$ = 0.5 m.

According to the agrotechnical requirements for the quality of pre-sowing treatment, there shouldn't be any soil lumps exceeding 10 cm in size, with lumps exceeding 2.6 cm being absent after treatment. The distance between the straps of the outer drum is recommended to be l_{s1} = 10 cm, l_{s2} = 2.6 cm. Then D_{k2} = 0.25 m and $D_{k1\ outer}$ = 0.5 m, their number being n_{s1} = 16 pcs on the outer drum and n_{s2} = 30 pcs on the inner one. The straps on the drums are arranged circumferentially, radially and regularly. The length of the outer drum straps protruding from the side discs is determined by the sowing depth equaling to 2 cm.

At the Institute of Agroecology, a branch of South Ural State Agrarian University, a two-drum roller was made according to the substantiated parameters.

The comparative assessment of tillage quality proved a tillage unit equipped with two-drum rollers to ensure:

the due structural composition of the seed layer: the content of lumps exceeding 10 mm was 28.4% of the total test piece, and there were no lumps exceeding 50 mm, with the content of lumps 2-5 mm in size being 45.9%;

a better evenness of the soil surface as compared to the one obtained with the traditional type of treatment: the mean-square deviation of the soil surface irregularities after the two-drum roller cultivator treatment was 1.7 cm, with the soil unevenness being 37% less if compared with the traditional type of treatment;

earlier and friendly sprouting of cultivated crops and increased yields: the yield of spring rape green mass obtained from the pieces of ground treated with the two-drum roller cultivator is 21.7% higher than the yield obtained from the pieces of ground treated traditionally.

The energy data of the operating unit according to GOST 52777-2007 were determined by the Ural Testing Center for Agricultural Equipment of South Ural State Agrarian University on the experimental field of the Institute of Agroecology. In the course of the experiment, the value of the two-drum roller traction resistance at the speed of 2.8 m/s was obtained. It amounted to 0.72 kN with the operating width b = 1 m, the theoretical and experimental data differing within the acceptable limits (10%).

4. Conclusion

The presented design model and analytical dependence made it possible to substantiate the main design parameters of a two-drum roller with energy saving parameters. The experimental studies proved the ability of the two-drum roller to provide the due structural composition of the seed layer, a better evenness of the soil surface as compared to the one obtained with the traditional type of treatment, earlier and friendly sprouting of spring rape and increased yields.

References

[1] Roberto Fanigliulo, Marcello Biocca, Daniele Pochi. Effects of six primary tillage implements on energy inputs and residue cover in Central Italy. *Journal of Agricultural Engineering*, 2016, 47 (3), 177-180.

[2] Zhang X.C., Li H.W., Du R.C., Ma S.C., He J., Wang Q.J., et al. Effects of key design parameters of tine furrow opener on soil seedbed properties. *Int J Agric & Biol Eng*, 2016, 9(3), 67-80.

[3] Dina Saber Salama, Sabreen Kh. Pibars, Yossry B. Abdelhay, Mohamed Y. Tayel, Gamal M. Nasr. Design a Combined Machine for Seedbed Preparation. *CIGR Journal*, 2018, 20 (1), 90-94.

[4] Paul Okoko. Draft and Power Requirements for some Tillage Implements operating in Clay Loam Soil. *CIGR Journal*, 2018, 20 (1), 95-102.

[5] Ritu Dogra. Effect of spade angle and spading frequency of spading machine on specific soil resistance and pulverization. *CIGR Journal*, 2017, 19 (1), 95-73.

[6] Kovalev N.G., Romanenko A.A., Mazitov N.K. Resource-saving Agriculture on the Basis of Domestic Machinery. *Sel'skohozyaystvennye mashiny i tekhnologii*, 2015, No. 1, 37-41 (in Russian).

[7] Rakhimov Z.S., Mudarisov S.G., Gabitov I.I., Rakhimov R.S., Rakhimov I.R., Alyabiev V., Tarkhova L.M., Farkhutdinov I.M., Mukhametdinov A.M., Gareev R.T. Resoning a Construction Diagram and Parameters of Tillers for Primary Cultivation. *Journal of Engineering and Applied Sciences*, 2018, 13 (S11), 8812-8818.

IOP Conf. Series: Materials Science and Engineering **582** (2019) 012027 doi:10.1088/1757-899X/582/1/012027

[8] Gabitov I.I., Mudarisov S.G., Gafurov I.D., Ableeva A.M., Negovora A.V., Davletshin M.M., Rakhimov Z.S., Khamaletdinov R.R., Martynov V.M., Yukhin G.P. Evaluation of the Efficiency of Mechanized Technological Processes of Agricultural Production, *Journal of Engineering and Applied Sciences*, 2018, 13 (S10), 8338-8345.

[9] Rahimov R.S. Parametry dvuhbarabannogo katka k orudiyam dlya obrabotki plasta mnogoletnih trav i zalezhi. *Traktory i sel'hozmashiny*, 2009, No. 7, 15-18 (in Russian).

[10] Mazitov N.K. [et al.] Importozameshchayushchiy kompleks mashin dlya vlagoakkumuliruyushchey energosberegayushchey tekhnologii obrabotki pochvy. *Mekhanizatsiya i elektrifikatsiya sel'skogo hozyaystva*, 2013, No. 1, 2 (in Russian).

[11] Mazitov N.K. [et al.] Itogi mnogoletnih sravnitel'nyh ispytaniy i vnedreniya novoy tekhniki dlya obrabotki pochvy i poseva. *Dostizheniya nauki i tekhniki APK*, 2016, 30 (8), 91-93 (in Russian).

[12] Polikutin N.G., Telichkina N.A., Rahimov R.S., Konovalov V.N. Pochvoobrabatyvayushchee orudie [Tillage tool]. Patent RF, no. 130188, MPK A01V29/00.

IOP Conf. Series: Materials Science and Engineering **582** (2019) 012028 doi:10.1088/1757-899X/582/1/012028

Material and technical support of the enterprises of the agro-industrial complex of the Omsk region management and certification of the technical component of the production processes in crop production

O V Myalo [a], S P Prokopov [b], V V Myalo [c], A S Soyunov [d], E V Demchuk [e]

Federal State Budgetary Educational Institution of Higher Education "Omsk State Agrarian University named after P.A. Stolypin 644008, Omsk, ul. Institutskaya Square

E-mail: [a] ov.myalo@omgau.org, [b] sp.prokopov@omgau.org, [c] vv.myalo@omgau.org, [d] as.soynov@omgau.org, [e] ev.demchuk@omgau.org

Abstract. The management of a modern enterprise of the agro-industrial complex should cover all aspects of the production of the main types of products produced. In this concept in modern economic conditions it is necessary to include not only economic indicators and ways of selling finished products, but also the use of modern technologies, production safety for the environment and people. Improvement of technologies for obtaining crop production is impossible without updating the equipment used for the mechanization and automation of production processes. For that reason, such a concept appeared in the legislation as certification of production technological processes, which pumps in the assessment of all technological process components, including the production means used, the quality of their technical condition and service. The article analyzes the current situation of the maintenance system, the structure and condition of the machine and tractor fleet of enterprises of the agro-industrial complex of the Omsk region. A method has been developed for processing information in terms of the agricultural enterprises revenue, and the share of work that the farm can transfer to a specialized technical service enterprise, taking into account its own profit, has been determined.

1. Introduction

Certification of technological processes and production is a documentary confirmation of the compliance of each component of the full functioning of the enterprise with the current standards. The procedure involves the finished products verification and technologies for its production, maintenance and rehabilitation work, test systems and technical control of the final result [1]. Since the beginning of 2017, many entrepreneurs and enterprises have heard of new certification - certification of technological processes. It is carried out in accordance with Article 215 of the Labor Code of the Russian Federation. As a rule, the state labor inspectorate comes to an unsuspecting enterprise, carries out an inspection, at the end of which a remark is issued stating that the technological processes in the organization do not comply with article 215 of the Labor Code of the Russian Federation, since they do not have a certificate. This note also indicates the date of the re-check. During this period it is necessary to eliminate all comments and pass the technological processes certification. If at the time of the re-verification the certification has not been passed, then a fine in the amount of 50,000 to 80,000 rubles will be issued for the legal entity.

In technological processes of agricultural production, the main means of performing technological operations are combines, tractors and agricultural machinery aggregated with them [2].

According to the Ministry of Agriculture and Food of the Omsk Region, the structure of the agricultural machinery park for 2018 is shown in Figure 1.

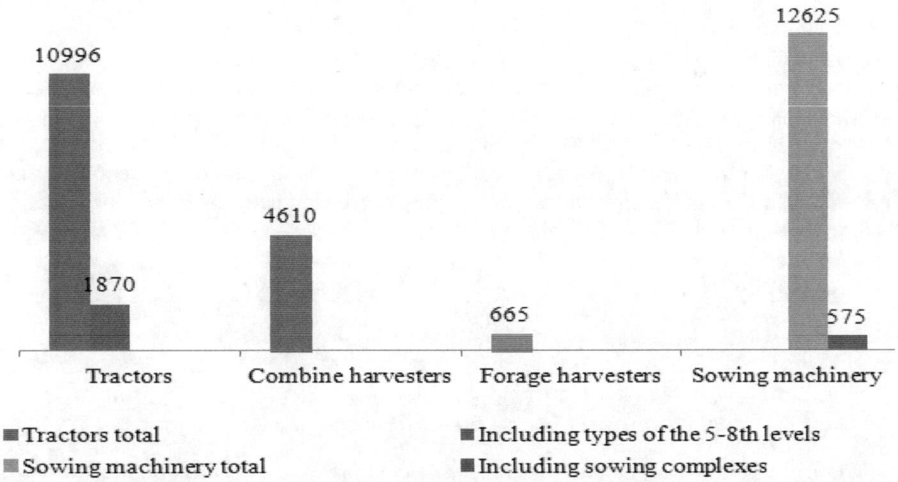

Figure 1. The park structure of agricultural machinery in the enterprises of the agro-industrial complex of the Omsk region in 2018

This technique is the main production means for the entire agro-industrial region complex and must meet the increasing requirements for production, as well as undergo appropriate certification of production processes, which implies not only a working condition, but also a corresponding maintenance level [3].

When considering the dynamics of the fleet structure over the past three years (2016 - 2018), it can be noted that the number of vehicles has changed slightly over the past year, mainly towards the reduction of existing machines (Fig. 2).

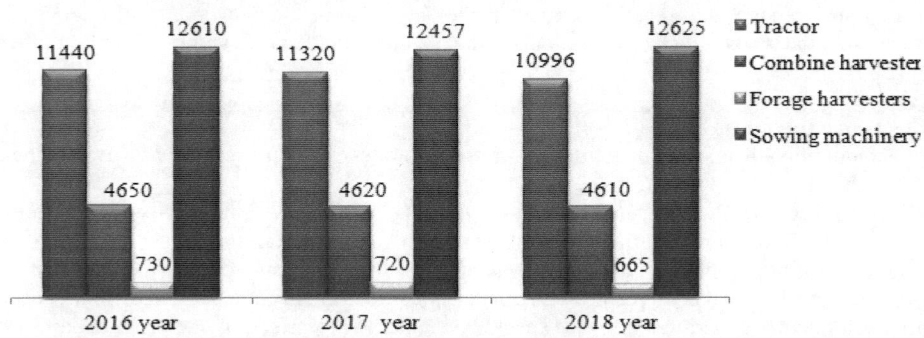

Figure 2. Structure dynamics of the machine-tractor park of the Omsk region for 2016 – 2018

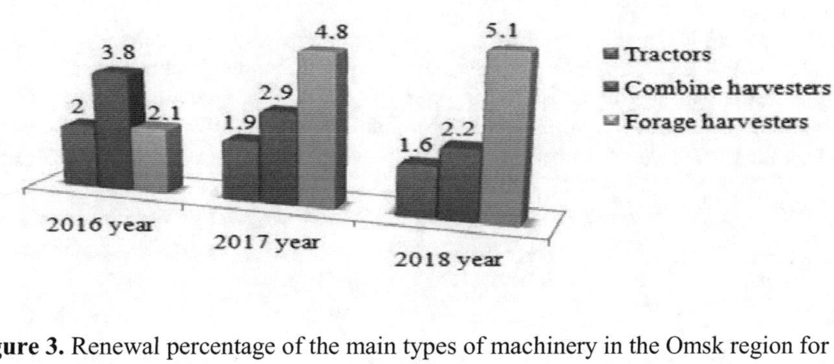

Figure 3. Renewal percentage of the main types of machinery in the Omsk region for 2016-2018,%

IOP Conf. Series: Materials Science and Engineering **582** (2019) 012028 doi:10.1088/1757-899X/582/1/012028

Fig. 3 shows the renewal percentage of the main types of machinery in the Omsk region, which on average over three years amounted to 2.9%. Thus, there is an increase in the service machinery life before it is written off, and this trend is not supported by either an increase in the quality of the new machinery or an improvement in the technical service itself, while the deterioration of agricultural machinery at enterprises of the agro-industrial complex is growing every year (Fig. 4).

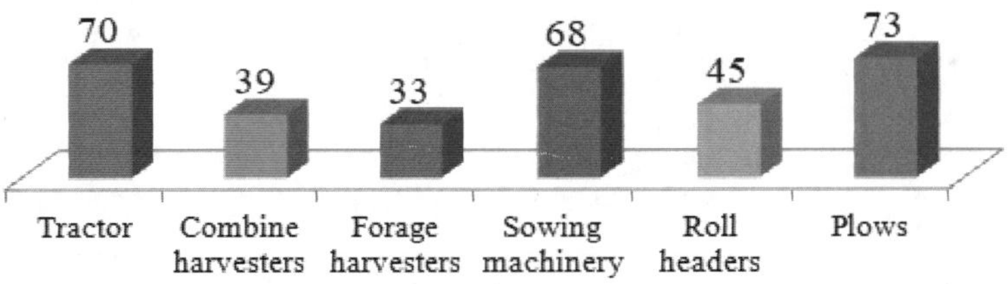

Figure 4. Wear of agricultural machinery by enterprises of the Omsk region for 2018,%

Based on the above data, it can be noted that the need for a radical vehicles fleet renewal for enterprises of the agro-industrial complex, which in modern economic conditions is practically impossible [4]. Taking into account the new requirements for certification of production processes, a number of agricultural enterprises will not actually be able to pass this test and receive certificates for compliance with the requirements of Article 215 of the Labor Code of the Russian Federation.

To maintain the equipment in working condition and meet new requirements, a new concept of technical service development is required, which is dictated not only by the change in the transport enterprises structure and the requirements of Article 215 of the Labor Code of the Russian Federation, but also by the crisis state of a significant part of the necessary material and technical base for machine -tractor park maintenance and repair. As a result, the enterprise of the agro-industrial complex is not able to carry out the whole range of operations of the planned maintenance and repair of equipment on its own [5].

2. Objects and methods

The methods of mathematical modeling and statistical processing of the results were used as the main research methods.

The profitability magnitude dependence of agricultural enterprises and the services needs formation in the field of technical service of agricultural equipment. Agricultural enterprises, having a certain amount of profitability in the form of revenue, can spend part of the technical service, including third-party organizations, which will affect the maintenance operations quality and volume, will directly affect the equipment reliability, and will allow to pass the legislation prescribed by the production processes at the enterprise [6, 7].

3. Research results

We have developed a method of processing information on the indicator of the agricultural enterprises revenue of the Omsk region. On the basis of the data obtained, a summary table was compiled of the farms revenue value in the Omsk Region per 100 hectares in ascending order, rub / 100 ha. In drawing up the methodology, the farms were impersonal, each farm was given a number, in order of increasing revenue rub / 100ga (Table 1).

Table 1. Agricultural enterprises revenue in rubles / 100 ha

Farm number	Revenue, thousand rubles	Farm number	Revenue, thousand rubles	Farm number	Revenue, thousand rubles
1	60.27	17	759.77	33	1079.62
2	68.88	18	782.56	34	1109.88
3	201.51	19	789.21	35	1124.56
4	243.26	20	792.24	36	1145.49
5	356.16	21	812.57	37	1146.19
6	388.84	22	814.01	38	1169.03
7	397.41	23	846.42	39	1171.82
8	450.30	24	857.39	40	1304.71
9	467.62	25	875.42	41	1356.63
10	491.41	26	893.07	42	1462.83
11	524.60	27	898.88	43	1586.02
12	567.06	28	975.89	44	1656.96
13	575.82	29	1033.72	45	1767.68
14	647.17	30	1044.00	46	1780.84
15	667.36	31	1071.67	47	1918.58
16	722.07	32	1077.87	48	2071.18
				49	2608.12

When processing the obtained data, a sample was taken from 49 farms, that is, the information is repeated n = 49, and the information is divided into N equal intervals, with an interval width n = 400. The number of intervals is determined from the expression:

$$N = \sqrt{n} \pm 1 \quad (1)$$

Further we accept

$$N = \sqrt{49} \pm 1 = 7$$

When developing the methodology, the interval size was adopted equal to 400 rubles per 100 hectares. Upon further processing, the presented data compiled a statistical series of revenues of the farms under consideration (Table 2).

Table 2 Statistical revenue range of agricultural enterprises

Revenue per 100 ha., rub / ha	The average value in the interval, ni	Test frequency, mi	Experienced probability, pi	Accumulated experimental probability, Σi
0-400	200	7	0.143	0.143
400-800	600	13	0.265	0.408
800-1200	1000	19	0.388	0.210
1200-1600	1400	4	0.082	0.292
1600-2000	1800	4	0.082	0.373
2000-2400	2200	1	0.020	0.394
2400-2800	2600	1	0.020	0.414

The experimental frequency shows how many agricultural enterprises in terms of revenue fall into one or another interval.

Experimental probability was defined as the experimental frequency ratio to the average value in the interval:

$$p_i = \frac{m_i}{n_i}, \quad (2)$$

where mi is the experimental frequency in the ith interval of the statistical series, ni is the average value in the interval.

Next, determine the average value of the company's revenue per 100 hectares according to the formula:

$$\bar{x} = \sum_{i-1}^{N} x_{ci} p_i, \tag{3}$$

where N is the number of intervals in the statistical series,

xci - the value of the middle of the ith interval,

pi is the experimental probability of the ith interval.

Having carried out the necessary calculations for the considered farms, the following results were obtained:

$$\bar{x} = 200 \cdot 0.141 + 600 \cdot 0.265 + 1000 \cdot 0.388 + 1400 \cdot 0.082 + 1800 \cdot 0.082 +$$
$$+ 2200 \cdot 0.02 + 2600 \cdot 0.02 = 934.694 rub/100ha$$

The dispersion characteristic of the revenue per 100 hectares was determined by the standard deviation:

$$\sigma = \sqrt{\sum_{i=1}^{n} (x_{ci} - \bar{x})^2 p_i} \tag{4}$$

After calculating the indicator obtained the following result:

$$\sigma = \sqrt{(200 - 934.694)^2 \cdot 0.143 + (600 - 934.694)^2 \cdot 0.265 + (1000 - 934.694)^2 \cdot 0.388 +}$$
$$\sqrt{+ (144 - 934.694)^2 \cdot 0.082 + (1800 - 934.694)^2 \cdot 0.82 + (2200 - 934.694)^2 \cdot 0.02 +}$$
$$\sqrt{+ (2600 - 934.694)^2 \cdot 0.02} = 525.88 \text{ rub/100ha.}$$

After carrying out the necessary calculations, the results were checked for outliers. Check for drop-down points carried out according to the rule $\bar{x} \pm 3\sigma$. If the extreme points of the information do not exceed the limits $\bar{x} \pm 3\sigma$, then all information points are considered valid [8].

In our methodology, the accuracy limits of the information were:

- lower bound: 934.694-3 · 525.88 = -642.9
- upper bound: 934.694 + 3 · 525.88 = 2512.33

The lowest enterprise profitability per 100 hectares is xmin = 60.27 thousand rubles, and the highest revenue of the enterprise per 100 hectares is xmax = 2608.12 thousand rubles, therefore both information points are valid and should be taken into account in further calculations.

The coefficient of variation (relative dimensionless quantity characterizing the dispersion of indicators) was determined by the formula.

$$v = \frac{\sigma}{\bar{x}}, \tag{5}$$

Since the variation coefficient in the calculations was in the range of 0.3 ... 0.5, the distribution law was chosen, which better coincides with the distribution of experimental information. In our case, this is the normal distribution law.

A distinctive feature of the differential function is the symmetric dispersion of the partial values of the indicators relative to the average value [9]. To determine the differential function in terms of the centered normalized function, the equation was used:

IOP Conf. Series: Materials Science and Engineering **582** (2019) 012028 doi:10.1088/1757-899X/582/1/012028

$$f(x)=\frac{A}{\sigma}f_o\left(\frac{x_{ci}-\bar{x}}{\sigma}\right), \tag{6}$$

where A is the length of the ith interval,
xci is the midpoint of the ith interval.
In addition, used the equation:

$$f_0(-x)=f_0(+x) \tag{7}$$

To determine the integral function $F(x)$ through $F_0(x)$ the equation used:

$$F(x)=F_0\left(\frac{x_{\kappa i}\bar{x}}{\sigma}\right) \tag{8}$$

where $x_{\kappa i}$ is the value of the end of the i -th interval.
The equation was also used.

$$F(-x)=1-F_0(+x) \tag{9}$$

The value of the differential and integral functions calculated according to the normal distribution law are given in Table 3.

Based on the obtained values $f(x)$ and $F(x)$ the graphs of the differential and integral distribution functions of the analyzed farms proceeds were constructed (Fig. 5).

Table 3. The value of the differential and integral functions calculated by the normal law

Revenue per 100 ha., rub / ha	The average value in the interval, n_i	*f(x)*
0-400	200	0.1141
400-800	600	0.251
800-1200	1000	0.304
1200-1600	1400	0.205
1600-2000	1800	0.007
2000-2400	2200	0.002
2400-2800	2600	0

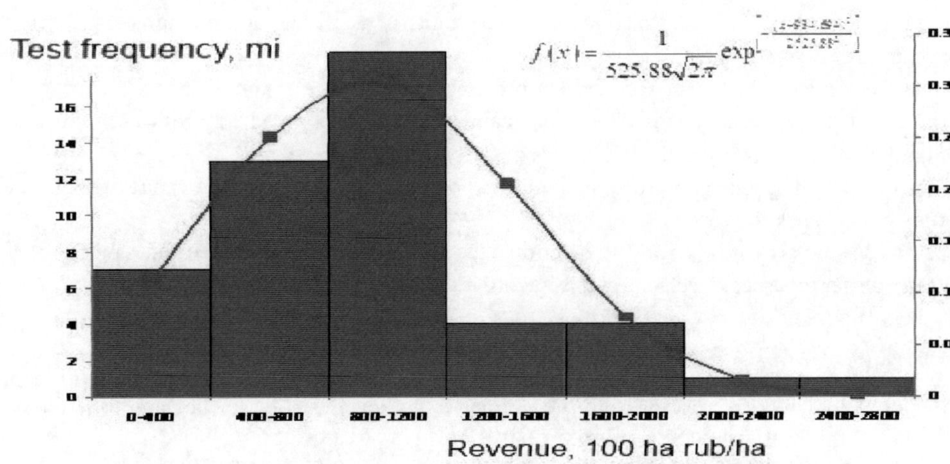

Figure 5. Graph of the enterprises profitability distribution by revenue

On the basis of the data obtained, it is possible to determine the maintenance work proportion that the farm can transfer to a specialized enterprise using the formula:

$$\varphi = 1 - \frac{z}{B} \tag{10}$$

where φ - the proportion of maintenance work;

z - is the cost of TO-2 and TO-3, when performing maintenance at specialized technical service enterprises;

B - revenue of the agricultural enterprise.

If the value, then the company can afford to use the services of specialized enterprises for carrying out complex maintenance of the second maintenance and the third maintenance for the entire fleet of vehicles, which will not only significantly improve the reliability of equipment and machine performance, but also to certify production processes in the enterprise in accordance with federal law. The corresponding system of technical service, using the services of specialized enterprises of technical service, can be formed only in the conditions of production of sufficiently large farms.

Taking into account the fact that at present the equipment of the material and technical base (MTB) of farms is low, mainly MTB is represented by equipment and equipment of the Soviet period, which is not suitable for maintenance and repair of modern equipment [10]. This regulation becomes another obstacle to updating an outdated fleet of machine-tractor units, apart from their high cost and the production processes of such enterprises of the agro-industrial complex cannot be certified according to article 215 of the Labor Code of the Russian Federation. The development strategy of the technical service of the machine-tractor park should be aimed at the development of the dealer system, which will assume not only warranty, but also post-warranty service of the machines.

4. Conclusion

Based on the research, we can draw the following conclusions:

1. At present, according to the requirements of Article 215 of the Labor Code of the Russian Federation, all farms of the agro-industrial complex must undergo certification of production processes, which implies that each component of the enterprise's full operation complies with the applicable standards, including the operability and reliability of the equipment used, and its quality service;

2. The machine and tractor fleet of the Omsk region is mainly represented by outdated equipment operating beyond the standard period of use, the average percentage of equipment renewal is 2.9%, therefore the wear of the fleet of machine and tractor units is increasing and the number of efficient machines decreases every year;

3. In the current conditions of the formation of technical service at the enterprises of the agro-industrial complex, the dealer system of technical maintenance can be formed only with a certain amount of profitability of the economy in the conditions of fairly large enterprises;

4. A method was developed for processing information on the indicator of the revenue of agricultural enterprises in the Omsk Region;

5. Using equation (10), it is possible to determine the proportion of maintenance and repair work that a farm can transfer to a specialized enterprise, taking into account its own profit.

References

[1] Ecological and economic indicators of new machines in the complex of promising agricultural technologies of crop production / I.I. Gureev // Achievements of science and technology of the agroindustrial complex - Moscow, 2014.№10. Pp. 61 - 64.

[2] Technical and technological support of agricultural producers in the transition to an innovative economy / A.V. Kuchumov, A.V. Belokopytov // Achievements of science and technology of the agro-industrial complex - Moscow, 2017. V. 31. №12. Pp. 78 - 81.

IOP Conf. Series: Materials Science and Engineering **582** (2019) 012028 doi:10.1088/1757-899X/582/1/012028

[3] Theoretical substantiation of machine-tractor fleet technical maintenance system on the example of Omsk region agricultural enterprises / O. V. Myalo, V. V. Myalo, S. P. Prokopov, A. P. Solomkin, A. S. Soynov - IOP Conf. Series: Journal of Physics: Conf. Series 1059 (2018) 012005

[4] Resource potential - the basis of sustainable agriculture and rural areas development / S.G. Golovina, S.V. Pugin // Achievements of the Science and Technology of the Agro-Industrial Complex - Moscow, 2015.T. 29 №1. P. 9 - 12.

[5] Improvement of technical service in the agro-industrial complex based on the assessment and analysis of the technological level of repair enterprises / A.I. Anosova, M.K. Buraev // Achievements of science and technology of agrarian and industrial complex - Moscow, 2014. №10. Pp. 65 - 68.

[6] Beshelev S.D. Mathematical-statistical methods of expert assessments / S.D. Bechelev, F.G. Gurvich. -2 ed., Pererab. additional -M .: Statistics, 1980. -263 p.

[7] Ensuring Machine and Tractor Aggregates Operability. / G.V. Redreev - IOP Conference Series: Materials Science and Engineering, Volume 142, Number 1.

[8] Modeling the Technological Process for Harvesting of Agricultural Produce / S.D. Shepelev, V.D. Shepelev, Z.V. Almetova, N.P. Shepeleva, M.V.Cheskidov, // IOP Conference Series: Earth and Environmental Science115(1), (2018) 012053

[9] Problems of using network information resources in the field of agriculture. Methodical recommendations / O.B. Sladkova, L.N. Pirumova, A.A. Pirumov // Achievements of the Science and Technology of the Agro-Industrial Complex - Moscow, 2016. T.29. Number 3. P. 59 - 64.

[10] Technique of agro-industrial complex - quality service / N.M. Ivanov, A.E. Nemtsev, V.V. Korotkikh // Achievements of Science and Technology of the Agrarian and Industrial Complex - Moscow, 2016. V. 30. №4. Pp. 81 - 82.

IOP Conf. Series: Materials Science and Engineering **582** (2019) 012029 doi:10.1088/1757-899X/582/1/012029

Application of Information Technologies for Automation of Railway and Cargo Owner Interaction

O H Strelko [a], H I Kyrychenko, Yu A Berdnychenko, O L Sorochynska, O Ya Pylypchuk

State University of Infrastructure and Technologies, 9 Kyrylivska Street, Kyiv, 04071, Ukraine

E-mail: [a] olehstrelko@gmail.com

Abstract. The results of formalization of the components in the process of calculation of the time of delay in the movement of wagons are presented. The formalization was carried out in accordance with the principles of system analysis with the purpose of next modeling and automation of the technological process of cargo transportation during the delay by the fault of a cargo owner. It was proposed to apply a methodology of the conceptual-logical mapping and project modeling by constructing models of the process of interaction between modes of transport.

1. Introduction

The problems of optimization of the interaction of railway with ports, industrial enterprises, border transit points include complex problems, which exist at the level of planning, coordination, realization of transportation and calculation of the performed work [1-6]. Solving such problems is associated with the development and implementation of information technologies in the management of transport processes on the basis of theory and methodology of a conceptual-logical mapping and project modeling [7] by construction of models of the process of interaction between modes of transport.

2. Aim of the work

The formalization of separate components according to the principles of system analysis is necessary for the next automation and modeling of the technological process of transportation. A detailed consideration of the components in the cargo transportation process with the analysis of costs, associated with a delay, remains urgent [8] and necessary to minimize the costs when changing the working schedule of enterprises and the effect on service conditions of the transport network. In such circumstances, the optimization of shipment planning can be made using the decision on the basis of a system for determination of a cargo traffic delay location.

The problems of the relevance in construction of a transport management system are discussed by scientists in numerous papers from different positions and considerations, but the development of a system coordinating the interaction and optimizing the management of the transportation process still requires significant theoretical grounds. In particular, the problem of coordinating the work at the junctions of railways and sea transport is problematic.

Interaction of the railway transport with the sea transport begins with the choice of the rhythm of shipment of cargoes destined to the port and the emergence of ships in the forecasted region. The work of ports, border stations of Ukraine is featured by handling powerful flows of transit wagons and cargoes arriving from the loading areas, located over a long distance, which complicates the forecasting and planning the operation of transport complexes. This results in considerable losses of time for using rolling stock and infrastructure of railways, causes arrhythmias in the operation of all transportation chains.

3. Investigation results

The degree of manageability in the railway subsystem of transportation management depends on numerous factors, such as largely stochastic provision of orders on transportation; volumes of transit flows forecasted on the basis of statistical and marketing studies; decision making by the dispatching unit regarding “leaving” of trains during delays in handling of cargoes in the traffic junctions, etc. The prerequisites for a quantitative evaluation of the railway subsystem conditions have been created, but the management of transportation based on the construction and use of optimization models as well as the integration of railway with industrial transport based on databases belongs to the future.

Multifactority in optimization of the management of sea transportation implies the dependence of ports functioning on the market situation, forecasting of arrival and servicing of ships at berths, navigational, climatic and weather conditions, rationality of internal organization, rhythm of railway trains approach, existence of differences in tariffs for handling and transfer of cargoes in different ports and other factors. Quantitative evaluation of the condition of the subsystem of sea transportation is problematic in connection with the lack of a single database, as in each port the own automated system operates, which is unique as to its construction and software. Specialists in the organization of sea transportation note that the activity of ports depends on other modes of transport and general logistics and marketing systems. Therefore, the development of the activities of sea ports, in their opinion, is related, in addition to the development of technical means, to the problems of integration, to the principles of logistics and system analysis, transport and automated information systems [9, 10].

Thus, the problems of interaction of railways with ports, industrial enterprises, border transit points exist at the level of planning, coordination, transportation and calculation of the performed work [8] or determining of a "culprit" of the unfulfilled work. Theoretically, the ways for solving these problems with the help of quantitative methods have been developed precisely in detail [11], but in practice it is difficult to use these methods.

Modern information systems allow performing an accurate quantitative evaluation of the state of objects and subjects of management due to the ability of a quick access to information about them [12] and then proceed to simulation of the technological process of transportation. In order to do that, according to the principles of a system analysis, formalization of separate components of the technological process of transportation is required. To understand the degree of detailing at the stage of understanding the process of interaction, let us consider the components participating in the automated (objective and transparent for both parties) determination of the time of delay in the movement of trains, wagons and cargoes caused by the stops of enterprises, including seaports. The processing of information flows in the automated system is carried out in the complex of workplaces available in the railway and enterprises, operating in a single information space, presented in the diagram of figure 1. Delaying of a cargo, a wagon, a train at the approaches to the destination station is predetermined by the interaction of circumstances, namely:

- stop of the enterprise and transfer of data about that to the railway,
- quick response by the dispatching unit to stop the train movement or form wagons in a train at a sorting station, i.e., quick-response ordet to the stop station,
- existing situation (availability of free tracks) at the stations of a probable train stop.

For the development of information provision to determine the time of start and end of delay in the movement of cargo flows, it is necessary to formalize the components of the operational process events.

The time of delay start (ta_{ds}) is calculated automatically, simultaneously the time of realization of three operational events is compared:

- order to delay a train or a group of wagons, to; such wagons include wagons which are handled at the stop station and have a destination to the station serving the user of services, who stopped receiving of wagons from the railway;
- arrival of a train to the stop station, t_a ;
- time of start of the enterprise stop, t_s.

In practice, all three events are carried out at different times. The 6 variants of staying on the time axis ts, ta, to are possible (figure 2). The time of a train delay at the approaches begins from a higher value on the time axis.

IOP Conf. Series: Materials Science and Engineering **582** (2019) 012029 doi:10.1088/1757-899X/582/1/012029

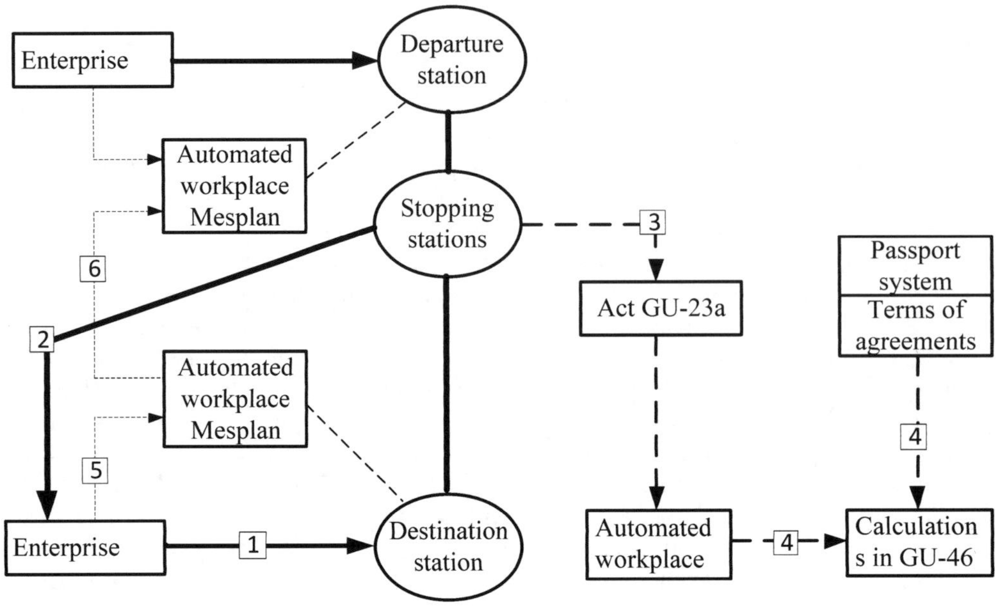

Figure 1. Automated interaction of information complexes: 1- notification about the stop of the enterprise; 2 - message to the destination station for the consignee about delaying of the train; 3 - drawing up an act GU-23a; 4 - calculation of payment for using of the wagon; 5 - correction of the plan of departure; 6 - correction of the loading plan in the departure stations.

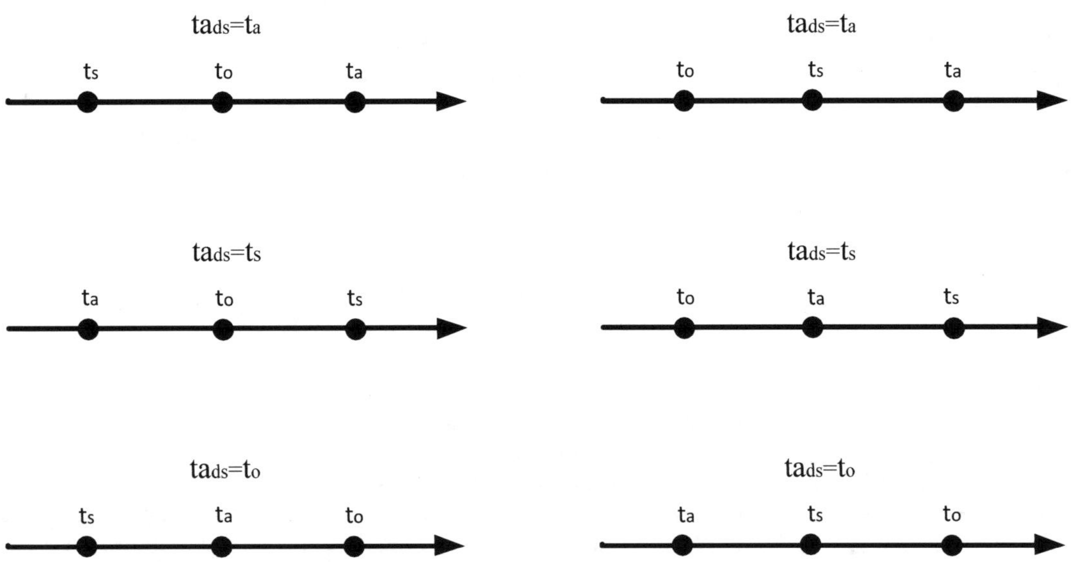

Figure 2. Variants of location of the time for operation fulfillment on the time axis

The time of end in delay (ta_{de}) is calculated automatically, simultaneously the time for the following operations is compared:

- order to resume the movement of trains or a group of wagons $t_к$;
- departure of a train from the stop station, t_d ;
- end of the enterprise stop, t_p.

In practice, all three events are carried out at a different time. The 6 variants of staying on the time axis tк, td, tp are possible (figure 3). The time of delay of a train at the approaches is finished at the moment of the lowest value (the first one on the axis).

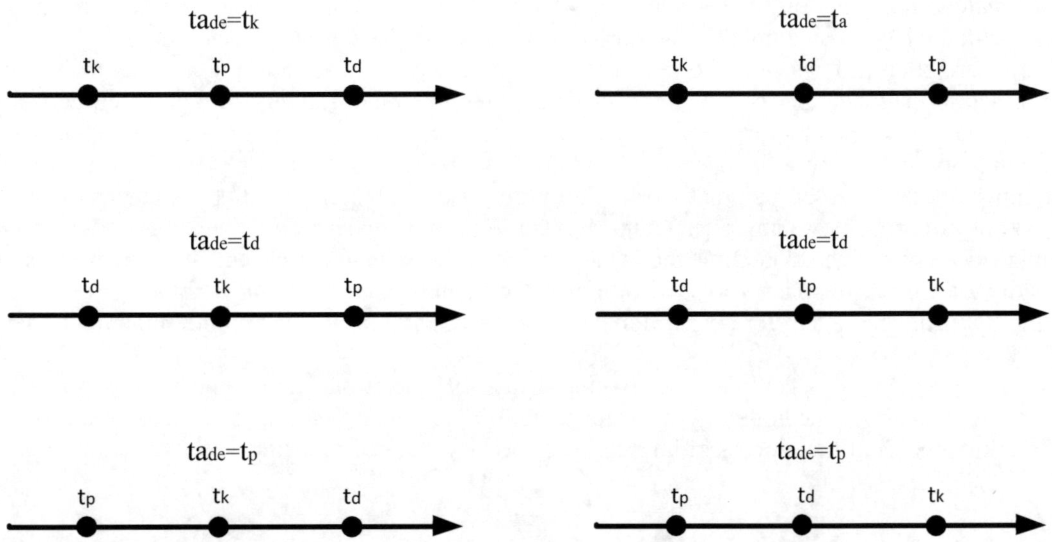

Figure 3. Variants of location of the time for operation fulfillment on the time axis

A cargo owner and the railway bear the minimum costs for idle time of wagons in detained trains and wagons in the case shown in figure 4.

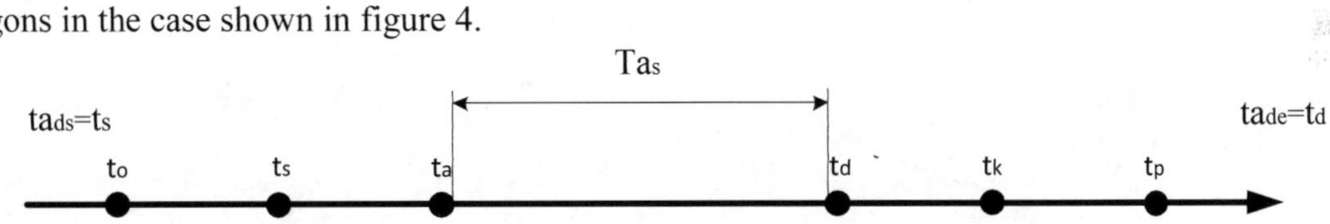

Figure 4. Taking into account the stopping time before payment

In order to reduce circumstances of uncertainty in planning of the next transportation to the destination station by a recipient, a Message about the delay of wagons is automatically made, indicating the events with a wagon, a group of wagons, a train:

- about the order to delay a train or a group of wagons, t_o;
- about arrival of a train to the stop station, t_a;
- about the start of the enterprise stop, t_s.

Data contained in the message about the delay of wagons:

- time of transferring the order and its number;
- time of arrival at the station of delay of a train, a group of wagons, a wagon;
- code and name of the enterprise;
- data on the cargo front (if available under the terms of the agreement);
- time of stop of the enterprise or the cargo front;
- time of start (determined in an automated way) of acting of the act about the delay at the approaches f. GU-23a;
- time of transfer of this certificate;
- numbers of a train and wagons with determined (as agreed with the user) cargo data.

The further detailing of the message depends on the set of constituent terms of the Agreement and takes into account the components of the Unified Technological Process of Work of the Station and Access Tracks. If necessary, the message for the user of services is detailed, indicating: an access track, an enterprise, a cargo front, a cargo area, a cargo, track, an individual customer.

Respectively, a message to the destination station about the end of a wagon delay at the address of the user of services is made, indicating the events with a wagon, a group of wagons, a train, namely:

- order for departure of a train or a group of wagons, ($t_к$) ,
- departure of a train from the stop station, (t_d);
- resuming of the enterprise work, (t_w).

Automated determination of the delay time is carried out in the information environment of the railway with the help of a complex of workplaces, available at the stations and enterprises.

The information provision of the technological process of transportation is a prerequisite for the development of optimization models of the systems of interaction of railway with ports, border stations and industrial enterprises. In this respect, it is relevant to develop a system of decision making by the dispatching unit in the railway concerning places of "leaving" trains during delays in handling of goods in the transport junctions. Stopping of work of the enterprises is connected with the change in the modes of movement of cargo flows, their delay and, as a consequence, the change in the plans of departure. To minimize the costs when changing the working schedule of enterprises and the effect on service conditions of the transport network and optimize the planning of shipment in such circumstances is possible by using the system for decision making on the basis of a system for determination of a cargo traffic delay location.

The development of a complex multifactorial task of operational planning is appropriate to be implemented using the methodology of a conceptual-logical mapping and project simulation [7] by constructing models of the process of interaction between modes of transport.

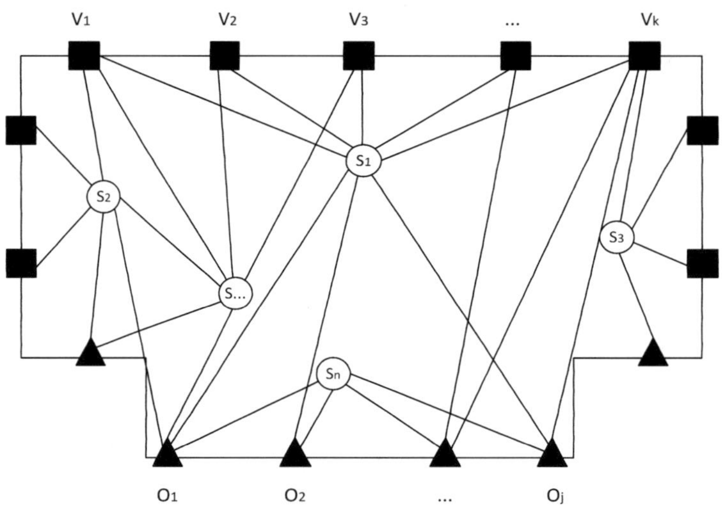

Figure 5. Scheme of information flows during planning of cargo shipment: S_1, S_2, ... S_n – systems of intersection and interaction of cargo traffic; V_1, V_2, ... V_k – cargo consigners; O_1, O_2, ... O_j – cargo consignees.

To construct the system for determination of the delay location (limitation in movement) of cargo flows (S_k) let us use the tools for system analysis.

The systems for determination of the delay location (limitation in movement) of cargo flows are the components of the system of transportation planning. In general, let us present the system of transportation planning as follows:

$$S_{pl} = \langle V, O, C, R \rangle \qquad (1)$$

where V - is the set of objects of cargoes shipment, $\{V_k\} \in V$; O - is the set of objects of destination and handling of goods, $\{O_j\} \in O$; R - is the relation between the objects, $\{R_i\} \in R$; C - is the set of objects of a cargo traffic delay, $\{C_m\} \in C$, which is the result of functioning of the system S_{pl} .

The further development of the task of decision making on "leaving" trains, wagons and operational change in the planning of shipment is advisable in the terms of the theory of sets, theory of possibilities for the ultimate creation of simulation model of the process of interaction of transport modes in the general transport information environment.

The Ministry of Transport and Communications of Ukraine started works on the development and implementation of the Program for improvement of the level of integrated information provision for the participants of the transport process. The program envisages the creation and development of a transport services market in Ukraine, further improvement of coordination of the work and interaction of all modes of transport on the basis of introduction of modern information technologies [13-14].

4. Conclusions

The need in optimization of costs for all the participants of transportation requires a complete formalization of constituents of the transport process in order to understand the complex interaction system for creating the system for supporting the decision making of persons in the future, who organize and manage the traffic.

Aggregation of factors, their mutual influence at the level of contacts of related modes of transport, presence of contradictory criteria in the subsystems predetermines the problem of optimization of interaction and management of transport and handling complexes. From the scientific point of view, this problem remains unsolved.

Proceeding from the abovementioned, the management of transport interaction as a single transport line requires scientific developments to solve a complex task: coordination of functioning of the transport complex and represents a scientific problem which is rational to be considered in compliance with the system approach in the methodology of conceptual-logical mapping and project modeling.

References

[1] Lyamzin A O 2008 Efficiency of the transport system of the industrial region in the port logistics chain of grain supply: author's abstract *Theses for Cand. of Tech. Sci.* (Kharkiv: National Academy of Municipal Economy) 20 p

[2] Strelko O, Kyrychenko H, Berdnychenko Y and Hurinchuk S 2018 Automation of Work Processes at Ukrainian Sorting Stations *International Journal of Engineering & Technology* **7**(2.23) 516-518

[3] Kyrychenko H, Statyvka Y, Strelko O, Berdnychenko Y and Nesterenko H 2018. Assessment of cargo delivery quality using fuzzy set apparatus *International Journal of Engineering & Technology* **7**(4.3) 262-265

[4] Okorokov A M 2015 Analysis of the prospects for the devel-opment of the container transport market in Ukraine *Transport Systems And Transportation Technologies* no10 pp 98-104

[5] Snezana T and Slobodan Z 2012 Development of Intermodal transport and logistics in Serbia *International Journal for Traffic and Transport Engineering* **2**(4) 380-390

[6] Dávid A 2013 Innovation of handling systems in the world container ports and their terminals *Proceedings of the 17th International Conference Transport Means 2013 (Kaunas)* (Kaunas: University of Technology) 250-253

[7] Petrashevsky O L, Redzyuk A M and Alekseenko O V 2009 Methodology of conceptual-logistic mapping and project modeling of the objectives of the traffic safety management system *NTU, Collection of Scientific Articles* **6**(10) 76-89

[8] Nikolashin V N 2008 Logistic Principles in the optimization of safety of the goods supply chains *Transport. Science, Technology, Management* no 10 (Moscow: VINITI RAS) pp 38-39

[9] Zaborsky L O 2008 Methodical fundamentals of organization of transport and technological processes in cargo delivery systems *Theses for Cand. of Tech. Sci.* (Odesa: Odesa National Maritime University) 20 p

[10] Denisov I R 2005 Role of Sea Ports in the System of Combined Transportation *Bulletin of Transport* no 1 pp 34-40

[11] Shapkin I N 2008 Organization of rail transportation based on discrete methods of management and a strict schedule of trains traffic *Transport. Science, Technology, Management* no 8 (Moscow: VINITI RAS) pp 2-8

[12] Budko T V and Lomotko D V 2007 Issues of organization of information interaction of railway and water transport, as well as logistic structures *Railway Transport of Ukraine* no 6 pp 62-65

[13] The Association Agreement between Ukraine, on one part, and the European Union, the European Atomic Energy Community and their member states, on the other part, the Verkhovna Rada of Ukraine, Kyiv, 2015. http://zakon2.rada.gov.ua/laws/show/984_011. Revised December 2017. Accessed December 18, 2017

[14] Marinov M, Zunder T, Mortimer P 2012 Rail Freight Services *World Transport, Policy & Practice* **18**(2) pp 30-38

IOP Conf. Series: Materials Science and Engineering **582** (2019) 012030 doi:10.1088/1757-899X/582/1/012030

Modeling a Stress Condition of Ore Massif in the Process of Bauxite Deposits Development in the Mines of «Sevuralboksitruda»

N I Kosukhin [1, a], D V Sidorov [1, b], L B Gil [2, c]

[1] Saint-Petersburg Mining University 199106 St. Petersburg, V.O. 21st line, house 2, Russia
[2] Tomsk Polytechnic University 634050 Tomsk, Lenina prospect 30, Russia

E-mail: [a] nikolay.kosuhin@uralkali.com, [b] sidorov-post@yandex.ru,
[c] gileno@mail.ru (corresponding author)

Abstract. The publication examines the problems of concentrator of elevated tension formation in pillars and ore massif. The article presents the results of modeling the stress state of interchamber pillars and ore deposits, using specialized geomechanical software PRESS 3D URAL. In the process of numerical experiments, it was found that the increase of natural flexibility in pillars leads to a decrease of normal stresses in comparison with the variant of a stiff scheme. If interchamber pillar involves different varieties of ores, there is a redistribution of stress on the site with more stiff ores. The article shows that the stress state control of interchamber pillars and ore massif in conditions of OJSC «Sevuralboxitruda» can be achieved by drilling relief holes.

Introduction

Forecasting the stress state of the rock mass is a very important task to ensure the safety of mining operations for many mining enterprises. [1-8]. This is due to the fact that at present depths of development in the zone of stope influence there is a significant change and redistribution of the stress field. A significant increase in tension at the approach of the front of the treatment works is experienced by areas with a lower capacity of the ore body and its greater strength. A significant tension increase in process of approaching the stopping front get areas with less thickness and greater strength of ore deposits The increase in normal stresses, acting in massif, in areas with more strength varieties of bauxite ore, determines their ability to accumulate significant elastic energy. At the same time, in the areas of the ore deposit, represented by less strength varieties of ores, as a rule, there is a process of active cracking, fracturing, which indicates a gradual withdrawal of latter from under the load and its redistribution to neighboring stronger areas. The situation becomes more complicated in the case of using room and pillar development systems. This problem is particularly relevant for mining enterprises, developing Severouralsk bauxite basin (OJSC «Sevuralboksitruda») in complex geological, mining and technical, geodynamic conditions.

Specialized geomechanical software PRESS 3D URAL [9-13] allows to solve spatial elastoplastic problems of geomechanics and can be used to identify the regularities of stress distribution in the mining-induced zones of ore deposit and pillars, as well as to determine the optimal parameters of preventive measures to prevent rock bumps and other geodynamical events.

Research objective

The aim of the study is to assess the stress state of the ore mass under different loading regimes of interchamber pillars in the process of underground mining of ore deposits of complex geological structure to improve the efficiency of mines operation at great depths.

IOP Conf. Series: Materials Science and Engineering **582** (2019) 012030 doi:10.1088/1757-899X/582/1/012030

Methods

Software PRESS 3D URAL allows to edit the source of mining and graphic documentation, to carry out numerical and analytical calculations, as well as to process the results. The software is developed in the system of visual programming Delphi Pro and allows with a high degree of automation to form a source database, including creation of model structural elements, initialize their properties: depth, thickness, physical-mechanical characteristics, tectonic dislocation characteristics and parameters of relief holes in study area.

Numerical-analytical block includes algorithms, which implement various integro-differential calculation methods, containing a significant amount of cyclic structures. The processor block of the PRESS 3D URAL software, allows to calculate stresses with the cumulative account of a wide range of mining-geological and mining-technical factors: mined-out spaces and pillars arbitrary in size and configuration, arbitrary location of pillars relative the boundaries of mined-out space, parameters of tectonic disturbances and thickness of relief slit. Setting of boundary conditions is carried out by specifying the values of loads, removed from the sole of excavation. The influence of this interaction on the mining-induced zone is taken into account by introducing the displacement parameter q, corresponding to the linear approximation of the stresses, acting on the excavation sole.

Processing unit of calculation results (stresses, strains, displacements, rock-bump hazard zones, allocation of rock-bump hazard zones within the ore deposit and destruction zones on the planes of tectonic fault, the parameters of relief holes) using technology of export results in a modern graphics visualization systems (AutoCad, Serfer, CorelDraw, etc).

To assess the effect of natural and man-made flexibility of ore interchamber pillars on their stress-strain state, numerical experiments were performed for the scheme of the chamber-column development system (with open developed space), presented in figure 1.

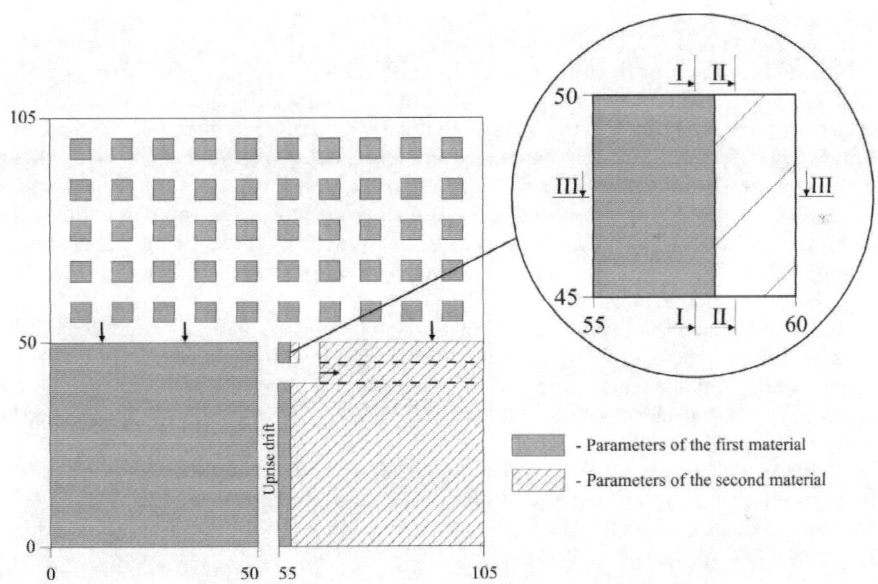

Figure 1. Basic design scheme

The initial data of the computational models, characterizing the possible combinations of the studied parameters, are given in table 1. To study the influence of the spatial configuration of ore deposit boundary part contour on its stress state, the conditions of model 1 were accepted.

Table 1. Initial data for computational models

Model	Parameters of the first material				Parameters of the second material				Note
	m_d, m	E_d, MPa	E_r, MPa	Δ_h, m	m_d, m	E_d, MPa	E_r, MPa	Δ_h, m	
1	0.5	$6.5 \cdot 10^4$	$3.0 \cdot 10^4$	0.00	0.5	$6.5 \cdot 10^4$	$3.0 \cdot 10^4$	0.00	Rigid scheme
2	6.0	$0.6 \cdot 10^4$	$3.0 \cdot 10^4$	0.00	6.0	$0.6 \cdot 10^4$	$3.0 \cdot 10^4$	0.00	Pliable scheme
3	6.0	$0.6 \cdot 10^4$	$3.0 \cdot 10^4$	0.00	2.0	$1.5 \cdot 10^4$	$3.0 \cdot 10^4$	0.00	Rigid-pliable scheme
4	6.0	$0.6 \cdot 10^4$	$3.0 \cdot 10^4$	0.00	2.0	$1.5 \cdot 10^4$	$3.0 \cdot 10^4$	0.01	Rigid-pliable scheme with relief holes
5	6.0	$0.6 \cdot 10^4$	$3.0 \cdot 10^4$	0.00	2.0	$1.5 \cdot 10^4$	$3.0 \cdot 10^4$	0.03	
6	6.0	$0.6 \cdot 10^4$	$3.0 \cdot 10^4$	0.00	2.0	$1.5 \cdot 10^4$	$3.0 \cdot 10^4$	0.05	

where m_d – thickness of ore deposits, m;
E_d – modulus of elasticity of ore Deposit, m;
E_r – modulus of elasticity of rock, m;
Δ_h – thickness of the loading slot, m.

Results

The results of numerical calculations of dimensionless additional stresses, acting in the boundary part of the ore deposit and interchamber pillar, are shown in figure 2. The results of numerical experiments for the conditions of models 1-6 are shown in figures 3-4. Analysis of the results of numerical evaluation of interchamber pillar stress state, shown in figures 3-4, allows us to make the following conclusions.

An increase in natural flexibility of interchamber pillar due to an increase in it's power and (or) a decrease in strength of the ore (figure 3, b) leads to a decrease in normal stresses, acting in it by 1,4 times compared to the variant of the rigid scheme (figure 3, a). In the case where the inter-chamber pillar is composed of different types of ores in strength (figure 3, c), there is a redistribution of stresses on the site with more rigid ores, while several unloading adjacent areas of the pillar. The value of additional loading on rigid section is an average of 20% (figure 3, b) compared with the variant, characterizing the uniform strength of interchamber pillar (figure 3, a). At the same time, the value of additional loading of rigid sections in the interchamber pillar will increase as width of rigid section decreases. A significant reduction of operating in pillar stresses can be achieved by giving the target man-made compliance (for example, by drilling discharge wells).

Giving the interchamber pillar additional man-made flexibility, leads to a significant decrease in stresses in it (figure 4). Control of the stress state of interchamber pillar can be carried out by changing the thickness of the man-made gap (parameters of relief holes). The results of the calculation show that the reduction of stresses in the areas of increased stiffness to the average level is achieved with the thickness of the man-made slot $\Delta_h = 0.03\ m$.

Significant practical experience, focused on the unloading of elevated stress concentrator in the mines of «Sevuralboksitruda» shows that when the thickness of the man-made gap is about 0.03 m and a diameter of relief holes amount 0.105 m, location of relief holes in the gap need to be done through two diameters of holes (0.2 m).

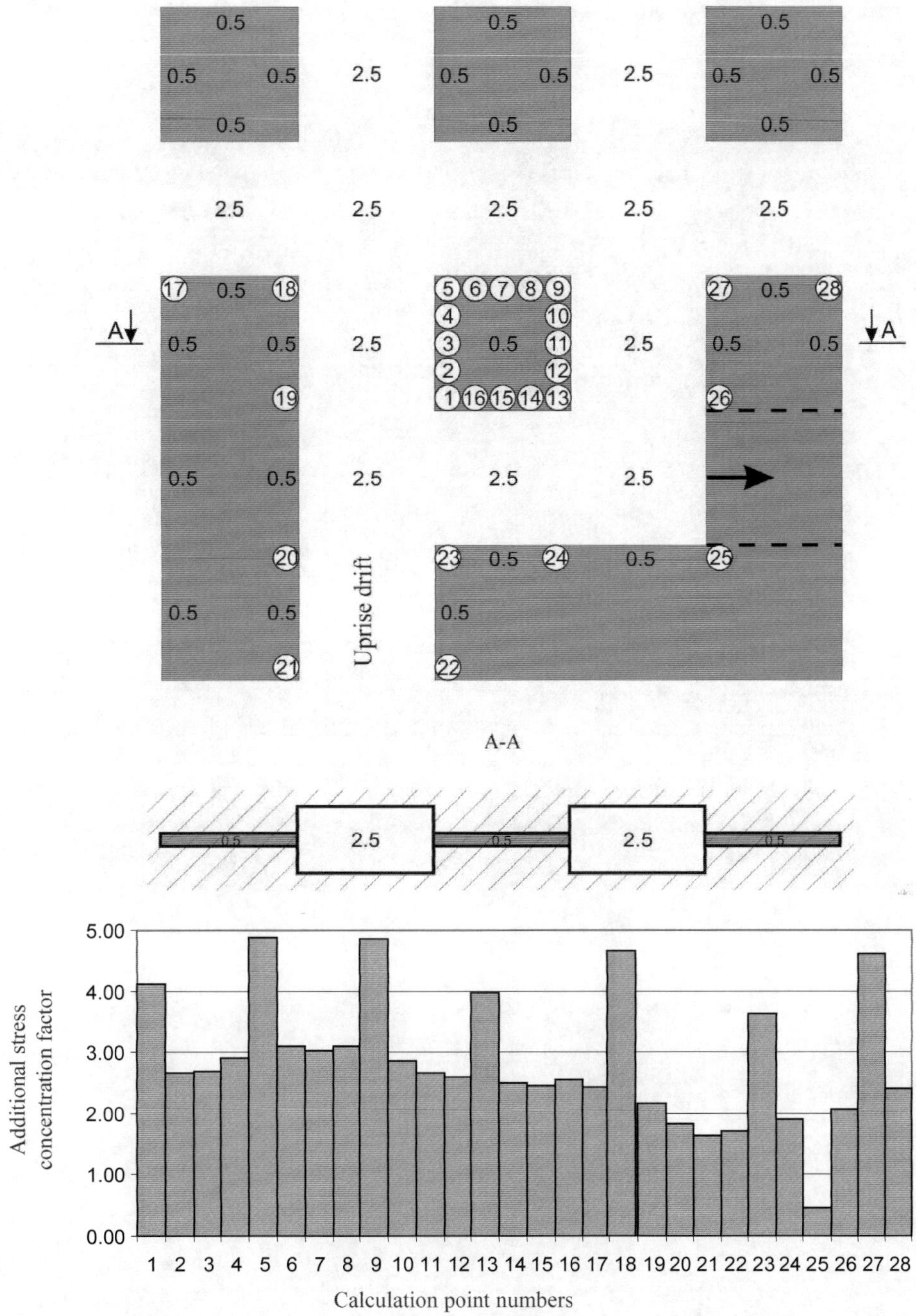

Figure 2. Results of stress state evaluation of ore deposit taking into account the spatial configuration of the boundary part of the deposit

IOP Conf. Series: Materials Science and Engineering **582** (2019) 012030 doi:10.1088/1757-899X/582/1/012030

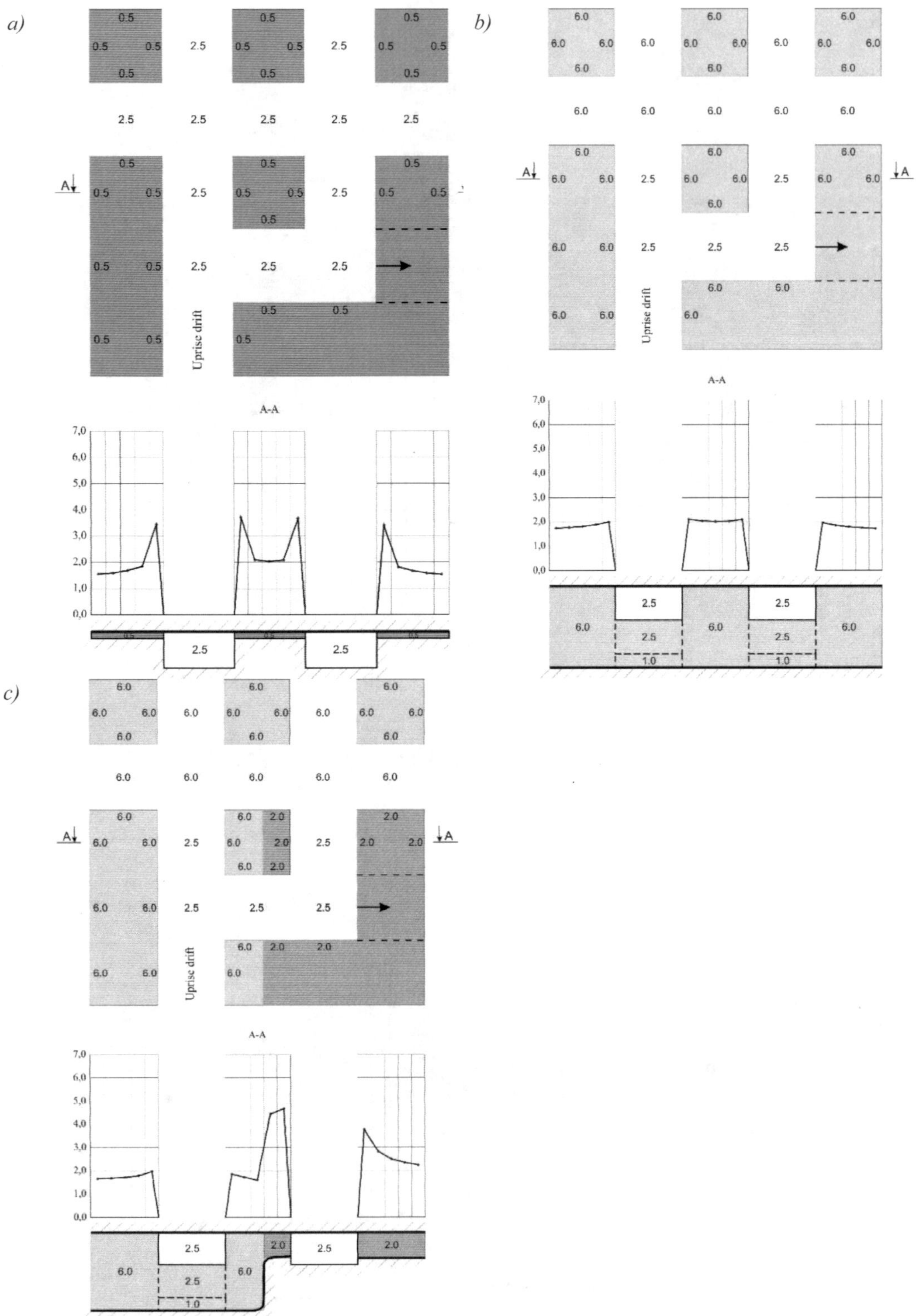

Figure 3. Results of stress state evaluation of the interchamber pillar for reference conditions of models 1-3: a – rigid scheme; b – pliable scheme; c – rigid-pliable schem

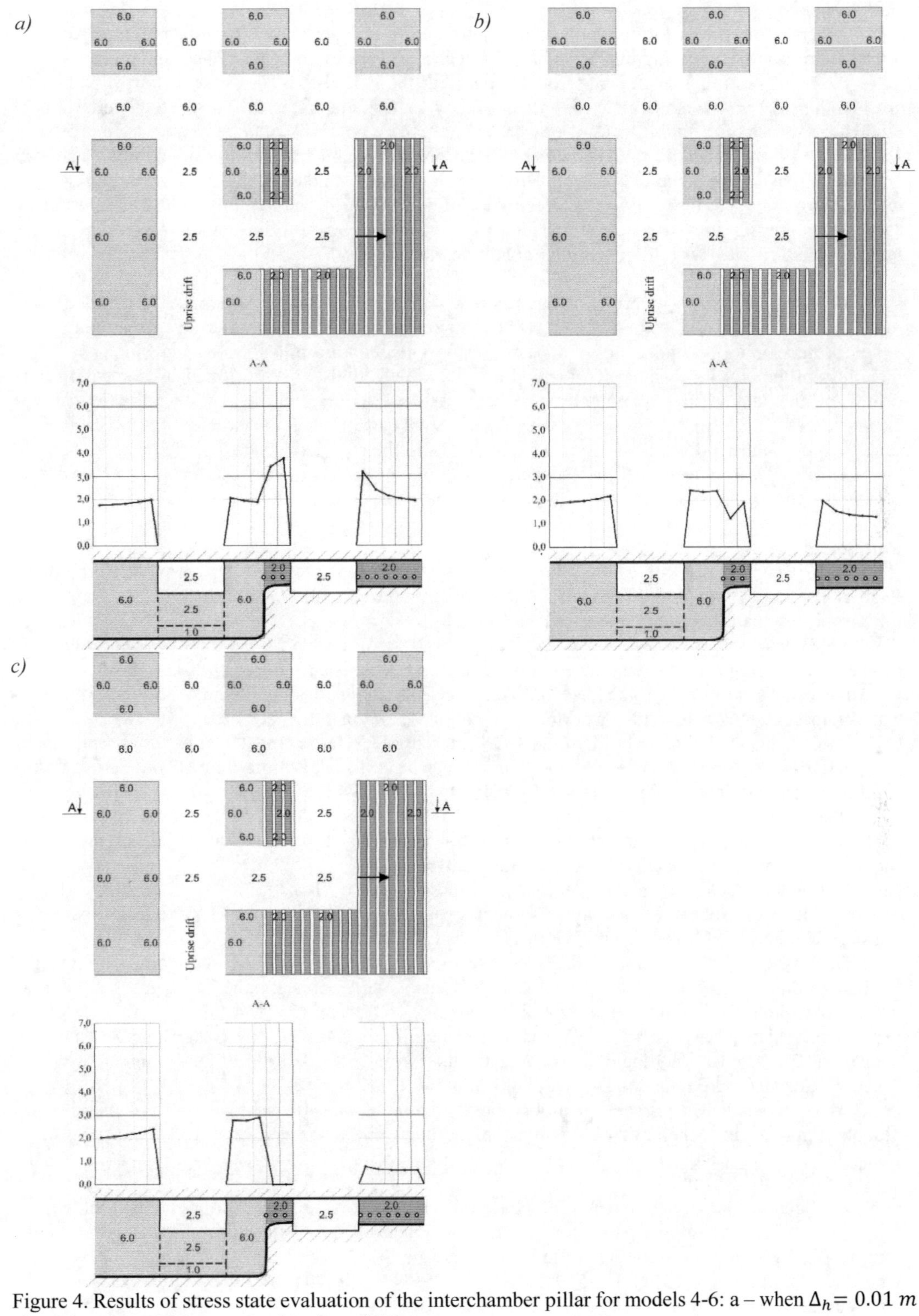

Figure 4. Results of stress state evaluation of the interchamber pillar for models 4-6: a – when $\Delta_h = 0.01\ m$; b – when $\Delta_h = 0.03\ m$; c – when $\Delta_h = 0.05\ m$

Conclusion

As a result, of the performed studies, the regularities of changes in the stress state of interchamber pillars were established, taking into account the influence of various geological and mining factors.

An increase in natural flexibility of interchamber pillar leads to a decrease in normal stresses, acting in it by 1.4 times compared to the variant of the rigid scheme. In the case where the interchamber pillar is composed of different types of ores in strength, there is a redistribution of stresses on the site with more rigid ores, while several unloading adjacent areas of the pillar. The value of additional loading on rigid section is an average of 20% compared with the variant, characterizing the uniform strength of interchamber pillar. Giving the interchamber pillar additional man-made flexibility, leads to a significant decrease in stresses in it. It is recommended to use relief holes to control the stress state of the interchamber pillars. For the operating conditions of mines of «Sevuralboksitruda» it can be recommended drilling relief holes every 0.2 m.

Similar qualitative regularities of stress-strain state of interchamber pillars were identified as a result of widespread mining experimental studies, carried out on ores of Dzhezkazgan and Mirgalimsay, which is developing a copper-nickel deposits, using chamber-and-pillar system development. The reliability of the developed modified method of integral equations was confirmed by comparing the results of the numerical experiment with the results of mining and experimental work (mine "Kalinskaya" OJSC "SUBR"), as well as with experimental data of laboratory studies of the stress-strain state of interchamber pillars, using equivalent materials in the certified laboratory of rock modeling of the Scientific research center of geomechanics and problems of mining production in St. Petersburg mining University.

References

[1] V.A. Eremenko, Kontrol geomexanicheskogo sostoyaniya geologicheskoj sredy` pri otrabotke Sheregeshevskogo mestorozhdeniya / V.A.Eremenko, A.A.Eremenko, V.N.Filippov i dr. // Gorny`j informacionno-analiticheskij byulleten. 12 (2007) 155–169.
[2] Yu.P Galchenko., V.A. Eremenko, A.V. Myaskov, M.A. Kosyreva, Solution of geoecological problems in underground mining of deep iron ore deposits // Eurasian Mining. 1 (2018) 35–40.
[3] A.P. Gospodarikov, Y.N. Vykhodtsev, M.A. Zatsepin, Mathematical modeling of seismic explosion waves impact on rock mass with a working / Journal of Mining Institute, 226 (2017) 405 – 411.
[4] N.I. Kosukhin, D.V. Sidorov, I.I Beloglazov, V.Yu. Timofeev, Assessment of stress-strain and shock bump hazard of rock mass in the zones of high-amplitude tectonic dislocations // IOP Conf. Series: Earth and Environmental Science, 224 (2019) 1-5, DOI: 10.1088/1755-1315/224/1/012014.
[5] T. Ponomarenko, I. Sergeev, Quantitative methods for assessing levels of vertical integration as a basis for determining the economic and organizational sustainability of an industrial corporation // Indian Journal of Science and Technology, 9 (20) (2016).
[6] I.Yu. Rasskazov, Geodinamicheskoe sostoyanie massiva porod Nikolaevskogo polimetalliche-skogo mestorozhdeniya i osobennosti proyavleniya udaroopasnosti pri ego osvoenii / I.Yu. Rasskazov, B.G. Saksin, V.I. Usikov, M.I. Potapchuk // Gorny`j zhurnal. 12(2016) 13–19.
[7] I.Yu. Rasskazov, Geomexanicheskaya ocenka uslovij razrabotki glubokix gorizontov polimetal-licheskogo mestorozhdeniya «Yuzhnoe» / I.Yu. Rasskazov, G.A. Kursakin, M.I. Potapchuk // Fiziko-texnicheskie problemy` razrabotki polezny`x iskopaemy`x. 5 (2012). 125–134.
[8] V.S. Koshkina, A.A. Serova, V.Y. Timofeev, Environmental-Toxicological Characteristics of Waters and Their Sources at Magnitogorsk With the Its Iron and Steel Industry //IOP Conference Series: Materials Science and Engineering. – IOP Publishing, 142(1) (2016) 012117.
[9] D.V. Sidorov, Prognozirovanie udaroopasnosti tektonicheski narushennogo rudnogo massiva na glubokix gorizontax Nikolaevskogo polimetallicheskogo mestorozhdeniya / D.V. Sidorov, M.I. Potapchuk, A.V. Sidlyar // Zapiski Gornogo instituta. 234 (2018) 604–611. DOI: 10.31897/PMI.2018.6.604.
[10] D.V. Sidorov, T.V. Ponomarenko, The development of a software suite for predicting rock bursts within the framework of a system for ensuring geodinamic safety of mining operations // 17th International Multidiscip-linary Scientific GeoConference SGEM 2017, www.sgem.org, SGEM 2017 Conference Proceedings, ISBN 978-619-7408-02-7 / ISSN 1314-2704, 29 June - 5 July, (2017), Vol. 17, Issue 22, 633-638 pp, DOI: 10.5593/sgem2017/22/S09.079.

[11] V.Y. Timofeev et al. The Mutual Effect of Reciprocally Moving Geokhod and Geological Environment Studied by the Discrete Element Method in Software PFC3D 5.00 //IOP Conference Series: Materials Science and Engineering. Vol. 142: Innovative Technologies in Engineering.—Bristol, 2016. – IOP Publishing, 142(2016) 12126.
[12] Z. Zhang, H. Shimada, T. Sasaoka, W. Kai, Study on the overlying strata movements and stability control of the retained goaf-side gateroad // Memoirs of the Faculty of Engineering, Kyushu University.76 (2) (2016) 1–17.
[13] S.V. Kovshov, D.A. Iconnicov, Growing of grass, radish, onion and marigolds in vermicompost made from pig manure and wheat straw // Indian Journal of Agricultural Research. – 51 (4)(2017).

IOP Conf. Series: Materials Science and Engineering **582** (2019) 012031 doi:10.1088/1757-899X/582/1/012031

Studying the Trajectory of Moving Particles in the Air-Screw Separator

S D Shepelev, M V Cheskidov, I Yu Novikova, F N Grakov, N A Kuznetsov

South Ural State Agrarian University, Russia

E-mail: mister.aspirant@yandex.ru

Abstract. When cleaning the heap of grain, the impurities reducing the grain value and affecting its safety, in particular, possible self-heating effect and further deterioration are to be released. For grain cleaning an air-screw separator is developed to separate particles moving along it helically. Computer simulation and experimental studies made it possible to detect the trajectory of moving particles and determine the optimal air flow rate.

Introduction

Cereals cultivation is one of the primary directions for agriculture. However, climatic conditions have a significant impact on the efficiency of grain harvesting and cleaning [1, 2]. When the agrotechnical period of grain harvesting exceeds, the grain quality and yield are significantly reduced [4, 5]. Choosing crops and varieties according to their maturing rates makes it possible to increase the agrotechnical period of grain harvesting and post-harvesting cleaning and involve fewer technical means [6, 7]. The most important stages in grain processing affecting its final quality and cost are pre-cleaning and drying [8]. When cleaning, the impurities reducing the grain value and affecting its safety, in particular, possible self-heating effect and further deterioration of grain are released from the grain heap. For separating light impurities such as parts of stems, husk, chaff, broken grain, dust the principle of mixture separation in the air flow is used according to the difference in aerodynamic properties and mass of particles [9, 10]. The heap for cleaning has non-uniform moisture; therefore, the particles have different terminal velocities. In the end, this leads to either the presence of whole grains in the waste, or the presence of trash in the grain after cleaning. It is possible to change the situation for the better by introducing another force into the system by complicating the motion trajectory of particles. A similar principle is used in the air-screw separator (Figure 1) where particles move along helically. However, at the moment, the motion trajectory of particles with different mass is not defined. In modern science there are a large number of mathematical methods and models for separating impurities due to centrifugal forces. ANSYS CFX is a high-performance tool for computational hydrodynamics, a combination of theoretical, experimental and numerical methods designed to simulate the flow of liquids and gases, heat and mass transfer processes, reactive flows, etc.

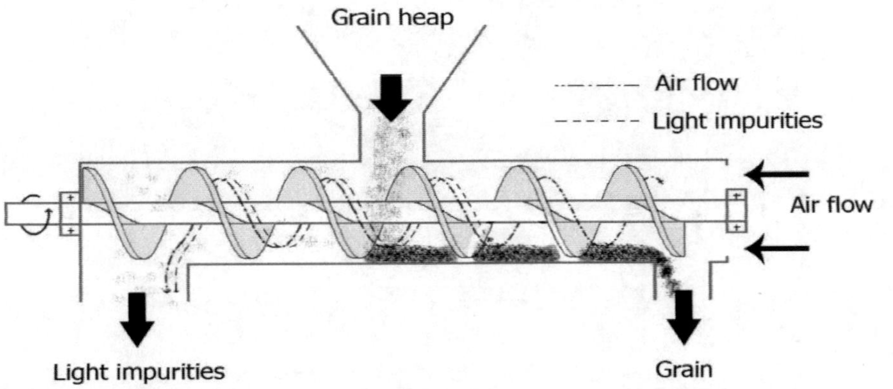

Figure 1 – The operation principle of the air-screw separator

Materials and Methods

To simulate the movement of particles in the air-screw separator, the boundary conditions for simulation were set and a part of the separator was modeled, representing one turn of a screw in a cylindrical body (Figure 2), the separator diameter being 300 mm, the screw pitch being 300 mm, the screw axis being 80 mm. An ideal gas at the temperature of 20^0C was chosen as the main medium. As a result of computer simulation, the trajectory of the air flow moving helically was determined.

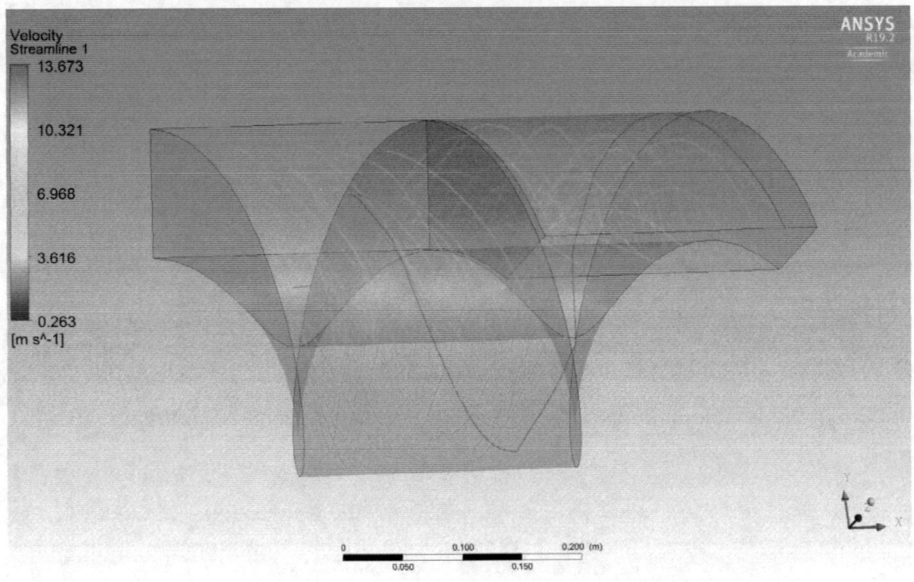

Figure 2 – The separator modeled with ANSYS CFX software

For successful separation, all light impurities should be removed from the initial grain heap, and whole grains in the waste should not exceed 1.5%. As broken grain is the fraction to be most difficult separated, its damage exceeds 50%. The operating mode of the separator, when the particles are removed by flowing air, is optimal. When the air flow rate in the modeled air-screw separator is 7.5 m/s, impurities are to be carried away by flowing air, while whole grains remain in the separator. The trajectory of moving impurities is presented in Figure 3.

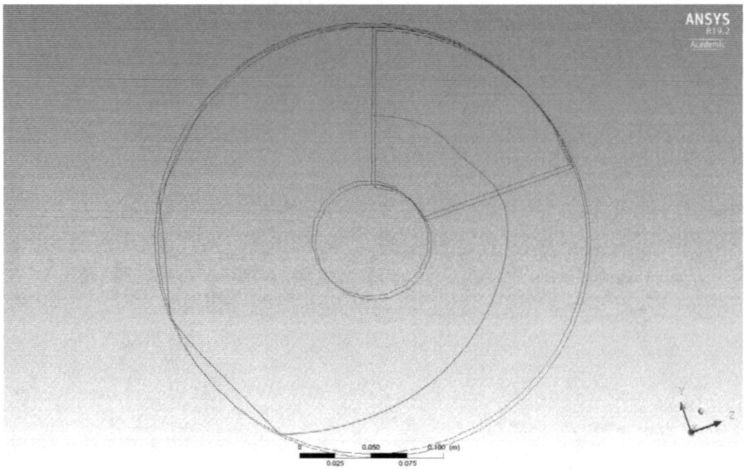

Figure 3 – The trajectory of moving impurities

To confirm the theory and studies on the trajectory of particles in the air-screw separator, a laboratory installation was made (Figure 4). The most part of the installation was made of translucent PET material (polyethylene terephthalate) to fix the experiment with the help of a camera.

To create the air flow, a 1.2 kW centrifugal fan with a regulating damper was used to change the air flow rate.

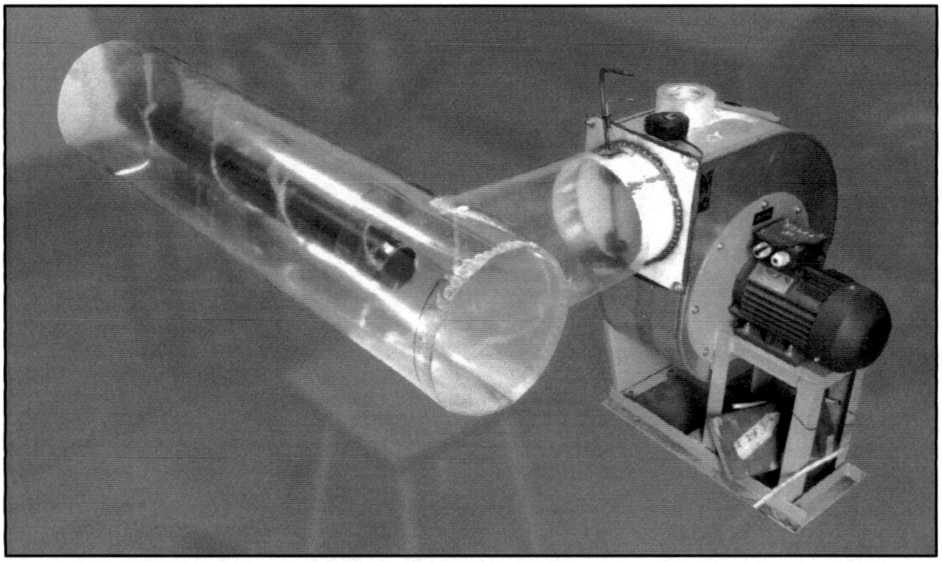

Figure 4 – Laboratory installation

Photo and video shooting was carried out with the camera in two directions: in the frontal and side ones. The video shooting frequency was 30 frames per second. Grain and light impurities were used for the experiments. The air flow rate was chosen according to the recommendations for wheat with 18% moisture. As a result of the experiment, according to the video analyzed, the location points of particles were marked to detect the trajectory of their movement.

Results and Discussion

The experimental studies showed that when the air flow rate is V=7.5 m/s, light particles are easily carried along by the air flow and removed from the separator, while the force of the air flow is not enough for whole grains to make a full turn in the screw channel of the separator (Figure 5).

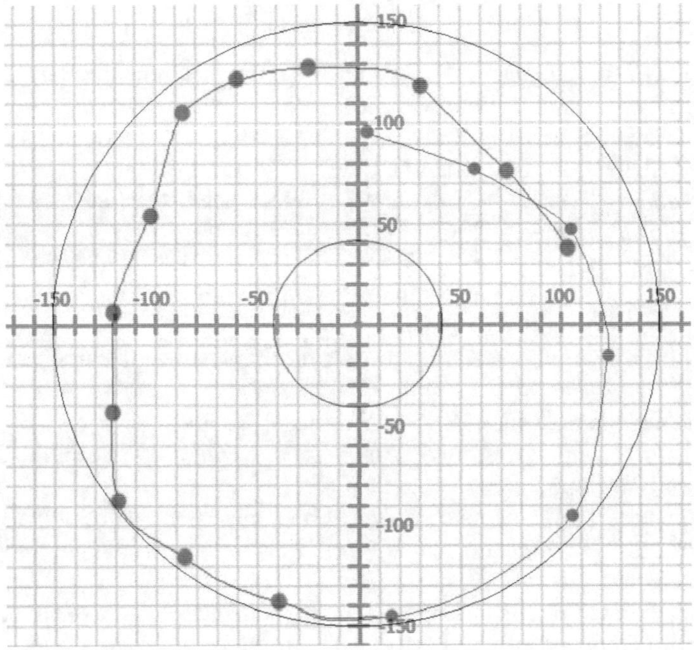

Figure 5 – The trajectory of moving impurities in the experimental installation

For further comparing the obtained results, the dependency graphs for the particles locating on the Y-axis and the experiment timing were constructed (Figure 6).

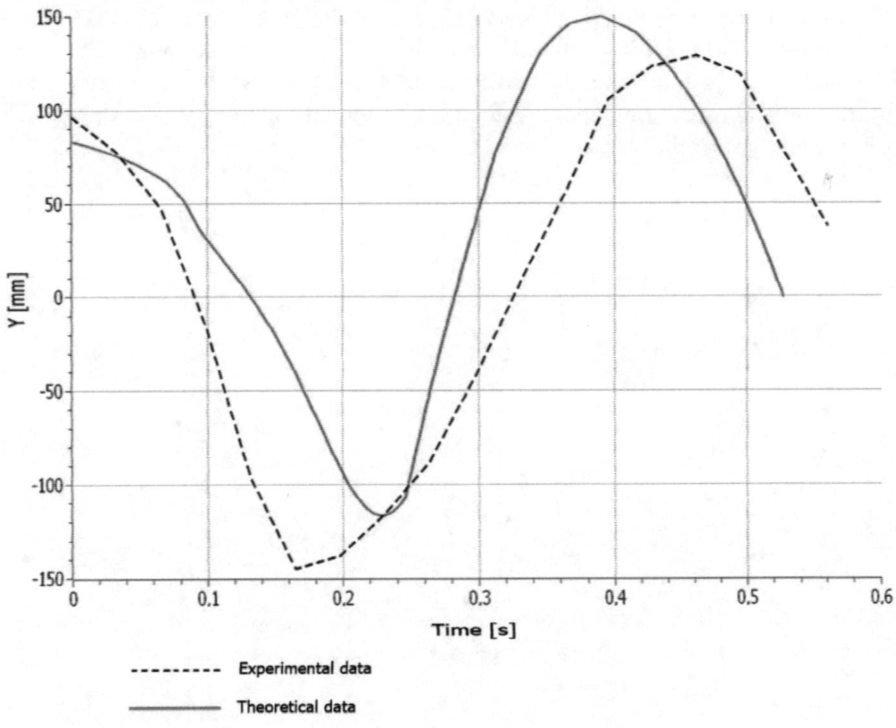

Figure 6 – The dependency graphs for the particles locating on the Y-axis and the experiment timing

Conclusion

Thus, the analysis of the data obtained makes it possible to say with confidence that the theoretical data on the air flow rate necessary for the separation process in the screw channel are confirmed by the experimental studies.

References

[1] Nawi N.M., Chen G., & Zare D. (2010). The effect of different climatic conditions on wheat harvesting strategy and return. *Biosystems Engineering*, 106(4), 493-502. doi:10.1016/j.biosystemseng.2010.05.015.

[2] Ziauddin A. M., & Liang T. (1986). WEATHER RISK MODEL FOR GRAIN DRYING SYSTEM DESIGN AND MANAGEMENT IN A DEVELOPING REGION. *Transactions of the American Society of Agricultural Engineers*, 29(5), 1434-1440. Retrieved from www.scopus.com.

[3] De Toro A., Gunnarsson C., Lundin G., & Jonsson N. (2012). Cereal harvesting – strategies and costs under variable weather conditions. *Biosystems Engineering, 111*(4), 429-439. doi:10.1016/j.biosystemseng.2012.01.010.

[4] Bolland M.D.A. (1984). Grain losses due to delayed harvesting of barley and wheat. *Australian Journal of Experimental Agriculture*, 24 (126), 391-395. doi:10.1071/EA9840391.

[5] Shinners K.J., Binversie B.N., Muck R.E., & Weimer P.J. (2007). Comparison of wet and dry corn stover harvest and storage. *Biomass and Bioenergy*, 31(4), 211-221. doi:10.1016/j.biombioe.2006.04.007.

[6] Shepelev S.D., Shepelev V.D., Almetova Z.V., Shepeleva N.P., Cheskidov M.V. Modeling the Technological Process for Harvesting of Agricultural Produce. IOP Conference Series: Earth and Environmental Science, 2018.

[7] Shinners K.J., Boettcher G.C., Hoffman D.S., Munk J.T., Muck R.E., & Weimer P.J. (2009). Single-pass harvest of corn grain and stover: Performance of three harvester configurations. *Transactions of the ASABE*, 52(1), 51-60. Retrieved from www.scopus.com.

[8] De Toro A. (2005). Influences on timeliness costs and their variability on arable farms. *Biosystems Engineering*, 92(1), 1-13. doi:10.1016/j.biosystemseng.2005.06.007.

[9] Gorial B.Y., & O'Callaghan J.R. (1990). Aerodynamic properties of grain/straw materials. *Journal of Agricultural Engineering Research*, 46(C), 275-290. doi:10.1016/S0021-8634(05)80132-5.

[10] Wold M.T., Kocher M.F., Keshwani D.R., & Jones D.D. (2011). Modeling the in-field logistics of single pass crop harvest and residue collection. Paper presented at the American Society of Agricultural and Biological Engineers Annual International Meeting 2011, ASABE 2011, 168-86. Retrieved from www.scopus.com.

The Structure of Training Program and Advanced Training of Enterprise Managers in Order to Ensure Environmental Safety

E N Abanina[1,a], Yu S Sergeenko[1,b], O V Devyatov[1,b], O Yu Ganyukhina[1,a], Yu M Nikitenko[2,b]

[1]Federal State Budget Educational Institution of Higher Education "Saratov State Law Academy", Volskaya st., 1, 410056 Saratov, Russia
[2] The Faculty of Law, Leningrad State University named after Alexander Pushkin, St. Petersburg sh., d. 10., Pushkin, 196605, St. Petersburg, Russia

E-mail: [a]elena-abanina@yandex.ru, [b]ulya.sergeenko@yandex.ru

Abstract. Large, medium-sized and small enterprises have violation of environmental requirements and negative impact on the environment quality through emissions, discharges, waste generation. The authors suppose that one of the causes of a high degree environmental pollution in the course of economic activity by enterprises is the lack of a system for training managers and specialists in specific courses with binding studying regulatory documents on environmental safety. The article analyzes the reasons for necessary development of a compulsory program of training and advanced training of business managers in order to ensure environmental safety. The dependence of the level and complexity of preparation on the level of environmental hazard of the enterprise is shown. In order to ensure environmental safety it is necessary to define the purpose and objectives of training programs for managers. At the end, the authors present the developed model structure of the training program and advanced training of enterprise managers in order to ensure environmental safety.

Introduction

Environmental safety of regions is imperative for sustainable development of the region, is the basis of preservation of natural systems, and maintain environmental quality. The basic meaning of environmental safety associated with the necessity of compliance with environmental regulations. The violation of such regulations affects the public health seriously and leads to the destruction of the stability of ecological systems [1].

In order to follow the goals of ensuring environmental safety and sustainable development, enterprises in many countries of the world are encouraged to develop environmental innovations actively and to exercise greening of production. It is about “developing any new or significantly improved products (goods or services), processes, and organizational changes or marketing solutions that reduce the consumption of natural resources (including materials, energy, water and land) and reduce emissions of harmful substances during the entire life cycle” [2]. Unfortunately, for example, in Russia, a small number of enterprises apply greening production tools (for example, creating and implementing a new waste management system) and are aware of the benefits of green manufacturing. Enterprises often complain that it is burdensome to follow changes in environmental requirements; especially it is difficult for them to understand which requirements they should comply directly [3]. Resolving the current situation can help the environmentally competent head of the company.

Russian legislation establishes the obligation of business managers to have training in environmental safety. Also, the obligation to provide training related to preparedness for emergency and other abnormal situations and response actions for persons managing an organization is provided for by ISO standards on environmental management [4].

However, at present in Russia and in many other countries there are no appropriate requirements for the managers training, as well as the procedure for the approval of training programs for their preparation.

The purpose of this work is to develop model structure of the training program and advanced training of enterprise managers in order to ensure environmental safety.

Tasks of the research:

- identify threats to the environment from individual objects of economic activity, depending on their danger;
- determine the complexity of the training programs for enterprise managers depending on the hazard levels of the facilities;
- define the goal and objectives of the general program for training managers to ensure environmental safety, regardless of the hazard level of enterprises;
- theoretically develop model structure of the training program and advanced training of enterprise managers in order to ensure environmental safety.

Results and discussion

It is indisputable that environmental safety is an integral part of general safety in the work performance on site that poses a potential hazard to the environment. Therefore, the manager must have the necessary level of training to ensure the security of his or her enterprise.

Such managers can be recognized as subjects of environmental danger - persons whose activities can create an environmentally dangerous situation. Of course, these categories of citizens, as a rule, already have a great deal of practical management experience and special education. They studied the basics of ecology, life safety, the fundamentals of law and economics. However, since they got their education, these managers have received new work skills, and faced with new features of work and difficulties associated with the necessity to ensure environmental safety. Therefore, competent management of enterprises is necessary to prevent environmental disasters, to preserve the lives of people on earth.

Table 1 – Correspondence of environmental threats to business objects depending on their danger

Threats	Potential consequences
energy facilities (for the provision of electric energy, gas and steam)	air pollution [5], thermal water pollution
hydraulic structures	water pollution, coastal destruction, extreme events (floods)
facilities for the operation of radiation sources, nuclear facilities, including nuclear plants	degradation of natural objects, population exposure, disease and death of the population
facilities for the oil extraction, natural gas, heavy metals, precious metals	1. soil contamination with high-tech material [6]; land degradation, water bodies
facilities for the transport, storage, destruction of chemicals, including weapons	chemical contamination or damage to people, animals and plants, earth pollution, water with dangerous chemicals
facilities that emit pollutants into the air	air pollution, subsequently disease of the population
facilities that discharge pollutants into water bodies	polluted drinking water, transboundary pollution, extreme events (floods) [7]
facilities involved in the storage, disposal, disinfection, industrial and consumption waste disposal	pollution and land degradation, large amounts of legal and illegal waste landfills [8], air pollution due to landfill gas emissions [9], air pollution from waste incineration [10]

Enterprises headed by managers are called objects of environmental hazard. These enterprises may have a negative impact on nature and man. Most of the enterprises, depending on the implementation of economic activity, can be sources of various threats to certain objects of nature and to the natural environment as a whole. Under the threat of environmental safety, we understand the types of economic or other activities that have a harmful impact on the environment and pose a danger to human life and health due to violations of environmental safety requirements.

The table shows the types of objects (enterprises) that pose a threat to environmental safety and potential consequences arising in case of violation of environmental requirements.

Based on the data in the table, we can conclude that every object of environmental hazard presents threats of varying degrees of danger to nature and to humans. Therefore, it is necessary to prepare managers of enterprises in the field of environmental safety. Moreover, for especially dangerous objects, it is necessary to conduct additional or special training and certification of managers (see Table 2).

Table 2 – Compliance of training programs for enterprise managers with environmental hazards

Threats	Training of managers
energy facilities (for the provision of electric energy, gas and steam)	training and certification in the field of industrial safety
hydraulic structures	training and certification in the field of industrial safety
facilities for the operation of radiation sources, nuclear facilities, including nuclear plants	training and certification in the field of atomic energy use, radiation safety
facilities for the oil extraction, natural gas, heavy metals, precious metals	training and certification in the field of industrial safety
facilities for the transport, storage, destruction of chemicals, including weapons	training and certification in the field of industrial safety, chemical safety, special training and certification on the organization and work safety
facilities that emit pollutants into the air	general environmental safety training
facilities that discharge pollutants into water bodies	general environmental safety training
facilities involved in the storage, disposal, disinfection, industrial and consumption waste disposal	general training and certification in the field of environmental safety, special training and certification on the organization and work safety

One of main reasons for the high level of environmental pollution at work is the lack of a system for training managers with binding studying regulatory documents on environmental safety. The head of the company should have a clear understanding of the system of environmental legislation and know the hierarchy of laws. (Figure 1).

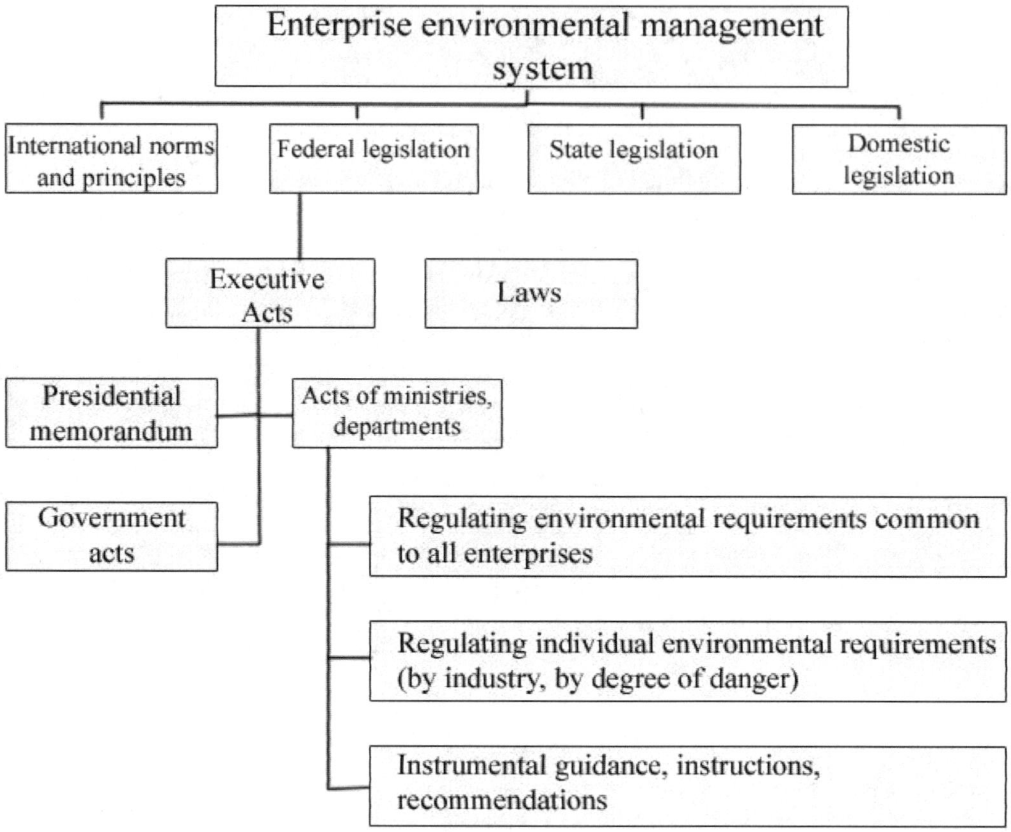

Figure 1. The system of regulatory documents on environmental safety

In this connection, in the course of general training of managers in order to ensure environmental safety, it is necessary to focus on the legal environmental issues necessary for a manager to perform his or her duties.

Before developing separate training programs for managers, it is necessary:

- to form up an obligatory minimum set of legal (economic, technical) knowledge for each profession;
- to establish a law on the state level about the admission of the manager to the management of a potentially dangerous object to the environment, that may be possible only after passing an exam before a special commission on the established minimum set of ecological knowledge;
- to enshrine at statutory level that the training of the head should be carried out at the expense of the enterprise, and at especially dangerous facilities - at the expense of the state;
- to hold legal responsibility (disciplinary (up to dismissal), administrative and other), for refusing to train, to pass an exam, and for unsatisfactory passing an exam.

Managers of enterprises who pose a small risk to the environment can receive training in the following general programs:

1. Environmental safety management systems.
2. Environmental protection.
3. Environmental management and audit based on ISO (International Organization for Standardization) standards.

Such general training programs will provide the minimum knowledge necessary to manage any enterprise in order to ensure environmental safety. In addition, training in such programs will not require large funding, since such training can be carried out both on the basis of educational organizations and private centers with qualified teachers in the field of law, ecology and economics, but without access to special knowledge (for example, in the field of atomic energy).

Russian companies have already successfully used this opportunity. For example, in such public corporation as System of Oil Pipelines in 2010–2012, 47 employees of the company (managers responsible for making decisions that could affect environmental safety) were trained in the environmental protection and environmental safety-training program. Besides training, professional development, a number of activities have been undertaken to improve the environmental culture of workers [11].

In order to ensure environmental safety, the purpose of the general program of training managers, regardless of the level of enterprises' risk, is to obtain specialized knowledge in a specific field of activity - ensuring environmental safety. The purpose of advanced training of managers in this area is to update the knowledge, principles and practices of management in accordance with the requirements for ensuring environmental safety.

Regardless of the level of hazard of enterprises, the main tasks of the general program for training managers to ensure environmental safety are:

- the formation of a holistic view of the environmental safety management system at the enterprise level;
- the development of the ability to make effective decisions in environmental emergencies;
- studying how to ensure economic growth with the lowest possible depletion of natural resources [12];
- training on the transition from traditional production with a potential environmental threat to environmentally responsible production [13].

Conclusions

Based on the identified data, we can offer the following model structure of the training program and advanced training of enterprise managers in order to ensure environmental safety.

The first block - environmental legislation (international norms and principles, federal legislation, state legislation, domestic legislation).

The second block - organizational and legal means of ensuring environmental security: the work organization on environmental protection; environmental impact assessment procedure of pre-planned, pre-project and project materials; environmental standards and regulations; environmental certification system; environmental audit; environmental management system; environmental planning; environmental monitoring procedures; environmental reporting; environmental control and supervision.

The third block - economic means of ensuring environmental safety: payment for a negative impact on the environment; ecologically oriented tax policy; environmental equipment depreciation; concessional lending for resource-saving and environmental programs; environmental insurance.

After the completion of training, you should pass an exam. It is necessary to improve the qualification of general and special programs once every three years. This will enable business managers to be always aware of changes in legislation and to maintain the necessary level of environmental safety in their organization.

The reported study was funded by RFBR according to the research project № 19-011-00416.

References

[1] Abanina E N, Timofeev L A, Agapov D A, Sorokina Yu V, Ustinova A N Systems of Environmental Security of Urbanized Territories Within the Framework of the Program of Ecological Development of Urbanized Territories *IOP Conf. Series: Earth and Environmental Science,* Volume 224 (2019) 012031. DOI:10.1088/1755-1315/224/1/012031

[2] Hojnik J., Ruzzier M., Manolova T. Eco-Innovation and Firm Efficiency: Empirical Evidence from Slovenia *Foresight and STI Governance*, Volume 11, N 3, pp. 103–111. DOI: 10.17323/2500-2597.2017.3.103.111

[3] Effectiveness of Regulation: Literature Review and Analysis, Report SC090028, Environment Agency, Bristol, UK.

[4] ISO 14001:2015 Environmental management systems - Requirements with guidance for use, IDT.

[5] Rybenko I A; Ermakova L A Modelling of CWS combustion process *IOP Conference Series: Earth and Environmental Science* Vol. 45 (2016) 012016. DOI:10.1088/1755-1315/45/1/012016.

[6] Damarmoyo K S, Handayani S, Utami S N H, Indarti S Soil physical properties and abundance

of soil fauna in conventional and organic rice field *IOP Conference Series: Earth and Environmental Science* Volume 215 (2018) 012009 doi :10.1088/1755-1315/215/1/012009
[7] Prykhodko M M, Romaniuk V V, Kukhtar D V, Rodzinska O V. A modern approach to monitoring the territories of solid waste landfills 2018 *17th International Conference on Geoinformatics Theoretical and Applied Aspects 17, Theoretical and Applied Aspects* URL: http://www.earthdoc.org/publication/publicationdetails/?publication=93296 DOI: 10.3997/2214-4609.201801829
[8] Weißhuhn P., Müller F., Wiggering H. 2018 Ecosystem vulnerability review: proposal of an interdisciplinary ecosystem assessment approach *Environ Manage.* 61(6):904-915. DOI: 10.1007/s00267-018-1023-8
[9] Sicard P., Agathokleous E. Araminiene V., Carrari E., Hoshika Y., Paoletti E., De Marco A. 2018 Should we see urban trees as effective solutions to reduce increasing ozone levels in cities? *Environ Pollut.* 243(Pt A) 163-176. DOI: 10.1016/j.envpol.2018.08.049
[10] Hamzah Z, Shuhaimi S N A Biochar: effects on crop growth *IOP Conf. Series: Earth and Environmental Science* Vol. 215 (2018) 012011 doi :10.1088/1755-1315/215/1/012011
[11] Bolshakov V.N., Mugu R.R. Improvement of the JSC «Severnye Magistralnye Nefteprovody» staff competence and awareness in the field of environment protection and environmental management *Science & Technologies: Oil and Oil Products Pipeline Transportation* Vol. 1 (2014). pp.91-93.
[12] Morgunov B. A., Bagin A. M., Kozeltsev M. L., Terentiev A. A. Problems of Environmental Safety of Russia in the Light of the "Green" Growth Concept *Ekologiya cheloveka [Human Ecology]* Vol. 4 (2017), pp. 3-11.
[13] Hojnik J., Ruzzier M., Manolova T. Eco-Innovation and Firm Efficiency: Empirical Evidence from *Slovenia Foresight and STI Governance*, Vol. 11, no 3, pp. 103–111. DOI: 10.17323/2500-2597.2017.3.103.111

IOP Conf. Series: Materials Science and Engineering **582** (2019) 012033 doi:10.1088/1757-899X/582/1/012033

Ecological Safety of Progressive Technologies in the Life Cycle

M V Grafkina, E Y Sviridova [a], A V Pitryuk

Moscow Polytechnic University,
Russia, 107023, Moscow, Bolshaya Semenovskaya, 38

E-mail: [a] evg_sviridova@mail.ru

Abstract. The article is devoted to the analysis of the implicative (according to the "What-If?" principle) interaction of all decisions on the organization and implementation of technological processes and providing them with material and energy resources, machines, units, equipment, technological equipment, technological structures and materials at various stages of the life cycle response in the environment, which allows you to identify and formulate recommendations for improving the environmental safety of technological processes in the life cycle.

Introduction

The development of advanced technologies is a response to the needs of the market for highly efficient production, constant updating of products and requirements for environmental protection and reducing environmental tensions. The technologies being developed must meet both technical, economic and environmental requirements, therefore the task of choosing the best option will always be multi-criteria. Recently, environmental requirements during the development of technical systems are becoming increasingly important.

The progressive and most significant modern technological processes include electron-beam, laser, membrane technologies, powder metallurgy, etc. Technologies have undoubted technical advantages, for example, the quality and accuracy of the products obtained. The performance of laser welding machines, the strength of the joints obtained is higher than that of modern automatic arc welding machines, and the use of laser welding avoids deformation of the welded parts. Technologies make it possible to increase the utilization rate of construction materials at the expense of the accuracy of the operations performed. Distinctive environmental features of advanced technologies are their focus on the best available technologies and, above all, their focus on combating emissions into the environment and reducing the consumption of material and energy resources, introducing closed (waste-free) technological cycles (regeneration of solutions used, energy recovery, recycling of materials used, rational water consumption), the introduction of modern systems for the cleaning of ventilating gases and waste gases in d.

However, at present, not all opportunities for minimizing the industrial environmental impact of technological processes in the life cycle are used effectively. Technological processes as well as industrial products have their own life cycle, and the organization of processes is more costly, both economically and from the standpoint of use and the constant need for significant amounts of natural, material and energy resources, the cycle stages take a considerable time period, therefore environmental consequences will have a larger scale, multiplicity, a certain inertia, and may eventually meet more stringent environmental requirements.

Analysis of implicative (according to the "What-If?" principle) interaction of all decisions on the organization and implementation of technological processes and providing them with material and energy resources, machines, units, equipment, technological equipment, technological compositions and materials at different stages of the life cycle with a response in the environment will allow to identify and formulate recommendations for improving the environmental safety of technological processes in the life cycle. When choosing the best solutions, you should use both qualitative and quantitative methods for evaluating alternatives.

IOP Conf. Series: Materials Science and Engineering **582** (2019) 012033 doi:10.1088/1757-899X/582/1/012033

Results and discussion

To analyze the environmental safety of technological processes, we can consider various stages of the life cycle, in this paper we will focus on the following:

• project design;
• stage of scientific work;
• implementation of the process and provision of it with all necessary resources;
• direct operation of the technical process (main and auxiliary operations);
• repair and adjustment processes;
• modification of technological processes;
• recycling, disposal or destruction.

The generalization of existing developments [1-5] to improve the environmental safety of technical systems allows us to formulate the main directions for reducing the negative impact on the environment that are applicable at all stages of the life cycle of the technological process include:

• reduction of consumption of raw materials, material and energy resources at all stages of the life cycle (including intermediate technological losses, losses during storage and transportation, etc.);

• reduction of the use of extremely hazardous and highly hazardous substances and materials (including taking into account the long-term prospects for restrictions on the use of certain substances and materials);

• improvement of the main and auxiliary technological operations in order to eliminate or reduce the sources of formation of pollutants, wastes and other negative factors affecting the environment;

• recycling (regeneration of technological compositions, recycling and reuse, recovery) of raw materials, materials, reagents, water, energy;

• processing and use of waste products by creating by-products based on the integrated use of raw materials and waste;

• coordination and cooperation with other industries (business partners) in the field of saving and saving raw materials and energy resources, using secondary resources, processing and using waste, transferring environmental information about processes, machines, aggregates of materials, etc.;

• reduction of situations with increased environmental risk and the risk of occurrence and development of emergency situations;

• preparation for environmental mitigation activities under conditions of emergency impact on the environment;

• environmental education and awareness of the organization's personnel to prevent negative environmental impacts;

• development of effective industrial environmental monitoring and control (including monitoring and control of the use of all types of resources, emission sources of polluting emissions, etc.);

• reflected in the policy, goals and objectives of the organization of the relationship of the main production and environmental activities of the enterprise;

• active use of various internal qualitative and quantitative environmental indicators, characteristics, including those developed independently by the enterprise;

• effective planning and organization of environmental activities in accordance with accepted goals and objectives;

• rational and effective use of all the opportunities and means at the enterprise (primarily cost-free and low-cost) to solve environmental problems;

• use of an independent performance measurement (environmental audit);

• development of a "green" accounting system for the movement of material resources in the course of technological operations, etc.

It is especially effective to think over all the environmental consequences of the technologies used, machines, units, energy, materials at the initial stages of the life cycle of the process - the design and scientific work concept, that is, before material and energy resources are involved in the process, that is, the material base is laid, possessing a huge multiplicative effect on the life-supporting resources of the earth.

Environmental safety analysis and assessment of alternatives should be carried out using the "What-If" method, which is used both to assess the reliability of technical systems and to assess environmental risks. The use of the method involves the assessment and analysis of the implicative relations of technical and environmental indicators of the technological process and allows you to predict the negative response in the environment when changing any process parameters. This approach will allow you to manage the indicators of environmental safety of processes and reduce the tension of a number of environmental problems even before they occur. This method is most often used to assess qualitative changes, but it also gives an idea of the deviations of indicators of specific physical quantities and can serve as a basis for deterministic assessment methods, and allows you to develop corrective effects not only from the standpoint of improving the functioning of the system but also environmental safety.

At the stage of implementing the process and providing it with all the necessary resources, a qualitative and quantitative assessment of the environmental safety of the process equipment when choosing alternatives can be carried out on the basis of the following criteria:

- indicators of mass and volume of equipment;
- indicators of the number of construction materials in processes and equipment;
- energy efficiency;
- reliability and service life of the equipment;
- application of eco-labeling of materials and products;
- volumes of consumables during operation, repair and maintenance of the process and equipment;
- the possibility of using recycled materials;
- the possibility of regeneration, recycling, recovery of material and energy resources.

These indicators primarily ensure the saving of material and energy resources, as well as an increase in the efficiency of their reuse.

The maximum task in the implementation of technological processes is the transformation into finished products of all the involved material resources. The assessment of the environmental safety of a technological process requires consideration not only of the environmental characteristics of the main process itself, but also of the characteristics of auxiliary processes that precede or follow it [2]. For example, the welding process requires preliminary preparation of the blanks: it is necessary to carry out a mechanical cleaning from rust and other contaminants, as well as to degrease and dry the parts being welded. The soldering processes require subsequent operations to clean up the flux.

To improve the environmental safety of the stages of repair, adjustment processes, modification of technological processes and recycling, it is necessary to provide for the possibility of optimal disassembly, unification of components, marking of materials. This will allow reuse of some nodes, provided they meet their quality requirements, and facilitate their disassembly at the end of the life cycle. In general, the same requirements for environmental safety of the life cycle as for technical products are applied to equipment of technological processes.

When analyzing the environmental safety of technological processes, it is necessary to strive for the collection of environmental information and, above all, quantitative information. The transition to quantitative assessment methods is most effective at the stages of implementation and operation of the process and the involvement of all the necessary resources, as well as carrying out repair and commissioning works and modification of technological processes. Here it is advisable to use the developments that are used in assessing the environmental safety of products.

The authors propose a number of criteria for assessing the environmental safety of machines and units used in the process, the choices and decisions on which are applied at earlier stages, and the environmental consequences are manifested at the stage of utilization and destruction, equipment that has served its life, equipment and accessories (Table 1).

The following criteria are proposed:

1. The coefficient of repeatability of grades of materials in the design:

$$K_{rg} = \frac{1}{M_m}, \quad (1)$$

where M_m – the number of brands of materials used in the product.

Reducing the number of materials in the design will improve the efficiency of sorting and subsequent processing (recycling) of materials.

2. The coefficient of disassembly of units of the product:

$$K_{du} = \frac{N_{sn}}{N_t}, \quad (2)$$

where N_{sn} – number of split nodes,

N_t – total number of nodes in the product.

Increasing this ratio makes the design more environmentally friendly, as permanent connections virtually eliminate the disassembly and restoration of components.

3. The coefficient of unification of the design:

$$K_u = \frac{E_u}{E}, \quad (3)$$

where E_u – number of unified assemblies in the structure;

E – total number of subassemblies in a structure.

4. The possibility of recycling materials and nodes will also depend on the economic indicator of the complexity of this process - T, which is defined as:

$$T = \sum T_i, \quad (4)$$

where Ti – the complexity of the disassembly, restoration, quality control and testing of the i - component of the machine, in hours.

Table 1. Criteria for assessing the environmental safety of machines and units

№	**The name of the criterion, units measurements**	**Designation**	**Formula for calculating**
1.	The coefficient of repeatability of grades of materials in the design	K_{rg}	$K_{rg} = \frac{1}{M_m}$
2.	The coefficient of disassembly of units of the product	K_{du}	$K_{du} = \frac{N_{sn}}{N_t}$
3.	The coefficient of unification of the design	K_u	$K_u = \frac{E_u}{E}$
4.	The complexity of the disassembly process, in hours	T	$T = \sum T_i$

Thus, when choosing alternative variants of machines, units, and tooling, it is necessary to evaluate not only their technical characteristics, but also environmental indicators in the life cycle. To obtain the necessary environmental information, it is advisable to establish interaction with suppliers of the necessary products.

Special attention should be paid to the environmental assessment of the materials used. On a geo-ecological and geopolitical scale, a certain imbalance between demand and supply of materials, especially metals, has already been noted. The work [1] characterized the criticality of 62 metals and metalloids in a three-dimensional “criticality space”, consisting of supply risk, environmental consequences, and vulnerability to supply restrictions. Factors contributing to the occurrence of extreme situations include a high geopolitical concentration of primary production, the lack of available suitable substitutes and political instability. For example, the results show that the restrictions for many metals important to developing electronics (for example, gallium and selenium) are largely related to the risk of supply. The growth of technological innovation over the past decades has been partly made possible by the fact that an increasing number of metals in the periodic table are used to perform specialized functions. However, concerns are growing regarding the reliability of the supply of some of these metals. The main reason for these concerns is the fact that many of these metals are extracted only as a by-product from a limited number of geopolitically concentrated ore deposits, which makes their stocks incapable of responding to rapid changes in demand. Dependence of accessibility of the associated metal on the production of base metals introduces a new aspect of the risk of supply for modern technologies [2]. In these works, the environmental safety of materials is mainly assessed using expert ball methods.

There are also methods for quantitative assessment of the environmental safety of materials [6-12]. The most frequently cited information is on atmospheric pollution during the production and recycling of structural materials, which is justified from the standpoint of assessing the overall scale of

IOP Conf. Series: Materials Science and Engineering **582** (2019) 012033 doi:10.1088/1757-899X/582/1/012033

the pollution of the biosphere, including the atmosphere, since it accounts for more than 80% of anthropogenic pollution. The generalization of these data allows us to estimate the emissions of harmful substances into the atmosphere during the production and recycling of various materials. As an example, in Figure 1. The results of generalizations on the negative effects on the atmosphere of certain substances released during the production of copper are presented. The estimated amount of emissions in grams per kilogram of material produced was estimated.

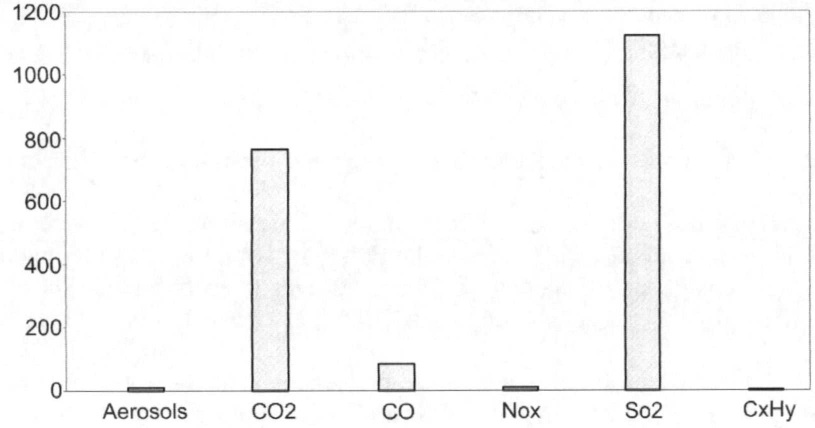

Figure 1. Emissions in the production of copper (g/kg)

It also summarizes the energy consumption in the production and recycling of steel, aluminum, copper and lead. The specific indicators of electricity consumption in kW per one kilogram of metal produced are shown (Figure 2).

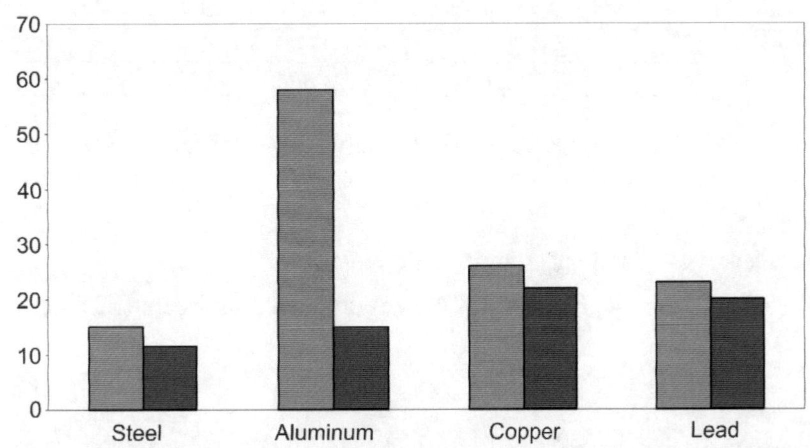

Figure 2. Energy consumption in the production and recycling of materials (kW/kg of material)

However, by absolute indicators of emissions of harmful substances into the atmosphere, as well as by the volume of consumption of water and energy resources in the production and recycling of construction materials, it is impossible to conduct a comparative analysis and draw a conclusion about the degree of complex (integral) negative impact on the components of the production environment and recycling of various materials.

The method developed by the authors, as well as a comparison of the results obtained with the results of other researchers, allows us to identify materials with the lowest environmental impact [7]. The method is based on formalized requirements for an integrated (integrated) environmental indicator of technical systems, which can be represented as a functional:

IOP Conf. Series: Materials Science and Engineering **582** (2019) 012033 doi:10.1088/1757-899X/582/1/012033

$$K(\varepsilon) = f\left[\varepsilon_A\left(\sum_{i=1}^{n} e_i\right)\right] = f\left[\sum_{j=1}^{m}\sum_{i=1}^{n}\omega_j\left[X(A_i)\right]\right], \quad (5)$$

where $\varepsilon_A\left(\sum_{i=1}^{n} e_i\right)$ – a vector characterizing the parameters of the natural-technical system that are manageable;

ε_A – the level of negative changes;

e_i – номенклатура показателей изменений, вызванных технической системой;

j, i – corresponding indexes of the nomenclature of the object of nature ($j = 1...m$) and technical indicators of the object ($i = 1...n$);

$\omega_o[X(A_i)]$ – technogenic impact throughout the life cycle of object X, characterized by technical indicators A_i.

This functional does not necessarily contain additive relations in the expression of the integral criterion. In this case, it only does not exclude the possibility of a linear approximation of the integral criterion from its constituent parameters e_i. Controllable parameters include those that characterize technical solutions to improve the environmental safety performance of a technical system in the life cycle.

Table 2 shows the results of a quantitative assessment of the environmental safety of some structural materials obtained by the method of the authors in comparison with other methods [8].

Table 2. Comparative table of environmental safety of materials

Materials	**Environmental indicator (Grafkina M.V.) [7]**	**EcoIndicator score / kg of material (Kozlov A.V.)[8]**
Aluminum	2975.84	0.564
Copper	22720.6	3.148
Plastic	1375.12	0.2983
Lead	4126.88	0.769
Steel	998.367	0.222

This method is applicable to the assessment of environmental safety of technological processes; the efficiency of calculations is enhanced by using data from "green" accounting. Calculation example:

$$\Delta\omega_{pr} = \omega_t - \omega_p, \quad (6)$$

where ω_t – level of technological impact of the process;

ω_p – level of engineering protection of the environment, taking into account the work of cleaning devices:

$$\omega_p = \sum_{i=1}^{n}\omega_{pi}, \quad (7)$$

where ω_{pi} – contribution to the reduction of anthropogenic impact from the use of methods, tools and devices for cleaning ventilation emissions, wastewater, reuse of waste production of technical systems.

$$\Delta\omega_{pr} < \omega_s, \quad (8)$$

where ω_s – maximum permissible emissions for the enterprise, waste generation standards and wastewater quality standards [13].

Quantitative methods for assessing the environmental safety of processes, equipment, and materials allow us to simplify the choice of developers, whose main efforts should be focused on the technical and functional characteristics of the processes.

To analyze and assess the environmental safety of technological processes in the life cycle (in different space-time coordinates), it is necessary to develop the skills of strategic environmental thinking and environmental responsibility for decision-making among process developers.

Conclusion

Despite the fact that advanced technologies are being created including in response to environmental protection requirements, not all opportunities are currently used to improve their environmental safety in the life cycle. It is most effective to conduct thinking about environmental aspects in the early stages of the life cycle of technological processes, before the involvement of material and energy resources in the process begins. Such an approach makes it possible to reduce and in some cases eliminate the negative impact on the environment. To analyze environmental safety at the early stages of the life cycle, it is advisable to apply qualitative assessment methods, whereas at subsequent stages the application of quantitative methods seems to be more effective and simplifies the choice of the necessary process components for developers.

References

[1] Graedel T.E., Harper E.M., Nassar N.T., Nuss P. Criticality of metals and metalloids BK Reck. Proceedings of the National Academy of Sciences. 2015, 112 (14). P. 4257-4262.

[2] Nassar. N.T., Graedel T.E., Harper E.M. By-product metals are technologically essential but have problematic supply. Science Advances. 2015, 1 (3), e1400180.

[3] Kharchenko S., Dorokhina E. System Analysis as the Best Way to Environmental Safety. Ecology and Industry of Russia. 2017, 21(1). P. 42-49. (In Russ.)

[4] Dewulf, J., Benini, L., Mancini, L., Sala, S., Blengini, G. A., Ardente, F., et al. Rethinking the area of protection “Natural Resources” in life cycle assessment. Environmental Science and Technology. 2015, 49. P.5310–5317.

[5] Gemechu, E. D., Helbig, C., Sonnemann, G., Thorenz, A., Tuma, Import-based Indicator for the geopolitical supply risk of raw materials in life cycle sustainability assessments. Journal of Industrial Ecology. 2015, 201(1). P.154–165.

[6] Lipunov I., Pervova I., Nikolaev I. On the Question of Recycling the Industrial Wastes. Ecology and Industry of Russia. 2015, № 19(4). P. 24-29. (In Russ.)

[7] Grafkina M.V., Sviridova E.Y., Sdobnyakova E.E. Improving ecological performance of design processes accounting for product life cycle European Research Studies Journal. 2017. 20 (2B). P.294-307.

[8] Luksho V.A., Kozlov A.V., Terenchenko A.S., Mirenkova E.A. Mathematical modeling of power units operating on natural gas, with a mono- and dual-fuel working process in total life cycle. International Journal of Mechanical Engineering and Technology. 2018, 9 (7). P. 1405-1415.

[9] Elshkaki, A., Graedel, T. E. Dynamic analysis of the global metals flows and stocks in electricity generation technologies. Journal of Cleaner Production. 2013, 59. P. 260–273.

[10] Graedel, T. E., Barr, R., Chandler, C., Chase, T., Choi, J., Christoffersen, L., et al. Methodology of metal criticality determination. Environmental Science and Technology. 2012. 46. P. 1063–1070.

[11] Nuss, P., Harper, E. M., Nassar, N. T., Reck, B. K., Graedel, T. E. Criticality of iron and its principal alloying elements. Environmental Science and Technology. 2014, 48(7). P. 4171–4177.

[12] Panousi, S., Harper, E. M., Nuss, P., Eckelman, M. J., Hakimian, A., Graedel, T. E. Criticality of seven specialty metals. Journal of Industrial Ecology. 2016, 20(4). P. 837–853.

[13] Trofimenko Yu.V., Grigoryeva T.Yu, Avenarius I.A. Estimation of Ecological Safety of Passenger Car. Ecology and Industry of Russia. 2004. P. 38-43. (In Russ.)

IOP Conf. Series: Materials Science and Engineering **582** (2019) 012034 doi:10.1088/1757-899X/582/1/012034

Environmental Monitoring as an Information System to Ensure Environmental Safety of a Region (as Exemlified in Kemerovo Region)

O B Sheveleva [a], E V Slesarenko [b], N D Vagina [c], N V Kudrevatykh [d], T F Mamzina [e]

T.F. Gorbachev Kuzbass State Technical University, 650000 Kemerovo, Vesenny st. 28,

E-mail: [a] shob.fk@kuzstu.ru, [b] sev.fk@kuzstu.ru, [c] vnd.fk@kuzstu.ru, [d] knv.fk@kuzstu.ru, [e] mtf.fk@kuzstu.ru

Abstract. the particularities of the environmental situation in the industrialized region with the predominant economic activity "Mining" are described in the article. Environmental monitoring of such basic components as the quality of atmospheric air and water, as well as investments into the environmental renovation and protection is recognized as an information system for ensuring the environmental safety of regions. As a result of the monitoring, the main discrepancies between the financial support of environmental renovation measures and the results of anthropogenic impact of coal mining enterprises on the environment were identified; attention was focused on the need to improve the efficiency of funds allocated for environmental protection and rational use of natural resources. The information system of environmental monitoring used by the coal-mining company SUEK, which today is a set of measures aimed at assessing the negative impact of the enterprise's activities on the environment of the region and eliminating its consequences, was considered. The system of environmental renovation measures developed and implemented by the Company, their effectiveness, along with adequate funding, allows the company to become one of the industry leaders in social accountability and corporate sustainability.

1. Introduction

Environmental safety can be considered as an integral and one of the most important parts of the concept of sustainable development of a coal-mining region, since the environmental risks occupy one of the leading places among the risks of backbone enterprises of such regions due to the particularities of the coal industry and its impact on the environment.

The Kuznetsk coal basin, located in Kemerovo region, is one of the largest coal deposits in the world, so the problem of the environmental safety state is extremely important for Kuzbass. The region possesses significant mineral deposits and a variety of natural resources. Several industries are developed on the territory of Kemerovo region - coal mining, metallurgy, mechanical engineering and power engineering. However, Kuzbass specializes primarily in coal mining - there are more than 380 mines and open cuts in the region.

The most important centers of the coal industry are the cities of Prokopyevsk, Mezhdurechensk, Belovo, Berezovsky, Kemerovo, Novokuznetsk, Leninsk-Kuznetsky, Novokuznetsky and Prokopevsky districts. There are more than two hundred mines and open cuts on the territory of the above cities and districts. There are two large coal basins - Kuznetsk coal and Kansk-Achinsk lignite - in the region. The share of Kemerovo region accounts for mining more than 75% of coking coal.

2. Research methodology

The information about the state of the natural environment (changes in weather, climate, the state of the animal and plant world around, etc.) has been actively used in human life since ancient times.

Recently, changes in the environment are increasingly occurring under the influence of the human factor. The study of anthropogenic changes is the basis of environmental monitoring, which is an information system for analyzing and predicting the state of the environment and the management actions developed on this basis that are necessary to resolve issues in the field of environmental safety. The environmental monitoring system consists of monitoring the state of the biosphere and the sources and the factors of anthropogenic impact on it.

In terms of the area covered by the analysis, monitoring is divided into global, national, regional, local, single-point. The global environmental monitoring system was developed globally at the First Intergovernmental Monitoring Meeting in 1974.

In Russia, considerable attention is paid to the environmental monitoring at the federal level - a number of legislative and regulatory acts have been developed:

- Federal Law "On Environmental Protection" No. 7-FZ dated January 10, 2002 (Chapter X. State Environmental Monitoring);
- Decree of the Government of the Russian Federation No. 681 of 09.08.2013 (as amended on November 11, 2018) "On state environmental monitoring and the state database of state environmental monitoring";
- Decree of the Government of the Russian Federation of the Russian Federation No. 477 of 06.06.2013 (as amended on July 10, 2014) "On the implementation of state monitoring of the state and environmental pollution";
- Decree of the Government of the Russian Federation of the Russian Federation No. 1410 of December 21, 1999 (as amended on June 06, 2013) "On the creation and maintenance of the Unified State Database on the state of the environment and its pollution" and others.

The regional level deserves special attention, since it is here where we can take into account the factors characteristic of the economies of specific territorial entities within the country, and therefore at this level specific problems can be considered and special programs for studying the environment state and the impact of anthropogenic factors on it can be developed.

3. Results and discussion

In order to identify the level of severity of environmental problems in Kuzbass, the comparative analysis of indicators of environmental monitoring of the quality of atmospheric air and water resources of Kemerovo region and the Siberian Federal District was conducted (Table 1). The indicators “Emissions of pollutants into the atmospheric air from stationary sources” and “Proportion of captured and treated atmospheric pollutants in the total amount of pollutants from stationary sources” are included in the State Program of the Russian Federation “Environmental Protection” for 2012-2020.

The amount of emissions of pollutants into the air from stationary sources in the Siberian Federal District in 2010–2017 varied from 5.5 to 6 thousand tons, and in the Kemerovo Region - from 1.3 to 1.5 thousand tons. Thus, the share of emissions of pollutants into the atmospheric air in Kuzbass accounted for 22.6-25.8%, i.e. almost a quarter of all pollutant emissions in the district. Moreover, the highest concentration of emissions was noted in 2017 (1,488 thousand tons), which led to an increase in the share of Kuzbass in the overall structure of emissions of the Siberian Federal District to 25.8%. According to this indicator, the Kemerovo region hold the second place (behind only the Krasnoyarsk Territory) among the 12 regions of the Siberian Federal District, which indicates the severity of this problem for our region. In 2017 compared to 2010, the amount of pollutant emissions into the atmospheric air in Kuzbass increased by 5.5% (from 1,411 thousand tons to 1,488 thousand tons), while the similar indicator in the District decreased by 1.8%. At the same time, the share of coal production in Kuzbass accounts for 61.5% of the total amount of pollutant emissions into the atmospheric air (915 thousand tons), i.e. the coal mining industry is the main air pollutant in the region.

According to the indicator “The proportion of captured and treated air polluting substances in the total amount of pollutants from stationary sources,” the Kemerovo Region is in 7th place among 12 regions of the Siberian Federal District. The percentage of captured and treated atmospheric pollutants in the total amount of pollutants in Kuzbass is comparable to that in the District. In 2017, there is a slight lag in the indicator for the Kuzbass compared to the District data - by 1%, which is not critical because the situation changes from year to year, with a slight change in the direction of increase and decrease.

Table 1. Indicators of environmental monitoring of the quality of atmospheric air and water

Indicators	*Years*							
	2010	*2011*	*2012*	*2013*	*2014*	*2015*	*2016*	*2017*
1. Emissions of pollutants into the air from stationary sources, kilotons								
Siberian Federal District	5868	5919	6017	5816	5570	5688	5605	5764
Kemerovo region	1411	1390	1360	1356	1332	1344	1349	1488
Share of Kemerovo region in Siberian Federal District, %	24.1	23.5	22.6	23.3	23.9	23.6	24.1	25.8
2. Captured air pollutants from stationary sources, kilotons								
Siberian Federal District	21998	23420	20896	19459	19469	18497	17401	18012
Kemerovo region	4823	6321	5334	4512	4636	4683	4234	4422
3. The proportion of captured and treated air polluting substances in the total amount of pollutants from stationary sources, %								
Siberian Federal District	78.9	79.8	77.6	77.0	77.8	76.5	75.6	75.8
Kemerovo region	77.4	82.0	79.7	76.9	77.7	77.7	75.8	74.8
4. Discharge of polluted wastewater to surface water bodies, million cubic meters								
Siberian Federal District	2218	2142	2077	1941	1750	1696	1654	1640
Kemerovo region	700	661	572	598	478	462	444	433
Share of Kemerovo region in Siberian Federal District, %	31.6	30.9	27.5	30.8	27.3	27.2	26.8	26.4
5. The volume of recycling and successively used water, million cubic meters								
Siberian Federal District	15739	16093	17052	16288	15997	16504	15978	15870
Kemerovo region	4524	4759	5114	5043	4766	4895	4891	4895
Share of Kemerovo region in Siberian Federal District, %	28.7	29.6	30.0	30.1	29.8	29.7	30.6	30.8

However, the situation looks quite different if we consider it in terms of absolute indicators, namely: 25.2% of non-treated air pollutants in 2017 in Kuzbass amount to 1,488 thousand tons of emissions per one region - Kemerovo region, while 24.2% of non-disposed substances in the District is 5764 thousand tons, which is 480 thousand tons per one region of the District (taking into account that there are 12 regions in the Siberian Federal District), i.e. emissions in Kuzbass are 3 times higher than the average in other regions of the Siberian Federal District.

The situation of polluted wastewater discharge into surface water bodies is being improved in the both in the Siberian Federal Distrcit and Kuzbass. Thus, in the Siberian Federal District, discharges for 2010–2017 decreased by 26.1%, and in the Kemerovo region - by 38.2%. However, despite the rapid decline in discharges in the region compared to the District, the Kuzbass and in 2017 ranks second in this indicator among the 12 regions of the Siberian Federal District, and it accounts for more than a quarter of the total polluted wastewater discharge in the District.

Within 2010-2017, the amount of investment in fixed assets aimed at environmental protection and rational use of natural resources in the Kemerovo region increased by 4.1 times. The largest share of investments throughout the analyzed period (except for 2010) is accounted for by investments in the protection and rational use of water resources (in 2017 it is 62.5% of the total investment in fixed assets). The amount of investments in this environmental protection area over the eight analyzed years increased by 1,596.1 million rubles, or 7.4 times.

Table 2. Composition and structure of investments in fixed assets, aimed at environmental protection and rational use of natural resources

Indicators	***Years***								
	2010	***2011***	***2012***	***2013***	***2014***	***2015***	***2016***	***2017***	***2018***
Investments in fixed assets, including	714.9	1161.6	1713.1	1404.4	2833.1	2058.1	1668.9	3150.1	2954.2
1) on protection and rational use of water resources	251.2	556.0	1459.1	1143.2	2025.7	1107.0	1255.5	2651.7	1847.3
- the same as a percentage of investments in fixed assets	35.1	47.9	85.2	81.4	71.5	53.8	75.2	84.2	62.5
2) on protection of atmospheric air	292.5	506.4	105.7	144.6	67.6	337.4	220.1	385.3	584.5
- the same as a percentage of investments in fixed assets	40.9	43.6	6.2	10.3	2.4	16.4	13.2	12.2	19.8
3) to other activities (including protection and rational use of land)	171.2	99.2	148.3	116.6	739.8	613.7	193.3	113.1	522.4
- the same as a percentage of investments in fixed assets	24,0	8,5	8,6	8,3	26,1	29,8	11,6	3,6	17,7

Return on such investments was expressed:

- in reducing the polluted wastewater discharge into surface water bodies by 38% or 267 million cubic meters;

- in the growth of savings in fresh water intake through the use of recycling and reverse water supply systems, which reflects the indicator of the volume of recycling and successively used water - there is a rapid growth in the Kuzbass (an increase of 8.2%) compared to the increase in this indicator for the SFD only 0.8% (Table 1).

Return on investments in fixed assets for air protection was less productive - with an increase in the amount of investments by 292 million rubles, or almost 2 times, emissions of pollutants into the air from stationary sources increased, and the proportion of captured and treated air pollutants decreased. It is quite possible to recognize such a situation as logical from the point of view of the formation of the structure of investments in fixed assets, since the share of investments in air protection in the total amount of investments decreased by 21.1%.

When analyzing the ratio of investments (table 2) and commissioning of facilities for environmental protection and rational use of natural resources (table 3), there is no obvious correlation between these indicators. So, the amount of investments in 2010 being 251.2 million rubles; the capacities of the commissioned wastewater treatment facilities were only 0.7 thousand cubic meters per day, and a year later, with the amount of investments exceeding only 2.2 times, the capacities of the commissioned facilities were 220 times more!

Table 3. Dynamics of commissioning of facilities for environmental protection and rational use of natural resources

Indicators	*Years*								
	2010	***2011***	***2012***	***2013***	***2014***	***2015***	***2016***	***2017***	***2018***
Wastewater treatment facilities, thousand cubic meters per day	0.7	154.5	34.8	110	69.9	124.4	22.9	94.0	189.8
Plants for trapping and treatment of harmful substances from waste gases, thousand cubic meters per hour	600	1036	471.7	322	-	75.7	27.6	16.0	30.0

The situation of efficiency of investments in the protection of atmospheric air the looks even more ambiguous - the amount of investment is increasing annually, while reducing the commissioning of facilities for capture and treatment of harmful substances from waste gases. At the same time, according to statistics, the type of economic activity "Mining" is the main source of emissions of pollutants into the atmosphere. The emissions figures are comparable with two other activities - "polluting sources" ("Manufacturing industries" and "Providing electric energy, gas and steam"), but the emissions for these two types of activity even in total terms are 2 times lower. As for all other types of economic activity, the exceedance of emissions in mining is hundreds of times higher! In this regard, it is completely natural that the investment in environmental protection measures in mining should be maximal.

However, according to statistics, the efficiency of disposal of harmful substances in the mining industry of Kuzbass is quite low - the proportion of captured and treated harmful substances from all sources of pollution in mining ranges from 43% to 50% in 2013-2017 years (Fig. 1), i.e. more than half of the harmful substances remain in the atmospheric air, adversely affecting the quality of life of people, an increase in morbidity and mortality, and a decline in the birth rate.

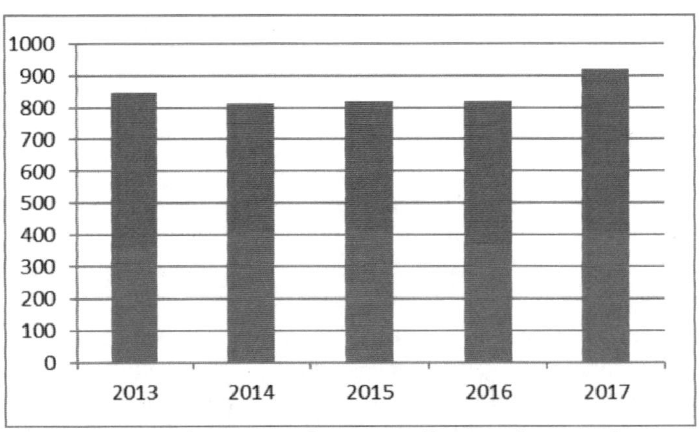

Fig. 1. Comparison of the total amount of waste pollutants from stationary sources and captured and treated substances in the field of mining activities in Kemerovo region in 2013-2017, kilotons

In 2012-2017, in Kuzbass, an increase in the current (operational) costs of environmental protection was by 3,980 million rubles or by 69.7%, including rising costs for air protection and climate change mitigation, amounted to 1,850 million rubles (or 2 times), wastewater treatment - 2452 million rubles (94.3%), waste management - 251 million rubles (51.4%). The remaining costs decreased to a critical level, for example, over the past five years, the cost of ensuring the radiation safety of the environment decreased by 17 times, and in 2015–2017 the costs of research and development activities to reduce negative anthropogenic environmental impacts, and biodiversity conservation and protection of natural territories.

Table 4. Dynamics of current (operational) environmental costs

Indicators	*Years*					
	2012	***2013***	***2014***	***2015***	***2016***	***2017***
Total, mln. Rub., including	5708	7579	10224	6858	7250	9688
1. air protection and climate change mitigation, mln. Rub.	1764	2097	3240	1920	2114	3614
- the same as a percentage of current environmental costs	30.9	27.7	31.7	28.0	29.2	37.3
2. wastewater treatment, mln. Rub.	2599	4129	5314	3543	4292	5051
- the same as a percentage of current environmental costs	45.5	54.5	52.0	51.7	59.2	52.1
3. other activities in the field of environmental protection, mln. Rub.	1345	1353	1670	1395	844	1023
- the same as a percentage of current environmental costs	23.6	17.8	16.3	20.3	11.6	10.6

The largest share in the structure of current (operational) expenditures on environmental protection in Kemerovo region over the past five years was attributed to wastewater treatment costs. So, in 2017, the share of these costs amounted to 52% of the total amount.

Thus, for the regions with the predominant location of the coal mining industry (including Kuzbass), the following main environmental problems are characteristic:

- topsoil disturbance, as well as its contamination by waste generated by the operations of mines and open cuts;
- negative impact on water bodies as a result of discharges of pollutants into them, as well as washing off surface pollution from the territories of mines, open cuts and dumps;
- plant cover depletion due to open cut mining, often causing complete destruction of soil and vegetation;
- increasing emissions of pollutants into the air from stationary sources, and reducing the proportion of captured and treated air pollutants;
- exacerbation of the sanitary and epidemiological situation, causing a deterioration in the health of the population, an increase in morbidity and mortality.

All these problems are often conditioned by the lack of incentives for the environmental activities of coal enterprises. It is more profitable for the economic entities to pay for the negative impact on the environment than to carry out reconstruction and modernization of production in order to reduce the harmful effects.

The experience of coal industry enterprises that are leaders in social accountability and corporate sustainability and implement a ramified and ambitious environmental policy, deserves special attention in this regard.

One of these companies that have been mining coal in Kuzbass since 2001 is JSC "SUEK". Coal is mined in 17 mines and 12 mines, which are located in seven regions of Russia. From 2005 to 2017, coal production increased by 39%. The volume of "SUEK" coal reserves is 5.61 billion tons - according to this indicator, "SUEK" ranks fifth in the world. SUEK's strategy provides for sustainable long-term competitiveness based on an integrated chain of production, transportation and marketing of coal, and provides for strategically important areas, including a special emphasis on sustainable development, including investment in environmental, public and social projects to improve well-being of the regions in which it operates.

In the Kemerovo region, JSC "SUEK-Kuzbass" mines coal in 2 open cuts and 9 mines. In order to improve the quality of coal, it is processed in 4 concentration plants in the region. The activities of JSC "SUEK-Kuzbass" are related to various factors that have an impact on the environment (Table 5), and since exceeding the permissible levels of impact on natural resources is fraught with fees for pollution in excess of the established limits, administrative liability, or suspension of activities, then to reduce environmental risks an environmental policy has been developed in the enterprise to reduce the impact of these factors.

Table 5. Lines of environmental policy of jsc "suek-kuzbass"

Factors affecting the environment	Types of monitoring	Monitored Indicators	Priority areas of environmental activities
air pollutant emissions	monitoring of air pollutant emission sources	air quality of the working area, atmospheric air at the industrial site and in the sanitary protection zone	gradual reduction of emissions of pollutants (above all, greenhouse gases and solids) into the atmosphere
discharge of pollutants into surface water bodies	monitoring of surface water bodies above and below the sewage discharge point	the quality of wastewater discharged from technological facilities and entering the treatment plant, treated wastewater, surface water	gradual reduction of discharges of polluted wastewater into water bodies
soil disturbance in the process of mining	waste disposal sites monitoring	waste disposal sites	arrangement of waste disposal sites to reduce the man-made impact on the environment

Following on the principles of the environmental policy, JSC "SUEK-Kuzbass" introduced an environmental management system certified in 2009 for compliance with the requirements of the international standard ISO 14001: 2004. Environmental indicators are monitored by accredited sanitary and hygienic and sanitary and industrial laboratories. The company has defined goals and objectives for meeting the standards for discharges and emissions of pollutants and mined-land reclamation; hazards were identified, environmental risks were estimated, measures aimed at minimizing these risks were developed, i.e. comprehensive and diverse work is underway to reduce the negative impact on the environment in the region.

However, despite fairly close attention to the issues of ensuring environmental safety at the federal level, the problem of the anthropogenic impact of industrial enterprises on the ecosystem, as well as improving the efficiency of the implemented environmental protection and recovery measures, remains one of the most pressing. Thus, a number of scientists [1, 2] recognize the threat of low social accountability and corporate sustainability as one of the most acute in addressing issues of improving environmental safety. The problem of accumulated waste from industrial enterprises, as well as their storage and disposal is widely investigated in the works of such authors as V. Anishchenko, V. Marhasova, A. Fedorenko, M. Puzyrov, O. Ivankov [3], A. Wolak, S. Hornik [4], M. Kwiatkowski [5]. The relationship between environmental protection and investment in innovation processes is presented in the works of C. Barra, G. Bimonte, L. Senatore [6]. The influence of coal mining enterprises on the level and quality of life is described in the works of K. Mustafina [7], V. Novotna [8] and many other scientists.

At the same time, the successful global and regional experience in improving the environmental safety of industrialized regions [9, 10] provides a rich set of techniques and methods for their implementation.

4. Conclusion

Despite the presence of coal mining companies such as SUEK, a rather impressive number of mining companies use dirty technologies, without considering the most efficient methods for the conservation of mined-out rock masses and land reclamation. Rent-oriented behavior does not allow their owners to spend on these goals more than is required by law. In such conditions, the presence of even 20% of coal mining enterprises in the industrialized region, focused on reducing the anthropogenic impact of their activities, is unlikely to be able to fundamentally resolve the situation. The following systematic and comprehensive works in this direction are required:

1) increase of standard requirements for already operating coal mining enterprises in terms of investments in environmental innovations;

2) reduction of standards requirements for emissions of harmful substances into the atmosphere, water, soil and strict monitoring;

3) tightening the rules for coal mining enterprises when issuing mining licenses, in particular, reducing the corruption component in this aspect. There are frequent situations when regulatory authorities warn coal companies in advance about inspections, measurements of harmful emissions are carried out within a specified time, so violations, as a rule, are not detected;

4) increase of the established distance (sanitary protection zones) between settlements and mining sites. Currently, the sanitary protection zone for coal mines and processing plants is 1000 m. However, often even such a short distance is not maintained;

5) introduction as a mandatory condition for the operation of coal mining enterprises of the gradual transfer of their production facilities to resource-saving, waste-free technological solutions, which allow minimizing the release of harmful substances into the environment;

6) development of a system of benefits for enterprises implementing measures to ensure the environmental safety;

7) extended application of environmental insurance;

8) use of higher tax rates for the production of environmentally hazardous products;

9) improving the efficiency of environmental monitoring by conducting more thorough environmental monitoring and clarifying indicators to assess the efficiency of the environmental safety of enterprises' production activities;

10) environmental certification of production of all enterprises of the coal industry taking into account the requirements of international standards;

11) changing the attitude of owners to environmental problems, promoting their social accountability and corporate sustainability.

References

[1] Zonova O.V., Nekhoda E.V. Social technologies for management: opportunities for coal-mining enterprise [M]. Coal in the 21st Century: Mining, Processing and Safety, 2016: 125-129.

[2] E. Ganebnykh, T. Burtseva, A. Petuhova, A. Mottaeva Regional environmental safety assessment, E3S Web of Conferences 91, 08035, 2019.

[3] V. Anishchenko, V. Marhasova, A. Fedorenko, M. Puzyrov, O. Ivankov Ensuring environmental safety via waste management. Journal of security and sustainability issues. file:///C:/Users/slekv/Downloads/ENSURING_ENVIRONMENTAL_SAFETY_VIA_WASTE_MANAGEMENT.pdf

[4] Wolak A., Hornik S. Ecological problems related to dangerous waste [M]. Conference: 18th IGWT SYMPOSIUM: Technology and Innovation for a Sustainable Future: a Commodity Science Perspective, At Rome, Italy, September, 2012.

[5] Kwiatkowski M. Energy, economic and ecological problems of waste management [J]. Archives of Waste Management and Environmental Protection, 2010, vol. 12, issue 4: 40-52.

[6] C. Barra, G. Bimonte, L. Senatore Innovation Processes and Environmental Safety, online file:///C:/Users/slekv/Downloads/dp_141.pdf

[7] Mustafina K. Social Issues of Ecological Problems [J]. AlmaTourism, 5, 2012: 109-117.

[8] Novotna V., Bedřich Půža, Jiří Hřebíček Modelling socio-ecological problems with delay. Case study on environmental damage [M]. Conference: International Environmental Modelling and Software Society (iEMSs) 8th International Congress on Environmental Modelling and Software Toulouse, France, Sabine Sauvage, José-Miguel Sánchez-Pérez, Andrea Rizzoli (Eds.) http://www.iemss.org/society/index.php/iemss-2016-proceedings.

[9] Chuturkova R., Simeonova A., Bekyarova J., Ruseva N., Yaneva V. Assessment of the Environmental Status of Devnya Industrial Region, Bulgaria [J]. Journal of Environmental Protection and Ecology, 2011, 12, 3: 805–813.

[10] Nurongguli Tuoheti, Maimaitiaili Abulaiti, Zafar Ahmed & Namaitijiang Maimaiti Anti-desertification Legislation and Ecological Problems in Xinjiang China [J]. Journal of Politics and Law, Vol. 6, 3, 2013: 160-168.

IOP Conf. Series: Materials Science and Engineering **582** (2019) 012035 doi:10.1088/1757-899X/582/1/012035

Automation of Processes in Dairy Cattle Production

I Ivanova [1], I Trotsenko [1] and V Trotsenko [2,a]

[1]Omsk state agrarian university named after P. A. Stolypin, 1 Institutskaya square, Omsk 644008, Russia
[2]Omsk tank-automotive engineering institute, 14 military town, Omsk 644098, Russia

E-mail: [a] iv.trotsenko@omgau.org

Abstract. The priority direction of modern animal husbandry is the creation of smart farms, which are managed through the use of automated systems and IT. However, not all farms use new technologies, so this article analyzes the efficiency of milk production in the conditions of the leading enterprises of the Omsk region, located in different climatic conditions, with traditional technology, and then introduced elements of a smart farm. In the course of the study in the conduct of traditional technology of milk production in large numbers were identified violations of animal identification, there was an increase in the duration of the open period and there were lower rates of purity and grade of milk. The introduction of elements of a smart farm at the same enterprises allowed to avoid errors in the identification of animals, to improve the analyzed indicators and to increase milk production.

1. Introduction

The global goal of animal husbandry is the satisfaction of world's population needs and the constant increase of availability for animal industry products providing the safety of products and the animals' health. Every year the consumption of animal products increases by 7%, thereby we can see the continuing need in the development of the existing technologies of livestock production [1, 2].

The production growth is due to intensification and automation of technological processes. Technical progress does not stay still and is actively implemented in all spheres of human activity, including dairy cattle breeding. The main reasons for the introduction of technical innovations in agriculture are the reduction of production costs and obtaining maximum profit [3, 4].

The main problem in the development of domestic livestock farming is the low competitiveness of livestock products on the market, which is due to the technical and technological lag in livestock raising in Russia from developed countries [5]. The development of the dairy production influences on the level of food supply in the country. The decision of this problem is the creation of production industry and sustainable development of agriculture by the use of modern scientific and physical infrastructure with the advanced management technologies [6].

The development of digital animal husbandry in the world began about thirty years ago and at present, modern technical solutions are comparable to robotics, since all technological processes are carried out without direct human participation [7, 8]. The dairy cattle breeding has reached a sufficient level of development and modern digital technologies are used along with the traditional processes of milk production in the conditions of the south of Western Siberia.

Thus, the use of modern automated technologies for the production of milk is relevant for animal husbandry in the region.

The aim of the research was to determine the efficiency of milk production when introducing modern automated herd management systems under the conditions of Omsk Region enterprises.

2. Research object and methods

The studies were conducted at the Institute of Veterinary Medicine and Biotechnology of the Omsk State Agrarian University.

The object of research was the technological processes of milk production in the leading enterprises of the Omsk Region. We selected the enterprises that have implemented the elements of a

IOP Conf. Series: Materials Science and Engineering **582** (2019) 012035 doi:10.1088/1757-899X/582/1/012035

smart farm and located in various natural and climatic zones of the Omsk Region. During the study, the main production indicators were analyzed for two periods of the enterprises' economic activities - before and after the introduction of the smart farm. Analysis of the production activities of enterprises was carried out for the period from 2000 to 2018. Automated herd management systems in the Omsk Region have been used since 2009. Thus, we identified two periods: 1 period - 2000–2009 – traditional milk production technologies; 2 period - 2010-2018 – milk production technologies with the using of a smart farm elements. The research scheme is presented in Figure 1. We also analyzed the changes in milk production indicators depending on the use of modern automated systems.

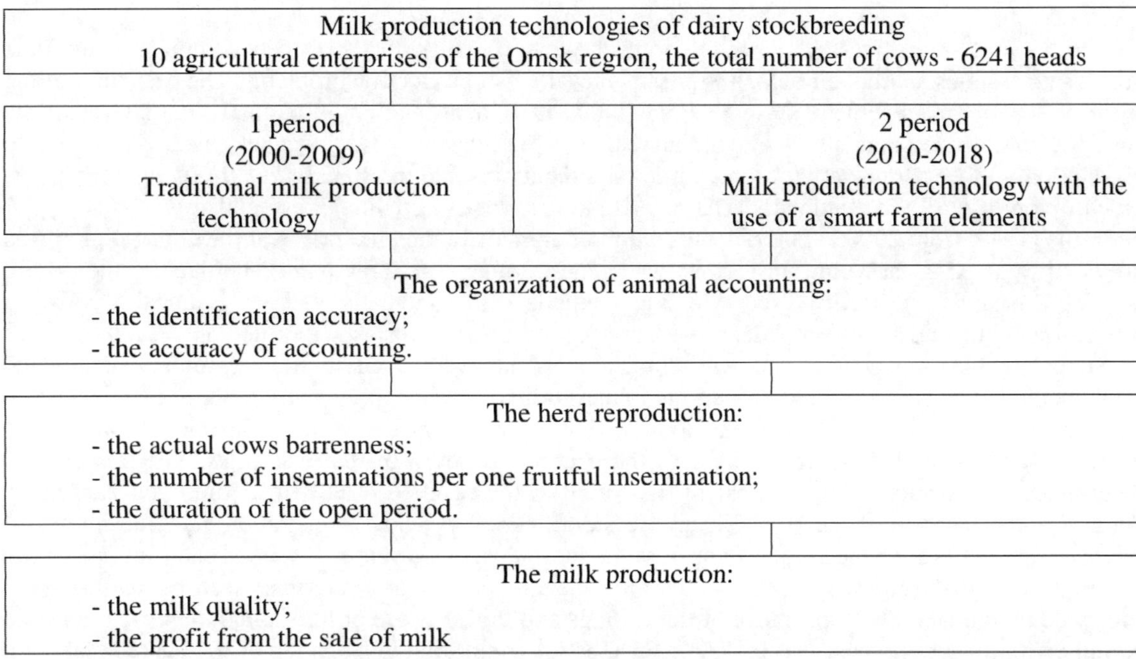

Figure 1. The scheme of the research

Livestock enterprises of the Omsk region are geographically located in different climatic zones and have a different status, so their needs are not the same. Some of the enterprises included in the research have the status of breeding reproducers or factories. In addition there are the commercial dairy complexes among them. Thereby they use the different software packages that are more suitable for specific production conditions. Breeding reproducers and factories use the software package of "Selex" Analysis and Information System for dairy cattle, but in commodity enterprises, software packages are different, for example, Data flow, Deli Master and others. The software is designed to keep primary and breeding livestock records. Nowadays in Russia, in order to ensure pedigree registration in dairy cattle breeding, automated accounting is required, which is determined by the requirements imposed on enterprises that have the status of a breeding factory or a reproducer.

3. Results and Discussion

There are red steppe and black pied cows for the production of milk in the enterprises of the region.

The priority direction in modern animal husbandry is a smart animal husbandry. Smart livestock is a branch of agriculture engaged in the breeding of farm animals, a feature of which is the introduction of new generation systems and technologies to automate the animal care in order to increase the number of products and reduce the cost of costs [9].

The development of the smart animal husbandry in Western Siberia is only gaining momentum and 2016-2018 years were called "years of testing", the results of which were predicted the trend of increasing "smart" systems being introduced into production.

The automated herd management systems are based on the principles of digitizing farm production processes, which are including the monitoring of the production performance, animal health status and well-being, nutrient balance in animal diets taking into account their physiological needs [10].

Modern information technologies, sensors and robots are being introduced into stockbreeding, making it possible to monitor each animal and arrange its life cycle as efficiently as possible so that the animals start making profit as soon as possible and longer than they were before the introduction of technology.

At present, in dairy cattle breeding, three keys to success can be distinguished: reliable accounting, the use of digital technologies, and competent control and interpretation of the data obtained by a specialist.

The basis of smart livestock is the equipment and sensors, methods of their communication, data processing systems. With their help, it is possible to solve all production problems. The basic unit of the automated herd management system is the individual identification of animals in the real time conditions. Information about a particular animal goes to a specialist every minute. Specialists take into account only those parameters for which it is possible to analyze the state of health, living conditions, time of mobility. All these indicators characterize the productive qualities of animals.

In the period from 2000 to 2009 identification of animals in the enterprises of the Omsk region was provided by tagging the young in various ways, tags, collars, and brandings. The data on the origin, development and productivity of cows were systematized mechanically, and zootechnical accounting was carried out using a cow card file, as well as other established forms were filled out.

Since 2009 a transponder and two duplicate ear tags are used to identify animals after the implementation of the automated accounting in the studied enterprises. A transponder is a sophisticated modern processor used to identify and monitor animal activity. The signal from the transponder is transmitted to the collector, which controls the health and phase of the sex cycles. The transponder determines the position of the animal in the room, at the result of that in real time you can easily determine the location of the cow. Electronic identification provides registration and traceability of animals (herds) based on the unique identifiers and individual electronic passports of each animal. These passports contain the entire event history, confirming zootechnical and veterinary well-being during the life cycle of animals. The appearance of the ear tags and the structure of the animal identification code comply with the requirements of the FAO (Food and Agriculture Organization of the United Nations). The ear tag shows a bar code that matches the animal's identification number, thus mobile scanners can be used to identify animals and record events in the field conditions.

It is important to avoid mistakes when identifying animals for the organization of reliable primary and pedigree zootechnical accounting. В таблице 1 представлена структура выявленных ошибок идентификации животных в зависимости от технологии ведения первичного зоотехнического учета. Table 1 shows the structure of the identified errors of animal identification, depending on the technology of conducting primary zootechnical accounting.

Table 1. The structure of the cows' identification

The type of error	1 period (2000-2009)	2 period (2009-2018)
The proportion of detected errors, %	27	3
Including:		
The duplication of the individual number	9	0
The loss of the individual number	13	0
The destruction of the numeration order	2	0
The incorrect record of origin	3	3

The proportion of errors in the management of zootechnical accounting, caused by identification errors in the traditional milk production technology, was at the level of 27% of the total cattle population in the studied farms. The loss of an individual number or a decrease of its quality, at which the possibility of its correct reading is lost, was the most common cause of cow identification errors. In the period from 2000 to 2009 errors due to the loss of the individual number accounted for 13%, which is by 4% more than due to duplicate numbers within the same enterprise. The incorrect record of the origin of the cows is due to the lack of reliable information about the ancestors of the cows. Violations in the assignment of individual numbers to animals, subsequently distorted the accuracy of the production indicators' accounting, thereby hampering the proper organization of dairy cattle breeding.

Because of the introduction of an electronic system for the animals' identification in the enterprise conditions, there were changes in some production indicators, as well as a reduction in the number of identification errors. This way for the period from 2009 to 2018 there is a significant reduction in errors for animal identification. The proportion of detected errors was 3% of the total number of cows in the studied enterprises. After the introduction of electronic identification of animals, there were no violations in the assignment of individual numbers, which is due to the automatic exclusion of two animals with the same numbers within the same herd. The system simply does not make a mistake in assigning an individual number to an animal. When entering data into the electronic identification system, data validation is performed. Loss of the individual number to animals is also excluded. In addition, to increase the reliability of zootechnical accounting, data entry uses a calendar, international reference books of allowable values when entering details: breed, animal species, gender, disease, vaccine, research result, etc.

With automated accounting, we detected 3% of violations in the records of the origin of cows. These errors are caused by the different spelling of the names of foreign stud bulls. The same producer can be recorded differently, due to the transliteration of their names, since there is no single system for translating the names of animals of foreign origin in the country.

The obtained data let us to conclude that the introduction of an automated system for zootechnical accounting and animal identification minimizes errors. There was a decrease in the proportion of cow identification errors by 24% for the period from 2000 to 2018.

Digitalization in zootechnical accounting minimizes the number of errors during data entry, as the role model is used to regulate user access to the system's functionality. It is allowed to create new roles and edit existing ones by the system administrator.

The modern accounting system in dairy cattle breeding is not only the indicators of milk yield and fat content, it is also the core of the genetic improvement of animals, the tool of a specialist for increasing milk production and improving of technologies.

A transponder located on every farm animal not only identifies, it also transmits indicators of cow activity. The obtained information is displayed on the monitor screen in an accessible form in the type of tables and graphs. Analysis of the obtained information let you to quickly identify deviations in the behavior and state of health of the animal.

Organizing the reproduction of a herd of cows and planning milk production, it is necessary to identify cows in hunting in a timely manner and to inseminate them fruitfully. Cultural dairy cattle breeding involves the use of artificial insemination of cows. The timely identification of cows in hunting and their transfer to insemination is the most important technological link in the production of milk. Table 2 presents the indicators of the reproduction of cows in enterprises for the period from 2000 to 2018.

Table 2. Reproductive qualities of cows

Indicators	1 period (2000-2009)	2 period (2009-2018)
The quantity of inseminations per one fruitful, doses	4.1±0.7	2.3±0.5
The actual barrenness, %	44.2±2.8	32.7±1.3
The duration of the open period, days	183.0±12.4	156.0±7.2

The analysis of the reproduction of a cows herd in the studied enterprises showed a significant positive trend after the introduction of a smart farm. The multiplicity of insemination of cows decreased, as evidenced by the number of insemination per fruitful one. On average, in enterprises of the Omsk region, before the introduction of a smart farm elements and an automated system for identifying cows in hunting, the number of insemination per fruitful was 4.1 times, which made it difficult to organize the reproduction of dairy cattle. There was the reduction in the multiplicity of insemination of one head in the studied enterprises on average to 2.3 times for the period from 2009 to 2018. Reducing the number of insemination per one fruitful can make the rational use of the reproductive function of cows and seed stocks of bulls. This reproduction indicator is mainly due to timely insemination. If the cow is properly determined in the hunt and inseminated at the optimum time, the insemination rate will be lower, that increases the efficiency of dairy cattle breeding.

IOP Conf. Series: Materials Science and Engineering **582** (2019) 012035 doi:10.1088/1757-899X/582/1/012035

The violation of reproduction leads to barrenness of the livestock, that is infertility of cows. Barren cows are unprofitable for the enterprise, since no offspring are received from them during the calendar year. In order to improve the efficiency of dairy cattle breeding, it is necessary to develop a plan of measures to reduce the actual productivity of cows and the prevention of infertility. Barrenness is caused not only by diseases of the cows, but also by technological errors in identifying cows in the hunt. Highly productive cows are characterized by a strongly pronounced dominant of lactation, which inhibits the reproductive function of the animal. Therefore, the animals with high milk production have the poor-expressed sexual hunting and, as the result, their open period is prolonged due to non-timely insemination. Thus, due to the automated system of reproduction control in the enterprises of the Omsk region, it was possible to reduce the actual barrenness to 32.7%. Skillful analysis of the received information let us to identify cows in the hunt in a timely manner and to inseminate them fruitfully. The duration of the open period is an important indicator of the reproductive qualities of cattle, since the synthesis of milk in the udder of a cow is not possible without receiving a calf from a cow. The open period is the period from calving to fruitful insemination. Figure 2 shows the trend of the open period from 2000 to 2018.

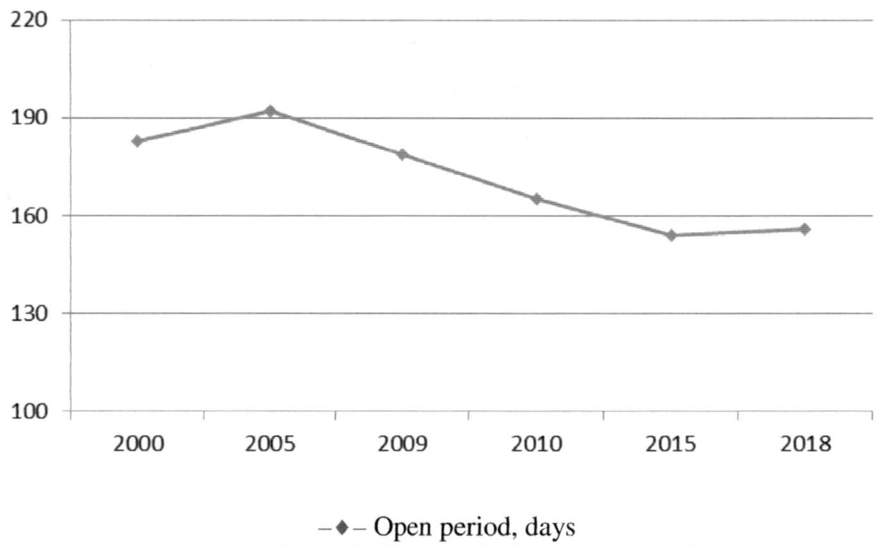

Figure 2. The trend of the open period

For the period from 2000 to 2018, the duration of the open period on average decreased by 27 days, or by 14.75%. The maximum values of the open period were noted in 2005, before the introduction of automated reproduction control systems. In 2015, the period from calving to fruitful insemination had minimal values, on average, the pregnancy in cows occurred on the 154th day of lactation.

Thus, we can conclude that the introduction of a smart farm leads to an improvement in reproduction rates in dairy cattle breeding by 52.3%.

At the same time, it is currently more cost-effective for enterprises to produce the highest quality milk. It is possible to improve the quality characteristics of the produced milk only by upgrading the milking equipment. Physical and chemical indicators of milk are the main criteria for determining its grade. The average values of milk quality indicators in the analyzed periods are presented in table 3.

Table 3. Physical and chemical milk characteristics

Indicators	1 period (2000-2009)	2 period (2009-2018)
Acidity, °T	17.6±3.21	17.2±0.15
Purity group	1…2	1
Density, kg/m	1027.0±2.4	1028.0±2.1
Mass fraction of true protein, %	2.7±0.04	2.8±0.03

Studying the obtained data, we can conclude that for the second period there were changes in the physical and chemical parameters of milk in the region. The purity of the milk produced in the conditions of modernized farms improved. All the milk produced in modern automated milking parlors belongs to the first purity group, which is undoubtedly due to the optimization of the milking technology and minimal contact of the raw milk with the environment.

The mass fraction of true protein in milk depends on the feeding conditions of the cows, as well as the breed and individual characteristics, therefore the difference between the periods before automation and after is not statistically significant.

To carry out the milking process the studied enterprises are equipped by the modern milking parlors with milking units such as a parallel or herringbone. The entire milking process is automated. The modern milking parlor of the smart farm provides comfortable milking, reduced stress for animals, and improved working conditions for machine milking operators.

The milking process is displayed on the PC monitor screen in the form of a graph and the use of color indicators that show statistical information and the quantity and quality of the milk produced.

Analysis of gross milk production, depending on the grade is presented in table 4.

Table 4. Milk produced, %

The grade of milk	1 period (2000-2009)	2 period (2009-2018)
The highest	42.0±4.8	56±4.1
The first	36.0±4.2	31±3.8
The second	22.0±3.3	13±3.6

The analysis of the grade of produced milk for the studied enterprises showed that in the period up to 2009, the share of top grade milk in gross production was 42%, and after the milking rooms were commissioned, the proportion of top grade milk was 56%, the quality of milk produced improved by 14%.

4. Conclusion

The increase of milk production is associated with the intellectualization of labor both based on promising means of automation and mechanization, and based on the organization of labor and production management using modern computer technologies.

Ensuring high growth rates of milk production and raising to a qualitatively new level of the economy of dairy cattle breeding are possible only by early transition to intensive production methods. In modern conditions of agricultural production, it is necessary to reduce the cost of human labor per unit of production. The introduction of digital technologies for smart farm, the automation of basic technological processes can reduce the load of the worker, thereby reducing the cost per unit of production.

Analysis of the economic efficiency of the use of modern herd management systems in the south of Western Siberia indicates the feasibility of modernizing domestic dairy cattle breeding. Nowadays the profitability of milk production in the studied enterprises is relatively low, since the cost of the equipment of a smart farm is quite high and pays off in a long time. Therefore, the main indicator of the effectiveness of the enterprise can be considered the specific volume of revenue from the sale of milk before the introduction of modern technologies and after, taking into account the price of selling milk.

The specific growth of profits from the sale of produced milk in the conditions of enterprises in the region amounted to +42%, and in certain production conditions, the specific profit was increased by 68%.

Thus, the introduction of modern automated herd management systems for milk production in the Omsk Region leads to an increase in key production indicators.

The result of the work of a smart farm was a reduction in the cost of fruitful insemination to 52.3% due to a reduction in the rate of insemination of cows, an increase of the young stock birthrate due to a decrease of the barrenness capacity by 34%. Due to the improvement of the quality indicators of milk and the reduction of labor costs, the use of modern technologies contributed to an increase in the specific profit from the sale of produced milk by 42...68%

In the current economic conditions, when a breed of cows was selected, a way of housing was determined, and a feeding ration was established, the role of mechanization and automation of technological processes increases in obtaining high-quality and competitive milk. Analysis of the production experience of two enterprises breeding the same breed of cattle located in different natural and climatic zones showed that the improvement of labor and the improvement of technology using modern equipment created the prerequisites for increasing the productivity of animals. This approach to the organization of dairy cattle in the conditions of the south of Western Siberia contributes to the preservation and development of relevant rural areas.

References

[1] Gupta B, Tripathi A, Joshi R, Pareek A and Singla-Pareek S 2015 Designing climate-smart future crops employing signal transduction components *Elucidation of Abiotic Stress Signaling in Plants: Functional Genomics Perspectives* vol 2. pp 393-413.

[2] Ortiz-Gonzalo D, Oelofse M, de Neergaard A, Vaast P, Albrecht A, and Rosenstock T 2017 Farm-scale greenhouse gas balances, hotspots and uncertainties in smallholder crop-livestock systems in Central Kenya *Agriculture, Ecosystems & Environment* vol 248. pp 58-70.

[3] Kipling R, Scollan N, Bannink A, Bellocchi G, Dalgaard T, Hutchings N, Kjeldsen C, Fox N, Topp C, Lacetera N, Sinabell F, van Oijen M and Virkajärvi P 2016 Modeling European ruminant production systems: facing the challenges of climate change *Agricultural Systems* vol 147. pp 24-37.

[4] Hovinen M, Siivonen J, Taponen S, Hänninen L, Pyörälä S, Pastell M and Aisla A 2008 Detection of clinical mastitis with the help of a thermal camera *Journal of Dairy Science* vol 91. № 12. pp 4592-4598.

[5] Pismennaja E, Loshakov A, Shevchenko D, Odincov S and Kipa L 2015 Comprehensive approach for evaluating the potential of the Stavropol agricultural territory *International Journal of Economics and Financial Issues* vol 5. № 3S. pp 113-120.

[6] Nagy P, Thomas S, Markó O end Juhász J 2013 Milk production, raw milk quality and fertility of dromedary camels (camelus dromedarius) under intensive management *Acta Veterinaria Hungarica* vol 61. № 1. pp 71-84.

[7] Aliev E, Bandura V, Pryshliak V, Yaropud V, Trukhanska O 2018 Modeling of mechanical and technological processes of the agricultural industry *INMATEH - Agricultural Engineering* vol 54. № 1. pp 95-104.

[8] Mottram T, Velasco-Garcia M, Berry P, Richards P, Ghesquiere J and Masson L 2002 Automatic on-line analysis of milk constituents (urea, ketones, enzymes and hormones) using biosensors *Comparative Clinical Pathology* vol 11. № 1. pp 50-58.

[9] Mottram T 2016 Animal board invited review: precision livestock farming for dairy cows with a focus on oestrus detection *Animal* vol 10. № 10. pp 1575-1584.

[10] Onteru S, Ampaire A and Rothschild M 2010 Biotechnology developments in the livestock sector in developing countries *Biotechnology and Genetic Engineering Reviews* vol 27. pp 217-228.

Technology of Poultry Manure Utilization as a Renewable Energy Source

M V Zapevalov [1], N S Sergeyev [1], G V Redreev [2], Yu B Chetyrkin [3], M S Zapevalov [1]

[1] South Ural State Agrarian University, Russia
[2] Omsk State Agrarian University named after P. A. Stolypin, Russia
[3] South Ural State University (National Research University), 454080, г. Chelyabinsk, Lenin Avenue, 75, Russia

E-mail: weerwg@mail.ru

Abstract. the aim of the research is to improve the efficiency of technologies for recycling poultry manure. The achievement of this goal involves the development of technological regulations for the processing of manure, research and development work on the creation of technological lines for the production of fuel briquettes and organic fertilizer, development of the technological process of processing of manure and its introduction into production. The technological scheme of the manure processing complex has been developed and substantiated, which includes: a receiving and storage station for receiving manure coming from poultry houses, and its dosed supply for further processing; a container for collecting liquid formed during the drying of manure; a technological line for the production of fuel briquettes and a line for the production of organic-mineral fertilizer. The capacity of the complex with a litter moisture content of 65–70% coming for processing is about 5 t/h. It is proved that the annual processing of manure at the round-the-clock operation of the complex will be about 43.0 thousand tons. At the same time, the production of organic-mineral fertilizers with a nutrient content of up to 32% will be about 1.5 thousand tons, and the fuel briquettes with a calorific value of up to 18.0 mJ/kg will be about 8.0 thousand tons. The estimated annual economic effect from the processing of litter on this complex is about 20.0 million rubles, the payback period of the complex is up to 1.5 years. This technology of poultry manure utilization allows to process all poultry manure removed from poultry houses in the stream, as a result of which there is no need for a litter storage. The production of organic fertilizer can solve the problem of increasing soil fertility. When the price of one ton of this fertilizer is 5thousand rubles and the contents until 320 kg of active ingredient, its use in the cultivation of crops would be economically profitable. As a result of the utilization of manure through this technology, the company receives energy, agro-chemical, economic, environmental and social effects.

Introduction

By the beginning of the twenty-first century, the world has intensified the research to find new, alternative energy sources. In particular, there was found the possibility of using solar radiation, wind energy, water energy of small rivers and reservoirs, geothermal energy, biomass energy, etc. Nowadays, the annual economic potential of alternative energy sources exceeds the production volume of all types of fossil fuels and is within 200 billion tons of conventional fuel. According to the forecasts of analysts, by 2030 the share of alternative energy sources in the total energy consumption will be up to 21% and will be equal to the share of the natural gas. At the same time, the great attention is paid to the use of biomass, primarily waste wood, pulp and paper industry, utilities and agriculture.

Any production is accompanied by the release of manure. At the same time, the higher the production culture, the more attention is paid to the recycling processes. Recycling should be understood as the processing of waste into a new and popular product. Agriculture is one of the main producers of organic waste, which include crop residues, animal manure, poultry manure, etc. Chelyabinsk region is one of the three leaders in Russia in the production of eggs and poultry. As a result of the modernization of existing poultry farms and the introduction of new ones, there is an annual increase in the number of

IOP Conf. Series: Materials Science and Engineering **582** (2019) 012036 doi:10.1088/1757-899X/582/1/012036

poultry. However, along with this, the yield of poultry manure is increasing, the annual volume of which is close to 1.5 million tons. At the cellular content of poultry, the moisture content of the litter is in the range of 65-70 % and belongs to the third class of hazardous substances, so it requires a special approach to its disposal [1–2]. Currently, the culture of the egg production and meat in poultry farms is very high, however, in most cases, the technology of litter disposal does not meet modern environmental and economic requirements. Very often, hundreds of thousands of tons of hazardous substances accumulate in the storage facilities, causing harm to the environment. When using such a litter as a fertilizer, which has a low amount of nutrients, while transporting over long distances, the cost of application is not paid off by increasing the yield of crops. The crop production needs an effective fertilizer. It is necessary that this fertilizer has a low cost, high nutrient content, a long period of action and it can be used by the existing system of machines [3–4].

An urgent problem in agriculture is energy supply. It is known that in the cost of agricultural production a large share of energy costs, which is mainly due to the high cost of energy. The constant increase in the cost of electricity, petroleum products, natural gas, solid fuel, leads to an increase in the cost of production, resulting in reduced profitability of all agricultural production. Agriculture, receiving a large amount of organic waste, for its own needs can produce solid and gaseous fuels, thermal and electrical energy.

This study is aimed at improving the efficiency of recycling processes of poultry manure. To achieve this goal, the following tasks were set and solved:

1. The experimental design work has been carried out and the technological line for the production of fuel briquettes and organic fertilizer has been created.

2. The technological process of manure processing has been developed for widespread introduction into production.

Materials and methods

The current level of agriculture development, both in Russia and abroad, requires new directions for the creation of waste-free production with a high sanitary and veterinary level. For later use of crop residues, animal manure and poultry manure, with the aim of preserving in it the essential nutrients, improve the physical and chemical properties, are used available for the enterprise methods of processing [5]. All known methods of processing can be divided into four separate methods: biological, physical, chemical and mixed. These methods contain 15 methods. Each of the methods can be carried out by 2-3 different technologies. Thus, currently there are more than thirty different technologies for processing livestock and poultry manure (Figure 1).

The largest amount of organic waste processing technologies are based on the biological method. The deeper understanding of the factors affecting the processes in industrial methods of biological processing, allow us to take into account the technological features of the choice and compliance with the optimal modes and parameters of the equipment [6].

The biological method of processing is based on aerobic, anaerobic, oxidative and biothermal methods, as well as the method using Californian worms. The implementation of the same method can be performed by different technologies, indoors or directly in the open air.If we take into account that biological processes are based on the laws of microbiologists, its application in the technology of manure and litter utilization allows to use more fully and with greater effect the laws of biological processes of biofermentation of various components in which the main ingredient is organic matter.

Without knowledge of the basics and features of composting technology, with a biothermal method of processing organic mixtures, the fermentation process can lead to unknown and even negative results. Instead of valuable organic fertilizers can be obtained, at best, ballast material, at worst - significant amounts of additional environmentally hazardous waste with a large number of seeds of various weeds and seedlings of pathogenic microflora.

One of the rational methods of processing organic waste is physical, which is based on changes in physical and mechanical properties, in particular and the removal of moisture. For this purpose, a variety of drying methods are used. The high-temperature drying deserves the greatest attention. The drying is carried out in cylinder-type dryers at a temperature of 800-9000C. During the heat treatment of manure or droppings, it is dehydrated to a moisture content of 12–14%, disinfection from pathogenic bacteria, viruses, helminth eggs, liberation from germinating seeds of weeds, down hair and feathers. The result is a dry substance containing essential nutrients (NPK) that can be used as an organic fertilizer

[7]. However, due to the fact that such a fertilizer contains a small amount of nutrients (3 – 4%), in order to provide plants with the necessary amount, it is required to make doses in a larger volume, which leads to an increase in transport costs. The cost of using this fertilizer is not paid off by increasing the yield of the cultivated crops [8].

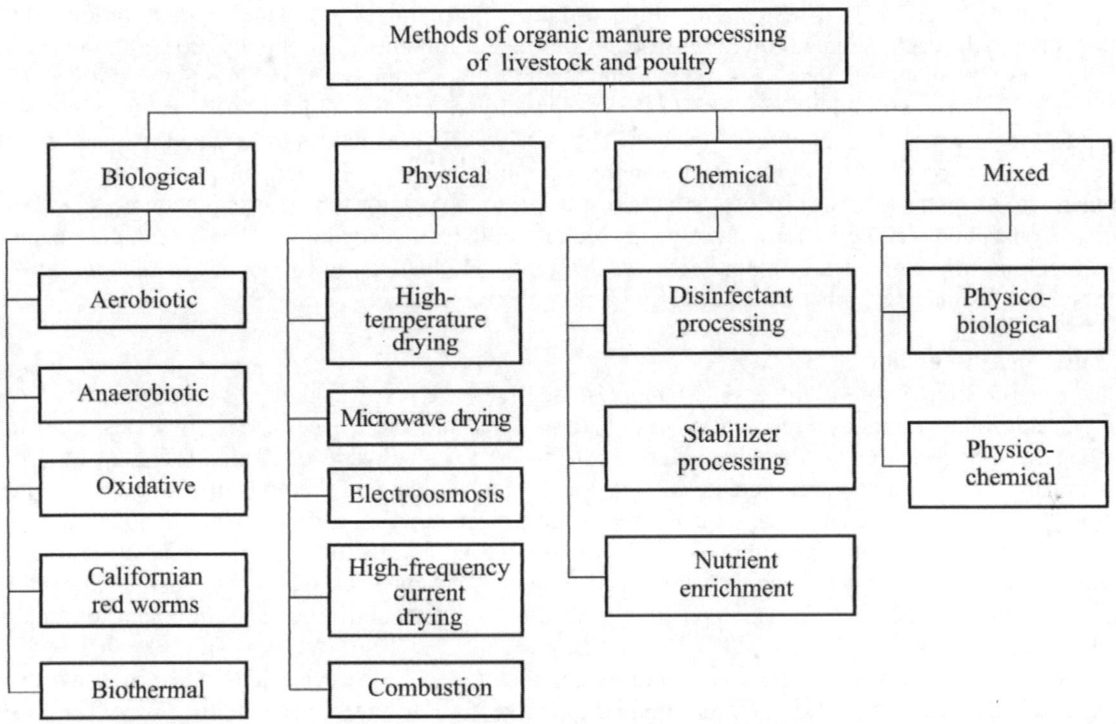

Figure 1 – Organic waste treatment methods

To solve the problems of poultry manure utilization and providing producers of the crop products with effective fertilizers in the South Ural State Agricultural University, the technology of processing of poultry manure with the production of complex organo-mineral fertilizer and fuel briquettes has been developed [9–10]. The scheme of the technological line of the complex for processing of the unshielded poultry manure with a capacity of 5.0 t/h is shown in Figure 2.

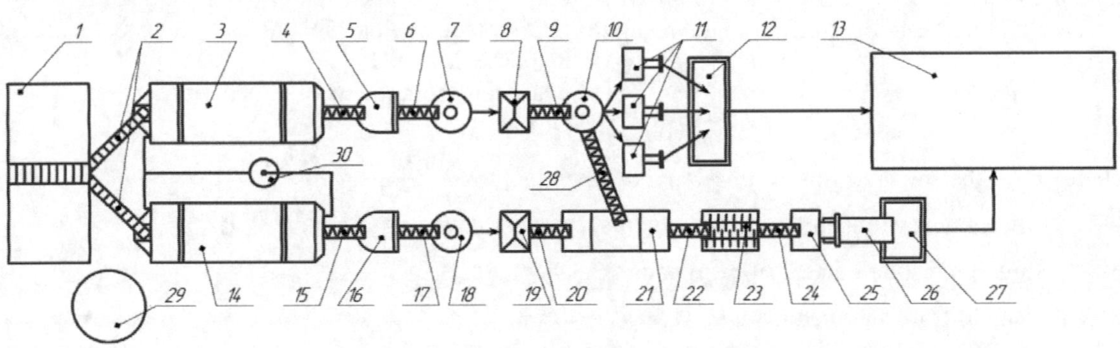

Figure 2 – Diagram of the technological line of the complex for poultry manure utilization
1 – receiving-storage station; 2 – scraper conveyor (TSN); 3 – reactor for drying manure; 4, 6, 9, 15, 17, 20, 22, 24, 28 – conveyor; 5, 16 – cooling column; 7, 10, 18 – storage tank; 8, 19 – shredder centrifugal rotary; 11 – piston press; 12 –briquettepacking table; 13 – warehouse 14 – reactor for manure gasification; 21 – dispenser; 23 – mixer; 25 – granulator; 26 – conveyer belt; 27 – pellet packing table; 29 – water tank; 30 – gas tank

The complex consists of the receiving-funded station, intended for reception of manure coming from the poultry houses, and batching it for further processing, containers for collection of liquids generated during drying of the manure, technological lines on manufacture of fuel briquettes, line for production of the organo-mineral fertilizers. All technological equipment is linked to a single process.

While removing manure from poultry houses load it in a dump vehicle, deliver to the facility for processing and overload in the emergency-funded station. From the receiving and storage station 3.0 t/h litter is fed to the drying and 2.0 t/h for gasification. When drying the litter, about 1200 kg/h of dry litter with a moisture content of up to 15 %. The dried manure is cooled, ground and fed into the storage tank. The part of the dry crushed manure is fed to the preparation of organo-mineral fertilizer, and the rest is used to the preparation of the fuel briquettes. The received fuel briquettes are packed and sent to the warehouse. As a result of the pyrolysis of manure, a combustible gas is formed, the part of which is used to maintain its own pyrolysis process, and the rest is fed to the reactor for the dried manure. As a result of pyrolysis, about 160 kg/h of ash is formed, which is cooled, crushed and fed to mix with dry manure. The resulting organic-mineral mixture contains 25 % of dried manure and 75 % of ash. The mixture is formed into granules, which are packaged and sent to the warehouse.

Results and discussion

The proposed technology of processing manure in the first stage involves the removal of moisture from it by drying. This occurs as a result of a combination of thermal and mass transfer processes occurring both at the surface and inside the wet manure, contributing to its dehydration from 70–75% to 14–15%. As a result, there are structural-mechanical, chemical, biochemical and rheological changes in the manure. The rate of these changes and the degree of the completion depends on many factors, the main factors are the method of supplying heat to the dried material and the drying mode. Since the manure contains a high percentage of organic matter, its drying prevents the occurrence of undesirable phenomena in it, such as self-heating, decomposition with the formation and release of ammonia, oily compounds, etc.

The poultry manure refers to a pasty dispersed material, so it is advisable to dry it by convective or contact methods. The most common method is convection drying, where the heating of the dried material is carried out by a gaseous drying agent (heated air, flue gases or a mixture thereof), in direct contact with the surface of the manure. The same agent is used to remove moisture vapor outside the dryer. The speed of the drying process of the litter with a gaseous agent depends on the intensity of the external and internal heat and mass transfer, since the amount of moisture supplied to the evaporation surface depends on it.With this method, the part of the thermal energy is not used for its intended purpose, but is lost along with the vapors. While drying the manure, a large amount of evaporating moisture is formed, which requires condensation. To do this, the dryer must be equipped with a reliable condensing unit. At the same time, the steam removed from the dryer has a sufficiently high temperature, so it can be used to preheat the manure in the receiving and storage station. This helps to reduce the energy consumption for its drying and reduce the time of the process.

The amount of moisture that is required to evaporate from the manure is defined as:

$$Q_{WED} = Q_{MDAY} \cdot (W_I - W_D) / 100 - W_D \tag{1}$$

where Q_{WED} - the amount of moisture evaporated per day, tons.

Q_{MDAY} - the amount of manure processed by the dryer per day, tons.

W_I - the initial moisture content of the manure, %.

W_D - the humidity of the dried manure, %.

For the purpose of expediency the dryer should work round-the-clock. With an hour of its capacity of 3.0 t/h, 72 tons of wet manure will be processed per day, while about 51.0 tons of water will evaporate. The daily amount of dried manure coming out of the dryer will be 21 tons.

The process of drying the litter is very energy-intensive and involves heating the material and evaporation of moisture. The heat consumption for evaporation of 1 kg of moisture is determined based on the heat and moisture content of the air:

$$q_{EVA} = \frac{I_{AE} - I_{AB}}{d_{ME} - d_{ME}}, \text{ kcal/kg} \tag{2}$$

where q_{EVA} - the amount of the heat to evaporate moisture, kcal/kg;

I_{AE} - the heat content of air at the exit from the dryer, kcal/kg (I_{AE}=230 kcal/kg);

I_{AB} - the heat content of air before heating in the dryer, kcal/kg (I_{AB}=10.25 kcal/kg);

d_{MB} - the moisture content of air before heating in the dryer, kg/kg (d_{MB}=0.01 kg/kg);

d_{ME} - the moisture content of air at the outlet of the dryer, kg/kg (d_{ME}=0.308 kg/kg).

Taking into account the data on the heat and moisture content of the air before heating in the dryer and heated, the amount of heat required to evaporate 1 kg of moisture is about 716 kcal/kg.

The required amount of the heat for heating 1 kg of evaporated moisture is defined as:

$$q_{HE} = G_{MD} \cdot \frac{100 - W_I}{W_I - W_D}(T_{MD} - T_{MW}) - G_W \cdot T_W, \text{kcal/kg} \tag{3}$$

where G_{MD} - the heat capacity of dried poultry manure, kcal/kg. deg. C (G_{MD} = 0.42 kcal/kg. deg. C);

T_{MD} - the temperature of dried poultry manure, deg. C;

T_{MW} - the temperature of wet poultry manure, deg. C;

G_W - the heat capacity of water, kcal/kg. deg. C (1 kcal/kg. deg. C);

T_W - the water temperature, deg. C.

Since before feeding the manure for drying in the receiving and storage station, its preheating is provided, in this regard, the temperature of the wet manure entering the drying drum will be about 300C, and the temperature of the dry manure when leaving the dryer is 1200C.The amount of heat to heat 1 kg of evaporated moisture is about 2.5 kcal/kg.

While drying the manure, the heat loss to the environment also occurs, the approximate amount of which can be determined by multiplying the amount of evaporated moisture by 10 kcal. In our case, the heat loss to the environment will be about 7.0 kcal/kg. Thus, the total heat consumption will be about 726 kcal/kg. With the efficiency of the dryer 95%, the total heat consumption for evaporation of 1 kg of moisture from the litter will be about 764.0 kcal/kg. the Total hourly heat consumption will be 1619680 kcal/h.

The amount of burned fuel (natural gas) to obtain this amount of the heat for drying the manure is determined by the formula:

$$G_{GAS} = \frac{q_H}{g_{GAS}}, \text{ m}^3 \tag{4}$$

where G_{GAS} - the amount of burned gas during the drying of manure, m³//h;

q_H - the hourly heat consumption, kcal/h;

g_{GAS} - the calorific value of gas, kcal/m³.

When the gas heat of combustion is about 8040 kcal/m³ , its consumption will be 201.4 m³/h.

The required volume of the drying cylinder is defined as:

$$V_{DC} = \frac{Q_{MEH}}{A}, \text{ m}^3 \tag{5}$$

where Q_{MEH} - the amount of moisture evaporated per hour, kg/h;

A - the tension of the cylinder for the evaporated moisture, kg(m³/ h).

With the cylinder voltage of 80 kg of evaporated moisture (m³/h), the volume of the dryer must be at least 26.5m³. Taking the cylinder diameter of 1.5 m, its length should be about 12 meters.

Taking into account these studies, a prototype of the dryer was made, which passed production tests (Figure 3).

Figure 3 – the prototype of the dryer

The proposed technology of manure processing allows to process up to 5.0 t/h of wet manure, while producing about 0.7 MW/h of thermal energy, 1150 kg/h of fuel briquettes and 210 kg/h of organic-mineral fertilizer. As a result of mixing dried poultry manure, with a content of 8.4 % of nutrients, with ash and a content of 41.3 % of nutrients, a complex of organo-mineral fertilizer is obtained, which contains about 320 kg of the active substance. The resulting organic-mineral fertilizer in the amount of nutrients is comparable to mineral fertilizer [11].

The analysis of the fertilizer price in the Chelyabinsk region and the content of the active substance shows that the average cost of one kilogram of the active substance is 41.5 rubles. Based on the cost of 1 kg of the active substance of the mineral fertilizer, the price of one ton of the organo-mineral fertilizer containing 320.0 kilograms of active substance should be about 13.0 thousand rubles. However, the calculations show that if 1 kg of the active substance of the fertilizer gives an increase in the yield of wheat grain 5.0–7.0 kg, then to cover the costs only for the purchase of fertilizer, the purchase price of grain should be about 8.0–6.0 rubles/kg. If we take into account the costs of storing this fertilizer, transportation, introduction into the soil, post-harvest processing of the increase in the crop from its application and the minimum profit, the purchase price of grain in this case should be 12-10 rubles/kg. Often, the purchase price of commercial grain is much lower than these figures, so the use of mineral fertilizers becomes economically unprofitable. Consequently, the price of mineral fertilizers in the domestic market is overstated, which hinders its use. In our calculations to determine the economic efficiency of poultry manureprocessing on the proposed technology, the price of 1 kg of the active substance of the resulting fertilizer was taken within 16-17 rubles. Based on this, with the content of nutrients in the organo-mineral fertilizer is 32 %, its price will be about 5.0 thousand rubles/ton.

The fuel briquettes on calorific value correspond to coal. The briquettes require a minimum of storage space, it can be automated feed during combustion. It can be used as a flue material in existing boiler plants without conversion, and combustion is relatively harmless to the environment. When burning fuel briquettes remains ash, which can be used as fertilizer. A significant share of consumption of fuel briquettes falls on European countries. In recent years, the interest in the fuel briquettes has also increased in Russia. Depending on the material from which the briquettes are made, its market price is 4.0–6.0 thousand rubles/ton. For comparison, the price of coal is 3.1–3.9 thousand rubles/ton. In determining the economic efficiency of the price of the fuel briquettes, we adopted 3.0 thousand rubles/ton.

When processing 120 tons per day of manure with a moisture content of 65-70 %, the annual production of the organo-mineral fertilizer will be about 1500 tons and fuel briquettes about 8000 tons. The revenue from the sale of these products will be about 31.5 million rubles. With the amount of direct costs of 11.5 million rubles, the estimated annual economic effect of the utilization of poultrymanure on the proposed technology is about 20.0 million rubles. With the cost of the complex 25.5 million rubles, the payback period is not more than 1.5 years. In the processing of poultry manure, the company gets the following effects:

1. Energy effect: the production of the fuel briquettes with qualities superior to classical energy resources (wood, coal, etc.);

2. Agrochemical effect: obtaining an effective complete organo-mineral fertilizer and restoration of natural soil fertility;

3. Economic effect: profit from the sale of fertilizers and fuel briquettes. Diversification of agricultural production. Reduction of expenses for purchase of mineral fertilizers for own needs. Reduction of costs for heating business premises due to the self-reliance of solid fuels;

4. Environmental effect: utilization of poultry manure in the stream, without the use of a repository. Reducing pollution of the environment and arable land;

5. Social effect: creation of new jobs. Increasing the attractiveness of living in rural areas and employment of rural population. The emergence of an additional source of income. Improvement of rural infrastructure, increase of literacy and communication of the population with territorial authorities.

Conclusion

1. By utilizing the hazardous substance, which is the manure, this method for the poultry farm solves the environmental problem, establishes the production of popular and competitive products. With the utilization of one production line of 120 tons per day of manure with a moisture content of 65-70 %, the annual production of thermal energy will be about 6000 MW, the fuel briquettes - 8000 tons, the organo-mineral fertilizer – 1500 tons, the implementation of which the company receives an economic effect of about 25 million rubles/year.

2. The production of the organic-mineral fertilizer can solve the problem of soil fertility increase. With the price of the organo-mineral fertilizer 5.0 thousand rubles/ton containing 320.0 kilograms of the active substance, its use in the cultivation of crops becomes economically beneficial.

References

[1] GOST 31461-2012 Poultry manure. Raw materials for the production of organic fertilizers. Technical conditions. – M. : STANDARTINFORM, 2013. – 11 p.

[2] Technologies and technical means for processing manure at poultry farms : scientific.-method. The manual / under the General editorship of V. I. Fisinin, V. P. Lysenko. – M.: OOO «NIPKTS Voskhod-A», 2011. – 296 p..

[3] Zapevalov M. V. Mechanized processes of preventive plant care. Technological and technical support for the use of organic fertilizer and chemical treatment of seeds. –Saarbrücken: Lap Lambert Academic Publishing, 2011. – 88 p.

[4] Govdya, V. V. Organizational and economic problems of efficiency increase of use of fertilizers and means of plant protection (On an example of Krasnodar Krai): dis. ... Dr. Econ. Sciences: 08.00.05 / Govdya Victor Vilenovich.– Krasnodar, 2001. – 354 p.

[5] Shchetkin, B. N. Methodology of environmentally safe processing of poultry manure into organo-mineral fertilizers and creation of devices for assessing the quality of its application to the soil in the cultivation of crops: dis. ... Dr. Techn. Sciences: 05.20.01 / Shchetkin Boris Nikolaevich. – St. Petersburg, 2004. – 350 p.

[6] Malinov, G. I. improving the efficiency of production lines of manure disposal and poultry farming in the dry animal feed by optimizing the composition and improve the tools: dis. ... Dr. Techn. Sciences: 05.20.01 / Malinov Gennady Ivanovich. – Petrozavodsk, 2003. – 342 p.

[7] Podkolzin, A. I. Evolution, reproduction of soil fertility and optimization of fertilizer application in agricultural landscapes of Central Ciscaucasia : dis. ... Dr. Biol. Sciences: 06.01.04 / Podkolzin Anatoly Ivanovich. – Moscow, 2008. - 379 p.

[8] Dabakhova, E. V. Scientific justification for the use of organic fertilizer of the poultry industry in the agro-ecosystem: dis. ... Dr. Agricultural Sciences: 06.01.04, 03.00.16 / Dabakhova Elena. – N. Novgorod, 2005. – 409 p.

[9] Zapevalov M. V., Kachurin, V. V. Improving the efficiency of recycling manure generated during the production of poultry products. Theoretical and scientific-practical magazine «News of the Orenburg State Agrarian University» №3 (71) 2018, 144-146 p.

[10] Zapevalov, M. V., Zapevalov S. M. Processing of poultry manure with obtaining electric and thermal energy, and a complex organo-mineral fertilizer / Vestnik CSAA, vol. 67/2,Chelyabinsk 2014, 45-49 p.
[11] Kapranov, V. N. Use of natural agrochemical means as sources of mineral nutrition of the field crops : dis. ... Dr. Agricultural Sciences: 06.01.04 / Kapranov, Vladimir Nikolaevich. – Nemchinovka, 2009. – 358 p.

IOP Conf. Series: Materials Science and Engineering **582** (2019) 012037 doi:10.1088/1757-899X/582/1/012037

Investigation of Moving Flat Hexagonal Disk Attachment

A S Soyunov [a], S P Prokopov [b], N A Zaripova [c], A Yu Golovin [d], R K Abdylmanova [e]

Federal State Budget Educational Institution of Higher Education "Omsk State Agrarian University named after P.A. Stolypin" 644008, Omsk, ul. Institutskaya square, 1

E-mail: [a] as.soyunov@omgau.org, [b] sp.prokopov@omgau.org, [c] na.zaripova@omgau.org, [d]ayu.golovin@omgau.org, [e]rkh.abdylmanova@omgau.org

Abstract. The article "Investigation of moving flat hexagonal disk attachment" reviewed the operation of flat hexagonal disk attachment used on discer for tilling the soil, exposed to wind erosion. The research of flat disk attachment is carried out in FSBEI HE OSAU named by P.A. Stolypin. The main laws are set and described and the quality evaluation criteria of cutting up weed vegetation δ in dependence upon angle of attack, cultivation depth and the distance between discs on the shaft are also represented. Using received dependences in practice it is not difficult to define the quality of weed vegetation cutting up and when required to make regulations of disc implement for quality weed vegetation cutting up. Arrangement is done by setting of basic parameters of stubble cleaner: angle of attack, cultivation depth and the distance between discs on the axle.

1. Introduction

Investigation of moving flat hexagonal disk attachment [1, 2, 3] is carried out in FSBEI HE OSAU named by P.A. Stolypin with the aim to determine the completeness of cutting up weed vegetation. The methods, described by G.N. Sineokov were used in this research [4, 5, 6, 7, 8, 9, 10] (fig. 1).

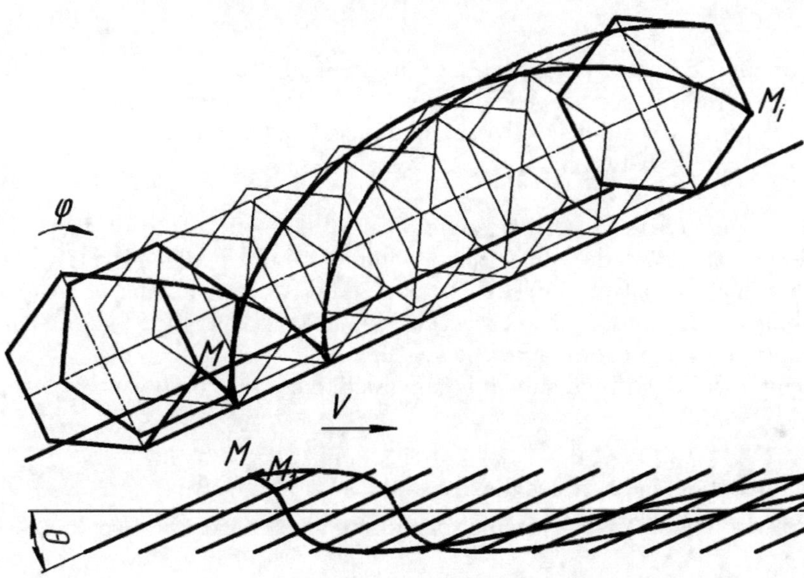

Figure 1 – Movement of a hexagonal disk in the soil

Thanks to the hexagonal form, the disk forms a spiral groove in the soil that favorably influences for retardation of a drain of subsoil and melt water on biases.

IOP Conf. Series: Materials Science and Engineering **582** (2019) 012037 doi:10.1088/1757-899X/582/1/012037

2. Theoretical investigations

During the operation of round disc attachments on different angles of attack on soil surface the soil trace looks like the furrow is formed. During the operation of hexagonal disks under smaller and bigger angles of attack, fitted on the shaft offset relative to each other on 30°, track is formed on soil surface from penetration of disk blade into and out the soil (fig. 2).

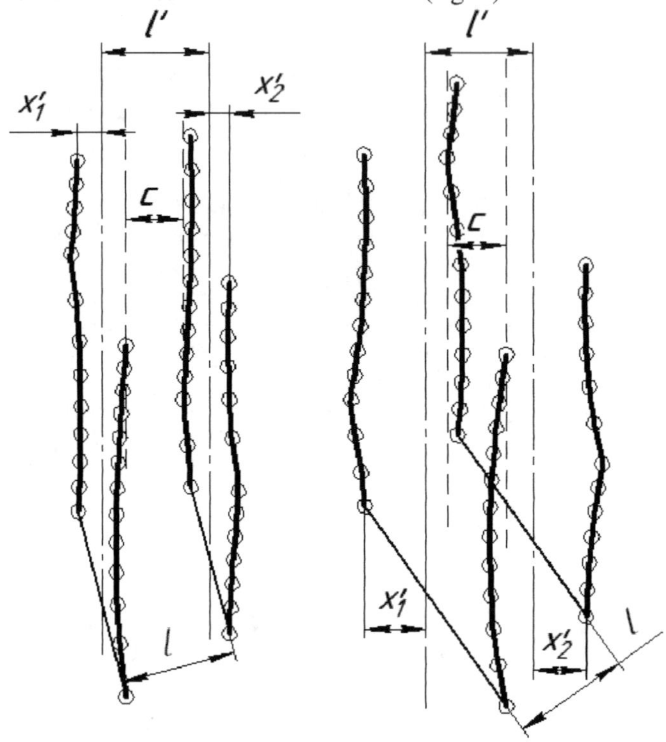

a b

Figure 2 – Disk trace on field surface: a –angle of attack θ=15°; b – angle of attack θ=35°; l – distance between disks on the shaft; l' – distance projection between disks on frontal surface; x'_1 и x'_2 – positions of crosspoint of disk blade from field surface; c – distance between traces of extreme disks

During the disks operation with smaller angle of attack 15° (fig. 2, a) on the field surface between traces of extreme disks undulating soil stripe of width c is formed, which is a flaw. During the disks operation with big angle of attack 35° (fig. 2, b) of flaw is absent owing to imaginary disk overlap, it promotes complete demolition of weed vegetation. That is why stripe width c can be used as the parameter of quality control of cutting up weed vegetation.

For quality control of cutting up weed vegetation it is suggested to use the coefficient δ

$$\delta = \frac{c'}{l'} \cdot 100\% , \quad (1)$$

Where c'– distance of cut plot of land between the disks, $c' = x'_1 + x'_2$ (fig. 3);

l' – distance projection between disks on the shaft on frontal surface, $l' = l \cdot \cos\theta$.

IOP Conf. Series: Materials Science and Engineering **582** (2019) 012037 doi:10.1088/1757-899X/582/1/012037

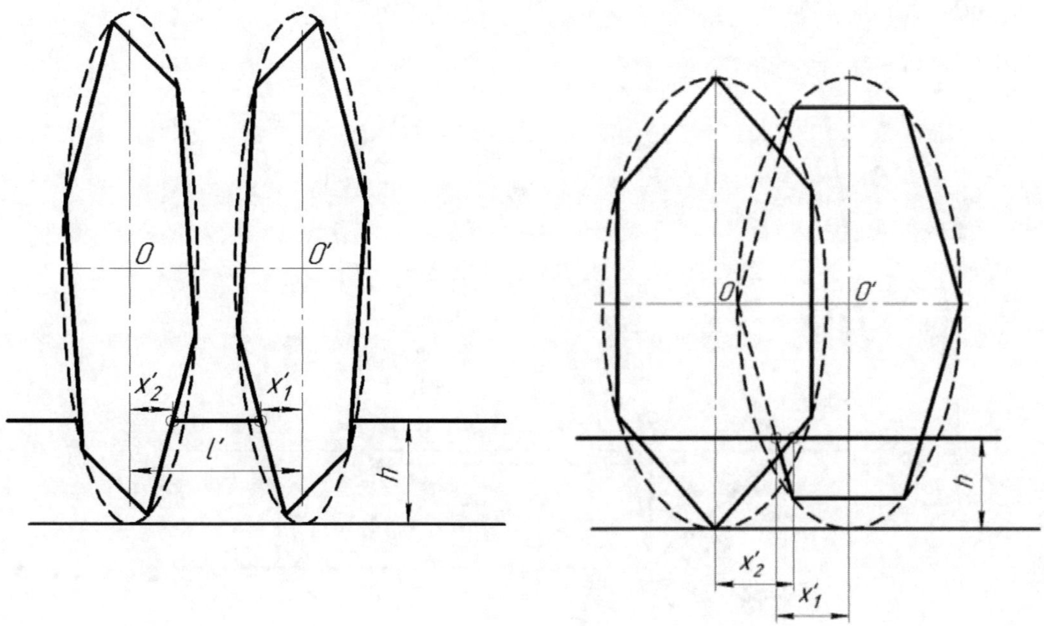

a b

Figure 3 – Projection of disks: a – under the angle of attack θ=15°;
b – under the angle of attack θ=35°; x'1 and x'2 – distance from disk axle to crossing of disk blade with the soil;
l' – distance between disks on the shaft; h – cultivation depth

Consider the parameters included in the equation (1).

The distance of nontreated plots of land by disks

$$c = l' - x'_1 + x'_2 \ , \tag{2}$$

where x'_1 and x'_2 – distance projection from disk axle to crosspoints of blade with field surface

$$x'_1 = x_1 \cdot \sin\theta; \quad x'_2 = x_2 \cdot \sin\theta \tag{3}$$

Nontreated plots of land is left under the smaller angle of attack between disks (fig. 3, a), but with increase of angle of attack imaginary overlap occurs and cultivation is done without flaws (fig. 3, b).

Meaning of the distances x_1 and x_2 is defined from geometric ratios of figure 3.

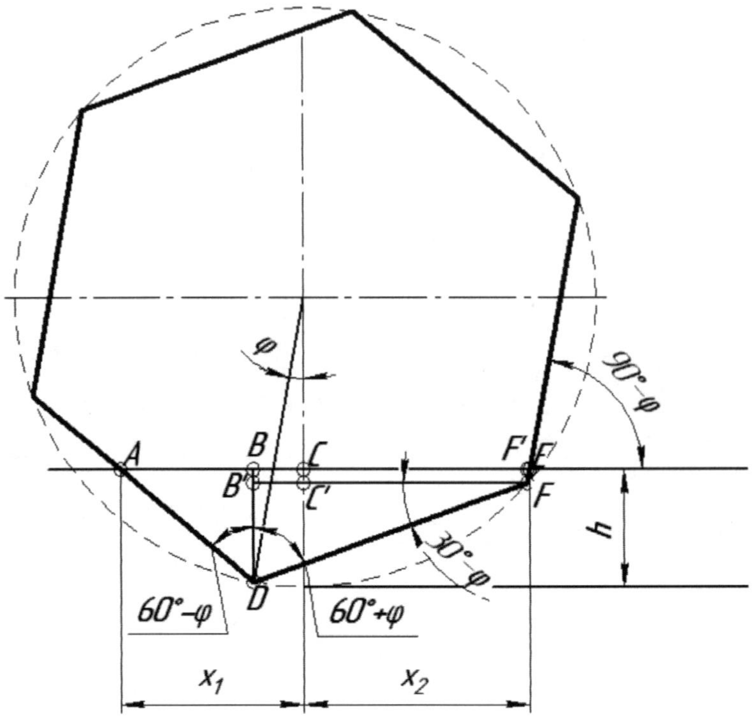

Figure 4 – Scheme of distance definition x_1 and x_2 before field surface crossing with hexagon face

From figure 4 the distance x_1

$$x_1 = AC = AB + BC\ , \tag{4}$$

where $AB = BD \cdot \mathrm{tg}(60^\circ - \varphi);$

$BD = h - R + R \cdot \cos\varphi;$

$BC = R \cdot \sin\varphi.$

Equation (4) takes the form of

$$x_1 = \left[h - R\ 1 - \cos\varphi\ \right] \cdot \mathrm{tg}(60^\circ - \varphi) + R \cdot \sin\varphi\ . \tag{5}$$

Distance x_2 is defined from the figure 3

$$x_2 = CE = CF' + F'E\ , \tag{6}$$

where $CF' = C'F = B'F - BC;$

$B'F = R \cdot \cos(30^\circ - \varphi);$

$CF' = R\ \cos(30^\circ - \varphi) - \sin\varphi\ ;$

$F'E = BB' \cdot \mathrm{tg}\varphi;$

$BB' = BD - B'D;$

$B'D = R \cdot \cos(60^\circ + \varphi)\ ;$

$F'E = \left[h - R\ 1 - \cos\varphi + \cos(60^\circ + \varphi)\ \right] \cdot \mathrm{tg}\,\varphi\ .$

Equation (6) takes the form of

$$x_2 = R\ \cos(30^\circ - \varphi) - \sin\varphi\ + \left[h - R\ 1 - \cos\varphi + \cos(60^\circ + \varphi)\ \right] \cdot \mathrm{tg}\,\varphi\ . \tag{7}$$

Insert (5), (7), (3) equation (2) and carry out calculations, received the table 1 and the schedule (fig. 5) of width of nontreated plot of land.

Table 1 – Dependence of nontreated plot of land width *c* (mm) upon cultivation depth *h* (mm), angle of attack $\theta°$ and distance between disks on axle *l* (mm).

l	160				170				180			
θ \ h	60	80	100	120	60	80	100	120	60	80	100	120
15	70.9	60.2	51.7	44.7	80.6	69.9	61.3	54.4	90.2	79.5	71.0	64.1
20	39.8	25.7	14.4	5.2	49.2	35.1	23.8	14.6	58.6	44.5	33.2	24.0
25	8.5	-9.0	-23.0	-34.3	17.5	0.1	-13.9	-25.2	26.6	9.1	-4.9	-16.2
30	-23.0	-43.6	-60.2	-73.6	-14.3	-35.0	-51.5	-64.9	-5.7	-26.3	-42.9	-56.2
35	-54.3	-78.0	-96.9	-112.3	-46.1	-69.8	-88.7	-104.1	-37.9	-61.6	-80.5	-95.9

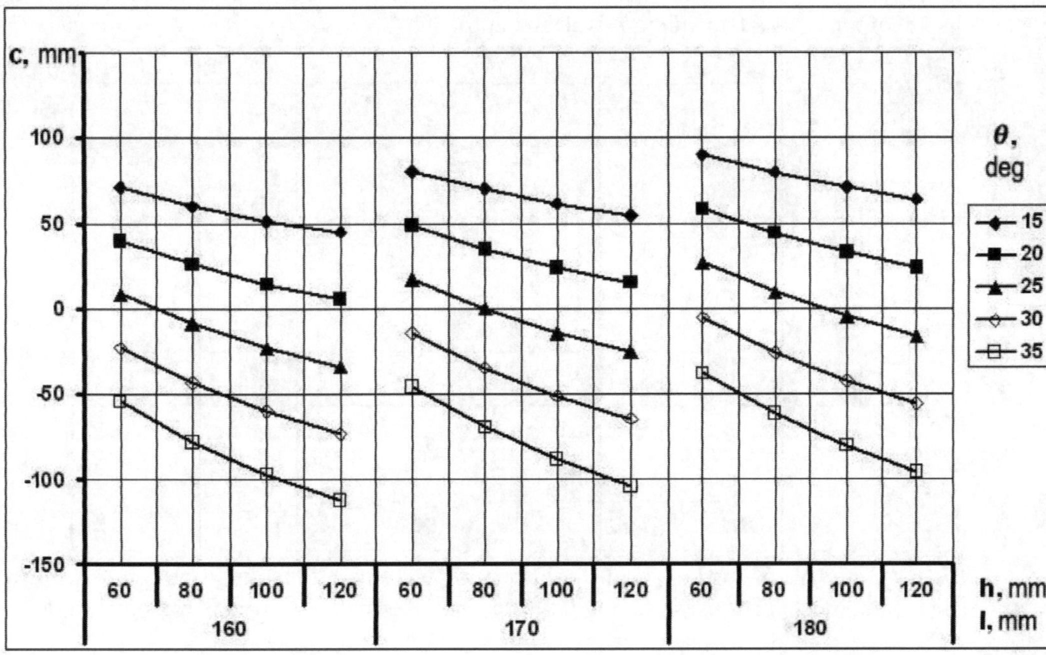

Figure 5 – Schedule of distance change *c* in dependence upon the distance of disks arrangement *l*, depth of movement *h* and angle of attack θ

Knowing the meaning of nontreated plot of land width *c*, consider quality coefficient of cutting up weed vegetation (1), calculations data are shown in the table 2.

Table 2 – Quality of cutting up weed vegetation in dependence upon depth of cultivation *h* (mm), distance between disks on axle *l* (mm) and angle of attack $\theta°$

	l	160				170				180			
θ	h	60	80	100	120	60	80	100	120	60	80	100	120
15	$\delta_{nontreated}$	45.9	39.0	33.4	28.9	49.1	42.6	37.3	33.1	51.9	45.8	40.8	36.8
	$\delta_{treated}$	54.1	61.0	66.6	71.1	50.9	57.4	62.7	66.9	48.1	54.2	59.2	63.2
20	$\delta_{nontreated}$	26.5	17.1	9.6	3.5	30.8	22.0	14.9	9.2	34.7	26.3	19.6	14.2
	$\delta_{treated}$	73.5	82.9	90.4	96.5	69.2	78.0	85.1	90.8	65.3	73.7	80.4	85.8

Continuation table 2

	l	160				170				180			
θ	h	60	80	100	120	60	80	100	120	60	80	100	120
25	$\delta_{nontreated}$	5.8	-	-	-	12.1	0.0	-	-	18.3	6.3	-	-
	$\delta_{treated}$	94.2	100.0	100.0	100.0	87.9	100.0	100.0	100.0	81.7	93.7	100.0	100.0
30	$\delta_{nontreated}$	-	-	-	-	-	-	-	-	-	-	-	-
	$\delta_{treated}$	100.0	100.0	100.0	100.0	100.0	100.0	100.0	100.0	100.0	100.0	100.0	100.0
35	$\delta_{nontreated}$	-	-	-	-	-	-	-	-	-	-	-	-
	$\delta_{treated}$	100.0	100.0	100.0	100.0	100.0	100.0	100.0	100.0	100.0	100.0	100.0	100.0

$\delta_{nontreated}$, $\delta_{treated}$ – respectively percentage ratio of treated and nontreated field surface.

Schedule built (fig. 6) according to the calculations (table 2), which clearly demonstrates the quality of cutting up of weed vegetations at distance between disks on the shaft 170 mm and parameters change of angle of attack and cultivation depth.

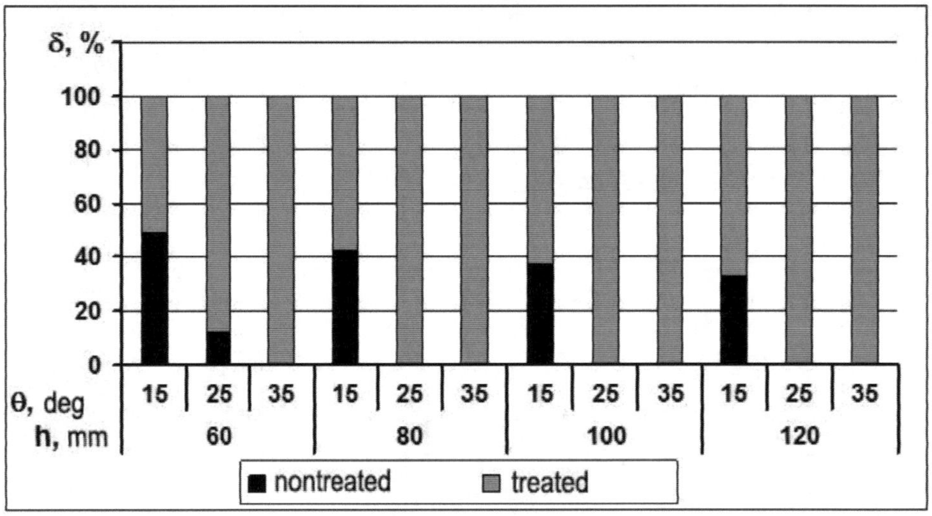

Figure 6 – Bar chart of working width usage of implement and the quality of cutting up weed vegetation at distance between disks on axle l =170 mm in dependence upon parameters: angle of attack θ, cultivation depth h.

From the schedule (fig. 6) can be seen, that at smaller angle of attack θ =15° across the working width of implement nontreated plots of land up to 50 % are observed and consequently noncut weed vegetation is left on these plots of land.

3. Conclusion

In the result of investigation the criteria of quality control of cutting up weed vegetation δ in dependence upon parameters: angle of attack, cultivation depth and distance between disks on axle is represented. According to the received dependences and schedules it is easy to define the quality of cutting up weed vegetation (in percentage) as well as the best parameters of stubble cleaner operation: angle of attack, cultivation depth, distance between disks on axle.

References

[1] Patent for useful model № 68838 RF, MPK A 01 B 15/18. Disk attachment of tillage implement / E. Yu. Kupriyan, I.D.Kobyakov, A.A. Degtyarev, V.V. Trotsenko; applicant and holder FSEI HPE Omsky SAU – № 2007119928/22; appl. 28.05.2007; publ. 10.12.2007, Bulletin № 34. – P. 18.

[2] Patent for useful model № 82508 RF, MPK A 01 13 23/00. Attachment of disk harrow / A.S. Soyunov, I.D.Kobyakov; applicant and holder FSEI HPE Omsky SAU. – № 2008144333/22; appl10.11.2008; publ. 10.05.2009, Bulletin № 13. – P. 12.

[3] Patent for useful model № 87859 RF, MPK A 01 B 35/20 (2006.01). Disk attachment of tillage implements / A.S. Soyunov, I.D.Kobyakov, E.I. Kobyakova; applicant and holder FSEI HPE Omsky SAU. – № 2009100697/22; appl. 11.01.2009; publ. 27.10.2009, Bulletin № 30. – P. 12.

[4] Sineokov G.N. Disk attachments of tillage implements: (theory and calculation): scien. publ. / G.N. Sineokov. – M.: Mashgiz, 1949. – 86 p.

[5] Sineokov G.N. Construction of tillage implements: scien. publ. / G.N. Sineokov. – M.: Izd-vo Mashinostroenie, 1965. – 308 p.

[6] Sineokov G.N. Theory and calculation of tillage implements: scien. publ. / G.N. Sineokov, I. M. Panov. – M.: Izd-vo Mashinostroenie, 1977. – 328 p.

[7] Influence of a form of disk edges on jamming of stalks / I.D. Kobyakov//Tractors and agricultural cars. – 2007. – No. 12. – P. 31-33.

[8] A research of interaction of a flat disk with the soil / F.M. Kanarev//Rational use of the machine and tractor park: науч. тр. / Cubic agricultural in-t. – 1971. – Issue 44. – P. 132-133.

[9] Research of process of cutting of the soil / I.D. Kobyakov//Achievements of science and technology of agrarian and industrial complex. – 2007. – No. 9. – P. 30-32.

[10] Strelbitsky, V. F. Disk tillage machines: науч. prod. / V.F. Strelbitsky. – M.: Mashinostroyeniye publishing house, 1978. – 135 p.

IOP Conf. Series: Materials Science and Engineering **582** (2019) 012038 doi:10.1088/1757-899X/582/1/012038

Elaboration of Theoretical Methods for Assessment of Isolated and Combined Physical Damage Effects of Technogenic Accident While Transporting Radiological Materials by Multimodal Transport

Aleksander Tumanov [1, a], Sergey Venevsky [2]

[1]The Peter the Great St. Petersburg Politechnic University, St. Petersburg, 195251, Polytechnicheskaya, 29, Russia
[2]Tsinghua University, Department for Earth System Science, Beijing, China

E-mail: [a] toumanov@mail.ru

Abstract. To enhance the safety of the transport of dangerous goods in tanks, the latter should improve the mechanical (physical) protection to prevent accidental damage to the boilers, such as objects containing dangerous goods directly. Damage to the boiler will almost inevitably leads to such consequences as decompression, the leakage of liquid radioactive waste and radiation pollution of the surrounding area. This article discusses scenarios accidents when overloading with marine railway transport; the explosion of gas cylinders on board the vessel. Calculation formulas are obtained to determine the speed at which the shards is provided by probabilistic the breakout a certain thickness of the tank walls. Received probability distribution function for the specified parameters shards and material barriers, which was determined by conducting a full-scale experiment. The results of theoretical studies, experiments used to improve designs of tanks carrying LRW and recommendations for improving the security of multimodal transport.

1. Introduction

In accordance with the transportation of dangerous goods terminology – radioactive material (RM) is a moving, including the use of various vehicles and its overload, since his departure from nuclear installation (item storage) and ending with the arrival of the nuclear installation (item storage). Radioactive material is a nuclear materials containing or capable to reproduce the fissile materials (substances) and radioactive substances (not related to nuclear material substances which emit ionizing radiation). As the RM are considered liquid radioactive wastes. Thus, the process of transportation includes the following stages of multimodal transport: loading RM onto a ship at the sea port (port facility); transportation of the vessel; transhipment of cargo RM from ship to rail transport; transportation by rail to destination.

To ensure the safe transport of radioactive materials by different modes of transport currently consists in the establishment of the necessary industrial regulatory framework reflecting the specificity of the sector and organizations effective departmental control. Analysis of foreign statistics showed that in the United States from 1997 to 2011 year happened 128 accidents with cargoes of RM (1 death neradiologicheskaja, 1 case of harm, damage, $2.2 million). With all dangerous goods (og) happened 5 052 170 493 accidents (serious, 138 death 2 825 cases of injury, damage to $635.58 million.

In the United Kingdom gives full statistics. So for 1958-2008 Gg. 913 occurred incidents during 2006-2008 Gg. 90 incidents occurred. Only three cases have been minor accidents, the rest is incorrect marking, documentation, malfunction, contamination, etc. In France for 1997-2007 biennium. happened 900 events (accidents, incidents, accidents). Adverse events occurred in all modes of transport (transportation, loading and unloading, the storage of packages while in transit, transshipment from one mode to another).

In the report, Rosatom in the Forum in Saint Petersburg in 2015, provides data on 33 accidents during transportation. Thus, securing transportation process maritime transport and railways, and especially when handling processes, security of transported goods, environmental security is a very important issue. The aim of the research is to increase the level of security at transportation and transshipment of liquid radioactive waste in multimodal transport. The object of the study are technical means of transport – tanks for transportation of RM in multimodal transport, and subject — processes for assessing the risk of transportation of liquid radioactive waste in tanks when accidental man-made explosions on floating nuclear thermal power stations (FNTPS).

Given that coastal and floating NAVAL and civilian spent fuel storage locations nuclear fleet is almost completely filled, the path to the burial of several thousand kilometres of marine and rail (fig.1).

Figure 1. Routes of transportation RM arising from multimodal transportation.

In [2-6] are the results of years of research. "The work took place in several stages. In the first phase, an analysis of real accidents during the carriage of dangerous goods and the development of scenarios of possible developments in emergency regimes. In the second phase, analysed structural features of tanks, influencing their behavior when emergency modes and defined the parameters of mechanical or thermal effects on the boiler (vessel) tank at the time of the loss of tightness (destruction) of the boiler.

The third phase consisted in the development of scenarios of the development of "settlement" emergency situations in which it is possible to save the boiler tightness that complemented the normative documentation in designing tanks for dangerous goods. In the fourth stage modeled various modes that can lead to accidents. The study dealt with the problems of strength, stability, thermal effects, thermodynamic processes, dynamic response, taking into account the dynamics of liquid cargo, etc. accompanied by Research experiments, model and full-scale samples. If this were verified and refined theoretical models were validated designs. Next, on the basis of the developed mathematical models for studies of different modes of emergency interventions. Fifth phase included the development of means of protection against accidental impacts. In this experimental and theoretical studies have been carried out on the basis of the developed methods.

It is known that "transportation RM carried out by the railway wagon-liternymi container trains consisting of locomotives, wagons, escorts for posting armed guards and attendants, cover wagons and carriages. Routes mainly run from NPP to and Krasnoyarsk (fig. 2). For transportation RM uses the following modifications to the cars.

2. Methods

2.1. Baseline data

Estimated assessment of radiation effects is made using the following source data [19]. Volumes of LRW are: drainage capacity to-4 PCs of 8 M3 = 32 M3; draining tanks d-4 PCs of 8 M3 = 32 M3; water tank decontamination unit overload-40 M3; Bucky excerpts-3 PCs for 5.49 M3 + 1.4 m3 (s) = 17.9 m3. The weight of the coolant I contour-20827 kg. The content of radionuclides in thermal liquid (I) at the time of circuit 3 phase is given in table 1.

Table 1. Maximum volumetric activity of LRW and its dose-forming radionuclide

Activity, 10^4 Bq/kg						
Corrosion products		Fission products				The sum
^{54}Mn	^{60}Co	$^{90}Sr+^{90}Y$	^{134}Cs	$^{137}Cs+^{137m}Ba$	$^{144}Ce+^{144}Pr$	
2	2	5	8	15	8	40

2.2. Construction of storage tanks

For liquid radioactive waste according to the normative requirements of security [1] capacity for LRW consists of an outer protective shell and placed in her tank. Shell is a vertical or horizontal cylindrical container sealed with bottoms. Material-stainless steel 12x18h10t. The gap between the walls of the tank shell and is filled with special sorbent to absorb leaking liquid radioactive waste in the event of damage to the tank or to seal formed holes. The tank is equipped with a mounting for mounting the nodes in the shell. Placed in the tank level sensors LRW injection systems/plum and mixing LRW, ventilation, alarm, outages LRW intensity required when waste level sensor, as well as heating (for preheating LRW and packaging systems in operation it offline in the cold season). At the bottom of the tank is the sump to collect remnants of LRW from pipelines after discharge.

2.3. Characteristics of LRW

LRW is a predominantly mineral solutions with a total salt content of not more than 500 g/l, including suspended matter (insoluble salt and sediments) with the total salt content of not more than 100 g/l density-no more than 1.25 t/m^3 pH-6-10, kinematic viscosity of no more than 2.5 centistokes. Volumetric activity waste shall not exceed 3,7*10^6 kBq/M^3 (10-4 CI/l) for α-and β-emitting radionuclides with the energy of the radiation is not more than that of 137Cs, and 3,7*10^3 kBq/M^3 (10^{-7} CI/l) for the rest of the α-and β-emitting nuclides. The design capacity of the dose from transported LRW, according to Sanpin 2.6.1.1281-03, does not exceed 2 mSv/h at any point on the surface of the package, and 0.1 mSv/h at any point 1 m from the packaging. Overall mass characteristics RM containers allow you to ship by rail rolling stock in sizes 02-VM GOST 9238-83, taking into account existing transportation and handling chain TUK. For example it is possible to transport in the wagon-container TC-13 with minor rework of the design of the car.

Emergency scenarios with LRW during transportation of the RM in scripted accidents lie exploration of possible hazardous effects on an object (packaging, container, etc.), as well as the surrounding Wednesday (exit and the spread of dangerous substances). As hazardous substances were considered.

When designing scenarios of accidents within the conservative approach considered the accident a third category.

Radiation effects Radiation accidents consequences in the form of receipts of LRW in surrounding Wednesday are possible with the following scenarios: partial destruction of transport packing container with LRW: fire and explosion defralgacionnom railway tank when overloading LRW from sea to rail fire in and subsequent explosion of oxygen cylinders on board the vessel during towing and parking; flooding or prolonged landing on the Rocky Shoal in storm conditions when towing along the Northern sea route.

2.4. Explosions of gas cylinders on board

Gas cylinders are sources of potential danger of explosion, which leads to the destruction of hull structures, failure of equipment delivered, and the emergence of a radiation accident. The explosion of cylinders may be caused by the following reasons. Violation of the integrity of the gas cylinder as a result of breakdown of any node (armature), damage or corrosion of the casing, improper operation can lead to an explosion of a cylinder at the working pressure. The probability of explosion of a cylinder (of the vessel) high pressure in this case is 10^{-7} 1/year. Exceeding the limit value of pressure in the event of a fire. For cylinders with nereagirujushhimi gas (air, nitrogen, argon) with the failure of the safety valve pressure of gas when heated increases until a blast. In cylinders with reacting gases (oxygen and acetylene) detonation may occur, resulting in a dramatic increase in pressure and explode.

IOP Conf. Series: Materials Science and Engineering **582** (2019) 012038 doi:10.1088/1757-899X/582/1/012038

Methods of evaluating the amount of destruction caused by the explosions of gas cylinders, is based on the provisions set forth in [10]. Affecting factors in the explosive destruction of the pressure vessels are shards, gas-dynamic Jet and overpressure in the emergency room as a result of the expiration of the gas from the vessel. Force the gas Jet, pouring from the vessel, with increasing distance quickly decreases and excess pressure in the emergency room because of the relatively large size of ship premises and the presence of "unloading" openings (doors, hatches, as well as the looseness of the damaged hull constructions) is not dangerous, as does not exceed the thresholds of excess pressure in the shock wave front (50 kPa). Thus, the major destruction of hull constructions and equipment are determined by influence of splinters. In the course of calculation is determined by the speed of fragments damaging the container, the equivalent thickness of a steel impenetrable barriers and the size of the area of devastation.

The study developed a model of accidents of the third category in which the impact of damaging factors on blast tanks with LRW resulted in holes in the wall of the tank. The challenge facing a researcher is to determine the minimum thickness of the wall of obstacles from a particular material, which may impair the integrity of the tank at the specified speed and configuration of the shard. For this purpose it is necessary to explore the mechanism of breaking through barriers and build a model of disruptive actions of the fragments in the explosion of gas cylinders.

Mechanism of breaking through barriers depends on many factors, among which prevail shard speed on approach to target and the ratio of the thickness of the barrier s to characteristic size fragment, for example, $d_0 = 2\ r_0$-r_0 (the radius of the spherical fragments equivalent mass). At high speeds the shard and $d_0 > s$ a powerful shock wave that occurs in hazard, reaches its opposite side and is reflected from it in the form of waves of rarefaction.

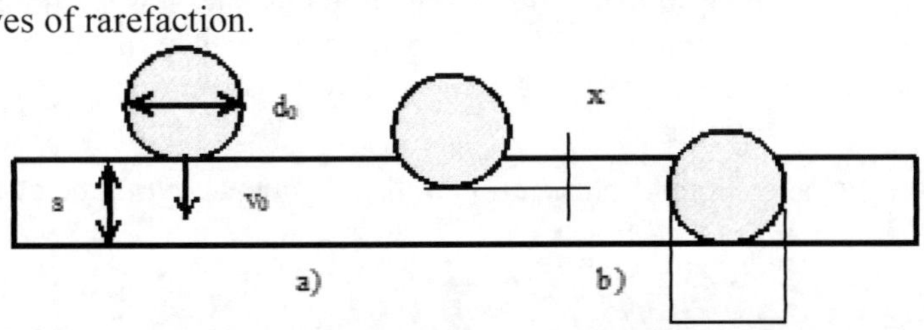

Figure 2. Mechanism of breaking through barriers fragment-the scoop on stages: a-introduction and b-ball movement with stopper

At moderate velocities of impact of a fragment with a metal barrier (100–1300 m/s) and s> r0, the process of collision is described by the deformation model of a certain volume of the obstacle material, which depends on the size (mass) of the fragment. The mechanism of the deformation of the barrier in the event of a splinter damage is shown in Figure 2.

Conventionally, it can be divided into two stages. At the first stage, the ball has an initial velocity v_0 embedded in the barrier, knocks out the tube and moves with it. The mass of the tube at the current time t> 0 is $\pi r^2 s \rho_M$. The material of the tube undergoes shear deformations, and the acting force R_τ varies linearly and is determined by the shear stress τ .

$$R_\tau = 2\pi r s \tau \qquad (1)$$

This force acts at the time of the introduction of the ball to a depth of r_0. In the future, the resistance force of the barrier will change according to the law (Figure 2, a, b)

$$R_\tau = 2\pi r_0 (s + r_0 - x)\tau \qquad (2)$$

in which x is the path traversed by the center of mass of the fragment from the moment of its collision with the obstacle. At the second stage, the motion of the "fragment – cork" system under the action of force (2) is considered, and the path of the center of mass of the fragment varies from r_0 to $r_0 + s$.

From the considered collision process, we conclude that the main part of the kinetic energy of the fragment is spent on the deformation of the material of the obstacle. This model does not take into account the energy loss of the fragment to the formation of shock waves in the barrier. Let us reproduce for our model calculation formulas based on the use of various assumptions. One of the first formulas was based on the assumption of the theory of plastic deformations that the specific deformation energy of the E_1 barrier does not depend either on the mass or on the shape of the striking body, nor on the thickness of the obstacle. The specific strain energy is determined only by the strength characteristics of the material of the obstacle, which is proportional to the magnitude of the destructive shear stresses of the material τ. This allows you to write the following relationship for a given material:

$$E_1 = \frac{E}{V} = \alpha k \tau = const \qquad (3)$$

where E is the energy expended on the deformation of the material of the barrier; V is the volume of the deformed material; k – is dynamic amplification factor, which takes into account the hardening of the material in the dynamic nature of shear deformations; α – is a matching factor.

If we assume that all the kinetic energy of a fragment with a mass m and velocity v is spent on the deformation of the material, the volume of which is equal to the volume $S \cdot s$ of the broken cork, then the actual specific energy (energy per unit volume of the destroyed obstacle) in this case will be $\frac{mv^2}{2Ss}$. The condition for breaking through the barrier will be determined by the ratio

$$\frac{mv^2}{2Ss} \geq E_1 \qquad (4)$$

In this expression, the impact area S of a fragment with a target is a random variable. Also, the random value is the depth of penetration of the fragment into the material x with thickness s. At the moment of breaking through $x=s$. We introduce a dimensionless quantity

$$D = \frac{x}{\bar{\bar{x}}} \qquad (5)$$

Where $\bar{x}$ – the average depth of penetration of the fragment. Using the designation

$$E_S = \frac{mv^2}{2\bar{\bar{S}}s} \qquad (6)$$

instead of (5) you can write

$$D \leq \frac{E_S}{E_1} \qquad (7)$$

2.5. Cumulative distribution function

When exposed to objects of the same type of damaging factor, each of them needs to achieve a certain effect of damage. However, due to the individual characteristics of the objects, to achieve a given effect of exposure, each of them will need a well-defined dose of the damaging effect. In this regard, it can be argued that the dose of exposure that causes a given, studied effect (response) of an object on the impact of energy and substances affects the factor is a continuous random variable.

Let us consider methods for describing a random variable dose, causing a given effect on a technical object and a person.

As we noted above, the most complete characteristic of a random variable is its distribution law. It can be specified in one of two forms: in the form of a probability distribution of a random variable (probability density or a differential distribution function of a random variable) and the probability distribution function of a random variable (a function distributions or integral distribution function of a random variable).

The differential function $\varphi(D)$ of the distribution of the random variable D – dose of a substance or energy causing a given effect to an object

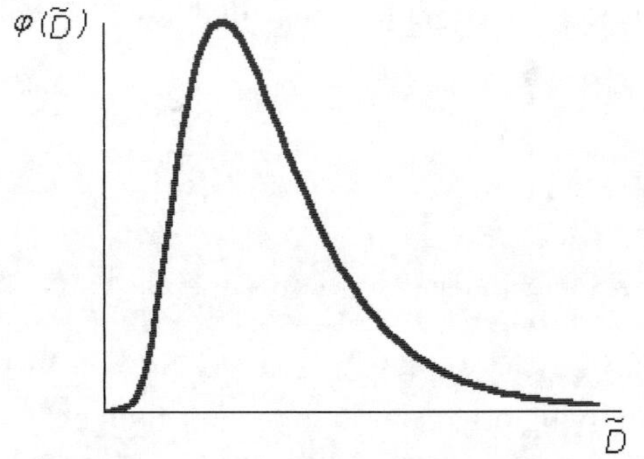

Fig. 3. The probability density of the random variable dose, causing a given effect of exposure

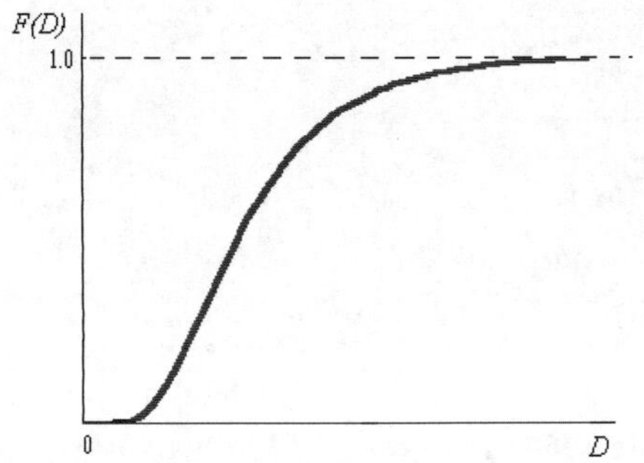

Fig. 4 The probability distribution function of the random variable dose, causing a given effect of exposure

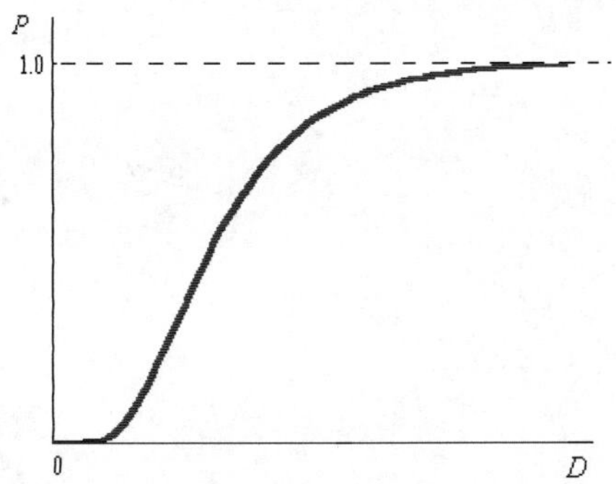

Fig. 5 Factor Impact Law

Factor Impact Law schematically shown in Figure 3. It characterizes the proportion of objects in which the dose of a physiologically active substance that causes a given effect of exposure was in the range of doses from D to $D+dD$

$$\varphi(\tilde{D}) = \frac{dN}{NdD}. \quad (8)$$

Cumulative distribution function F (D) of this random variable is schematically represented in Fig. 4. In this case, based on the definition, the distribution function F (D) is the probability that the random variable dose, causing the object to have a given effect, will be less than a certain value, i.e. $F(D) = P(\tilde{D} < D)$.

Then (Fig. 5), when a certain dose D of a physiologically active substance is applied to an object with a probability P equal to $F(D)$, the dose of the striking impact factor , which causes the object to have a given effect (the effect of a given force), will be less effective.

Therefore, with a probability of $P = F(D)$, an object will manifest effects not lower than a given one (not lower than a given force).

The correlation between the probability of manifestation of effects not lower than the specified and the magnitude of the dose of a substance or energy acting on an object will be called law of factorial exposure.

In the most general case, it has the form:

$$P = \int_0^D \varphi(\tilde{D}) d\tilde{D} \tag{9}$$

The specific type of factorial effect of exposure is determined from experimental data using special tests.

Currently, the literature discusses the application of the following laws [16–19, 27, 36, 39–41] to describe the results of the impact of damaging factors on living organisms and technical objects.

The logarithmically normal law – the density function of the distribution of the random variable dose, as applied to the problem in question, has the following form

$$\varphi(\tilde{D}) = \frac{1}{\sqrt{2\pi} D \sigma_{ln D}} e^{-\frac{(ln \tilde{D} - ln D_{50})^2}{2\sigma_{ln D}^2}} \tag{10}$$

Law of factorial exposure in this case can be written as follows:

$$P = \int_0^D \frac{1}{\sqrt{2\pi} D \sigma_{ln D}} e^{-\frac{(ln \tilde{D} - ln D_{50})^2}{2\sigma_{ln D}^2}} d\tilde{D} = \frac{1}{\sqrt{2\pi}\sigma_{ln D}} \int_0^D e^{-\frac{(ln \tilde{D} - ln D_{50})^2}{2\sigma_{ln D}^2}} dln\tilde{D} \tag{11}$$

Similar to normal density, for a log-normal density distribution of a random variable dose causing a given effect, the factorial effect using the Laplace function, the double Laplace function and the error integral can be written as follows:

using the Laplace function

$$P = \Phi\left(\frac{1}{\sigma_D} ln \frac{D}{D_{50}}\right) = \Phi\left(\frac{\sqrt{k}}{\sqrt{2}} ln T\right) \tag{12}$$

using the double Laplace function

$$P = 0.5\left[1 + F\left(\frac{1}{\sigma_D} ln \frac{D}{D_{50}}\right)\right] = 0.5\left[1 + F\left(\frac{\sqrt{k}}{\sqrt{2}} ln T\right)\right] \tag{13}$$

using the error integral (Crump function)

$$P = 0.5\left[1 + erf\left(\frac{1}{\sqrt{2}\sigma_D} ln \frac{D}{D_{50}}\right)\right] = 0.5\left[1 + erf\left(\sqrt{k}\, ln T\right)\right], \tag{14}$$

где $T = \frac{D}{D_{50}}$ – dimensionless quantity (effect) characterizing the amount of D_{50} of a substance or energy acting on an object;

$\sqrt{k}$ – law parameter

$$\sqrt{k} = \frac{1}{\sqrt{2}\sigma_{ln D}} \tag{15}$$

The parameters of the logarithmically normal factorial influence law have a statistical and physical interpretation.

Statistical interpretation (statistical meaning) of the parameters of logarithmically normal law of factorial exposure.

D_{50} – the median value of the random value of the dose that causes a given effect on an object (a given response);

$\sigma_{\ln D}$ – the standard deviation of the natural logarithm of the random variable dose, causing the object to a given effect of impact (a given response).

Physical interpretation of the parameters of the logarithmically normal:

D_{50} – the dose of a substance or energy, under the influence of which, with a probability of 0.5, the object will experience a response (effect of impact) not lower than a predetermined force;

$\sigma_{\ln D}$ or $\sqrt{k}$ – characterizes the homogeneity (heterogeneity) of similar objects: the smaller $\sigma_{\ln D}$ (more $\sqrt{k}$), the more homogeneous objects of the same type are to the effects of a substance and vice versa, the larger $\sigma_{\ln D}$ (smaller $\sqrt{k}$), the more homogeneous objects of the same type are to the effect of this substance.

Weibull distribution – is the density function of the random variable dose. Weibull distribution has the following view to our problem.

$$\varphi(\widetilde{D}) = \lambda\gamma\widetilde{D}^{\gamma-1}e^{-\lambda\widetilde{D}^{\gamma}} \tag{16}$$

In this case impact factor law of factorial exposure will look like

$$\mathrm{P} = \int_0^D \lambda\gamma e^{-\lambda\widetilde{D}^{\gamma}}\, d\widetilde{D} \tag{17}$$

We replace the variable $u = \lambda\widetilde{D}^{\gamma}$, and get the tabulated integral

$$\mathrm{P} = \int_0^u e^{-u} du \tag{18}$$

Where

$$\mathrm{P} = 1 - e^{-u} = 1 - e^{-\lambda D^{\gamma}} \tag{20}$$

The last expression we write as follows

$$\mathrm{P} = 1 - e^{-\left(\frac{D}{D_{50}}\right)^{\gamma} ln2} \tag{21}$$

The parameters of the law D_{50} and γ have the same meaning as the parameters of the log-normal law. There are other, less common models of factorial laws of exposure to chemical and other factors of origin.

Cumulative distribution function *F(D)*of this random variable is schematically represented in Figure 6. In this case, on the basis of the definition, distribution function *F(D)*there is a probability that a random variable *D*, calling the object specified effect will be smaller than a certain size. For our task in the light of the conditions of breaking through barriers (4) observations about the randomness of x values, and therefore *D* as well as symbol (5) the probability of breaking through barriers by a p_b can be defined as the conditional probability that a random variable (D) will a value that is not greater than D_1, i.e. $p_b = p\,(D \le D) = F\,(D_1) =$ Dp, where $F\,(D_1)$ is the distribution function of the random variable (D). According to the definition of the function $F\,(D_1)$ is an integrated distribution law of relative values holes (D) made by shrapnel. So, with probability $P = F\,(D_1)$ the object will show no effects below the set (not below the specified force) [9, 11].

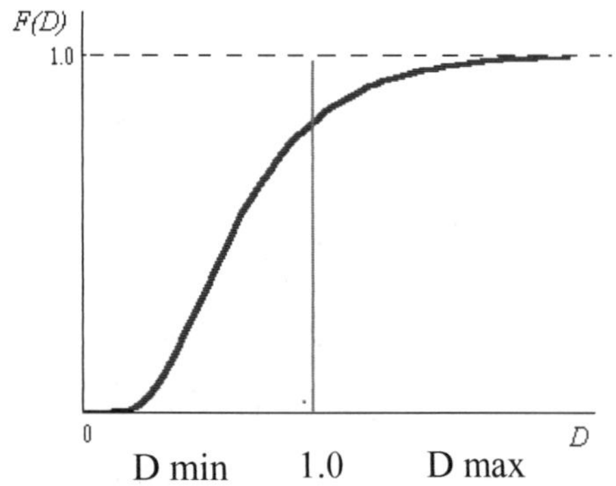

Figure 6. Distribution law of relative values holes

In the framework of the study, experiments were carried out, the essence of which consists in shooting identical fragments at a special barrier, determining the relative depths of holes in them and building static dependencies F (D1). Both methods should lead to the same dependency, although the second one is less cumbersome.

All the above makes it possible to obtain a calculation formula for determining the velocity of the fragment v_{np}, at which the breaking through of the obstacle with thickness s is provided.

$$v_{\text{пр}} = \sqrt{\frac{2E_1 s\Phi}{m^{1/3}}} \quad (22)$$

It is also possible to obtain a calculation formula for determining the limiting thickness $s_{\text{пр}}$ of an obstacle penetrated by a fragment of a given mass m and velocity v

$$s = \frac{m^{1/3} v^2}{2E_1\Phi} \quad (23)$$

Writing the formula (23) for an obstacle with a thickness of $s1$ and $E_1 = \alpha k_1 \tau_1$ and another obstacle with a thickness of s' and $E^{\cdot} = \alpha k^{\cdot} \tau^{\cdot}$ taking the ratio of the right and left sides and resolving with respect to s′, we get the formula

$$s^{\cdot} = s_1 \frac{k_1 \tau_1}{k^{\cdot} \tau^{\cdot}} \quad (24)$$

to find the so-called equivalent thickness s´. Relation (24) shows that it is possible to replace an obstacle with a thickness s1 of this material with an equivalent thickness s′ with a barrier of some other material considered as a reference one. Most often, duralumin is considered as a reference material; for any other obstacle (for example, steel), the so-called duralumin equivalent (24) is found, using the dependence.

3. The experiment

Short description of methodology of the experiment purpose constraints and assumptions, and object of research are determined, the estimated ratio. Figure 7 presents installation diagram for ballistics testing of samples.

3.1 The participants: Peter the Great St. Petersburg Politechnic University, Research-and-production Association of special materials

3.2 Procedure

3.3 This section describes how to complete the pilot session algorithm: procedure for holding the experiment and experimental data processing, etc.

3.4 Variables independent variables (factors). This paragraph provides the names of all IRS, their type (intra-group or intergroup), as well as the number of levels. Dependent variables. Specifies one or more of the RFP, as measured during the experiment. If the RP uses the derived index which is not measured directly during the experience, the way in which it is specified (that is, it is measured, but the indirect method).

Independent variables:

- speed of fragment.
- weight of the fragment.

IOP Conf. Series: Materials Science and Engineering **582** (2019) 012038 doi:10.1088/1757-899X/582/1/012038

- mathematical expectation square section

Dependent variable-depth

- implementation in this barrier

Table 2 summarizes the quantitative and qualitative data of the experiment without the elements of interpretation.

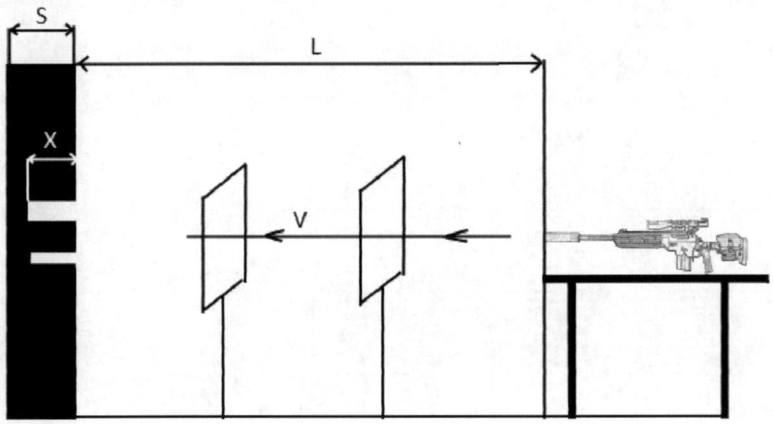

Figure 7. Installation diagram for ballistics testing of samples

Table 2. Fragment of test results.

Serial number	Submunition weight, m, g	Submunition speed, v, m/s	Material barriers	Deformation constraints, depth, x, mm	Deformation of the obstacles, the diameter of the midsection, mm
1	0.49	143	dural	x_1	S_1
2	0.49	152	dural	x_2	S_2
3	0.49	148	dural	x_3	S_3
….					
30	0.49	145	dural	x_{30}	S_{30}
1	0.49	147	Sculptural clay	x_4	S_1
2	0.49	143	Sculptural clay	x_5	S_2
3	0.49	145	Sculptural clay	x_6	S_3
…					
30	0.49	147	Sculptural clay	x_{30}	S_{30}
1	0.49	144	Pine Board	x_7	S_1
2	0.49	150	Pine Board	x_8	S_2
3	0.49	143	Pine Board	x_9	S_3
30	0.49	148	Pine Board	x_{30}	S_{30}

The statistical processing of the sample showed that the distribution function of the probability of the introduction of fragments into an obstacle can be represented in the form of a log-normal law and the Weibul law. The statistical processing of the sample showed that the distribution function of the probability of the introduction of fragments into an obstacle can be represented in the form of a log-normal law and the Weibul law. The verification of the hypothesis about the supposed distribution law was carried out using the Fisher significance criterion. The audit showed that the statistics of the logarithmically normal law are better than those of the Weibul law. Figure 6 shows the graph of the function after the statistical processing of the sample using Microsoft Office spreadsheets..

F(h)=0.5[1+erf(korKln(h/he))]
lnh=lnhe+(1/korK)argerf(2F-1)

h	k	F	2F-1	argerf(2F-1)	lnh	b	a			
4.02	3	0.111111	-0.77778	-0.86312307	1.391282	0.35208	1.693615	he=	5.44	mmm
5.75	16	0.592593	0.185185	0.16561798	1.7492	0.001638	0.001398	korK=	2.840	
7.05	23	0.851852	0.703704	0.73850854	1.953028	0.999957	0.002584	$ско_{lnh}$=	0.249	
8.5	26	0.962963	0.925926	1.26300271	2.140066	46183.23	2			
						0.308398	1.34E-05		1.000	
									<0.01	

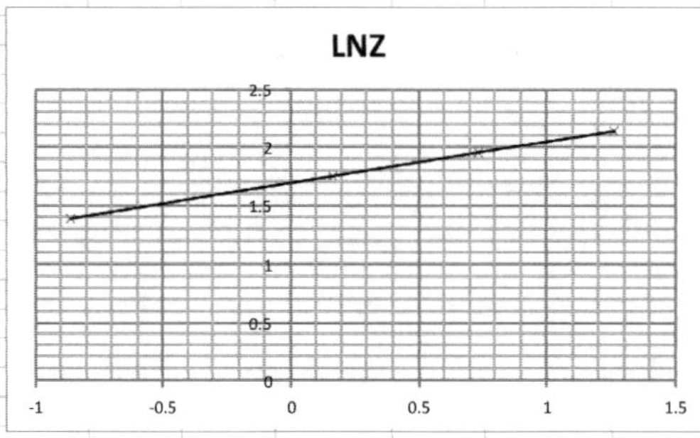

Figure 8. Lognormal probability distribution function of a random variable implanting fragments into an obstacle

4. Results

The development of a standardized tank module for multimodal transport of the RM, which can perform various tasks such as transporting a wide range of liquid radioactive waste by different modes of transport with opportunity safe overload, taking into account not only the rules for calculation and design of tank wagons [5, 6, 7, 8], but also taking into account the dependence of thickness of the tank from the probability of penetration by shrapnel. More detailed models and equipment for conducting the experiment can be found in the [12-20].

The main results obtained:

- accident scenarios considered in transfer the marine railway transport;
- мathematical model for determining the speed and weight of vlim shard m at which break through obstacles thickness s, and limiting the thickness of slim barriers easily by a given mass m and velocity v;
- received probability distribution function for the specified parameters shards and material barriers, which was determined by conducting a full-scale experiment.

5. Discussion

The results were discussed at scientific seminars and were presented during national and international conferences.

6. Conclusion

The study examined the construction of tanks and the requirements of the normative documents applicable to transportation LRW rail and sea transport. Emergency scenarios are considered: partial destruction of transport packing container with LRW in transfer of marine to rail transport; the explosion of gas cylinders on board the vessel. Received мathematical model for determining the speed and weight of v_{lim} shard m at which break through obstacles thickness *s,* and limiting the thickness of s_{lim} barriers easily by a given mass m and velocity *v*.

Calculation formulas are obtained to determine the speed at which the shards is provided by probabilistic the breakout a certain thickness of the tank walls. Received probability distribution function for the specified parameters shards and material barriers, which was determined by conducting a full-scale experiment. The results of theoretical studies, experiments used to improve designs of tanks carrying LRW and recommendations for improving the security of multimodal transporte.

Work is performed within the framework of the state job support (project No. 13.8874.2017)

References

[1] Federal rules and regulations in the field of atomic energy. No. 7. Collecting, processing, storage and conditioning of liquid radioactive waste. Security requirements. NP-019-2000. Moscow 2000

[2] Filippov V.N., Popov V., Ponomarev V.M., Shebeko Y.N., Bespalko S.V. 2017 *J. Transport, science and technology*, **1** 21-27.

[3] Popov V.G. Sukhov, F.I., Petrov S.V. 2012 *J. World of transport*,. **6 (44)** 150-155.

[4] Ponomaryov V.M., Shevchenko A.I. 2005 *J. Transport, science and technology* **2** 40-46.

[5] Specialized tanks for transport of dangerous goods. V.N. Koturanova and V.N. Filippov-M.: publishing standards, 1993.

[6] Filippov V.N., Skuratov A.E. 1999. General technical requirements for fittings wagons tanks for dangerous goods. The safety of trains. Proc. Of The MIIT. M.,

[7] Rules for calculation and design of railways wagons of gauge 1520 mm (non-self-propelled) with amendments and additions 2004 г.-m.: GosNIIV-VNIIZHT, 1996. -319 p.

[8] Shebeko Yu N., Bolodian I.A., Easy V.N., Navzenya V.Yu., Kostyuhin A.K., Zamishevski P.M., Tokarev E.D. A. 2000 J. Journal of Hazardous Materials **1-3** 43-56.

[9] Karmishin A.M., Kireev V.A. Mathematical methods of Pharmacology and Toxicology: monograph/ed. 2-e, revised. and extras. (M.: MGTU named after n.e.Bauman, 2008. (333 p.: ill.

[10] Moody f. 1971 Maximum consumption of single-component two-phase mixture. Proceedings Of ASME. Heat transfer **2** 47-56

[11] Tumanov A.Yu. Gumenyuk V.I. Scientific methodological fundamentals of risk assessment natural and man-caused emergencies at hazardous technical objects: monograph. Spb.: Polytechnic., 2018. -135 p.

[12] G. Petrov and A. Tarmaev "Modeling of railway vehicles movement having deviations in the content of running parts". Proceedings of the International Conference «Aviamechanical Engineering and Transport» (AVENT 2018). Series: Advances in Engineering Research, Atlantis Press, 2018, vol. 158, pp. 410-415. DOI:10.2991/avent-18.2018.79

[13] Tumanov A., Gumenyuk V., Tumanov V. Development of advanced mathematical predictive models for assessing damage avoided accidents on potentially-dangerous sea-based energy facility. IOP Conference Series: Earth and Environmental Science 19. Сер. "Energy Management of Municipal Transportation Facilities and Transport, EMMFT 2017" 2017. С. 012027.

[14] Burlov, V., Lepeshkin, O. Modeling the Process for Controlling a Road Traffic Safety System Based on Potentially Active Elements of Space and Time (2017) Transportation Research Procedia, 20, . DOI: 10.1016/j.trpro.2017.01.02

[15] Pykhtin, K., Simankina, T., Sharmanov, V., Kopytova, A. Risk-based approach in valuation of workplace injury rate for transportation and construction industry (2017) IOP Conference Series: Earth and Environmental Science, 90 (1), статья № 012065, . DOI: 10.1088/1755-1315/90/1/012065

[16] Bardin, A., Korotkov, A., Kazarnovskij, V. Limit of fire resistance of a steel transport structures. Simplified calculation methods (2017) IOP Conference Series: Earth and Environmental Science, 90 (1), № 012207, . DOI: 10.1088/1755-1315/90/1/012207

[17] Korshunov, G.I., Polyakov, S.L., Shunmin, L. Assurance of reliability and safety in liquid hydrocarbons marine transportation and storing (2017) IOP Conference Series: Earth and Environmental Science, 87 (6), № 062009, . DOI: 10.1088/1755-1315/87/6/062009

[18] A. Ustinov, O. Zybina, L. Tanklevsky, V. Lebedev, A. Andreev Intumescent coatings with improved properties for high-rise construction (2018) E3S Web of Conferences,33 (1), статья № 02039 DOI: 10.1051/e3sconf/20183302039

[19] Burlow, V.G., Grachev, M.I., Shlygina, N.S. Adoption of management decisions in the context of the uncertainty of the emergence of threats (2017) Proceedings of 2017 20th IEEE International Conference on Soft Computing and Measurements, SCM 2017, статья № 7970510, pp. 107-108. DOI: 10.1109/SCM.2017.7970510

[20] Bolodian, I., Melikhov, A., Tanklevskiy, L. Fire safety arrangement of inhabited pressurized compartments of manned spacecraft (2017) Acta Astronautica, 135, pp. 92-99. DOI: 10.1016/j.actaastro.2016.10.003

Probabilistic Approach to the Separation of Grain Material

U K Sabiev, A Yu Golovin, P V Chupin, A S Soyunov [a], R H Abdylmanova
FSBEI HE Omsky SAU, Omsk, Russian Federation

E-mail: [a] as.soyunov@omgau.org

Abstract. Possible ways to improve operating elements of grain cleaning machines due to mobilization factors directly influence on qualitative and quantitative aspects of separation are presented. In relation to sieve grain cleaning machine the technological process of grain mixture separation on the sieve is presented as a block diagram. Some variants of mathematical description (models) consideration of grain material separation process are also given. The dependencies between parameters of grain material and uniformity of seed distribution on sieve area are given. Calculation formulas and schemes of definition probability of a particle entering the hole are presented. It is established, that in the range β from 0 to 30° probability of particle passage increases from 0.315 to 0.548, that is 1.74 times.

1. Introduction

The Russian Federation is one of the largest producer of grain crops. According to the Government program of agricultural development and regulation of agricultural production, raw material and food in 2013 – 2020 years, gross yield of cereals and legumes on the farms of all categories must reach 135 million tons and provide stable grain export 40–50 million tons in a year.

Qualitative and quantitative grain safety depends upon providing the enterprises of agro-industrial complex with modern equipment for the post-harvest treatment, its technical level and effective usage. Foreign grain cleaning machines of different models have proven themselves well for rural manufacturers but not every company can afford to purchase equipment as the price level of foreign machines and other costs associated with their operation in the current economic environment is not favor of buying machines of foreign brands.

Post-harvest treatment of grain is carried out by production grain cleaning lines through which the grain pass after combine harvesting. Characteristics of freshly cleaned grain heap substantially affect the quality of grain cleaning machines operation. That is why agricultural enterprises of our country need innovative grain cleaning machines for post-harvest treatment of grain. The problem is particularly acute in the conditions of the Western Siberia, where the harvesting and post-harvesting treatment of grain is carried out under high humidity.

Undertaken review and analysis of literature show that at present time in creating new ways of grain cleaning and development of progressive attachments for grain cleaning machines there is no cardinal change for the better [1,2,3,4,5].

The efficiency of grain cleaning machines operation depends largely upon the ratio regimes of sieve mill movement close to optimal (passport). Study the influence of parameters of the sieve mill fluctuations and the detection of dependencies between the parameters of grain material and uniformity of seed distribution on sieve area on separation process of grain material allows to identify ways for increase the productiveness of grain cleaning machines, to reduce terms and to decrease the cost of grain treatment.

From the main ways of increasing the productiveness machines you can select the following: definition of dependencies between the parameters of grain material and uniformity of seed distribution on sieve area, theoretical substantiation of increasing rate movement of grain on the surface, complication of the fluctuation law of the sieve mills, detection of particles interaction on the sieve surface.

IOP Conf. Series: Materials Science and Engineering **582** (2019) 012039 doi:10.1088/1757-899X/582/1/012039

2. Definition of dependencies between the parameters of grain material and uniformity of seed distribution on sieve area.

The main task of theoretical study reduces to dependencies detection between the parameters of grain material and uniformity of seed distribution on sieve area with the aim to define rational performances of sieve mill operation.

Theoretical studies of the proceeding phenomena, namely technological process of separation is presented in the form of investigation object, which is characterized by input and output parameters [6].

In relation to sieve grain cleaning machine the technological process of grain mixture separation on the sieve is presented as a block diagram in figure 1.

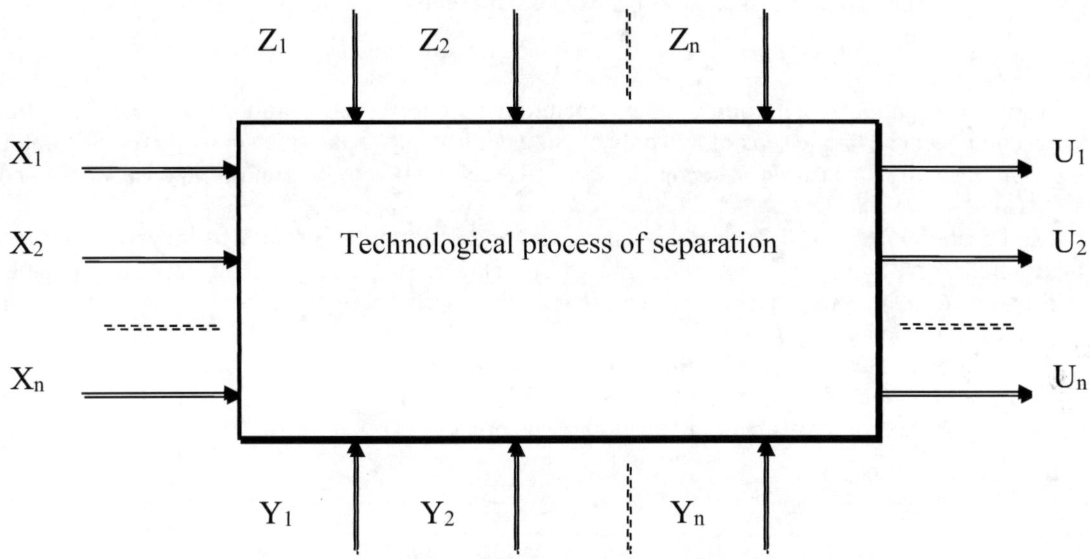

Figure 1 – Block diagram of technological process

Parameters group U is output data characterizing the technological process of separation. In relation to sieve grain cleaning machine it will be the productiveness, completeness of division, power input, seed injury, reliability etc.

Parameters groups X, Y and Z are input parameters. Parameters group X characterizes the properties of grain material received on the sieve: content of grain mixture, sizes, configuration, moisture, mass of particles, delivery etc.

Parameters group Y characterizes construction parameters of the separator. They include: sizes, sieve material, form and sizes of hole, form and sizes of jumpers, installation angle of the sieve etc.

Parameters group Z matches to the kinematic regime of separator drive (law of vibrations, oscillation frequency and amplitude etc.).

The task of theoretical investigation is to establish patterns of relationships between input and output parameters of object.

If input influences on the object are not changed over time and space and output parameters are also unchanged, the process is called stationary.

If only one input influence is changed over time and space, the process of corresponding change of output object parameters is called dynamic, at the end time of transition process it became stationary.

Mathematical model of technological process is made for theoretical study of statistical and dynamic characteristics of object.

For mathematical description of investigated model processes at probabilistic approach the most efficient is the mathematical apparatus of substance kinematic theory. Herewith separate grains of bulk material interprets as a specific molecules associated with some interaction forces.

The scientists, working on kinematic theory of separation are V.M. Tsetsinovsky, E.A. Nepomnyashy, A.A. Rassadin and others [7].

V.M. Tsetsinovsky obtained generalized equation of separation kinematic (1):

$$\varepsilon = 1 - \exp(-\int_0^t P_S dt)\,, \qquad (1)$$

where ε – completeness of division;

$P_S = f(t)$ – function of the separation length t, defining the intensity of division process in dependence upon the working regime of separator and physical components properties of the mixture.

It is established that the value P_s for some separators (sieve, trier, pneumatic separator) at relatively low loads and constancy of initial mixture content does not depend on time t and for this separator is constant value, that is
$P_s = k$. In this case the equation kinematics (1) can be represented as follows:

$$\varepsilon = 1 - e^{-kt} \qquad (2)$$

From the equation (2) it follows that separation process is asymptotic, complete extraction separated component in simple separator is only possible with a long separation $t = \infty$. In real separators the separation length is measured in seconds and that is why complete separation practically does not occur.

The equation kinematics (2) applies to the cleaning conditions of grain mixture on the sieve, at low loads on the sieve and the movement of grain in a thin layer. That is why it can be used in separation of small parties of grain, or by laboratory classifier, as the coefficient k is defined by resistance to the passage of small particles from sieve side.

E.A. Nepomnyashy considered the separation as a random markov process and used it for description Kolmogorov – Fokker – Plank's equation for multidimensional density of probability which is converted by the author to differential equation of the process (3):

$$\frac{\partial \omega}{\partial t} = \frac{b}{2} \cdot \frac{\partial^2 \omega}{\partial z^2}\,, \qquad (3)$$

where z – generalized particle coordinate;

$\omega(t_0\,, z_0;\, t, z)$ – the density of probability distribution defining the probability that random value $z(t)$ at time t will take the meaning $z < z(t) < z + dz$, if at time t_0 it was equal to $z(t_0)$;

b – the constant coefficient which is the measure of movement disorder of particles under random influence and defines the intensity of mixing.

The task of this study is to define the qualitative dependencies of coefficients, characterized screening of small particles from parameters of process and to check pattern by comparing with reliably established physics of the phenomena.

Herewith it is established that coefficient dependency of screening from layer height is very weak. It is quite natural since according to E.A. Nepomnyashy's opinion, this coefficient defines the passage rate of grain through the sieve and must not depend on layer thickness of grain on the sieve. So, the author admits that this coefficient is proportional to difference of hole sieve sizes and grains, initial grain concentration and depends on living area of sieve. Nature of obtained dependency of mixing coefficient from layer height does not correspond its physical content. This phenomena is explained by the extraction value depending on clogging of sieve holes during the experiments.

M.V. Kuzmin suggested another model of separation process in structural-logical approach. It means that state of the process is taken logically possible outcomes of which can be any particle of the mixture (figure 2.).

The author comes to decision that the intensity of selection material is characterized by indicator:

$$P_4' = \frac{\lambda_3}{2} t^2 e^{-\lambda t}\,, \qquad (4)$$

where t – selection time;

λ – value defining outcome of event in corresponding direction.

Maximum meaning P_4' is taken when $t = 2 / \lambda$, that is for intensification of separation process it is necessary to increase the meaning λ.

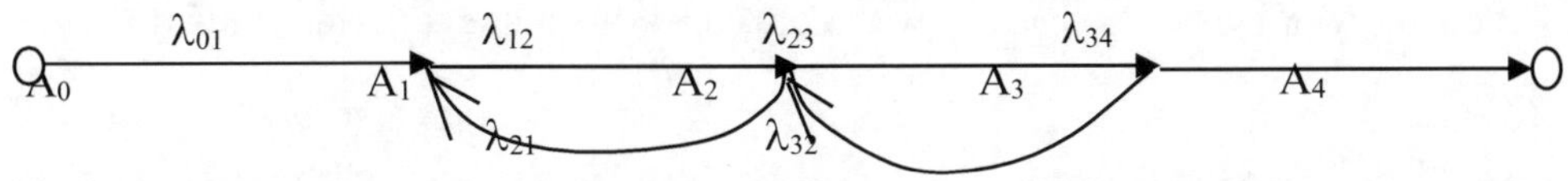

Figure 2 – Scheme of structural-logical approach in relation to separation process of grain (at the threshing unit, straw rack, sieve, trier), suggested by M.V. Kuzmin: A_0 – the grain with ear; A_1 – the grain without ear in a heap layer; A_2 – the grain is on the sieve surface; A_3 – the grain is near sieve hole; A_4 – the grain passed through separating attachment.

Despite the author's logic reasoning, this theory is difficult to use since it is unknown what methods you can find the numerical value λ.

B.D. Papin's research is of particular interest as it is based on queuing theory. The subject of queuing theory is finding probabilistic characteristics of queuing theory system [7].

Transitions from one state to another occur under the influence of random flow of applications $[\lambda(t)]$ and release flow $[\mu(t)]$. Flow of applications is the simplest and has exponential law of timing:

$$f[\lambda(t)] = \lambda \exp(-\lambda t), \tag{5}$$

where λ – intensity of flow applications;

t – current time

Release flow is also the simplest:

$$f_1[\mu(t)] = \mu \exp(-\mu t), \tag{6}$$

where μ – the density of release flow individual for each type of separators.

By applications the author means particles amount of passage fraction taking part in separation process. Release flow - particles amount of passage fraction, for one reason or another were not able to pass through the holes. The author mathematically described the character of objective relation of probabilistic characteristics of different separation process in high and low concentrations of grain flow on operating elements.

Thus, any model is an abstraction of real process that must possess possibility to use for prediction the process where the experimental data are absent; possibility to highlight the main characteristics of the process.

The conclusions obtained during the analysis of process model must not contradict known confirmed by practice theories but must expand the knowledge in the field of study. Choice of investigation method and mathematical model depend upon the laboriousness of theoretical study and reliability of obtained results.

3. Probability penetration of particles into the sieve oblong hole

Standard sieve has oblong holes located on the long side in direction of longitudinal sieve axis which corresponds with direction of grain flow movement.

Particles of grain heap entering the sieve located on it at random. Moreover two variants are possible. The first – when the grain falls on the hole area, then at a certain position of the particle regarding the hole, it can pass through hole.

The second situation, when the grain hits the jumper. In this case, during vibrations of the sieve regarding grain movement will be parallel to the longitudinal edges of the holes and if grain center of gravity is located on the jumper area, then it can't get in hole zone. Thus, the particles located on the jumpers do not participate in separation. They can get hole zone in the result of random process of interaction with other particles or in the case of special sieve construction, for example with jumpers having cross – section as a triangle or circle.

Therefore, in order to pass through the sieve hole, the particle must be in hole zone and located regarding holes edges definitely. Considering these events as independent, probability passage of particle through the sieve hole will be equal to:

$$P = P_\gamma \cdot P_\beta , \tag{7}$$

where P – probability passage of particle through the sieve hole;

P_γ – probability of «pass» position of particle on hole edges;

P_β – probability of interaction particle with hole edges due to the trajectory of particle movement on the jumper.

For definition of particle position in which its passage through the sieve hole is not difficult, we take into consideration the design scheme (figure 3). We consider the grain as a form of ellipsoidal oval where large axis equals to the grain length a, and small axis equals to the grain thickness b, central profile is the oval from conjugate radius arcs R_1 and R_2. Radiuses value R_1 and R_2 at given sizes of large and small oval axis can be calculated and defined graphically [8]. Grain center of gravity coincides with its geometric center O.

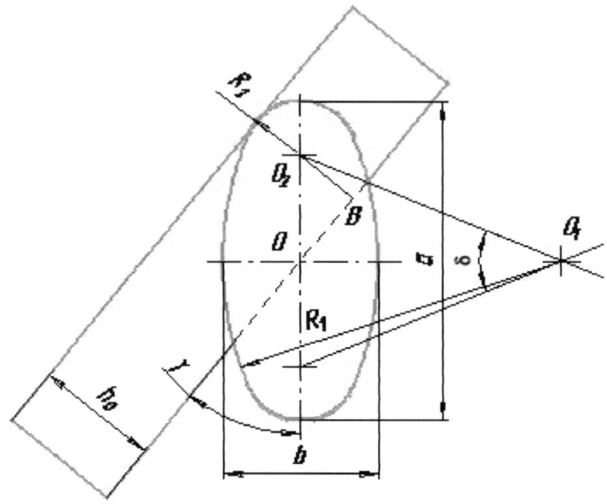

Figure 3 - Design scheme of mutual position of grain and sieve hole

For the accepted conditions limit position of grain regarding sieve hole according to the passage condition through it when grain center of gravity will be located on one longitudinal edge of the hole and the second longitudinal edge of the hole will be tangent to the curve forming ellipsoidal oval. The angle between the edge hole and the long axis of particle describe as γ.

Expressing sides of triangular OBO_2 through the hole sizes and the grain we will obtain:

$$\sin\gamma_{ПР} = \frac{2\cdot(h_0 - R_2)}{a - 2\cdot R_2}, \tag{8}$$

where $\gamma_{ПР}$ – limit angle when the grain passage into sieve hole is possible;

a – the size of long axis of particle;

h_0 – the width of hole;

R_2 – small radius of curvature at the top of particle.

Obtained ratio defines the maximum angle $\gamma_{ПР}$ between longitudinal axis of grain and edge of sieve hole according to the passage. So, grain passage into the hole is possible provided that:

$$0 \le \gamma \le \pm\arcsin\frac{2\cdot(h_0 - R_2)}{a - 2\cdot R_2}. \tag{9}$$

At angles $\gamma_{ПР}$ larger, than defined ratio (10), theoretically separating capacity of the sieve will approach zero at any direction of moving particle.

For the grain of sizes a = 5.2 мм, b = 2.1 mm [8] and radiuses of curvature surface R_1 = 2.5 mm and R_2 = 0.7 mm, as well as the sieve of the hole sizes 2.2×20 mm and the jumper width 2.5 mm according to the formula (9) critical angle is calculated $\gamma_{ПР}$, in which grain passage in hole sieve is possible:

$$0 \le \gamma_{np} \le arcsin\frac{2(2.2-0.7)}{5.2-2\cdot 0.7} = 52°$$

IOP Conf. Series: Materials Science and Engineering **582** (2019) 012039 doi:10.1088/1757-899X/582/1/012039

If we know the statistic characteristics of real particles position on sieve surface regarding its longitudinal axis, then probability of «pass» position of particle regarding hole edge will be equal:

$$P_\gamma = \sum_i P_{\gamma_i} = \frac{N_\gamma}{N}, \tag{10}$$

where i – the number of independent events in a range γ_i;

P_{γ_i} – probability of each events in a range γ_i;

N_γ – the number of particles on control section of the sieve located in a range γ_i;

N – the total number of particles on control section of the sieve.

According to the analyses of statistic characteristics of grain on control section of the sieve probability of favorable position of grain regarding hole sieve under the term of grain passage through the hole is calculated.

At limit value of the angle γ = ± 54°, probability of grain passage in hole sieve located on the hole edge amounts

$$P_\gamma = \sum_{-54^o}^{+54^o} P_{\gamma i} = 0.089 + 0.092 + 0.159 + 0.144 + 0.099 + 0.092 = 0.675$$

For definition of probability of particles movement in the direction of longitudinal axis of the sieve it will meet longitudinal edge of the hole located at an angle to the direction of grain movement, we use design scheme (fig. 4) where fragment of the sieve consisting of one hole and one jumper is depicted.

The position of grain is considered when the long axis of grain is parallel to the hole edge (γ = 0).

The justice of such assumption is confirmed by the analyses results of statistic characteristics of grain distribution on the sieve, as the mathematical expectation of random value M_γ approximately equals to the arithmetic mean of observed value γ_{cp} and amounts: $M_\gamma = 1.269° \approx \gamma_{cp} \approx 0$.

For definition of probability P_β we use ratio of elementary sieve area *ABB'A'* (fig. 4), formed by area of one hole and by area of one jumper and part of elementary area where the particle is located in the hole zone in pass position or where it may be in pass position at moving on jumper in direction of hole edge [9].

If $0 \le \beta \le arctg\frac{h_\Pi}{l}$, where h_Π – width of jumper, l – hole length; from figure 2 we will write geometrical ratio:

$$P_\beta = \frac{S_{ABDC} + S_{CDK}}{S_{ABB'A'}}$$

after transformation we obtained:

$$P_\beta = \frac{h_O}{h_O + h_\Pi} + \frac{l \cdot tg\beta}{2(h_O + h_\Pi)} \tag{11}$$

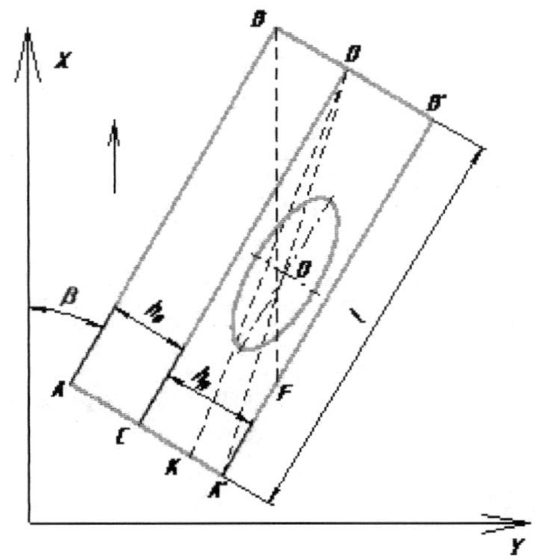

Figure 4 – Design scheme of definition P_β. The arrow indicates the direction of particle moving

If $arctg\frac{h_П}{l} < \beta \leq \frac{\pi}{2}$, from figure 4 we will write geometrical ratio:

$$P_\beta = 1 - \frac{S_{BB'F}}{S_{ABB'A'}}$$

after transformation we obtained:

$$P_\beta = 1 - \frac{h_П^2 \cdot ctg\beta}{2 \cdot l \cdot (h_O + h_П)} \quad (12)$$

thus:

$$P_\beta = \begin{cases} \dfrac{h_O}{h_O + h_П} + \dfrac{l \cdot tg\beta}{2(h_O + h_П)} \\ with \quad 0 \leq \beta \leq arctg\dfrac{h_П}{l} \\ 1 - \dfrac{h_П^2 \cdot ctg\beta}{2 \cdot l \cdot (h_O + h_П)} \\ with \quad arctg\dfrac{h_П}{l} < \beta \leq \dfrac{\pi}{2}. \end{cases}, \quad (13)$$

The probability P_β is calculated at initial values: for sieve – $h_o = 2.2$ mm, $h_п = 2.5$ mm, $l = 20$ mm; for grain – $a = 5.2$ mm, $b = 2.1$ mm.

According to the equation results (7) we obtained the dependence $P = f(\beta)$, which is shown in picture 5.

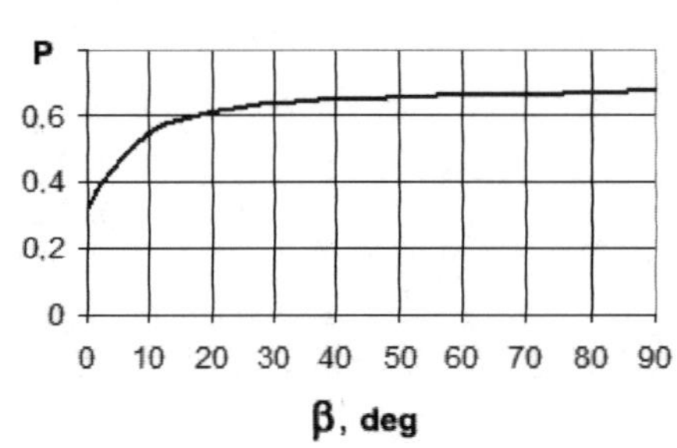

Figure 5 – Dependence of probability particle passage through hole sieve from the angle of inclination of the hole

The graph follows that angle of inclination of the sieve hole to the longitudinal axis helps to increase probability of particle passage in hole sieve. Probability increases at small angle of inclination of the holes.

In a range β from 0 to 30° probability of particle passage increases from 0.315 to 0.548 that is 1.74 times [7].

4. Conclusions

1. Block diagram of technological process of separation in the form of mathematical model of investigation object characterizing input and output parameters is presented

2. It is established that dependence of increasing the probability of particle passage in oblong sieve hole at small angle of inclination of the holes. It is defined that in a range β from 0 to 30° probability of particle passage increases from 0.315 to 0.548.

References

[1] Experimental registrations of plain sieve operation, making transverse fluctuations / U.K. Sabiev, E.V. Demchuk, A.Yu. Golovin, I.V. Skysanov. В сборнике Mechanical Science and Technology Update (MSTU-2018) IOP Publishing IOP Conf. Series: Journal of Physics: Conf. Series 1050 (2018) 012074 doi :10.1088/1742-6596/1050/1/012074.

[2] Aguirre R. Continuous – flowing portable separator for cleaning and upgrading bean seeds and grains / R. Aguirre, A. E. Garay // Agr. Mech / Afr. And Lat. Amer. – 1999. – V. 30, №1. – P. 59 – 63.

[3] Boumans, G. Grain Handling and Storage. – Amsterdam; Oxford; New-York; Tokyo, 1985. – 608 p.

[4] Choszcz, D. The determination of optimum parameters of various kinds of the sieve basket motion of winnowing machine on the cleaning process of mustard seeds / D. Choszcz, S. Konopka, K. Wierzbicki // Rocz. AR Poznaniu. Rol. – 1996. – № 49. – P. 13 – 22.

[5] Wang Guo-Xin Luoyang gongxueyuan xuebao / Wang Guo-Xin, Shi Qing-Xiang, Ni Chang-An, Liu Shi-Duo, Ji Jiang-Tao // J. Luoyang Inst. Technol. - 2000. – V. 21. – №1. – P. 7 – 10.

[6] Zilberangel, A.V. Theoretical research of particle moving trajectory on probability of its interaction with edges of oblong hole of the sieve / A.V. Zilbernagel, U.K. Sabiev // Bulletin of Altaj state agrarian university. - 2011. - № 9 (83). - P. 78-81.

[7] Golovin, A.Yu.Substantiation of constructive- regime parameters of flat sowing sieve, making circular motions: diss. .....cand. techn. science : 05.20.01. : defended 14.12.2018 / Golovin Alexandr Yurjevich. - Barnaul, 2018. - 118 p.

[8] Golovin, A.Yu. Trajectory of grain movement on the sieve, making circular motions / A.Yu. Golovin, U.K. Sabiev, A.Yu. Soyunov // Bulletin of Omsk stare agrarian university. - 2017. - № 4 (28). - P. 204-210.

[9] Zilberangel, A.V. The effect of particle position regarding the edge of oblong hole of the sieve on its limit rate / A.V. Zilberangel // Bulletin of Omsk state agrarian university . - 2003. - № 2. - P. 122.

IOP Conf. Series: Materials Science and Engineering **582** (2019) 012040 doi:10.1088/1757-899X/582/1/012040

Determination of Separator Constructive Parameters

V S Koval [1, a], **A V Chernyakov** [1, b], **A P Shevchenko** [1, c], **M A Begunov** [1, d], **E V Demchuk** [2, e]

[1] Tarsky branch of FSBEI HE Omsk SAU, 646532, Russian Federation, Omsk Region, Tara, ul. Tyumenskaya, 18
[2] FSBEI HE Omsk SAU, 644008, Russian Federation, Omsk, ul. Fizkulturnaya 1

E-mail: [a]vs.koval@omgau.org, [b] chernyakovavforester@mail.ru, [c]ap.shevchenko@omgau.org, [d]ma.begunov@omgau.org, [e]ev.demchuk@omgau.org

Abstract. The article describes the two swinging sieves separator with increased guide activity. The kinetics of two sieves oscillatory mechanism is examined on the basis of the deterministic approach to the process study. In the course of theoretical research, a theoretical model was obtained, which defines the sieve movement. This model allows to calculate the kinematics of both separator sieves and its design parameters.

Introduction

Post-harvest grain processing is the final stage in its production. The post-harvest processing and grain storage accounts for more than one-third of the costs associated with its production [1]. Timely post-harvest grain processing harvested by combine harvesters contributes to high harvesting rates, prevents grain spoilage and its quality. One of the final stages and the most crucial part of it is the secondary treatment. Since the 50s of the last century, the seed grain share is 20...25% of its total volume [2]. The seeds of the most stringent requirements of GOST on purity, germination and other indicators.

At present, due to the increased contamination of the fields occupied with crops, grain-cleaning machinery load has increased. The seeds are passed through grain cleaning machines several times to bring the grain to the sowing conditions, while processing costs and grain injury are increasing [3]. And this, in turn, leads to an increase in the prime cost of the final sales unit and a decrease in the machine productivity.

One of the most common separating elements, both for primary and secondary grain cleaning, are punch sieves [4, 5]. Of these, the most applicable are sieves with rectangular openings that separate the seeds by thickness. These sieves have much higher specific productivity values than sieves with round holes. The authors of many works come to the conclusion that cylindrical sieves are the most productive, and swinging sieves injure grains less than others. The most loaded sieves in seed cleaning machines are considered to be sowing and sorting, on which small and frail spines are separated. To increase the productivity of these, and others, is solved by those who are standing in the camp, in two ways. The first way is an extra strong one, connected with an increase in the battery size, the power and other similar aspects. The second way is a promoted one, includes a study for achieving the highest level of competitiveness possible through the mobilization of important factors influencing the quantitative and qualitative machine quality. The realization of the second way is kept due to the fact that there is not enough study of grain separation process on the sieve.

Analyzing various separator designs for cleaning grain material, we can conclude that today some of the most promising are machines capable of combining the advantages of both flat-sieve and machines with a curved sieve surface.

Such a combination of advantages is possible with the use of a grain cleaning machine with two cylindrical swinging sieves with elongated holes located at an angle α to the movement of the grain material.

This machine will allow you to increase the penetration degree through the net by improving the grain management through the holes.

Theoretical investigations of the grain cleaning process were carried out on the sieves with rectangular openings located at the angle to the longitudinal sieve axis.

We have developed a separator with swinging working organs which openings have a straightforward shape and are arranged under a specific angle α to seed movement.

Such a configuration allows to increase screening sieve ability by location improving of grains regarding the holes [7].

2. Object and research methods

The kinematic mechanism design of sieves suspension and its drive was made for a laboratory sample (Fig. 1).

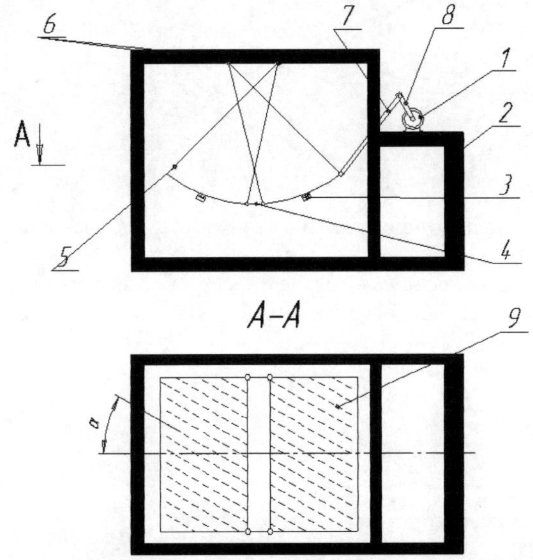

Figure 1 - Laboratory sample diagram of a machine for the grain separation with oscillating tools: 1 - electric drive; 2 - frame; 3 - sieve cleaners; 4 - jumpers; 5 - sieve pendants; 6 - mounting sieve suspensions; 7 - connecting rod; 8 - crank; 9 - sieve; α – inclination hole angle regarding the sieve.

Grain cleaning separator, contains a frame 2, on which with the sieve hangers help 5 concave sieves 9, cross-sectioned in a cross-section, are mounted, made in the form of a cylindrical surface. To the sieve 9 pivotally attached jumper 4, connecting the sieve. The mountings of the sieve hangers 6 lie in the same plane and are shifted regarding each other, yoke, and they have a longitudinal inclination to the horizon. Under the sieves 9 brushes 3 are installed.

In the movement of the sieve 9 are driven from the motor-reducer 1 by means of the crank 8 and the connecting rod 7, are connected to each other by a hinge. The movement from the first sieve to the second is transmitted through the jumper 4. The sieve 9 oscillates about its axis.

Grain cleaning separator works as follows: the source material received at the beginning of the first and second sieves, under the action of oscillatory motion begins to move across the sieves, and due to the mill and holes inclination - along the sieve above, freely falling on the sieve below. Having reached the extreme point, the movement of the sieves is repeated in the opposite direction. Due to the fact that the axles of fastening the suspensions of sieves are displaced regarding each other in opposite directions, during the movement, the sieves move in the horizontal plane between themselves. This makes it necessary for them to connect with the help of the first pin 4, which allows to move between the sieves and also connects them [8]. The cleaning of the holes will be done from the entrances of the discovered grain by means of brushes, which are placed in the upper mill part.

The advantages of such a separator are the structure simplicity, in which the particle moves under the influence of a number of forces: inertia, centrifugal and gravitational. Moving a particle from the sieve to the sieve and back promotes intensive mixing, and therefore actively prevents the separation of the divided mass, and allows the weevils in the upper layer to get to the sieve. The grain pile moves with a thin layer, and due to the arrangement of the longitudinal holes at an angle, the grain material is more intensively oriented on the sieve, which shortens the residence time of the small fraction on the sieve.

Previously, the kinematics of the lever mechanism as applied to the two-grid separator scheme have already been studied [9] on a crank mechanism with a beam, identifying it with a hinged four-link mechanism (Fig.2)

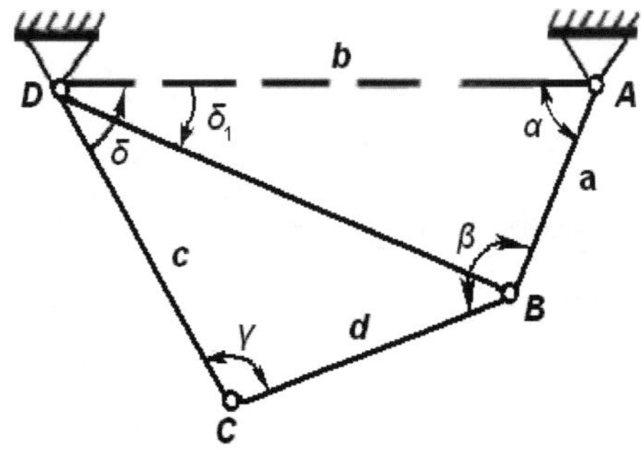

Figure 2 - Scheme of the crank mechanism with the rocker arm.

The functional dependence of the angle δ of the deflection of the rocker from the angle α, if the initial link is AB: where a, b, c, d are the lengths of the AB, AD, CD and BC links, respectively;

$$\delta = arctg\frac{b \cdot \sin\gamma}{c - b \cdot \cos\gamma} + \delta_1, \tag{1}$$

$$\cos\gamma = arctg\frac{b^2 + c^2 - a^2 - d^2}{2b \cdot c} + \frac{a \cdot d}{b \cdot c}\cos\alpha_h, \tag{2}$$

Angular velocity of the rocker arm:

$$\omega_3 = \omega_1 \cdot \frac{a}{c} \cdot \frac{\sin\beta}{\sin\gamma}, \tag{3}$$

where:

$$\beta = 360^0 - (\alpha_h + \gamma + \delta),$$

Angular rocker acceleration:

$$\varepsilon_3 = -\omega_1^2 \cdot \frac{a \cdot d}{b \cdot c} \cdot \frac{\cos\beta \cdot \sin\delta - \frac{a}{c} \cdot \sin\alpha_h \cdot \sin\beta \cdot ctg\gamma}{\sin^2\gamma}, \tag{4}$$

A four-link hinge mechanism was considered (Fig. 3):

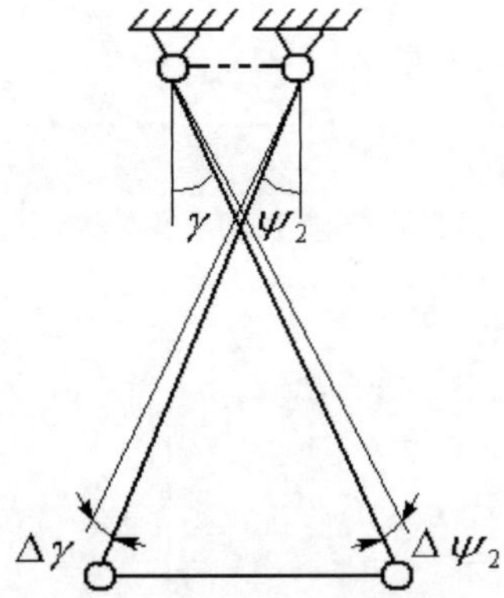

Figure 3 - Scheme of the hinged four-link mechanism.

The change dependence in the angles γ and ψ2 will be as follows:

$$\gamma = 90 - \delta, \tag{5}$$

Considering that the angles corresponding to the neutral position are equal $\gamma = \psi_2$, then their arbitrarily small angular displacements are also equal $\Delta\gamma = \Delta\psi_2$, so get $\psi_2 = \gamma - \Delta\gamma - \Delta\psi_2$, or

$$\psi_2 = \gamma - 2\Delta\psi_2, \tag{6}$$

where $\Delta\gamma$, $\Delta\psi_2$ – small angles changes γ, ψ_2

Taking into consideration $\Delta\psi_2 = \omega\Delta t$ and equation (7) we get

$$\psi_2 = 90 - arctg\frac{b\left(\sqrt{1-\left(arctg\frac{b^2+c^2-a^2-d^2}{2b\cdot c}+\frac{a\cdot d}{b\cdot c}\cdot\cos\alpha_h\right)^2}\right)}{c - b\cdot arctg\frac{b^2+c^2-a^2-d^2}{2b\cdot c}+\frac{a\cdot d}{b\cdot c}\cdot\cos\alpha_h} + \delta_1, \tag{7}$$

The speeds and accelerations of both sieves are the same, since the sizes of their suspensions are the same and the sieves are rigidly interconnected by a jumper that transmits an oscillatory motion from the first to the second.

To find the motion parameters of the first and second cylindrical swinging screens, we use the technique used in the theory of mechanisms and machines [10]. In it, structural, kinematic and kinetostatic research is studied in a graphanalytic way for certain positions of the links of the mechanism of angular velocity, peripheral velocity, acceleration and angular acceleration, as well as the force effect at certain points.

The kinematic scheme of the proposed technical solution (Fig. 1) for simplicity was divided into three parts: a crank and a connecting rod with rocker arms and two hinged four-link mechanisms (a sieve mill) (Fig. 4).

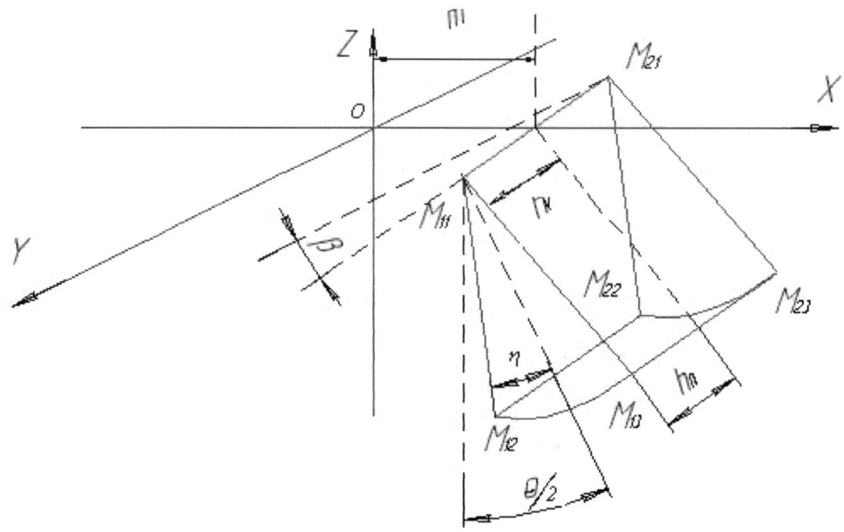

Figure 4 - Kinematic sieve scheme.

The sieve mill orientation in space can be determined as follows:

- assigned the sieve angle in the horizontal plane β;
- the angle between the sieve hanger mounts (lines $M_{11}M_{12}$, $M_{11}M_{13}$, $M_{21}M_{22}$, $M_{21}M_{23}$) from the axis of symmetry (η);
- assigned sieve mill deflection angle (θ), then the angle change of the sieve symmetry axis from the vertical axis ($\frac{\theta}{2}$);
- the angles of deflection of the sieve suspensions are calculated (lines $M_{11}M_{13}$, $M_{21}M_{22}$, $M_{21}M_{23}$) from vertical plane:

Angle of deflection for left-sided drives:

$$l = \frac{\theta - \eta}{2}, \tag{8}$$

- deflection angle for right-sided drives:

$$P = \frac{\theta + \eta}{2}, \tag{9}$$

- the radius r1 is calculated for different values of the arms (upper - h_v and lower - h_n) :

$$r_1 = \sqrt{r^2 - (h_v - h_n)^2}, \tag{10}$$

Next, the deviation angle value is calculated relative to the vertical position when the deviation β=0;

$$f_i = \frac{h_n - h_v}{r_1}, \tag{11}$$

- coordinates values are determined M_{11}, M_{12}, M_{13}, M_{21}, M_{22}, M_{23}.

Expressions to determine the value of the coordinates of points M_{11}, M_{12}, M_{13}, M_{21}, M_{22}, M_{23} will take the following form:

The point M_{11}:

$$\begin{cases} x_{11} = m; \\ y_{11} = h_v \cdot \cos(\beta); \\ z_{11} = -h_v \cdot \sin(\beta); \end{cases} \tag{12}$$

The coordinates of the points M12, M13 are calculated depending on the coordinates of the point M11.

$$\begin{cases} x_{12} = x_{11} - r \cdot \cos(\beta - f_i) \cdot \sin(l); \\ y_{12} = y_{11} - r \cdot \sin(\beta - f_i); \\ z_{12} = z_{11} - r \cdot \cos(\beta - f_i) \cdot \cos(l); \end{cases} \quad (13)$$

$$\begin{cases} x_{13} = x_{11} - r \cdot \cos(\beta - f_i) \cdot \sin(p); \\ y_{13} = y_{11} - r \cdot \sin(\beta - f_i); \\ z_{12} = z_{11} - r \cdot \cos(\beta - f_i) \cdot \cos(p); \end{cases} \quad (14)$$

In the same way we determine the values of the coordinates of points M21, M22, M23.
Similarly, we obtain the coordinates of points M21, M22, M23.

3. Results of the study.

Based on the above, the equations of sieve mill motion are derived. Taking the value of the sieve inclination angle β, the distance between the oscillation axes of the sieve suspensions (+ m for the right and –m for the left), the sieve mill angle deflection regarding the symmetry axes $+\frac{\pi}{k}$ for the right and $-\frac{\pi}{k}$ for the left, where k- coefficient of deviation, you can define various kinematic parameters. Based on the obtained mathematical model, a simulation model was developed. This model allows you to change the design parameters, namely: the suspensions length parameters and the values of the sieves deflection angle.

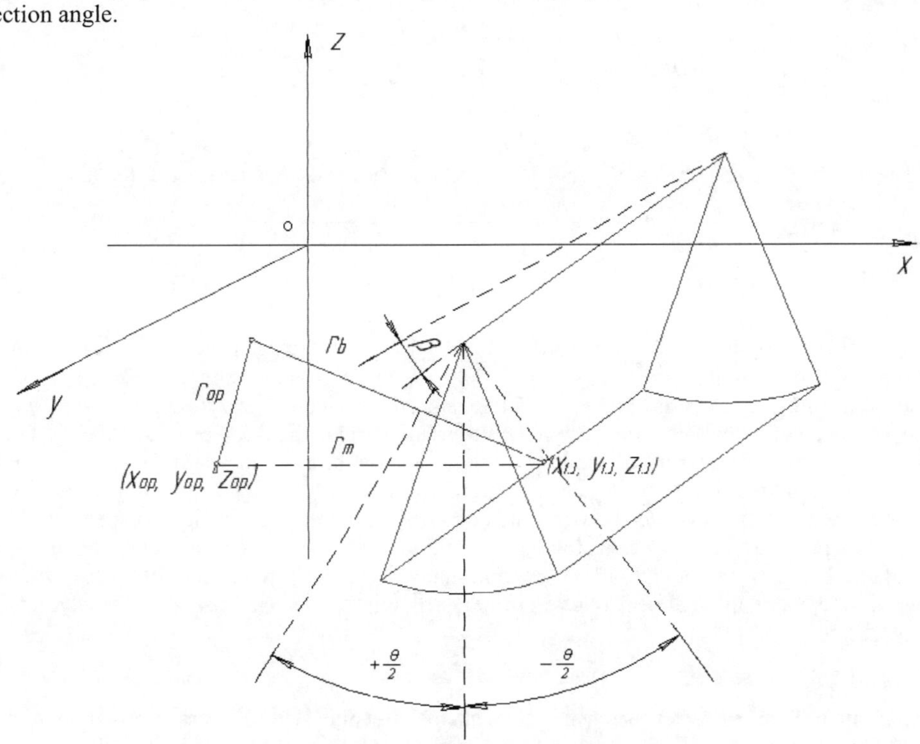

Figure 5 - Sieve drive diagram.

The connecting rod and crank lengths, sieve suspensions and jumpers are calculated using a mathematical model.

For the implementation of the above, the coordinate values of the connecting rod fixation points to the lattice with its maximum and minimum deviations are calculated ($x_{max}, y_{max}, z_{max}$), ($x_{min}, y_{min}, z_{min}$).

$$\begin{cases} x_{\max} = m - r \cdot \cos(\beta - f_i) \cdot \sin(-\frac{\theta}{2} + \frac{\eta}{2} + \frac{\pi}{k}); \\ y_{\max} = h_v \cdot \cos(\beta) - r \cdot \sin(\beta - f_i); \\ z_{\max} = -h_v \cdot \sin(\beta) - r \cdot \cos(\beta - f_i) \cdot \cos(-\frac{\theta}{2} + \frac{\eta}{2} + \frac{\pi}{k}); \end{cases} \quad (15)$$

$$\begin{cases} x_{\max} = m - r \cdot \cos(\beta - f_i) \cdot \sin(\frac{\theta}{2} + \frac{\eta}{2} + \frac{\pi}{k}); \\ y_{\max} = h_v \cdot \cos(\beta) - r \cdot \sin(\beta - f_i); \\ z_{\max} = -h_v \cdot \sin(\beta) - r \cdot \cos(\beta - f_i) \cdot \cos(\frac{\theta}{2} + \frac{\eta}{2} + \frac{\pi}{k}); \end{cases} \quad (16)$$

Thus, the minimum and maximum values of the connecting rod fixation points to the lattice mill are determined regarding the crank rotation axis.

$$\begin{cases} r_{\min} = \sqrt{(x_{\min} - x_{op})^2 + (y_{\min} - y_{op})^2 + (z_{\min} - z_{op})^2}; \\ r_{\max} = \sqrt{(x_{\max} - x_{op}) + (y_{\max} - y_{op}) + (z_{\max} - z_{op})^2}; \end{cases} \quad (17)$$

crank length value:

$$r_{op} = \frac{r_{\max} - r_{\min}}{2}, \quad (18)$$

rod length value:

$$r_b = r_{\min} + r_{op}, \quad (19)$$

The distance between the crank rotation axis and the sieve suspension is determined by:

$$r_m = \sqrt{(x_{13} - x_{op}) + (y_{13} - y_{op}) + (z_{13} - z_{op})^2}, \quad (20)$$

4. Conclusion.

The article substantiates the relevance of the grain post-harvest processing issue in modern agriculture, and, as a result, a grain separator model with a two-sieves design and installed sieve with increased orienting activity was proposed. This design allows you to carefully carry out the grain cleaning process without possible injury to it, however, using centrifugal forces. A simulation model of a two-sieves separator was obtained, which allows to calculate the coordinates of fixing the connecting rod points to the lattice mill with its maximum and minimum deviations, the crank length and the connecting rod length, the distance between the crank rotation axis and the sieve suspension. For the laboratory sample, the following values were obtained: crank length (150 mm), connecting rod (380 mm), and also sieve hangers lengths (350 mm) and jumpers (30 mm). In the case of an industrial scale separator development, these dimensions will be adjusted, but the mathematical dependencies used will remain the same.

References

[1] Scarification of seeds as an increasing element of perennial legume grasses productivity / A.P. Shevchenko, A.V. Bankrutenko, V.S. Koval, M.A. Begunov, E.V. Demchuk // Journal of Physics: Conference Series 9. Сер. "IX International Scientific Practical Conference "Innovative Technologies in Engineering"" 2018. С. 012011..

[2] The determination of optimum parameters of various kinds of the sieve basket motion of winnowing machine on the cleaning process of mustard seeds / D. Choszcz, S. Konopka, K. Wierzbicki // Rocz. AR Poznaniu. Rol. – 1996. – № 49. – P. 13-22.

[3] Sieve wire-welded construction with round bridges / I.V. Shevtsov // Agrarian Bulletin of the Urals. - 2007. - №2 (38). - pp. 46-52.
[4] Carousel seed cleaning device / Yu.A. Medvedev // Machinery and equipment for the village. - 2017. - № 6. - p. 32-33.
[5] Self-cleaning of flat sieves from stuck grains / P.N. Lapshin, N.O. Moskovchenko // Mechanization and electrification of agriculture. - 2009. - № 7. - p. 4-5.
[6] Grain Handling and Storage / Boumans G. // Amsterdam; Oxford; New-York; Tokyo, 1985. – 608 p.
[7] Continuous – flowing portable separator for cleaning and upgrading bean seeds and grains / R. Aguirre, A. E. Garay // Agr. Mech / Afr. And Lat. Amer. – 1999. – V. 30, №1. – p. 59-63.
[8] Patent for utility model 79012 of the Russian Federation MPK7 A 01 F 12/44, B 07 B 1/28, B 07 B 1/38; Application No. 2008109352/22 Claimed March 11, 2008; Publ. 20.12.2008 // Inventions. Useful models. - 2008. - № 35 - p. 697.
[9] Determination of design parameters and modes of operation of swinging sieves with a cylindrical working surface / R.A. Zverkov // Siberian Journal of Agricultural Science - 2007. - № 10 (178). - p. 80-84.
[10] Connection of experimental and theoretical indicators of sieve work / V.F. Evtyagin // Collection of scientific papers, Omsk Agricultural Institute. - Omsk, 1992. - p. 45-48.

IOP Conf. Series: Materials Science and Engineering **582** (2019) 012041 doi:10.1088/1757-899X/582/1/012041

Management of Repair Service Structure of Industrial Enterprise When Transferring Repair Equipment for Outsourcing

A V Shimohin, G V Redreev, V M Pomogaev, N A Zaripova, O M Kirasirov

Omsk State Agrarian University named after P.A. Stolypin
644008, Omsk, Institutskaya square st., 1

E-mail: weerwg@mail.ru

Abstract. in modern conditions, when the equipment of domestic industrial enterprises is significantly worn out, it becomes important to use outsourcing repairs as a way to improve the reliability of equipment by ensuring the quality of repair. The decision to outsource is made on the basis of different criteria using different approaches, some of which have certain drawbacks. For example, a decision based on a revenue change might be wrong because it is not related to outsourcing. Multi-criteria approaches are ineffective due to mutual compensation of deteriorating and improving criteria. The algorithm of decision-making on outsourcing of repair of the equipment and the choice of service providers on the basis of comparison of costs of outsourcing and insourcing is presented.The widespread structure of repair service of the industrial enterprises is considered and its reorganization in the conditions of repair outsourcing application is offered. The criteria and method of their calculation for the possibility of evaluating the effectiveness of repair outsourcing equipment. To assess the effectiveness of the reorganization of the repair service, the ratio of repair time, the number of failures, the number of defects, the repair costs are taken. For the detection of improvement or deterioration in the quality of the repair after the use of outsourcing, you must use the product of these relations.

1. Introduction

In conditions of greater wear and tear of the equipment of most domestic industrial enterprises, one of the ways to improve the reliability of the equipment by ensuring the quality of repair can be outsourcing. At the same time, many specialists pay attention to the expediency of developing the company's own maintenance service, noting its inherent strengths, noting its inherent strengths: efficiency, controllability, knowledge of the peculiarities of the equipment operation [1].

The development of our own repair services, departments may be associated with the concept of insourcing. The concept of insourcing is treated differently [2].Insourcing is the expansion of the division's activities for additional utilization of existing capacities or assets. Insourcing can also be associated with the transfer of the project to an employee or department of the enterprise instead of transferring it to an external contractor [3-5]. However, it should be noted that the effectiveness of insourcing depends on the suppliers of parts, personnel and senior managers, which remain the same.

The decision-making mechanisms for outsourcing include the following approaches. A single-criteria approach is used for economic efficiency [2]:

$$E_{OUT} = \frac{И_{BO} - И_{AO}}{R} \cdot 100\%, \tag{1}$$

whereE_{OUT} – the cost-effectiveness of outsourcing, %;

$И_{BO}$ – the enterprise costs in the internal performance of the function before outsourcing, in monetary units (currency unit);

$И_{AO}$ – the enterprise costs after the implementation of outsourcing, currency unit;

R – the enterprise revenue, currency unit.

The author I.L. Rudaya [6] notes the disadvantage of this approach is that the effect of outsourcing is calculated from changes in revenue, which may not be associated with the use of outsourcing.

In general, multi-criteria approaches can be expressed as follows [2]:

$$E = \sum_{i=1}^{n} Wi(KA_i - KB_i), \tag{2}$$

where$E$ – the outsourcing effect;

n – the criteria number;

Wi – the weight of*i-criteria*in the overall assessment;

KAi – the value of*i-criteria* after outsourcing;

KBi – the value of *i-criteria* before outsourcing.

The disadvantages of this approach may be due to the fact that the deterioration of some criteria can be offset by improving the values of others. As a result, the final value may indicate a positive effect, although the values of the most important criteria will be worse.

Whilerepair outsourcing, in addition to saving by reducing the cost of the repair process equipment, it is also necessary to take into account the efficiency of equipment use.

2. Results

The evaluation of the criteria for the effectiveness of repair outsourcing should be carried out continuously. The analysis of the literature [7-9] have shown that the quality of services in repair of equipment may be determined by parameters such as:

- the repair time;
- the number of failures;
- the number of defects that reduce the performance of the equipment.

All three parameters should be kept to a minimum. To minimize the factors that have a negative impact on the use of repair outsourcing, it is also necessary to introduce new functions in the activities of the repair service of the enterprise – joint with the service provider preparation of the repair schedule, analysis of statistical information, analysis of the quality of services parameters.

2.1 Mechanism of decision-making on outsourcing

As a criterion reflecting the quality of the repair, you can take the availability parameter [2], which reflects all the losses at the stops (D), including any unplanned stops: equipment failures, stops due to a shortage of the raw materials or lack of storage space. To analyze the quality of repair, we will consider only the stops due to equipment failure. Consider the indicator - the planned production time:

$$PPT = POT - PSD \tag{3}$$

where PPT – the planned production time;

POT – the total operating time of the enterprise;

PSD – the time of scheduled equipment stops for scheduled repairs and maintenance.

After you calculate the planned production time, you can determine the actual operating time for the past period:

$$OT = PPT - D \tag{4}$$

whereOT – determine the actual time loss time of the enterprise to stop,

D – the equipment stop time due to failures.

The availability criterion is calculated according to the following dependence:

$$A = \frac{OT}{PPT}, \tag{5}$$

whereA – the criterion of equipment availability.

Thus, the equipment can be classified according to the availability criterion, using the Pareto principle, that is, for further analysis, choose 20% of the equipment with the lowest indicators of the availability criterion. For the selected equipment, it is necessary to consider the issue of transferring its repair to outsourcing. The first step is to select the equipment that is used in the main production, that is, directly in the manufacture of products. Next, it is necessary to analyze the causes of failures, and leave under consideration the transfer of repair to outsourcing the equipment in which the causes of failures was low-quality repair.

Then, for each repair of the selected equipment, the possibility of outsourcing is considered, for this purpose, the possibility of performing repairs on the basis of the organization is estimated. But it should be borne in mind that the implementation of quality repairs on the basis of the organization is possible in the presence of competent personnel and equipment to obtain a quality result of the repair. In the absence of equipment, a comparison of the repair costs carried out by the organization, taking into account the cost of purchasing equipment and services of the outsourcing company.

If the purchase of new equipment costs the company more for a certain period than the outsourcing services, the efficiency of the process outsourcing is obvious. If the company does not have the necessary competencies to carry out repairs, then outsourcing the process can also be an effective solution.

If the decision to outsource is made, the calculation of equipment repair costs based on the use of outsourcing is calculated according to the formula:

$$Cp = a + b + p \qquad (6)$$

where C_p – the costs associated with the repair of equipment under outsourcing, currency unit;

a–the costs associated with the wages of repair service employees, in terms of outsourcing, currency unit;

b – the cost of materials required for repair service to perform repairs, outsourcing, currency unit;

p – the costs for the services of the outsourcing company, currency unit.

The calculation of equipment repair costs on the basis of insourcing is calculated by the formula:

$$Cp' = a' + b' + d \qquad (7)$$

whereCp' – the costs associated with the repair of equipment in the conditions of insourcing, currency unit;

a' – the costs associated with the wages of repair service employees, currency unit;

b' – the cost of materials necessary for the repair service to perform repairs, currency unit;

d – the depreciation of equipment, currency unit:

$$d = \sum_{n=1}^{w} C_n, \qquad (8)$$

whereC_n – the amount of depreciation for n-equipment, currency unit;
w – the amount of purchased equipment.

The decision to outsource occurs when the following conditions are met:

$$Cp` > Cp \qquad (9)$$

The algorithm of decision-making on equipment repair outsourcing can be described by the following block diagram (Figure 1):

2.2 Service provider choice

After deciding to outsource the repair and choosing the repair process that is outsourced, the next important issue is the choice of a service provider.

To do this, you need to answer the following questions:

Does the supplier provide equipment repair services?
Does the supplier carry out the type of repair that the organization has chosen to outsource?
Does it carry out repair of equipment that the organization has chosen to outsource?
Did the supplier carry out repair of the equipment unit, which the organization chose to outsource?

If the organization has chosen to outsource the repair of a particular model of equipment, it is preferable to have a supplier with experience with the same models of equipment. In the absence of suppliers with experience of repair of this model, it is necessary to choose suppliers offering as a whole current or capital repairs of the similar equipment.

If the organization is looking for vendor for repair of certain components, a preferable supplier having experience of repair of these sites. The mechanism for selecting service providers is shown in Figure 2.

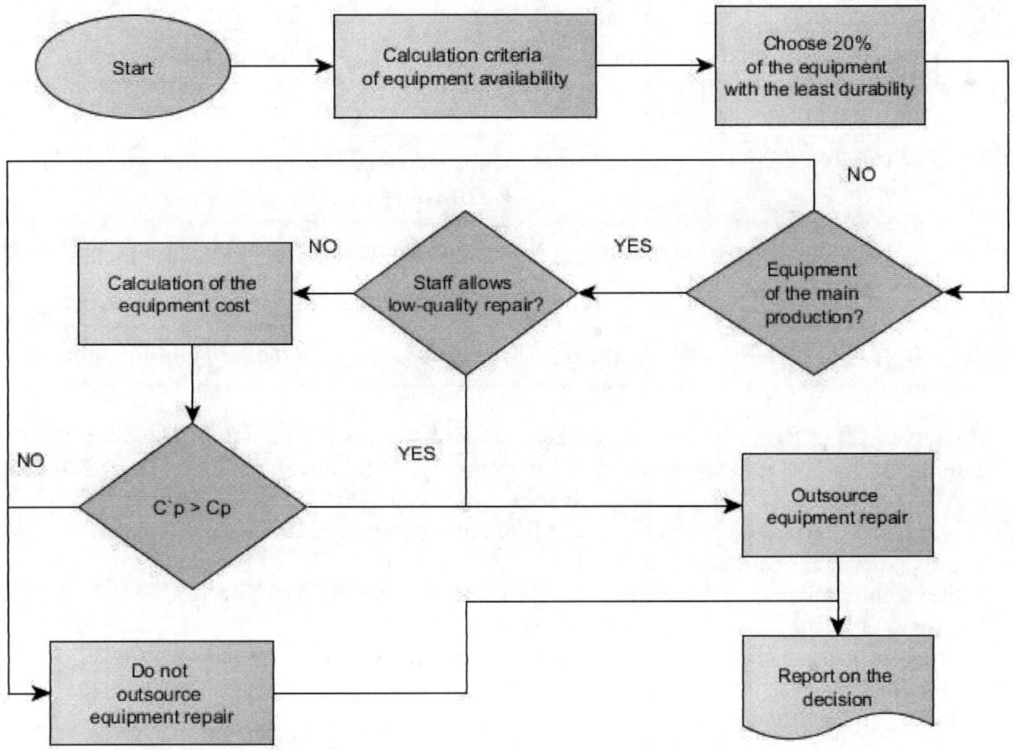

Figure 1 – Algorithm of decision-making on equipment repair outsourcing

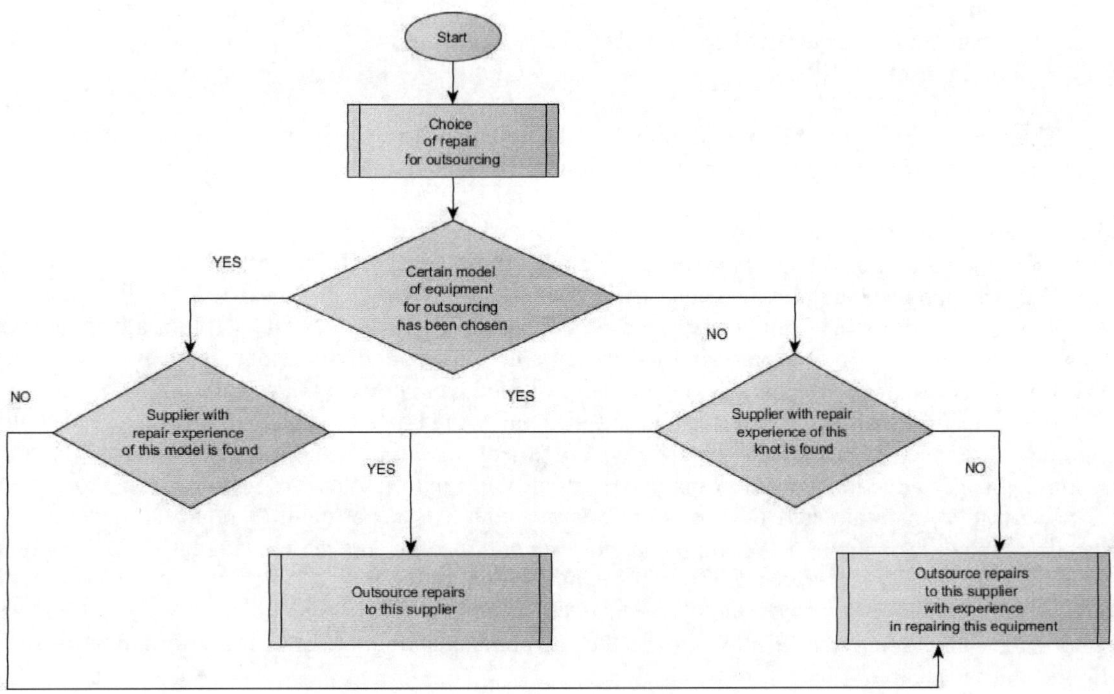

Figure 2 – Algorithm for the choice of service providers

2.3 Statistics of repair services in Omsk region

The formative evaluation of the proposals of the industrial service in the Omsk region (Table 1).

Table 1 – Enterprises – suppliers of industrial equipment repair services in the Omsk region

Enterprise	Location	Type of service
The Federal state unitary enterprise «Omsk production Association «Irtysh»	Omsk	Repair of metal-cutting equipment
		Inspection and preparation of equipment punch list
ZAO«Omsk CTO AT»	Omsk	Equipmentmodernization
		Overhaul of lathe groupmachine tools
OAO Factory «Trud»	Novosibirsk	Modernization and overhaul of machines

ZAO«Omsk CTO PO» also has an experience and the necessary equipment for repair of the following units of the equipment: devices of numerical program control NS-31, 2R22, 2S42-65, MS2109, MS2105, MicroDAT, electric drives «Razmer 2M-5-21», «Razmer 2M-5-21\11», «KEMROS», «KEMTOK», «KEMRON», «KEMEK» and other similar drives.

And any experience with equipment:

– lathemachine 16B1TT1S1, 16K20F3, 16A20F3, 16K30F323, 16M30F3141, R755F341, 16D16AF1-01, F-11;
– millingmachine AG-250, AG-400, NR5-38, MANO MN-500S, LF66, 6R13F3, WF-5N, GF2171S5, FP 7, SFP-13M;
–cutting machine SFP-1R;
– machining centers GDW-400, W400;
– coordinate boring machines IR2637PMF4, 24K40SF4, 2637GF1, 2620;
– heat cutting machine «Grant PPlKC-2,5»;
– grinding machine W3-205F3;
– hole-punching presses KO 126P, OCKO 126P;
– molding machine NR-5;
– galvanic line AG-44.

Thus, it is possible to choose suppliers on experience in repair with the certain model of the equipment, a type of repair (current, capital) or the node of the equipment.

2.4 Reorganization of repair service

In practice, the structure of the repair service includes the mechanical department, mechanical repair shops (MRS), shop mechanic service, equipment and material warehouses[7, 9-10].

The most widespread is the centralized organization of repairs, one of the main advantages of which is that the planning of repairs and its implementation are under one management. It increases reliability of performance ofrepair scheduleand improves maneuverability of repair crews.

In connection with the use of outsourcing, the following changes in the organizational structure of the repair service are proposed: to create thedepartment of statistics and management analysis, this structural change is due to the need to analyze the quality of repair in outsourcing conditions, for which it is necessary to maintain statistics on certain parameters. The department of repair management is created on the basis of the existing planning and schedule preventive bureau and schedule productivebureau with preservation of repair planning function on which load fell in connection with transfer of part of repair to outsourcing. To reduce the load of the chief mechanic in each unit, a Manager is appointed, whose functions are to analyze the incoming information and obtain solutions together with the chief mechanic.

Then the specialized teams are created according to the type of equipment, which includes specialists who are familiar with the relevant equipment of the enterprise. For example, several teams carry out repair of machines, other teams - the repair of particularly complex equipment and one team - the repair of crane equipment. The described recommendations are suitable for a mixed structure of repair service, which is the most common in industrial enterprises. The structure of the repair service after the proposed reorganization is shown in Fig. 3.

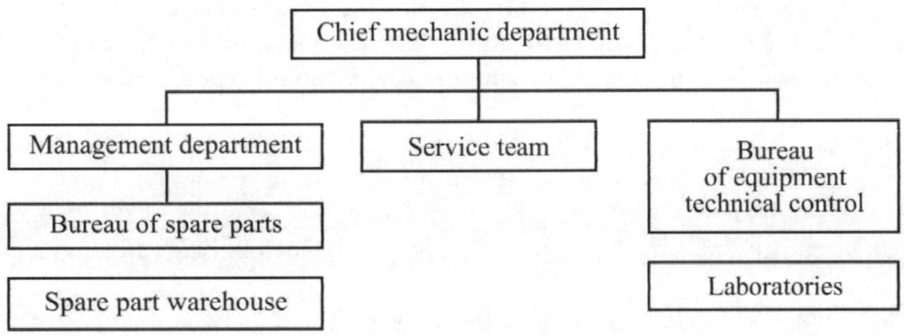

Figure 3 – Structural diagram of the repair service after the proposed reorganization

The criteria required to evaluate the effectiveness of equipment repair outsourcing were described above.

For the basic value of the repair time should be taken calculated value according to the standards. Then the time of repair performed by the outsourcer should not exceed the value according to the standards, or other values set by the enterprise.

To calculate the values of these criteria it is proposed to use the following relations:

$$K1 = \frac{t_r}{t_s}, \quad (10)$$

$$K2 = \frac{n_{f1}}{n_{f2}}, \quad (11)$$

$$K3 = \frac{n_{d1}}{n_{d2}} \quad (12)$$

wheret_r ,t_s - the repair time before the reorganization, and according to standards;

n_{f1}– the number of failures before the reorganization of the repair service;

n_{f2}– the number of failures after the reorganization of the repair service;

n_{d1}и n_{d2} - the number of defects that reduce the equipment performance before and after the reorganization of the repair service.

The degree of economic efficiency can be estimated as follows:

$$K4 = \frac{\text{Э}_a}{\text{Э}_p} \quad (13)$$

where$\text{Э}a$ - actual repair costs while using equipment repair outsourcing;

$\text{Э}p$- planned indicator of repair costs when using equipment repair outsourcing.

At $K4 \leq 1$ it is possible to talk about the effectiveness of outsourcing. The process of using equipment repair outsourcing can be considered successful at the following values:

$$K1 \leq 1,\ K2 \leq 1,\ K3 \leq 1 \text{ и } K4 \leq 1 \quad (14)$$

or with small deviations (5-10%).

For the detection of improvement or deterioration in the quality of the repair after the use of outsourcing, use following dependency:

$$K = K1 \times K2 \times K3 \times K4 \quad (15)$$

3. Conclusion

The mechanism of decision-making on outsourcing of equipment repair on the basis of availabilitycriterion and the mechanism of the choice of the supplier is offered. To improve the organization of repair service in the conditions of outsourcing, the criteria for evaluating the effectiveness of outsourcing equipment repair: repair time, the number of failures, the number of defects that reduce the performance of equipment; control of which will allow to make changes to the terms of the outsourcing contract. The presented dependencies can be used to detect improvement or deterioration of repair quality after outsourcing.

IOP Conf. Series: Materials Science and Engineering **582** (2019) 012041 doi:10.1088/1757-899X/582/1/012041

After the proposed reorganization, the structure of the repair service is proposed, in accordance with which it is advisable to create a department of statistics analysis, management and service teams specializing in the repair and maintenance of equipment of a certain type.

References

[1]Redreev, G.V Perfection of technical service of manufacturing plants equipment / Redreev G.V., Shimokhin A.V., Kiyko P.V. In the collection: Journal of Physics: Conference Series 9. «IX International Scientific Practical Conference «Innovative Technologies in Engineering» 2018. C. 012008.

[2]Chernova E.G. Integrated economic education: a typology and dynamics of organizational-legal forms/E.G. Charova// Vestnik of St. Petersburg University: Economics.- №1.2010.- 59-69 p.

[3]Tetter V. The Research Institute of Technology, Control and Diagnostics of Railway Transport / V. Tetter // TRANSPORT PROBLEMS. 2009. №2. 67-72 p.

[4]Laktionova, O.E. Financial Management of enterprises on the terms of the outsourcing contract/O.E. Laktionova//Finance and business.-№3.-2017.-134-142 p.

[5]Williamson, O.I. (2010). Outsourcing: transaction costs and supply chain management. *Russian management journal*, 8(1), 71-92 p.

[6] Smirnov, S. A. Concept of quality costs (Cost of Quality): problems and prospects of application / S. A. Smirnov// Bulletin of St. Petersburg University: Economics.-№3.-2012.- 146-156 p.

[7]Morozova, A.S. Comparative characteristics of strategies for entering foreign markets /A. S. Morozova//Science and technology.-№3.-2009.-72-78 p.

[8]Konorov, D.G. Enterprise restructuring. Change of organizational structure / G. D. Konokov // Technical journal «Automotive industry».-№3.-2013.

[9]Sivacheva, E.A. Outsourcing in the system of international economic relations/E. A. Sivacheva//Bulletin of St. Petersburg University: Economics. – №1. – 2008. – From 135–139 p.

[10] Technical support of energy conservation and saving technologies / V.A. Eviev, G.M. Muchkaeva // studyguide: St. Petersburg State University, Kalmyk State University named after B. Gorodikov. – Elista, 2010.

IOP Conf. Series: Materials Science and Engineering **582** (2019) 012042 doi:10.1088/1757-899X/582/1/012042

Improvement of Adhesion Strength of the Sprayed Material

S A Kornilovich, G V Redreev, D A Vorobyov, S V Zakharov

Omsk State Agrarian University named after P. A. Stolypin
644008, Omsk, Institutskaya square st., 1

E-mail: weerwg@mail.ru

Abstract. The effective way of enhancing resistance to deterioration of the parts is the change of surface properties due to coating with the required properties. Since the deposition process takes place at an elevated temperature, it is necessary to ensure the absence of phase and structural transformations in the metal, mixing of materials, changes in the chemical composition of the coating. The use of plasma spraying at the transport management facility showed a sufficiently high quality of the coating. However, the metallographic analysis showed the presence of discontinuities at the interface in the form of pores, oxide film and sections of decarburized steel. In the case of the process of melting–spraying, the overheating areas of the base metal are formed, which leads to disadvantages similar to the deposition process. To ensure the minimal thermal impact on the part, the thermal field calculations are required. For this purpose, the time of local melting of the part by a single-phase plasma jet was calculated. The temperature field of a part in which heat is transferred by thermal conductivity, the solution of the differential equation of thermal conductivity is described. The boundary conditions of three genera are formulated. The first kind is the distribution of temperature on the enclosing surfaces. The second kind – the task on the surface of the part of the heat flux density. The third kind – setting the temperature of the medium, which acts as a plasma jet, washing the surface of the part, and the intensity of heat exchange. The mathematical model of melting of the product is constructed. The formula for calculating the reflow time is derived. The temperature distribution expressions for the molten and solid zones are obtained by the solution of the problem with moving boundaries by the Fourier's method. On the basis of theoretical calculations it is proposed to increase the strength of the sprayed layer with the base to apply the process of plasma spraying with preliminary local melting by periodic exposure to a single-phase plasma jet.

1. Introduction

Spraying of the material is a productive and relatively simple process that provides a small heating of the part, the possibility of regulating the chemical and phase composition of the coating within a wide range and high wear resistance.

The essence of the material deposition as a coating process is to heat the material, its dispersion, transfer of the material by a plasma jet and impact on the coated surface, deformation of the particles and their fixation on the basis of (the area of the coated surface of the item).

According to the scientific conclusions of academics V.I. Chernoivanov and E.S. Karakozov [1], the main conditions for the formation of high-quality coatings are: the heating of the part should not cause phase or structural transformations in the metal; mixing of the base and coating materials should be minimal; there should be no reactions that cause a change in the chemical composition of the coating compared to the starting material; in the zone of the compound should not develop relaxation processes that can change the structure and phase composition of the metal [2]. These requirements correspond to a method of applying a coating by plasma deposition of a metal layer on the recoverable, as well as manufactured parts [2-3].

The experience of using plasma spraying at a car repairing plant when restoring parts of fixed joints using the nickel-aluminum powder PN85YU15 as a substrate showed a positive quality of coatings. In this case, the coating of the surface of the parts of movable joints without applying a substrate was obtained of poor quality.

IOP Conf. Series: Materials Science and Engineering **582** (2019) 012042 doi:10.1088/1757-899X/582/1/012042

A metallographic analysis of the coating state and base metal after plasma spraying of the central factory of «Transmash» laboratories showed the following: nothing was found at the interface that could lead to the formation of an oxide film, as well as areas of the decarburized steel to pure ferrite; the microstructure of the sprayed layer is a heterogeneous-layered structure with the participation of a spherical shape, there are areas of austenitic-martensitic and bainitic-austenitic structures. Martensite has been significantly increased. The areas after the plasma deposition of the layer had different hardness from HRC 35-40 to HRC 50-60 (measured by microhardness).The adhesion strength of the coating did not exceed 1...6 daN/мм^2, which is an order of magnitude lower than the strength of the base metal. After subsequent melting, the strength of the coating layer is much higher – 30...40 daN/мм^2. However, the zone of thermal influence increases, at a depth up to 2.5 mm, the overheating areas of the base metal are formed. There is a process similar to the process of surfacing with its inherent disadvantages.

In cases of using wire instead of a powder made of carbon steel without a substrate of nickel-aluminum alloy, a significant disadvantage of plasma spraying is the low adhesion strength of the coating to the substrate. Increasing the adhesion strength of the coating layer with the base is possible when spraying the material on the pre-melted surface of the part [4].

Pre-melting is carried out periodically in the order of «melting–spraying». Melting is carried out by a plasma jet without feeding the sprayed material. After melting, the supply of the sprayed material is resumed. To simplify the article provisionally called the plasma jet without the sprayed material as the «single-phase», and with the spraying material – the «two-phase». The repeated experiments performed at the Department of Technical Service, Mechanics and Electrical Engineering of Omsk State Agrarian University showed that the depth of the melted layer should be minimal - no more than the thickness of the film held on the surface due to tension forces. Outwardly, such a fusion corresponds to the physical state of «fogging». With an increase in the penetration depth of the liquid metal is blown by a plasma jet to the sides, a crater is formed, in addition, the part is heated unnecessarily, which is incompatible with the essence of the deposition process of the items.

The process of deposition requires a minimum thermal effect on the part, without any noticeable structural changes. The maximum heating of the part should not exceed the high tempering temperature [4-7]. At the same time, the adhesion strength of the sprayed layer increases with increasing temperature on the surface until it melts. The determination of the state prior to the melting point (time t) -x is possible by calculating the thermal field [5-8]. The most General description of the thermal field is the expression of the dependence $t = f(x,y,z,\tau)$ as a differential equation.

2. Results

The calculation of the local melting time of a part by a single-phase plasma jet is made in the following sequence.

2.1.Description of the temperature field of the sprayed part

The description of the temperature field of the part in which heat is transmitted by thermal conductivity can be done by solving the differential equation of thermal conductivity [2]. To simplify the derivation of the equation in the case of a non-standard thermal conductivity process in a stationary homogeneous and isotropic body, it can be assumed that the physical parameters of the body are constant, the deformation of the object under consideration associated with a change in temperature is small and can be neglected, the relative movement of body parts is absent.

To derive the equation, select an elementary parallelepiped with edges *dx, dy, dz* in the volume of the heated part, placing it relative to the rectangular coordinate system, and assuming that there is no internal heat source (Figure 1):

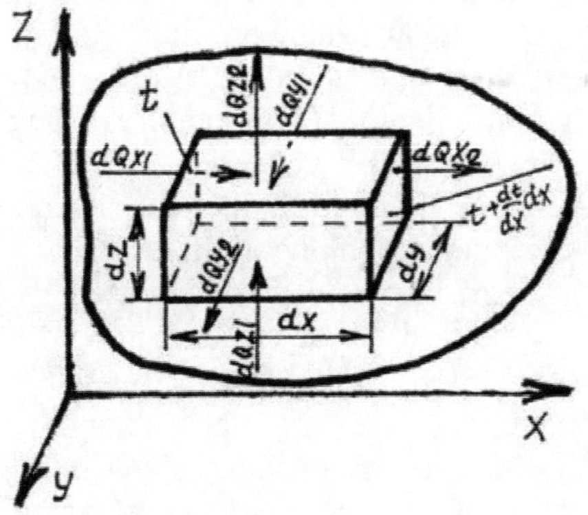

Figure 1 – Elementary volume of the heated metal to the solution of the differential equation

The heat flux vector through the elementary parallelepiped is decomposed along the coordinate axes. Considering its component in the direction of the X axis, we denote the elementary amount of heat obtained by the left side of the parallelepiped from the plasma jet in contact with it, by dQx1, and the heat transferred by the adjacent layer by the opposite face is dQX2.

According to Fourier's law, taking the temperature of the left face equal to *t*, the expression for dQx1 can be written as follows:

$$dQ_{x1} = -\lambda \frac{\partial t}{\partial x} dydzd \, , \tag{1}$$

where λ – the thermal conductivity

The change in body temperature in the direction of the x-Axis is indicated as $\frac{\partial t}{\partial x}$, on the segment dx the change in temperature will be equal to $\frac{\partial t}{\partial x} dx$, so the temperature of the opposite right side of the parallelepiped can be represented as a sum

$$t + \frac{\partial t}{\partial x} dx \tag{2}$$

Accordingly, we write the expression for dQ_{x2} as;

$dQ_{x2} = -\frac{\partial t}{\partial x}(t + \frac{\partial t}{\partial x} dx)dydzd$ or after opening the brackets we will have:

$$dQ_{x2} = -\frac{\partial t}{\partial x} dydzd - \frac{\partial^2 t}{\partial x^2} dxdydzd \tag{3}$$

The amount of heat dQ_x, which went to the heating of the parallelepiped during dt (from the component transmitted in the direction of the x axis), can be obtained by subtracting from the right side of equation (1) the right side of the equation (3):

$$dQ_x = -\frac{\partial t}{\partial x} dydzd + \frac{\partial t}{\partial x} dydzd + \frac{\partial^2 t}{\partial x^2} dydzdxd = \frac{\partial^2 t}{\partial x^2} dydzdxd \tag{4}$$

If we make similar arguments about the other two components of the heat flux vector along the y and Z axes, we can write:

$$dQ_y = \frac{\partial^2 t}{\partial z^2} dxdydz, \tag{5}$$

$$dQ_z = \frac{\partial^2 t}{\partial y^2} dxdydzd \, , \tag{6}$$

To obtain a complete picture that reflects the qualitative and quantitative characteristics of a particular problem, it is necessary to mathematically set this problem, determined by the conditions of its uniqueness in the form of geometric conditions (shape and size of the part), physical (the value of the coefficients of heat and thermal conductivity), as well as boundary [3]. The composition of the boundary conditions to the heat equation includes time (initial temperature distribution in the volume of the part) and boundary, determining the heat transfer at the boundaries of the part. These conditions together determine a single state, and in this sense can be called the conditions of uniqueness, and the problem solved with their help - the boundary value problem [2,4].

The initial temperature distribution in the part can be different. For example, at the initial moment of heating or cooling (t = 0) the part has the same temperature throughout its mass:

$$t(x,y,z) = t_c \tag{7}$$

In general, the initial temperature distribution in the part is given by the equation t = f(x,y,z) (at t = 0). Boundary conditions can be formulated in one of three ways.

1. The boundary conditions of the first kind are determined by the temperature distribution on the enclosing surfaces as a function of the position of the surface point and time. In conditions of the first kind, the problem of pure heat conduction is considered. This is the only possible mechanism of heat transfer through the wall from one of its boundary surfaces to another.

2. The boundary condition of the second roll consists in setting the heat flux density on the part surface as a function of time and coordinates. These tasks are characteristic of the heat exchange process.

3.The boundary conditions of the third kind consist in setting the temperature of the medium T_c, (plasma jet) washing the surface of the part and the intensity of heat transfer (heat transfer) between this medium and the surface of the part. The process of heat transfer is described by Newton's – Richman law, according to which the amount of heat given or received by the body from the environment is proportional to the surface area of the body, the temperature difference of the body t_b and the environment t_c and the duration of the process:

$$Q= \alpha(t_e - t_b)\, F\tau \tag{8}$$

or

$$Q = \alpha(t_e - t_b)F \tag{9}$$

If we take in equation (8) F and τ equal to one, then a = $Q_T/(t_e - t_b)$ represent a value numerically equal to the amount of heat exchanged per unit time unit surface of the part with the medium of the unit temperature difference between them.

In general, the heat transfer coefficient reflects the combined effect of convection and thermal conductivity of the medium, and therefore depends on many factors. It is obvious that the heat flux given by a single surface of the part medium should be equal to the heat flux supplied by the thermal conductivity of the internal volumes of the part to its surface, i.e.

$$\alpha(t_e - t_b) = -\lambda(\frac{\partial t}{\partial n})_c \text{or} -(\frac{\partial t}{\partial n})_c = -(\alpha/\lambda)(t_e - t_b), \tag{10}$$

This expression is the most general record of the boundary condition of the third kind [2-3].

2.2. Mathematical model ofthe sprayed part melting

The mathematical model can be built on the example of melting the product according to the scheme (Figure 2).

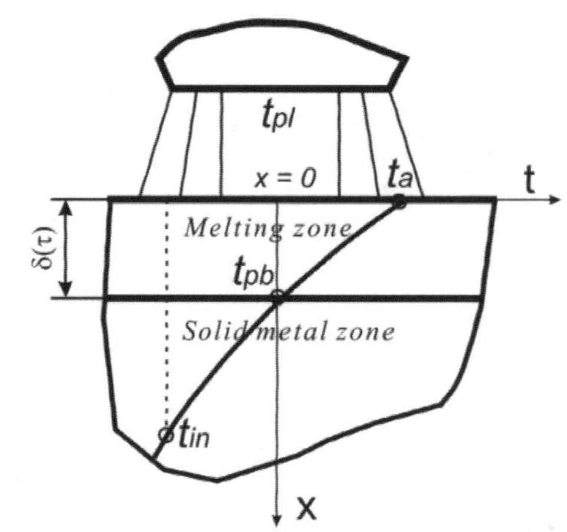

Figure 2 – The model of the reflow coating part: t_{in} – the initial temperature; t_{pb} – the temperature in the phase boundary (solid and molten); t_a – the ambient temperature; $\delta(\tau)$ - the coordinate of the phase boundary; t_{pl} – the plasma temperature

Imagine that the item has a sufficiently large thickness, such that the temperature perturbations on one surface will not reach the other surface during the period of time. The body is initially in a solid state at a given temperature $t_{in}<t_{pb}$. In the simplest case, it can be assumed that, starting from a certain point in time t = 0, a constant temperature $t_a>t_{pb}$ (a boundary condition of the first kind) is established on the bounded surface and during the period under consideration, which leads to the formation of a melted layer.

The lower boundary of this variable-coordinate layer is the interface (liquid and solid) and maintains a constant temperature t_{pb}. On this surface, the heat of the phase transition is released, equal to the melting heat of the material (for steel – 272.1 kJ/kg).As the phase boundary progresses, the solid zone is heated and melted, but at a sufficient distance ($x\rightarrow\infty$) the temperature retains the initial value t_{in}. Obviously, the heat transfer occurs only in the direction of the X-axis. We denote the thermophysical parameters in the molten layer λ_p , C_p, α_p , and in the solid $\lambda_{m,}C_{m,}$ α_m. In the simplest case, the density of the molten and solid material will take the same:
$\rho_p = \rho_m = \rho = Const.$

It is required to find the temperature field in the molten and solid $t_m(x,\tau)$ zones, as well as the law of displacement in time of the interface $\delta(\tau)$ and the velocity of this boundary ($d\tau/d\delta$ – the melting rate).

2.3. Mathematical formulation of the problem. Formula derivation for calculating the reflow time
The differential heat equation for different conditions can be written as:

a) for the molten zone

$$\frac{\partial t_p}{\partial} = \alpha_p \frac{\partial^2 t_p}{\partial x^2} \text{ where } 0 < X \le \delta(\tau);\ \tau > 0 \qquad (11)$$

б) for the solid zone

$$\frac{\partial t_m}{\partial} = \alpha_p \frac{\partial^2 t_m}{\partial x^2} \text{ where } \delta(\tau) \le X \le \infty;\ \tau > 0 \qquad (12)$$

Initial condition: where $\tau = 0,\ t\,(x,\,0) = t_{н;}\ \delta(0) = 0$.

External boundary conditions: where $X = 0;\ \tau > 0;\ t(0,\,\tau) = t_0 = Const$; where $X \rightarrow \infty;\ \tau = 0;\ t\,(\infty,\,\tau) = t = Const.$

It is also necessary to formulate conditions on the internal mobile interface. These two conditions. The first condition is the equality of the temperature of the molten and solid zones. The second – the equation of heat flow from the molten and solid zones. To formulate it mathematically, we write the heat balance equation on the phase boundary for the elementary time interval $d\tau$. Due to the simultaneity of the problem, the area of the calculated surface can be any. We will calculate on 1 m^2. The amount of heat coming from the molten zone should be equal to the sum of the heat withdrawn from the phase boundary to the solid zone and the heat of the phase transition absorbed at the boundary:

$$dQ_m = dQ_s + dQ_{pb} \qquad (13)$$

According to Fourier's law:

$$dQ_m = -(\frac{\partial t_p}{\partial x})_{+\delta} d; dQ_m = -(\frac{\partial t_m}{\partial x})_{-\delta} d\ , \qquad (14)$$

At this time $(\frac{\partial t_p}{\partial x})_{+\delta}$ and$(\frac{\partial t_m}{\partial x})_{-\delta}$ – the temperature gradient at the interface in the molten and solid zones.

During $d\tau$, the interface will move by the value of $d\delta$, so, based on 11 m^2 of the surface, the volume will melt: $dV = Ld\delta$, and the amount of heat of the phase transition will be:

$$dQ_{pb} = H_{pb} \times dV = H_{pb} \times d\tau \times L \qquad (15)$$

The equation of the heat balance after transformations at division by $d\tau$ takes the form at $x=\delta(\tau)$:

$$-(\frac{\partial t_p}{\partial x})_{+\delta} = -(\frac{\partial t_m}{\partial x})_{+\delta} + H\frac{d\delta}{d\tau} \qquad (16)$$

Here we introduce the designation $H = G{\cdot}L{\cdot}\rho$. The value H can be interpreted as the heat of the phase transition unit volume.

The solution of the problem with moving boundaries by Fourier's method leads to the temperature distribution expressions:

a) in the molten zone

$$\frac{t_m(x,\tau)-t_a}{t_a-t_{pb}} = \frac{erf\frac{x}{\sqrt{a_p\cdot\tau}}}{erfP}, \tag{17}$$

where *erf* – the tabulated Gauss error function;

б) in the solid zone

$$\frac{t_s(x,\tau)-t_a}{t_{pb}-t_a} = \frac{1-erf\frac{x}{\sqrt{a_p/a_m}}}{1-erfP(P\sqrt{a_p/a_m})} \tag{18}$$

The law of moving phase boundary:

$$\delta(\tau) = 2P\sqrt{a_p\cdot\tau} \tag{19}$$

The velocity of the phase at time

$$\frac{d\delta}{d\tau} = (2P\sqrt{a_p\cdot\tau})^{\frac{1}{2}} \tag{20}$$

In the expressions P – dimensionless quantity determined from the transcendental equation [8].

The problem is simplified if the solid zone initially has a phase transition temperature $t_s = t_{pb}$. While the heat transfer in the solid zone is not observed in the balance equation at the interface $dQ_T = 0$. If we assume that the temperature field in the molten layer at any time corresponds to the stationary one (the process is quasi-stationary), then the temperature distribution in the molten layer δ(τ) is linear, as in the infinite plate thickness δ at the stationary thermal conductivity [8-9], that is, to determine the temperature at any point of the molten layer, you can use the equation

$$\frac{t_a-t_p(x,\tau)}{t_a-t_{pb}} = \frac{x}{\delta(\tau)} \text{ where } 0 \le x \le \delta(\tau) \tag{21}$$

Transforming the formula for the speed of phase movement through the phase transition criterion, we will have the formula for the speed of movement of the phase boundary [7–11]:

$$\frac{d\delta}{d\tau} = \frac{\sqrt{\lambda_p(t_a-t_{pb})}}{2\Phi\tau} \tag{22}$$

The expressions for the velocity dependence are valid for the equality $t_s = t_{pb}$ and for the linear temperature distribution in the molten layer they are an important special case of Stefan's solution [8–10].

It is known that at moderate speeds in a very thin layer of the melt quasi-stationary approximation is quite acceptable. Physically, this corresponds to the assumption that the heat removed from the molten layer during its heating from the phase transition temperature t_{pb} to some average specified temperature at a given time is very small compared to the heat that went to the phase transition of the layer from the solid state to the molten state.

In this solution, the boundary condition on the external heated surface x=0 is written as a boundary condition of the first kind. For engineering practice, the case is more interesting when the specified temperature is not the surface temperature, but the temperature of the heated medium, in this case the plasma jet t_{pl}, i.e. the boundary condition of the third kind. In this case, in the quasi-stationary approximation for the law of the phase boundary displacement, we obtain

$$\delta(\tau) = \sqrt{\frac{2\lambda_p(t_{pl}-t_{pb})\tau}{H}} - \frac{\lambda_p}{\alpha_{pl}}, \tag{23}$$

where α_{pl} – the heat transfer coefficient of the plasma.

Since the thickness of the molten layer cannot be less than zero, then at δ(τ)=0 we find the time before the melting:

$$\tau_0 = \frac{\lambda_p H}{2\alpha_{pl}^2(t_{pl}-t_{pb})}, \tag{24}$$

3. Conclusion

Thus, under the boundary conditions of the third kind, in contrast to the boundary conditions of the first kind, the maximum value of the velocity of the phase boundary corresponds to the condition $\tau=\tau_0$ and has a finite (and not infinitely large) value, then with increasing τ the velocity decreases, so that at $\tau\rightarrow\infty$, $d\delta/d\tau=0$.

The practical application of the plasma spraying process with preliminary local melting by periodic exposure to a single-phase plasma jet lasting until the moment of melting increases the adhesion strength of the sprayed layer with the base.

References

[1] Chernoivanov, V. I. Formation of coatings on working surfaces by electrocontact surfacing / V. I. Chernoivanov, E. S. Karakozov, B. A. Molchanov, R. A. Latypov. Welding production, 1986, №4, 16-18 p.
[2] Restoration of details of machines: Reference book / F. I. Panteleenko, V. P. Lyalyakin V. P. Ivanov, V. M. Konstantinov; edited by V. P. Ivanov. – Moscow: mechanical engineering, 2003. 672 p.
[3] Makarenko N.D. Electrochemical hardening and restoration of parts of tribosystems: Scientific publication. – Omsk, Omsk scientific Bulletin, 2004. 254 p.
[4] Kornilovich, S.A. Improvement of quality of technological process of repair of agricultural machinery on the basis of the analysis of its accuracy and stability. / Thesis for the degree of doctor of technical Sciences. – St. Petersburg, 2000.
[5] Husay, A., Morigan O. Surfacing and spraying / Per. Japanese V.N.Popov; edited by V. S. Stepanov, N. D.Shestikova. – Moscow: Mechanical Engineering, 1985. 239 p.
[6] Lykov A. V., Theory of thermal conductivity. – Moscow: High school, 1967. 599 p.
[7] Malakhovsky V. A. Plasma processes in welding production. – Moscow: High school, 1988. 70 p.
[8] Podobrany, A. N. Calculation of transient temperatures in plasma spraying of wear-resistant coatings. / A. N. Podobrany, V. M. Rogozhin / the mechanical engineering Bulletin. 1995. 35-39p.
[9] Rykalin, N. S. Calculations of thermal processes in welding. / N.S. Rykalin – M.: Mechanical Engineering, 1951. 269 p.
[10] Sudenkov, E. G., Rumyantsev S. Yu. Restoration of parts by plasma metallization. / E. G. Sudenkov, S. Yu. Rumyantsev/ – M.: Higher school, 1980. 38 p.
[11] Kudinov V. V., Ivanov V. M. Plasma deposition of refractory coatings. – Moscow: Mechanical Engineering, 1981. 192 p.

IOP Conf. Series: Materials Science and Engineering **582** (2019) 012043 doi:10.1088/1757-899X/582/1/012043

Improving the Efficiency of the Power Take-Off Drive

G V Redreev, A N Sorokin, A A Luchinovich, V D Chervenchuk, S D Popov

Omsk State Agrarian University named after P.A. Stolypin
644008, Omsk, Institutskaya pl. St., 1

E-mail: weerwg@mail.ru

Abstract.power tools that are used in various industries have power take-off drives for aggregating various machines and mechanisms. In the agro-industrial complex, one of the most common power tools are tractors manufactured by the Minsk Tractor Plant. The drive of power take-off in these tractors is carried out at the expense of the planetary mechanism. The wheel and the solar gear of the planetary mechanism are connected to the drums, alternately stopping the drums is carried out on and off the drive. The brake tape of the firing drum connected to the sun gear is most susceptible to wear, as the drum is loaded with a torque of at least one third of the torque to the drive of the aggregated machine.Due to the design features of the existing control mechanism, the turn–on time is 1.5-2 times less than recommended for devices of this kind. As a result, there is an intense wear of the friction lining of the inclusion drum tape.Regulation of the control mechanism of adjusting screws is carried out untimely, as a result, the durability of the tapes is very low.To improve the durability and efficiency of the power take-off drive, a drive-on design is proposed by electromagnets.Applying force to the tape by electromagnets at four points, instead of one in the previous design, will improve the braking efficiency of the inclusion drum.It is possible to include electromagnets not at the same time, but with some time offset, with a gradual pressing of the tape, starting with the portion of the tape closest to the axis of its suspension. This improves the fit of the tape to the drum inclusion.In the transition from a mechanical drive to an electrical device, the control operation is eliminated, which reduces maintenance time, significantly increases the life of the brake belt and increases the efficiency of the power take-off drive.

1. Introduction

MTP tractors are widely used in the cultivation of crops as part of units with machines with active working bodies. The technological process is implemented by the energy supplied from the power take-off shaft (PTO) of the tractor (Fig. 1).

A planetary mechanism is used to change the torque. Turning on and off the PTO drive is carried out by two belt brakes. The pulley of one of the brakes is connected to the planetary gear driver; the pulley of the other is connected to the solar gear (Fig. 2).

2. Power take-off drive analysis

The most loaded in PTO is the brake tape of the inclusion drum (see13 in Fig. 1) connected to a sun gear. Due to the fact that the periodic adjustment of the PTO control mechanism is not made in time, the stop time of the drum switching is large enough, prolonged slippage of the drum in the brake tape leads to increased wear of the friction lining. The durability of the tape is significantly reduced, and in the practice of agricultural enterprises can be 2-3 months with the duration of the season of field work for at least 6 months.

One of the reasons for the increase in wear during slippage of the drum 13 (Fig. 1) is the design features of the inclusion mechanism. When turning the control roller, 9 double-shouldered levers act on the adjusting screws 10 and 11, loosening one of the brake bands and pulling the other. During the activation time of $t_{вк}$, the frequency of rotation of the drum carrier $\mathbf{n_h}$ increases from zero to a certain value, the frequency of rotation of the drum of the sun gear $\mathbf{n_a}$ decreases from a certain value to zero. At the same time, the activation time is rather small and does not exceed 1 s, which leads to increased

IOP Conf. Series: Materials Science and Engineering **582** (2019) 012043 doi:10.1088/1757-899X/582/1/012043

wear of the friction lining of the activation drum tape. For shoe brakes of hoisting machines having a similar design, the recommended braking time is 1.2–2 s.

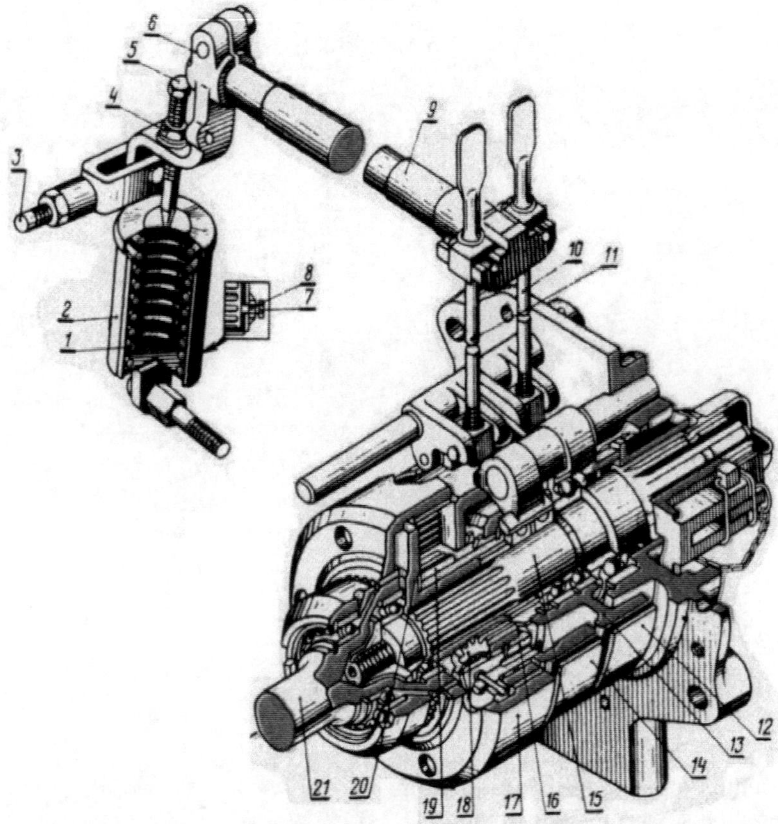

Figure 1 – PTO of MTP tractor 80/82 [1]
1 – spring; 2 – a glass; 3 – thrust; 4 – lock nut; 5 – stop bolt; 6 – control roller lever;7 – locking bolt; 8 – a nut; 9 – control roller; 10 and 11 – adjusting screws;12 and 14 – brake bands; 13 – drum inclusion; 15 – shaft; 16 – satellite axis; 17 – ring gear; 18 – satellite; 19 – sun gear; 20 – carrier assembled with brake drum;21 – ring gear shaft

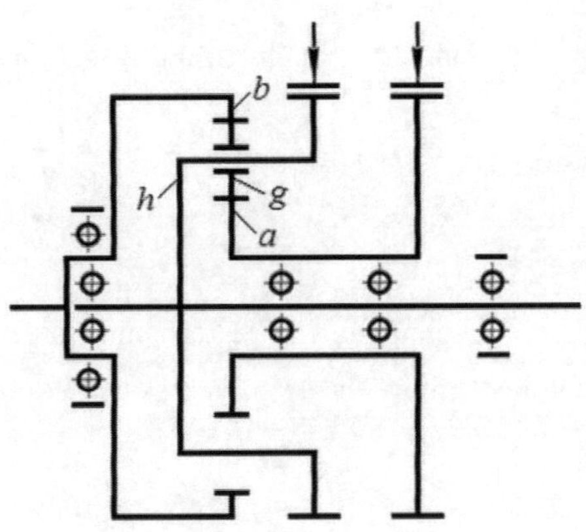

Figure 2 - Kinematic scheme of PTO
a - sun gear, b - ring gear, g - satellite, h– carrier

IOP Conf. Series: Materials Science and Engineering **582** (2019) 012043 doi:10.1088/1757-899X/582/1/012043

2.1 Determination of kinematic and power characteristics of PTO

Consider the calculated PTO scheme and define its individual kinematic and power characteristics (Fig. 3):

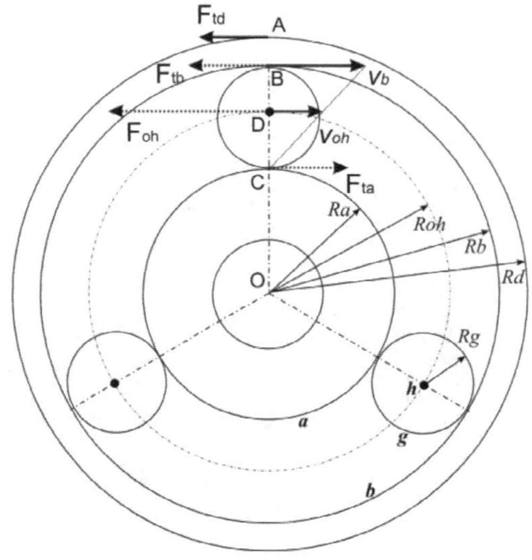

Figure 3 - Calculation PTO scheme

Designations on the scheme:
Ra – radius of the dividing circle of the sun gear
Roh – radius of the circle axes of the satellites
Rb – radius of the dividing circle of the ring gear
Rd– radius of the inclusion drum
Rg – radius of the pitch circle of the satellite
Ftd – circumferential force on the inclusion drum
Ftb – circumferential stress on the ring gear
Foh – circumferential force on the carrier on the circumference of the satellite axes
Fta – circumferential force on the sun gear.

The circumferential force on the $F_{tб}$ start drum will be determined:

$$F_{td} = \frac{T_a}{R_a}, \text{N} \quad (1)$$

where T_a is the torque on the sun gear, Nm

$$T_a = T_b\left(u^a_{bh}\eta^a_{bh} - 1\right), \text{Nm} \quad (2)$$

where u^a_{bh} is the gear ratio of the planetary gear from the ring gear to the carrier when the sun gear is stopped

$$u^a_{hb} = \frac{\omega_h}{\omega_b} = \frac{n_h}{n_b} = \frac{1}{u^a_{bh}} = \frac{z_b}{z_b+z_a} \quad (3)$$

where z_b and z_a are the number of teeth of the ring and sun gear, respectively.

$$u^a_{hb} = \frac{63}{63+30} = 0.677$$

η^a_{bh} – efficiency of the planetary mechanism in the transmission of torque from the ring gear to the carrier when the sun gear is stopped. According to various sources [2–4] η^a_{bh}=0.99–0.96.

Hence $T_a = T_b \times (0.677 \times 0.975 - 1) = -0.34T_b$

$$T_a = \frac{30 \cdot 0.34 \cdot N_{PTO}}{\pi n_{PTO}} \text{Нм} \quad (4)$$

The frequency of rotation of the shaft n_{PTO} coincides with the frequency of rotation of the carrier with the drum and is (for different types of PTO) 545 or 1000 r/min. The drive power N_{PTO} of various agricultural machinery ranges from 5 to 37 kW [5].

$$T_b = \frac{30 \cdot 0.34 \cdot 37000}{3.14 \cdot 545} = 220.5Nm$$

The magnitude of the circumferential force on the drum with its radius of *0.08 m* will be:

$$F_{td} = \frac{220.5}{0.08} \approx 2760N$$

Based on the amount of circumferential force, the tape performance is calculated.

2.2 Tape parameter determination

In accordance with the calculations, the friction force F_{TP} between the tape and the inclusion drum must be at least 2760 N. Taking into account the safety factor [6], the friction force F_{FF}^{S} must be at least 4140 N. force of pressing the tape to the drum inclusion is determined by the dependence:

$$F_{np} = \frac{F_{FF}^{S}}{fe^{f\alpha}}, N \tag{5}$$

where f is the coefficient of friction between the friction strip and the inclusion drum is 0.2–0.4; α – angle of the drum tape, rad. For tape design is used $\alpha = 5.76$ rad.

With the minimum value of the friction coefficient, the value of the clamping force will be:

$$F_{np} = \frac{4140}{0.2e^{0,2 \cdot 5,76}} = 3120N$$

One of the reserves to reduce the clamping force of the tape is to increase the angle of coverage of the drum tape. However, this requires a change in the design of the tape. The most rational is to change the method of clamping the tape.

3. The design of the drive electromagnets

We have developed a design for controlling the activation of the power take-off mechanism using electromagnets (Fig. 4) [7]:

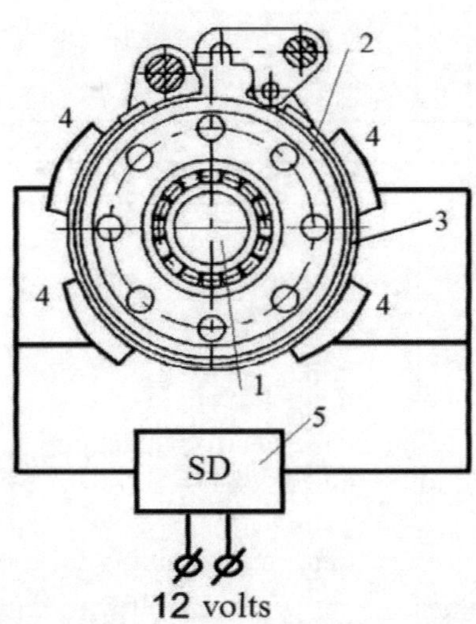

Figure 4 - Mechanism of the inclusion of a drive with electromagnets
1 - shaft; 2 - inclusion drum; 3 - tape; 4 - electromagnets;5 - switching device

The clamping force of the tape, with four electromagnets, will be 780 N per one magnet. The PTO design allows you to mount electromagnets on brake bands and install connectors for electrical wires on the rear wall of the PTO housing. The switch-on device is configured so that the drive is turned on for an optimal period of time.

In an electromagnet, a large part of the magneto motive force of the magnetization winding is applied to the air gap with a friction strip of the tape. Therefore, at any other point of the magnetic circuit the bulk energy density of the magnetic field will be less by several orders, i.e. with an acceptable approximation it is possible to assume that all the energy of the magnetic field of the winding goes to the work of attracting the tape with an electromagnet to the drum or its braking.

The bulk density of the energy of the magnetic field is [8]:

$$\frac{dW}{dV}=\frac{BH}{2}=\frac{\mu_0 H^2}{2}, \tag{6}$$

where $B-$ the magnetic induction, T; $H=\frac{B}{\mu_0}$ – magnetic field strength, $\frac{\text{A}}{\text{m}}$; $\mu_0=4\pi\cdot 10^{-7}\frac{H}{m}$ – absolute magnetic permeability of the air gap; $dV = Sdx$ – the elementary volume of the air gap, m^3, in which the magnetic energy is concentrated, J; S – cross-sectional area of the magnetic circuit, m^2; dx – infinitely small length of the magnetic circuit air gap, m; F – electromagnet traction force.

If the magnetization of the electromagnet core does not reach the saturation point, then the magnetic field strength in the air gap $H \approx Iw/x$, where $Iw-$ the magneto motive force (MMF) of the electromagnet winding, A; $x-$ the thickness of the tape friction layer ($x \approx 4$ mm).

Then

$$\frac{dW}{dV}=\frac{Fdx}{Sdx}=\frac{F}{S}=\frac{\mu_0 (Iw)^2}{2x^2}. \tag{7}$$

Thus, the winding of the DC electromagnet coil should provide MMF

$$Iw = x\sqrt{\frac{2F}{\mu_0 S}}. \tag{8}$$

We will calculate the electromagnet coil voltage that is powered from a constant voltage source at nominal voltage $U = 12$ V (the minimum voltage at which it can work $U_{\min}=$ 10.8 V); the design of such a coil is shownbelow (Fig. 5).

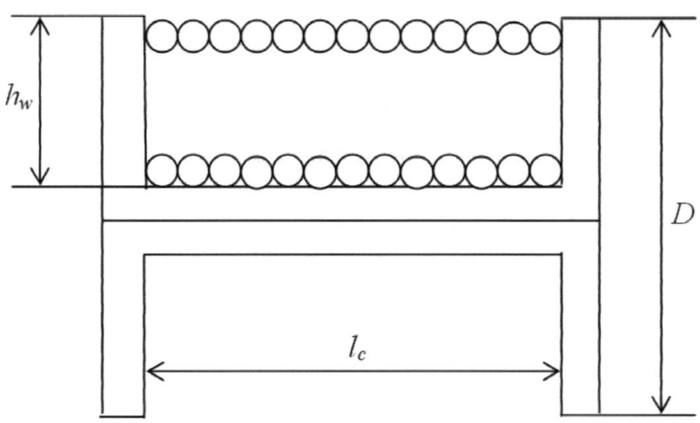

Figure 5 - Electromagnet coil in section:
h_w - winding thickness; D - coil diameter; l_c - coil length

Its geometrical dimensions are determined taking into account the PTO construction and correspond to the following values: h_w= 10 mm; $D = 40$ mm; l_c = 45 mm. In this case, the cross-sectional core area

$S=\pi\frac{(D-2h)^2}{4}=$ 3.14·10^{-4}m^2, and the required MMF according to (7) will be equal to

$$Iw = 4\cdot 10^{-3}\sqrt{\frac{2\cdot 258.4}{4\cdot 3.14^2\cdot 10^{-11}}}\approx 4576A$$

We calculate the average coil length (Fig. 5) $l_a = 2\pi r_a = \pi(D-h_н) = 3.14(40-10) = 94.24mm$

Further, from Ohm's law $I = \frac{U_{\min}}{R_c}$, where $R_c = \rho \frac{l_a}{q_{cs}} w$ – the active coil resistance, and $\rho = =$ $1.8 \cdot 10^{-11}$ Ohm · m – the copper resistivity, we find the wire cross section

$$q_{cs} = \frac{Iw\rho l_a}{U_{\min}} = \frac{4576 \cdot 1.8 \cdot 10^{-11} \cdot 94.25}{10.8} = 0.719\,\text{mm}^2$$

From where the wire diameter $d_w = 2\sqrt{q_{cs} / \pi} = 0.957$ mm.

The closest copper winding wire produced to this section has a copper cross section $q_{cs} = 0.712$ мм^2 and a nominal copper diameter $d_w = 0.95$ mm. The wire of the PEV-1 with such a cross section has a wire diameter taking into account the insulation $d_1 =$ 1.01 mm, the maximum allowable temperature for continuous operation $\tau = 105$ ° C, which corresponds to class Y of the heat resistance of insulating materials [9].

With the stacking factor $k_s = 0.85$, the filling factor of the winding space with copper will be equal to $k_{fs} = k_s \frac{q_{cs}}{d_1^2} = 0.599$.

We find the number of windings w. By definition, the fill factor of the winding space with copper

$$k_{fs} = \frac{q_{cs}}{h_w \cdot l_c} w, \text{ from which } w = k_{fs} \frac{h_w \cdot l_c}{q_{cs}} = 375.$$

Next, we find the current value: $I = Iw/w =$ 12.2 A, the coil active resistance $R_c = \frac{U_{\min}}{I} = 0.885$ Ohm, the power consumption $P_c = I^2 R_c =$131.8 W and if the heat transfer coefficient is k_{ht} = (8–14) W/m^2, according to the Newton formula, we calculate the excess coil temperature over the external environment

$$\Delta\tau = \frac{P_c}{k_{ht} S_{cool}} = \frac{P_c}{k_{ht} \pi D (l_c + D/4)} = 1702.67°\text{C}.$$

According to the results of calculations, an electromagnet of this design does not satisfy the requirements for the thermal resistance of wire insulation. Itisnecessarytochange the electromagnet core (Figure 6):

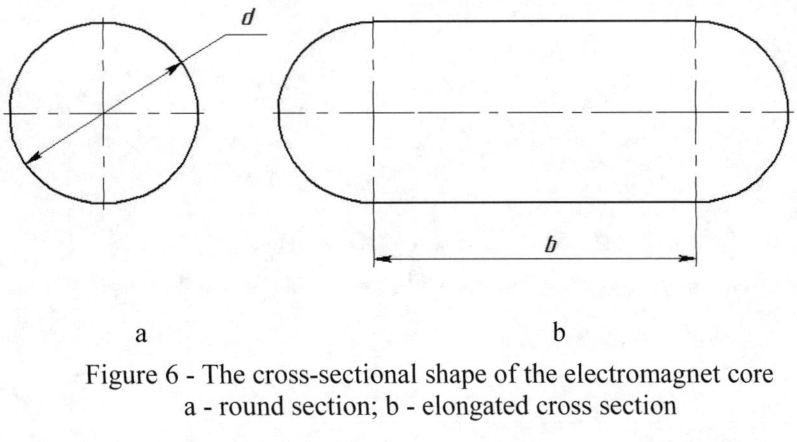

Figure 6 - The cross-sectional shape of the electromagnet core
a - round section; b - elongated cross section

With the dimensions of the transverse core rectangular part of the elongated shape *d*x2*d*, the coil temperature Δτ will exceed the temperature of 200° C, which corresponds to the class C of the insulation heat resistance [9–10].

IOP Conf. Series: Materials Science and Engineering **582** (2019) 012043 doi:10.1088/1757-899X/582/1/012043

4. Conclusions

To increase the braking efficiency of the drum of the planetary power take-off mechanism can be due to the application of force on the tape by electromagnets at four points, instead of one in the previous design. It can be used as a finished electromagnets, specially made according to the results of our calculations. The proposed design makes it possible to include electromagnets not simultaneously, but with some time offset, with a gradual pressing of the tape, starting with the portion of the tape closest to the axis of its suspension. The first is the upper left (see Fig. 4) electromagnet, latest – top-right. This improves the fit of the tape to the drum inclusion. In the transition from a mechanical device the drive enable to the electric device eliminates the operation of regulation, which reduces maintenance time and significantly improves the durability of the brake band.

References

[1]Ksenevich, I.P. Tractors of MTP-80 and MTP-82 / I.P. Ksenevich, S.L., Kustanovich, P.N. Stepanyuk and others; Under the general editorship of I.P. Ksenevich. - 2nd ed., revised and added. - M.: Kolos, 1984. - 254 p.

[2]Ivanov, M.N. Machine parts / M.N. Ivanov, V.A. Finogenov. - 11th ed., revised - M.: Higher school, 2007. - 408 p.

[3]Planetary gear. The directory/Under the editorship of V.N. Kudryavtsev and Yu. Kardasheva. - L.: Mechanical Engineering, 1977. - 536 p.

[4]Alexandrov, M.P. Brakes hoisting machines / M.P. Alexandrov; 3rd ed., added and revised - M .: Mechanical Engineering, 1976. - 383

[5]Regulatory reference material for the planning of mechanized operations in agricultural production: Collection. — Moscow: FGNU «Rosinformagrotekh», 2008. - 316 p.

[6]Alexandrov M.P. Lifting and transport machines. - 5th ed., revised and added – M.: Higherschool, 1979. - 558 p.

[7]Pat. №185594 Russian Federation. IPC B04K 17/28. Power take-off shaft / G.V. Redreev and others; patent holder of FSBEI HE «Omsk State Agrarian University. P.A. Stolypin». - №2018120959; declared. 06.06.2018; published 11.12.2018, bull. No. 35. – 2.

[8]Zilberman, G.E. Electricity and magnetism / G.E. Zilberman – M .: Science, 1970. – 384 p.

[9]Ogonykov, V.G., EvtushenkoYu.M. Electrical insulation materials and insulation systems for electrical machines. In 2 books. / V.G. Ogonykov, Yu.M..Evtushenko – M .: Publishing House MEI, 2012. – 304 p. (Book 2).

[10] Asanovich E. Z., V. A. Kolganova Highly heat-resistant electrical insulation. – M.: Energy atom published, 1988. – 264 s.

IOP Conf. Series: Materials Science and Engineering **582** (2019) 012044 doi:10.1088/1757-899X/582/1/012044

The Classification of Methods of Defining of Calculated Mechanical and Physical Characteristics of Composite Materials

V I Kucheryuk, O L Umanskaya [a], N A Krivchun

Tyumen Industrial University, 625000, Russia, Tyumen, Volodarskogo Street, 38

E-mail: [a] olgaumanskaya@yandex.ru

Abstract. A short description with the analysis of the choice of the equivalent characteristics of the material with the further classification of the casting method is represented in this article. The choice and definition of elastic characteristics of composite materials depend on a number of factors: they are the geometry of the estimated element of the construction, the stress-strained state and the mathematical model of this state, the calculation method, the final precision of the solution. The technical methods of calculation of such elements as rods, shells and plates are analytically connected with the resolving equation set-up. One of the most widely used approaches to the solution of practical tasks is to transform heterogeneous materials to homogeneous ones, either isotropic or orthotropic. The most common variants of the choice of physical and mechanical characteristics of the composite material are discussed in this article. One limitation is introduced, that is the work in the elastic strain range and the fulfillment of the conditions of the strain compatibility of the reinforced matrix. These limitations are in accordance with the security of the work of the constructions. A high-quality formulae analysis depending on the final task is given in the article. The given formulae analysis for defining the effectiveness of physical and mathematical characteristics of composite rods, plates and shells enables to make calculations for more complicated tasks. A structural scheme of methods of defining of characteristics of composite materials is given.

1. Introduction

Composite materials are a heterogeneous system. The material consists of two or more components, which possess different mechanical and physical properties. The distinguishing feature of these materials from homogeneous ones is the ability to manipulate their qualities while creating them.

The creation of new composite materials is on the increase. The spheres of science and production, which use composite materials are developing in the most dynamic way. A detailed classification of these materials is necessary to make the right choice of the composite material for every particular case.

It is known from the mechanics of deformed elastic systems that their statics and dynamics can be described as a system in the matrix form:

$$A\frac{\partial^2 u}{\partial t^2} + B\frac{\partial u}{\partial t} + Cu + F = 0, \tag{1}$$

where, u – is a displacement vector, A–a mass matrix, B- a deformation matrix, C – a rigidity matrix, F– an external load vector. One of the types of calculations for composite systems is turning a heterogeneous system into a homogeneous one, which is described by the system (1).

The calculation of constructions made from composite materials using both analytical and numerical methods is connected with the defining of the physical and mechanical characteristics of the material.

The choice and definition of elastic characteristics of composite materials depend on a number of factors: they are the geometry of the estimated element of the construction, the stress-strained state and the mathematical model of this state, the calculation method, the final precision of the solution.

IOP Conf. Series: Materials Science and Engineering **582** (2019) 012044 doi:10.1088/1757-899X/582/1/012044

2. Materials and Method

The variants of the structure of composite materials based on the matrix type and the reinforcement type are given in the following reference materials [1,2]. Laminates, disperse-reinforced and fibrous composite materials are most widely used. The disperse reinforcement of the matrix can be achieved with the help of the powder which is represented as macroscopic and microscopic particles. The main advantage of fibrous structures is their ability to produce fibers as monocrystals, which possess high strength. Composite materials, which have the gradient distribution of the powder according to the given law constitute disperse-reinforced composite materials. For example, cermet is a structure, which represents the reciprocal penetration of one material into another one.
The reinforcement of the matrix of the dimensional shape (figure 1) provides the dimensional rigidity of the composite material. The matrix material additionally enlarges the stability of the compression elements of the matrix. Plate-rod reinforcement allows to optimize the distribution of the material for plates and shells working in the conditions of simple and combined stress.

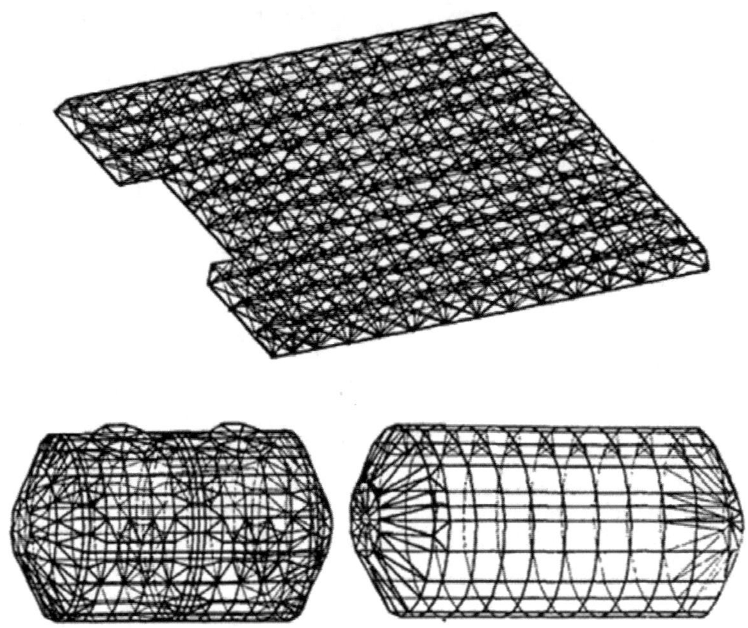

Figure 1. Dimensional reinforcement of composite materials

A considerable number of variants of turning a heterogeneous system to the homogeneous one can be represented in the structural scheme of the calculated model (figure 2).

Averaging elastic constants of the material. In 1887 Voigt introduced the idea of averaging by volume taking into consideration the homogeneity of the generalized displacement, and in 1929 Reiss introduced the idea of averaging according to the homogeneity of the united strains area [1].

Nemirovskiy Y.V. and Reznikov B.S. [3] applied a structural approach while solving the task of the mechanics of deformable solids in order to define the elasticity component and temperature rigidity of the material. Tensor components are expressed with the help of mechanical characteristics of the elements of the composition, the reinforcement structure and other macroscopic parameters. It is postulated that longitudinal forces are accepted by reinforcing fibers and shear forces are accepted by the matrix. The strains in the matrix and reinforcement elements follow Duhamel-Neumann's law (subject to postulation):

$$\sigma_{c}^{\alpha\beta} = \frac{E_c}{1-\nu_c^3}(e_{\alpha\alpha} + \nu_c e_{\beta\beta}) - \frac{a_c E_c}{1-\nu_c}\theta, \quad \alpha \neq \beta, \tag{2}$$

$$\sigma_{c}^{12} = \frac{E_c}{1+\nu_c} e_{12}, \quad \sigma_c^{\alpha 3} = 2G_c e_{\alpha 3}^{c}, \quad \sigma_{ak} = E_{ak} e_{ak} - a_{ak} E_{ak} \theta, \tag{3}$$

where σ, e, E, G, ν– strain, deformation, Young's modulus, shear modulus, Poisson's ratio; *c,a* – indices for the bonding material (matrix) and reinforcing elements; k – a set of fibers; 1,2,3 – Cartesian coordinates; α, β – Gaussian coordinates; θ– temperature changes.

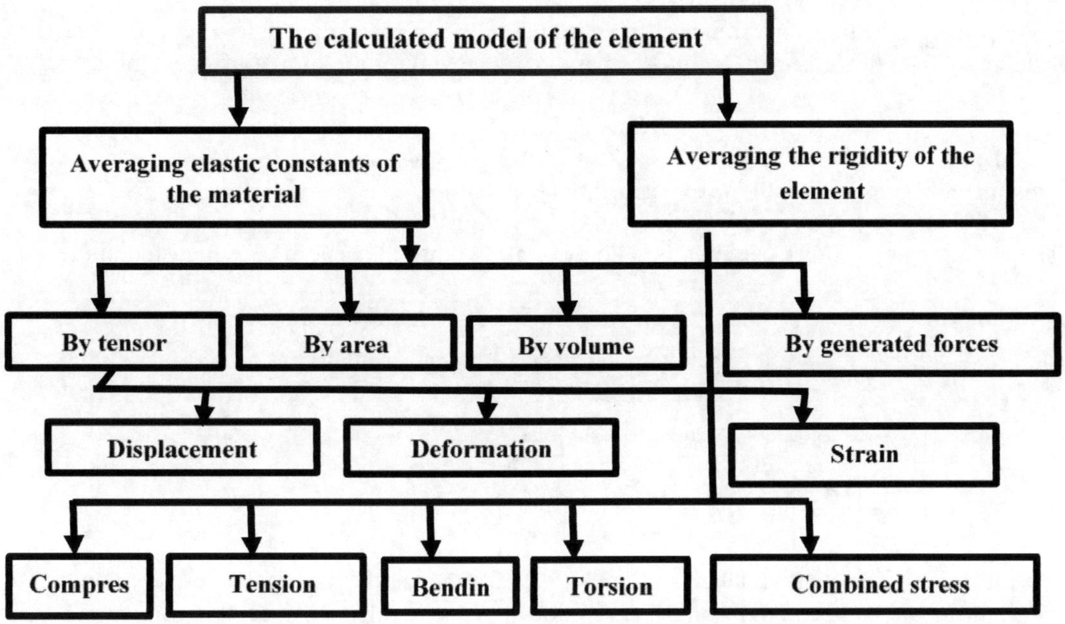

Figure 2. The structural scheme of the calculated element

In micromechanics the defining of the effective elasticity modules is achieved with the help of averaging by volume the values of components of the tensors of strains and deformations under definite boundary conditions [4]. Effective modules C_{ijkl} are found from the equations of the generalized Hooke's law which connect strains σ and linear elastic deformations ε:

$$\sigma_{ij} = C_{ijki} \cdot \varepsilon_{ki} \tag{4}$$

The simplest variant of the macromechanical approach is averaging by the cross-section area A or by volume V:

$$E_{\text{К}} = \frac{E_{\text{М}} A_{\text{М}} + E_{\text{В}} A_{\text{В}}}{A_{\text{М}} + A_{\text{В}}}, \tag{5}$$

where indices «М» and «В» correspond to the matrix and the reinforcing fiber of the composite material. In [5] the formulas of the longitudinal elasticity for the unidirectional continuous fiber composite (orthogonal directions -axes X, Y) are given:

$$E_{XK} = E_{\text{В}} V_{\text{В}} + E_{\text{М}}(1 - V_{\text{В}}), \tag{6}$$

$$E_{YK} = \frac{E_{\text{В}} E_{\text{М}}}{E_{\text{М}} V'_{\text{В}} + E_{\text{В}}(1 - V'_{\text{В}})}, \tag{7}$$

where $V_{\text{В}}$, $V'_{\text{В}}$ - is the ratio of the fiber cross-section areas to the composite and the ratio of the fiber length to the length the matrix of the studied element.

We got the formula (6) from the condition of the strain compatibility, Hooke's law and the representation of the longitudinal force as a sum of strains in the matrix and fibers. We got the formula (7) when studying the singled out element from the force equation in the fiber and the matrix, the implementation of Hooke's law, the representation of the relative deformation of the element as a sum of deformations of the fiber and the matrix.

In the works [6,7] they took into account the position of fibers at the angle α (for unidirectional reinforcement, $\alpha = 1$; for orthogonal reinforcement, $\alpha = 0.5$ and for random reinforcement $\alpha \cong 3/8$):

$$E_{\text{К}} = \alpha E_{\text{В}} V_{\text{В}} + E_{\text{М}} V_{\text{М}} \tag{8}$$

This formula goes well with the experiment [1].

The simultaneous averaging by strains and deformations is described in the work [8].

The averaging by rigidity parameters with the plates and shells bending is suggested in the work [1].

For this variant the cylindrical rigidity which we obtained by singling out a beam-strip looks as

$$\begin{aligned} D_1^{\text{пр}} &= \frac{E_M^c a}{1-\nu_M^2}\frac{z_0^3}{6} + \frac{E_M^p a}{1-\nu_M^2}\left(\frac{h^3}{3} + \frac{z_0^3}{6} - \frac{z_0 h^2}{2}\right) + \\ &\frac{a\pi d^2}{4t_B}\left\{\left[\left(z_0 - \frac{d}{2}\right)^2 + \left(z_0 - k_B - \frac{3d}{2}\right)^2\right]\left[\frac{E_B}{1-\nu_B^2} - \frac{E_M^c}{1-\nu_M^2}\right] + \right. \\ &\left. + \left[\left(h - \frac{3d}{2} - k_B - z_0\right)^2 + \left(h - \frac{d}{2} - z_0\right)^2\right]\left[\frac{E_B}{1-\nu_B^2} - \frac{E_M^p}{1-\nu_M^2}\right]\right\}, \end{aligned} \tag{9}$$

where E_M^p, E_M^c, E_B – are longitudinal elasticity modules for the matrix (a stretched area or compressed areas for multimodulus materials) and fibers ; $\nu_{\text{M}}, \nu_{\text{B}}$ – are coefficients of the transversal strain deformation of the matrix and the fibers; h – the thickness of the plate (the shell) a – the width of the beam-strip; d – the diameter of the fibers; t_B, k_B– the distance of the fibers in the plate surface and in the thickness correspondingly ; z_0– the position of the neutral axis of the cross-section of the beam-strip with the equivalent material which is found from the condition of equaling zero relatively the neutral axis of the statical moment of the equivalent area of the cross-section.

Having singled out a beam-strip in the orthogonal direction, we get the cylindrical rigidity in the same way $D_2^{\text{пр}}$.

Studying the torque of the beam-strip we get the rigidity which takes into consideration the torque:

$$D_3^{\text{пр}} = G_M \frac{ah^3}{12} + \frac{a\pi d^2}{2t_B}\left[\left(\frac{h}{2} - \frac{d}{2}\right)^2 + \left(\frac{h}{2} - k_B - \frac{3}{2}d\right)^2\right](G_B - G_M), \tag{10}$$

where G_M, G_B– are shear modules of the matrix and the fibers.

If necessary the equivalent mechanical characteristics can be defined using the known formulas for the equivalent orthotropic plates:

$$D_1^{\text{пр}} = \frac{E_1^{\text{пр}} h^3}{12(1-\nu_1^{\text{пр}}\nu_2^{\text{пр}})}, D_2^{\text{пр}} = \frac{E_2^{\text{пр}} h^3}{12(1-\nu_1^{\text{пр}}\nu_2^{\text{пр}})}, \ D_3^{\text{пр}} = \frac{G^{\text{пр}} h^3}{12}, D_1^{\text{пр}}\text{E}_2^{\text{пр}} = D_2^{\text{пр}}\text{E}_1^{\text{пр}}, \tag{11}$$

$$D_1^{\text{пр}}\nu_2^{\text{пр}} = D_2^{\text{пр}}\nu_1^{\text{пр}}.$$

It is suggested to integrate the averaging by rigidity parameters with the multilayer plates and shells bending over the thickness of the packet [9,10].

In the monograph[11] G.L. Gorynin and Y.V.Neirovskiy when studying bending and torque of laminated rods and plates base on averaging the displacement dots along the cross-section area F:

$$u(z) = \frac{1}{F}\sum_i \int_{F_i} u_i dF. \tag{12}$$

Averaged constants depending on the position of the *x*-layer and its elastic modulus E_i are included into the resolving equations:

$$c_0 = \sum_i \int_{F_i} x E_i dF / \sum_i E_i dF. \tag{13}$$

After we have chosen the model of the material, it is advised to choose the casting method of a heterogeneous system to a homogeneous one in order to calculate the elements of a composite system. An analytical approach to solving the task (1) is reduced to the preliminary defining of equivalent matrixes A, B, C through elastic constants of the components of the composite system.

3. Results and Discussion

A structural scheme of defining methods of physical and mechanical characteristics of the material depending on the different qualities of the material is shown in figure 3.

Calculation methods of composite constructions and the corresponding formulas are described in details in the works listed in the reference list of the given article.

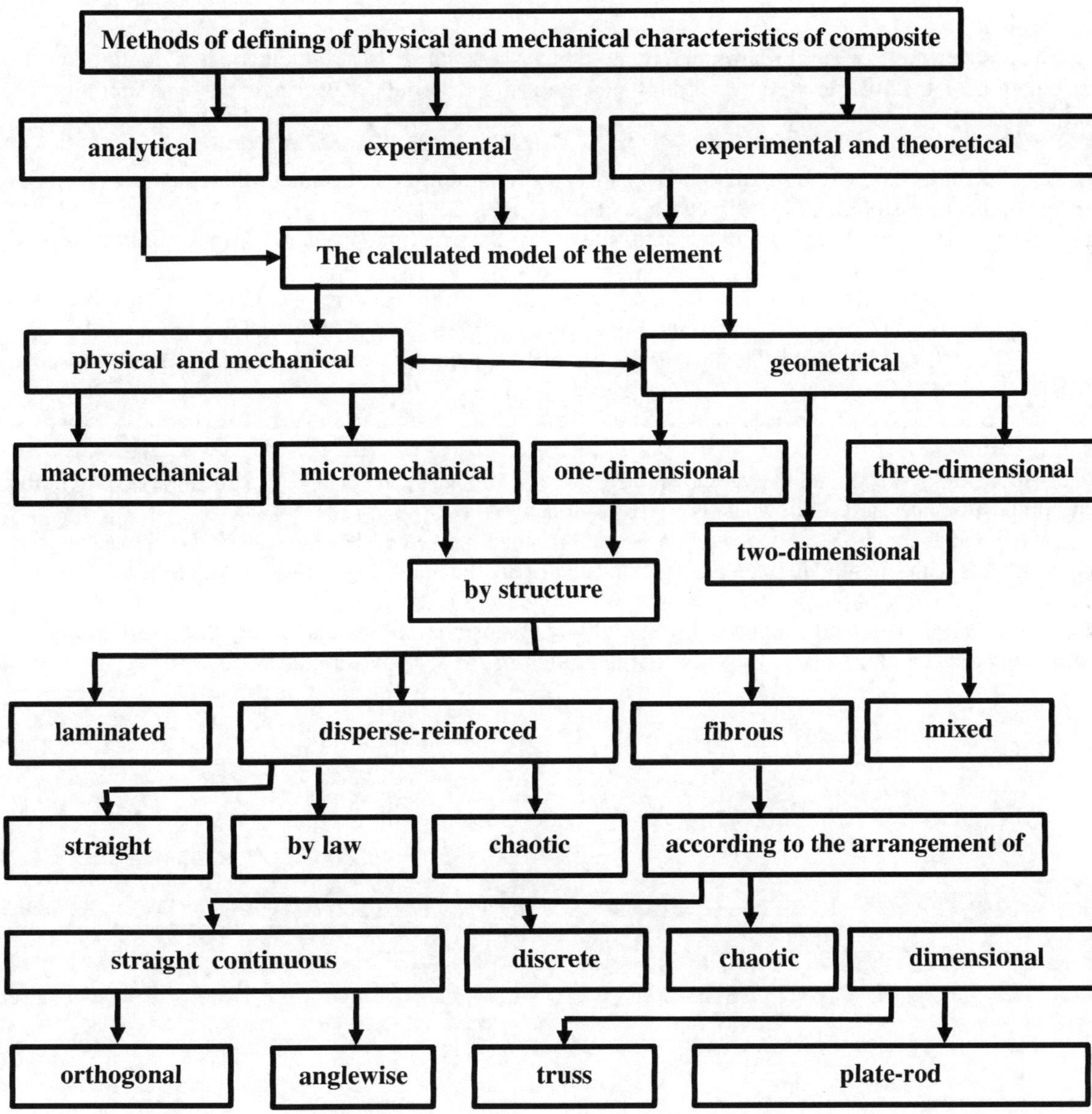

Figure 3. A structural scheme of methods of defining physical and mechanical characteristics of composite materials

IOP Conf. Series: Materials Science and Engineering **582** (2019) 012044 doi:10.1088/1757-899X/582/1/012044

4. Conclusion

According to many researchers the experimental check of the suggested methodologies and formulas is impeded by the fact that as being different from homogeneous materials, the heterogeneity of composite materials and especially their manufacturing technology makes the results of strain and deformation calculations different. Besides, multimodulus qualities of materials from which they are produced influence greatly the calculations of composite materials. Depending on the aim, it is essential to have a high-quality analysis of formulas. Averaging by volume is well-suited for the central compression or stretching of the elements for disperse-reinforced materials with the homogeneous distribution of the reinforcement. For bending elements formulas [1] are more suitable which take into consideration the position of the reinforcing fibers, multimodulus qualities and the position of the neutral axis and the plane. The given analysis of formulas defining the effectiveness of physical and mechanical properties of composite rods, plates and shells enables to make calculations for more complex problems

References

[1] Kycheryuk V and Krivchun N and Umanskaya O 2018 Mathematical modeling of the calculation of the stiffiness characteristics of composite materials. Key Engineering Materials 771, p.82-87

[2] Pobedrya B 1984 The mechanics of composite materials (Moscow: MGU)

[3] Nemirovskiy Y and Reznikov B 1986 The strength of the elements of the constructions made from composite materials (Novosibirsk: Nauka)

[4] Bataev A and Bataev V 2006 Composite materials: structure, production, application (Moscow: Universitetskaya kniga)

[5] Sendetski G 1978 The mechanics of composite materials under the editorship (Moscow: Mir)

[6] Karpinos D Composite materials. Encyclopaedia 1986 (Kiev: Naukova Dumka)

[7] Fudzi T and Dzako M The mechanics of the destruction of composite materials 1982 (Moscow: Mir)

[8] Bolotin V and Novichkov Y 1980 The mechanics of multilayer constructions (Moscow: Mashinostroenie)

[9] Yakubovskiy Y and Gulyaev B and Kolosov V and Krivchun N and Yakubovskaya S Bending of composite plates and shallow shells 2016 (Tyumen: TIU)

[10] Yakubovskiy Y and Krivchun N and Umanskaya O and Kolosov V 2016 The accounting for materials with strength differential effect in calculating the durability in order to raise the exploitation effectiveness. IOP Conference Series Materials Science and Engineering

[11] Gorynin G and Nemirovskiy Y 2004 Dimensional problems of bending and torque of laminated constructions.The asymptotic disintegration method (Novosibirsk: Nauka)

IOP Conf. Series: Materials Science and Engineering **582** (2019) 012045 doi:10.1088/1757-899X/582/1/012045

Linear Model of Information System of Industrial Processes in Crop Production

G V Redreev, P V Kiyko, E V Demchuk, N V Shchukina

Omsk State Agrarian University named after P. A. Stolypin
644008, г. Omsk, Institutskaya square st., 1

E-mail: weerwg@mail.ru

Abstract. the timeliness of information received during the implementation of any production process, as well as its qualitative composition, significantly affect the efficiency of this process. This is particularly important for agricultural production, as field operations are highly time-sensitive, both in terms of start-up and duration. The real possibility of a significant increase in the efficiency of production processes in crop production appeared at the current level of information technology and the spread of digitalization of various fields of activity. On the basis of existing methodological approaches to the system representation of objects of reality, we have developed an information system of the production process in crop production. The elements of the system are the amount of knowledge of specialists in the cultivation of crops, information about the composition of cultivation technologies, information about the objects processed in the implementation of production processes, recorded in any form replenished special knowledge, knowledge about the goals achieved in the implementation of production processes. Elements of the system are interconnected, developing in a certain way and, at the same time, limiting each other. The formalization of the relationships between elements of the information system allowed us to obtain ten equations with fifteen unknowns. To solve this system of equations, a set of simultaneous linear equations is presented. Five sets of simultaneous equations can be defined, depending on which of the relationships of the elements of the information system of the type f_i will be broken. The obtained solutions will allow to quantify the volume of elements of the information system, which makes it very promising to use information technology management of production processes in crop production.

1. Introduction

The efficiency of any production process depends on the timely receipt of the information and its qualitative composition [1-3]. This is especially true for production processes in crop production, where the time factor is especially important. At the same time, in addition to the operational information which is necessary for the current correction of the field operations, the knowledge accumulated during the implementation of production processes in previous years is needed. This can be facilitated by fixing the personal experience of agricultural production specialists. At the present level of information technology development, the positive effect of the use of previously accumulated knowledge can be very significant.

2. Results

On the basis of the conceptual design of the technical service system, taking into account the methodological recommendations [4], we have developed an information system that adequately describes the production processes [5–6]. For the crop industry, the information system of the production process is shown in Figure 1.

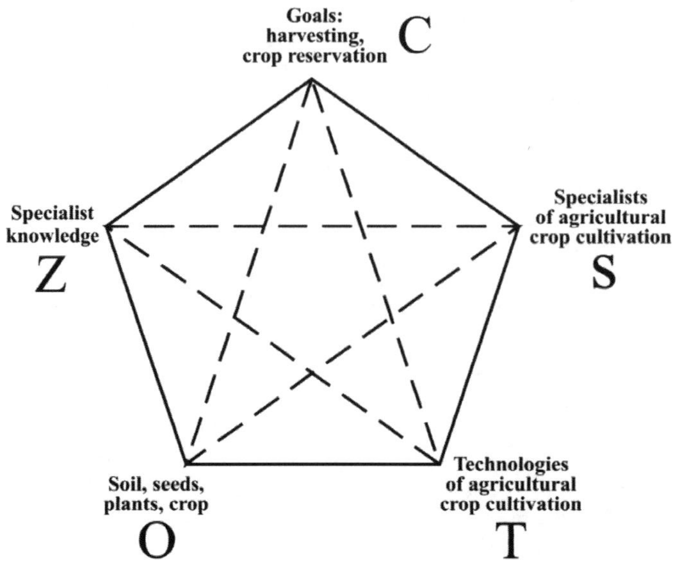

Figure 1 – Information system of the production process in crop production

The elements of the system are:
– the amount of knowledge of specialists in the cultivation of crops, hereinafter referred to as «Specialists (S)»,
– the information on the types of crop technologies, hereinafter referred to as «Technologies (T)»,
– the information about the agricultural objects processed during the implementation of production processes - soil, seeds, plants, crops, hereinafter referred to as «Objects (O)»,
– the special knowledge recorded in any (paper, electronic) form, hereinafter referred to as «Knowledge (Z)»,
- the knowledge of the goals achieved in the implementation of production processes, hereinafter referred to as «Goals (C)».

We consider the relationship between elements of the information system, taking into account that the main element of the system are «Specialists (S)».

1. Specialists (S) determine which technology will be selected (T).
2. The implemented technologies (T) affect the objects (O).
3. The information about the objects (O), changing under the influence of the technology (T), adds the knowledge (Z).
4. The knowledge (Z) forms and develops concepts about the goal (C), which is possible to achieve in the future.
5. The developing goals (C) expand the orientation of specialists (S) when choosing the technologies (T).

The formalization of these relations can be expressed as a combination of (1) dependencies:

$$\begin{cases} T = f_1(S) \\ O = f_2(T) \\ Z = f_3(O) \\ C = f_4(Z) \\ S = f_5(C) \end{cases} \tag{1}$$

In turn:
6. Specialists (S) limit which exposure objects (O) will be selected..
7. While changing, the objects (O) limit the goal achievement (C).
8. The goals (C) limit the choice of technology (T).

9. The technologies (T) unilaterally increase the knowledge (Z).
10. The knowledge (Z) orients specialists (S) in a certain direction.

$$\begin{cases} O = \varphi_1(S) \\ C = \varphi_2(O) \\ T = \varphi_3(C) \\ Z = \varphi_4(T) \\ S = \varphi_5(Z) \end{cases} \tag{2}$$

The dependencies of both sets are a mathematical model of the information system of the production process in the crop production.

Ten dependencies contain 15 unknowns: five elements of the information system, five functionals of the type f_i, five functionals of the typeφ_i.Determining the impact of the system elements on each other, you can describe the existing relationships by a set of simultaneous equations. Then the dependences of the type f_i, highlighted in Figure 1 by solid lines, are represented as the following system of simultaneous equations:

$$\begin{cases} T = a_1 + b_1 \cdot S \\ O = a_2 + b_2 \cdot T \\ Z = a_3 + b_3 \cdot O \\ C = a_4 + b_4 \cdot Z \\ S = a_5 + b_5 \cdot C \end{cases} \tag{3}$$

And the type dependencies highlighted in Figure 1 with dashed lines are described by the following system of simultaneous equations:

$$\begin{cases} O_t = \alpha_1 + \beta_1 \cdot S_t + \gamma_1 \cdot T_t \\ C_t = \alpha_2 + \beta_2 \cdot O_t + \gamma_2 \cdot Z_t + \gamma_2' \cdot Z_{t+1} \\ T_t = \alpha_3 + \beta_3 \cdot C_t + \gamma_3 \cdot S_t \\ Z_t = \alpha_4 + \beta_4 \cdot T_t + \gamma_4 \cdot O_t \\ S_t = \alpha_5 + \beta_5 \cdot Z_t + \beta_5' \cdot Z_{t-1} + \gamma_5 \cdot C_t \end{cases} \tag{4}$$

Each system consists only of endogenous (dependent) variables. Specialists in the methodology and theory of science point out to the circularity of the system relationships. In other words, we get a system consisting of identities.

For our case:
-S-T-O-Z-C- etc.,
-S-O-C-T-Z- etc.

3. Discussion

Instead of exogenous variables, some models may use lagged variables, that is, endogenous variables obtained in the previous period. Thus, the goals can depend not only on the knowledge (Z_t) available at a given time, but also on the knowledge of the previous years (old experience) (Z_{t-1}), and can also be predetermined by knowledge (Z_{t+1})which will be obtained later.It is rational to choose variables that can be regulated as exogenous variables. Now, following the proposed project of the information system, interpreting each equation of the system from different points of the pentagram, you can get the solution of the whole system [7 – 8].

Choosing the technology as an exogenous (independent) variable, we break the ring of interconnections of the system elements and obtain a system of simultaneous equations in which the first equation specialists limit the objects and affect the technology. In the second equation, the objects limit the goals and affect the knowledge that is available at a given time (Z_t), and the knowledge that will only be obtained (Z_{t+1}). In the third equation, the goals limit the technology and influence the specialists; in the fourth equation, the technology limits knowledge and affects the objects. And in the last equation we see that the knowledge of previous years (old experience) (Z_{t-1})as well as the real knowledge limits the specialists and affects the goals. Now we examine our system for identification to determine the method by which this system can be solved. Identifiability is uniqueness correspondences between the reduced and structural forms of the model [7, 9].

From the standpoint of identifiability, the structural models can be divided into three types: identifiable, unidentifiable, over-identifiable.

Necessary condition for identifiability:
D + 1 = H – the equation is identifiable;
D + 1 < H – the equation is unidentifiable;
D + 1 > H – the equation is over-identifiable,
where H– the number of endogenous variables in the equation,
D– the number of exogenous variables that are contained in the system, but are not included in this equation.

Sufficient condition of identifiability:
the equation is identifiable if it is possible to obtain a matrix from the coefficients of the missing variables in other equations of the system, the determinant of which is not zero, and the rank of the matrix is not less than the number of endogenous variables in the system without one.

A model is considered identifiable if each equation of the system is identifiable. If at least one of the equations is unidentifiable, then the whole model is considered unidentifiable.

If the model is accurately identifiable, then we use the indirect least squares method, if over-identifiable, we use the two-step least squares method.

According to the necessary condition of identifiability we obtain:

Endogenous variables – O_t, C_t, Z_t, S_t;
Exogenous variables – T_t, Z_{t-1}, Z_{t+1}.
First equation: $O_t = \alpha_1 + \beta_1 \cdot S_t + \gamma_1 \cdot T_t$
$H = 2(O_t, S_t) D = 2(Z_{t-1}, Z_{t+1})$obtain$2 + 1 > 2$; the equation is over-identifiable.
Second equation: $C_t = \alpha_2 + \beta_2 \cdot O_t + \gamma_2 \cdot Z_t + \gamma_2' \cdot Z_{t+1}$
$H = 3(C_t, O_t, Z_t) D = 2(T_t, Z_{t-1})$obtain$2 + 1 = 3$; the equation is identifiable.
Third equation: $T_t = \alpha_3 + \beta_3 \cdot C_t + \gamma_3 \cdot S_t$
$H = 2(C_t, S_t) D = 2(Z_{t-1}, Z_{t+1})$obtain$2 + 1 > 2$; the equation is over-identifiable.
Fourth equation: $Z_t = \alpha_4 + \beta_4 \cdot T_t + \gamma_4 \cdot O_t$
$H = 2(Z_t, O_t) D = 2(Z_{t-1}, Z_{t+1})$obtain$2 + 1 > 2$; the equation is over-identifiable.
Fifth equation: $S_t = \alpha_5 + \beta_5 \cdot Z_t + \beta_5' \cdot Z_{t-1} + \gamma_5 \cdot C_t$
$H = 3(S_t, Z_t, C_t) D = 2(T_t, Z_{t+1})$obtain$2 + 1 = 3$; the equation is identifiable.

Since, in addition to identifiable equations, we also have over-identifiable equations, the whole model will be over-identifiable. Consequently, our system can be solved by the two-step least squares method [7, 9 - 11].

Thus, we have considered only the case when lagged variables affect endogenous variables, and since the ring can be «broken at any place», then any element of the system can be chosen as an exogenous variable, thereby obtaining five independent models which are easily amenable to analytical and empirical solution using application program packages, for example, «STATISTICA», «EViews», MS Excel [12-14].

The functional structure of the entire information system of the production process in crop production consists of a set of interrelated functional blocks that implement a set of management functions for all processes. It is customary to distinguish between informal and formal management systems. The informal management system is based largely on non-formalized procedures and decision-making methods, which, in turn, are divided into determinant and non-determinant. A formal management system, by contrast, is based on formalized procedures and decision-making methods.

The determinant methods are used when all the conditions of the problem situation are known, i.e. worthwhile the goals have no uncertainties. These methods are divided into direct, answering the question:

«What crop can be obtained if highly qualified specialists will use modern technologies of agricultural crop cultivation, with a large amount of knowledge?»

IOP Conf. Series: Materials Science and Engineering **582** (2019) 012045 doi:10.1088/1757-899X/582/1/012045

And the reverse answer to the question:

«What kind of knowledge of specialists is needed, what modern technologies need to be applied, what objects do you need to choose (soil, seeds, plants), what specialists to choose in order for the goal to reach its extreme (minimum or maximum) value?»

Using the economic terms and mathematical models, we see that direct methods allow us to determine the performance indicator W when choosing one of the remaining four of the above basic concepts, and the inverse methods allow us to refine or supplement the concepts with information in which the indicator W will take its extreme (minimum or maximum)) value [3,4].

Among the inverse methods there are:

– linear methods, when the dependence $W = W(x, k)$has the form of a linear function, where k is the set of elements of the information system of the production process in crop production; – nonlinear methods when the dependence $W = W(x, k)$ has the form of a nonlinear function;

– brute force used in those cases, the information system contains a small number of concepts, and for each solution$x_i$of the whole set is an indicator of the effectiveness of W, and then of all the optimal (minimum or maximum), depending on the formulation of the problem.

But, as practice shows, in most cases, the information system, in addition to the known elements k and solutions x may be some uncertainties (unknown, or require additions and clarification of concepts). To solve such problems, non-determinant methods of substantiation of solutions are used, which can be classified into: stochastic, adaptive, expert evaluation methods.

When using non-determinant methods, it is assumed that the efficiency criterion cannot be calculated because it depends on unknown parameters. Thus, it can be concluded that with the help of non-determinant methods the optimal solution cannot be obtained, and it is impossible to determine the element of the system that allows to find a solution close to the optimal one. We conclude that these methods in practice do not give good answers to the questions.

We will not exclude the use of non-determinant methods, because in practice their use is still quite wide – from bad decisions you can always choose the best. The elementary case of application of such methods is the probabilistic nature of unidentified elements$k$, which are random variables whose numerical characteristics are unknown. In this case, the numerical characteristics are understood as the expectation, variance and distribution law. Here we use stochastic decision support methods which are subdivided into stochastic methods without constraints and stochastic methods with constraints. The most simple are stochastic methods without restrictions, which are used in cases where the situation allows to replace random variables with their mathematical expectations [1,3,5].

For example, if it is necessary to determine the minimum number of specialists in the cultivation of crops, the mathematical expectation of technologies for the cultivation of crops – for example, the number of modern equipment for sowing grain and harvesting-can be considered as a criterion of efficiency. Let it be 50 units, then, from the point of view of the goal of obtaining a total crop, nothing terrible will happen if one field will work 10 units, and the other 40. And a completely different situation can arise when different fields include different soil and landscape conditions, different varieties of grain, different sowing time, then as a criterion of efficiency to consider the expectation of 10 and 40 units will not be enough, a small number of units of equipment for one field can not compensate for the large downtime of equipment for another. In such cases, a limit is usually introduced, i.e., with a mathematical expectation of 20 pieces of equipment. And the methods solving such problems are called stochastic methods with restrictions. We must not forget that stochastic methods are used only in repetitive situations. The adaptive models can provide some assistance in case of non-random uncertainty [3,5].

Having considered determinant and non-determinant decision-making methods, we focus on the linear function of inverse determinant methods. To obtain the maximum yield in the presence of specialists, technologies, objects and knowledge, we will make a mathematical model:

Table 1 Result maximization model

	Product 1	Product 2	Product	Product 4	Product 5	
Specialists	a_{11}	a_{12}	a_{13}	a_{14}	a_{15}	b_1
Technologies	a_{21}	a_{22}	a_{23}	a_{24}	a_{25}	b_2
Objects	a_{31}	a_{32}	a_{33}	a_{34}	a_{35}	b_3
Knowledge	a_{41}	a_{42}	a_{43}	a_{44}	a_{45}	b_4
Harvesting	c_1	c_2	c_3	c_4	c_5	max

$$\begin{cases} a_{11} \cdot x_1 + a_{12} \cdot x_2 + a_{13} \cdot x_3 + a_{14} \cdot x_4 + a_{15} \cdot x_5 \leq b_1 \\ a_{21} \cdot x_1 + a_{22} \cdot x_2 + a_{23} \cdot x_3 + a_{24} \cdot x_4 + a_{25} \cdot x_5 \leq b_2 \\ a_{31} \cdot x_1 + a_{32} \cdot x_2 + a_{33} \cdot x_3 + a_{34} \cdot x_4 + a_{35} \cdot x_5 \leq b_3 \\ a_{41} \cdot x_1 + a_{42} \cdot x_2 + a_{43} \cdot x_3 + a_{44} \cdot x_4 + a_{45} \cdot x_5 \leq b_4 \end{cases} \quad (5)$$

$$f(x) = c_1 \cdot x_1 + c_2 \cdot x_2 + c_3 \cdot x_3 + c_4 \cdot x_4 + c_5 \cdot x_5 \rightarrow max$$

We can consider an example of minimizing the use of objects with the growth of knowledge, goals, specialists and technologies.

Table 2 – Resource minimization model

	Product 1	Product 2	Product 3	Product 4	Product 5	
Knowledge	a_{11}	a_{12}	a_{13}	a_{14}	a_{15}	b_1
Goals	a_{21}	a_{22}	a_{23}	a_{24}	a_{25}	b_2
Specialists	a_{31}	a_{32}	a_{33}	a_{34}	a_{35}	b_3
Technologies	a_{41}	a_{42}	a_{43}	a_{44}	a_{45}	b_4
Object using	c_1	c_2	c_3	c_4	c_5	min

$$\begin{cases} a_{11} \cdot x_1 + a_{12} \cdot x_2 + a_{13} \cdot x_3 + a_{14} \cdot x_4 + a_{15} \cdot x_5 \geq b_1 \\ a_{21} \cdot x_1 + a_{22} \cdot x_2 + a_{23} \cdot x_3 + a_{24} \cdot x_4 + a_{25} \cdot x_5 \geq b_2 \\ a_{31} \cdot x_1 + a_{32} \cdot x_2 + a_{33} \cdot x_3 + a_{34} \cdot x_4 + a_{35} \cdot x_5 \geq b_3 \\ a_{41} \cdot x_1 + a_{42} \cdot x_2 + a_{43} \cdot x_3 + a_{44} \cdot x_4 + a_{45} \cdot x_5 \geq b_4 \end{cases} \quad (6)$$

$$f(x) = c_1 \cdot x_1 + c_2 \cdot x_2 + c_3 \cdot x_3 + c_4 \cdot x_4 + c_5 \cdot x_5 \rightarrow min$$

These tasks can be easily solved with the help of the built-in function «Search for solutions» of the MS Excel spreadsheet processor [14].

4. Conclusion

The use of information technologies in management is not always effective because of the underdevelopment of theoretical ideas about the information processes taking place in production systems. The information system developed by us, as well as one of the possible options for formalizing the relationships of the elements of the information system, make it possible to use information technologies in the management of production processes in the agricultural production. Although the main element of the information system is «Specialists(S)», significant potential opportunities for improving the efficiency of production processes are contained in the development of the element of «Knowledge (Z)» system. With a constant increase in the volume of knowledge, the problem of structuring this volume and developing methods for converting them and providing them to «Specialists (S)».

References

[1] Korneev I.K., Mashurtsev V.A. Information technology in management. M .: Infra-M, 2001. 158p.
[2] Alekhina G.V. Information technology in economics and management. - Moscow: Moscow International Institute of Econometrics, Informatics, Finance and Law, 2004. - 236 p.
[3] Titorenko G.A. (ed.) Information Technology Management / Ed. prof. G. A. Titorenko. - 2nd ed., Ext. - M .: UNITY-DANA, 2003. 439 p.
[4] Razumov, V.I. Categorial-systemic methodology for the training of scientists: Textbook / Emb. Art. A.G.Teslinova / V.I. Razumov. - Omsk, Omsk State. University, 2004. 277p.
[5] Redreev, G.V. Ensuring the efficiency of technology in the implementation of flow processes in agriculture / G.A. Okunev, G.V. Redreev / Omsk State Agrarian University, No. 2 (30) Omsk, 2018. 125-132 p.
[6] Redreev, G.V. Information System of Machines and Tractors Fleet Technical Service. /G.V. Redreev, A.A .Luchinovich, E.I. Ustiyantsev and A.S. Laskin//. IOP Conference Series:Journal of Physics, Volume 1059, conference 1 http://iopscience.iop.org/article/10.1088/1742-6596/1059/1/0120033-1.
[7] Teaching methodology of econometric modeling with the help of interactive teaching methods Kiyko P.V., Shchukina N.V. International Journal of Economic Research. 2017. T. 14. 7. P. 59-75.
[8] Kiyko P.V., Shchukina N.V. Innovative aspects of studying econometrics -http: //e-journal.omgau.ru/ Electronic scientific methodical journal of the Omsk State Agrarian University, 2016, № 4 (7).
[9] Kiyko P.V. Mathematical modeling as a system-forming factor in the implementation of interdisciplinary connections of mathematics and special disciplines in teaching future economists: monograph / P.V. Kiyko: Ministry of agriculture of the Russian Federation, Omsk State Agrarian University - Omsk: Publishing house of FGOU VPO OmGAU, 2010. 116 p.
[10] Meshcheryakova N. A. Mathematical models of physical phenomena and teaching students of technical University their construction. Extended abstract of dissertation. ... PhD. in Pedagogy. Leningrad. 1974. 178 p.
[11] Burmistrova, N.A. Computer technologies in the process of professionally directed training in mathematics of future bachelors of the direction «Economics» / N.A. Burmistrova, N.I. Ilyin // Bulletin of the Laboratory of Mathematical, Natural Science Education and Informatization: a peer-reviewed collection of scientific papers. M., GBOU VPO MGPU, Voronezh: «Scientific Book», 2012. Vol. 3. 99–102p.
[12] http://statsoft.ru/products/
[13] http://www.eviews.com/home.html
[14] https://www.microsoft.com/ru-ru/

IOP Conf. Series: Materials Science and Engineering **582** (2019) 012046 doi:10.1088/1757-899X/582/1/012046

Practice of Use of Professional Standard While Developing Program of Professional Retraining on Ecological Safety

S K Belyakin, N K Smirnova, V A Krivobokova

Kurgan state university, 640002, Kurgan, 62Proletarskaya street, Russian Federation

E-mail: bgd@kgsu.ru

Abstract. the article considers the practice of the use of the professional standard "Specialist on ecological safety (in industry)" while realizing professional retraining of specialists and organizations on ecological safety. On the basis of this standard the program of professional retraining is developed. The curriculum and the fund of rating means are given. The material may be interesting for the people, providing professional training, retraining of specialists on ecological safety.

Introduction

In the 21 century local and global ecological problems are becoming sharper and sharper due to the increase of the influence of human economic activities on the environment. The growth of human pressure on the environment leads to reserve depletion of natural resources, loss of natural connection between the man and the nature, decline of inartificial values, and as a consequence, deterioration of moral and physical health of people.

The necessity of providing ecological safety, while modernizing economy and in the process of innovative development, conditions the elaboration of Cornerstones of state policy in the sphere of ecological development of the Russian Federation for the period to 2030, in which it is stated that in 40 constituent territories of the Russian Federation more than 54 per cent of the urban population is under the impact of high and very high level of atmospheric air pollution. The volume of aqueous wastes, discharging into surface water bodies without purification or without proper purification, remains high. Practically in all the regions the tendency of deterioration of land and soil state holds. The processes, leading to the exhaustion of cultivated lands and their removal out of economic turnover, are developing intensively. Desert advancing is going on at this or that degree in 27 constituent territories of the Russian Federation on the square of more than 100 mln hectares. The quantity of wastes, which is not recycled, but is sent to landfill sites, is increasing. Moreover, the conditions of waste storage and landfilling fail to meet the requirements of ecological safety [1].

The solution of problems of environment protection depends on ecological mindset of broad population and the qualification of specialist-ecologists a lot.

The state policy is defined as an array of economic, legal and organizational measures, aiming at maintaining the environment ability to satisfy the needs of the existing and not to threaten the satisfaction of needs of the future generations [1, 2, 3], and it is focused on the creation of favourable environment, improvement of living conditions and population health, guarantee of ecological safety. The important task of realization of the country policy in this sphere is the preparation of specialists of high qualification, they should manage ecological risks effectively, do the corresponding economic calculations, organize efficient policy in the sphere of environment protection.

Object of study

Professional standard is a characteristic of qualification, necessary for the worker to carry out a certain type of professional activity (paragraph 2 article 195.1 Labour code of RF) [4]. This notion of professional standard was introduced into the Labour code at the end of 2012, because the characteristics of positions, included into Unique qualification references, did not correspond the modern situation at the labour market.

IOP Conf. Series: Materials Science and Engineering **582** (2019) 012046 doi:10.1088/1757-899X/582/1/012046

The decree of the Ministry of labour and social security of the Russian Federation [5] ratified the professional standard "Specialist on ecological safety (in industry)", which set up claims to the education and training, experience of practical work and special conditions of the specialist admission. It presents the functional map of the type of the professional activity and the main general cultural and professional competences of the graduates.

Results and their discussion

Nowadays in Kurgan state university scientific-educational center "Ecology, labour protection and life safety" of the department "Ecology and life safety" realizes additional educational program of professional retraining "Ecological safety (in industry)". This program implies lectures and practical classes, consulting in the process of teaching and preparing for final assessment, unsupervised work of listeners, training on probation, conducting intermediate and final assessment.

The program of professional retraining is developed in accordance with laws and regulations:

- Federal Law of the Russian Federation dated 29 December 2012 № 273-FL "Of education in the Russian Federation" [6];
- Order of organization and realization of educational activity on additional professional programs, ratified by decree of Ministry of education and science of the Russian Federation dated 1 July 2013 № 499 [7];
- Decision of Ministry of labour of the Russian Federation dated 21 August 1998 г. № 37 "Of the statement of Qualification reference of positions of managers, specialists and other staff (with changes on 27 March 2018)" [8];
- Federal state educational standard of higher education in the field of study 20.03.01 "Technosphere safety" (bachelor degree level), ratified by decree of Ministry of education and science of the Russian Federation dated 21 March 2016 № 246 [9];
- Professional standard "Specialist on ecological safety (in industry)", ratified by of Ministry of labour and social security of the Russian Federation dated 31 October 2016 № 591н [5].

The aim of the realization of professional retraining program is listeners' formation of professional competences, necessary to provide the organizations of industry segments with environment protection activity with the purpose of development of the complex of organizational and technical measures, directing to ecological safety, minimization of the negative effect of economic activity in industry on the environment.

The sphere of professional activity is throughout professional activity in the industry in the sphere of providing environment protection activity.

Professional standard [4] gives detailed characteristics of generalized labour functions, possible names of positions, requirements to education and training, as well as to the experience of practical work, which have become the bases of program of professional retraining in the Kurgan state university. The description of labour functions, included into professional standard (functional map of the type of the professional activity), is presented in 1.

According to the names of generalized labour functions of a specialist on ecological safety, it seems reasonable to master the following professional competences (PC):

PC-1 – ability to carry out control of meeting requirements in the sphere of environment protection;

PC -2 – ability to develop activities on environment protection and provision of ecological safety and paperwork in accordance with stated requirements;

PC -3 – ability to develop activities on economic regulation and personnel staff management in the sphere of environment protection.

IOP Conf. Series: Materials Science and Engineering **582** (2019) 012046 doi:10.1088/1757-899X/582/1/012046

Table 1 - Functional map of the type of the professional activity

Generalized labour functions			Name of labour function
Code	Name	Qualification level	
A	Realization of control of requirements fulfillment in sphere of environment protection	5	Control of adherence of technological regimes of environment protection objects of organization, analysis of their work, control of normative state of environment in the area of organization location
			Preparation of necessary materials for carrying out production ecological control
			Performance of technical tests of equipment, providing ecological safety of enterprise , and determination of efficiency of equipment work
			Control of waste accumulation, utilization, neutralization and emplacement in organization
B	Development of activities in organization on environment protection and provision of ecological safety and paperwork in accordance with stated requirements	6	Performance of ecological analysis, involving enlargement and reconstruction of existing productions, as well as new technologies and equipment
			Performance of work on production of new goods with improved ecological characteristics
			Performance of production ecological control and paperwork of event fulfillment of activities on environment protection
			Record keeping of values, characterizing environment state, of data of ecological monitoring and other paperwork
			Ecological paperwork preparation in accordance with stated requirements in the sphere of environment protection and provision of its up-to-date reconsideration
			Development and introduction of activities, aiming at meeting requirements in the sphere of environment protection, prevention of occurrence of emergency situations of natural and technogenic character
C	Development of activities in organization on economic regulation and personnel staff management in sphere of environment protection	7	Development and economic justification of plans of introducing new equipment and technologies, providing minimization of organization impact on environment
			Development of activities on economic regulation of environment protection measures of organization
			Fulfillment of justified calculations of ecological risks with the view of foreseeing of impact of business activity of organization on environment
			Organization of organization of reasons and consequences of wastes and interception of harmful substances, preparation of offers on prevention negative consequences
			Managing personnel staff of organization department on providing ecological safety

The program of professional retraining, realizing at the Kurgan state university, provides the achievement of the seventh level of qualification in accordance with the requirements of professional standard [4]. The accomplishing of this program of professional retraining will allow fulfilling labour functions and taking corresponding positions (table 2).

IOP Conf. Series: Materials Science and Engineering **582** (2019) 012046 doi:10.1088/1757-899X/582/1/012046

Table 2 – Positions, corresponding to qualification levels

Qualification level	Names of possible positions
5	Technician-environmental expert Technician–laboratory assistant Technician of department of ecological safety
6	Engineer on ecological safety Engineer on environment protection Engineer - laboratory assistant Specialist in the sphere of environment protection
7	Chief of the department of environment protection Chief of the department of ecological safety Group leader (in industry)

People, having and/or getting secondary technical and/or higher education, can be admitted to master the additional educational program of professional retraining, as well as acting specialists: technicians-environmental experts, technicians–laboratory assistants, technicians of department of ecological safety, engineers on ecological safety, engineers on environment protection, engineers - laboratory assistants, specialists in the sphere of environment protection, chiefs of the department of environment protection, chiefs of the department of ecological safety, group leaders (in industry).

Credit value of study comprises 540 academic hours with the account of all kinds of in-class and unsupervised work of the listener. The form of study is part-time. It is possible to realize the program of profession retraining in a distance-type form of education. The study runs for 4 months; 13 academic weeks. 5 weeks is given to fulfill final assessment work.

In table 3 the curriculum of the additional educational program of professional retraining «Ecological safety (in industry)" is presented.

Table 3 - Curriculum of the additional educational program of professional retraining «Ecological safety (in industry)"

Names of subjects (courses)	Total number of hours	In-class		Unsupervised work	Form of control
		Lectures	Practical classes		
Legal regulation of ecological safety	**36**	6	6	60	graded test
Ecological requirements to organization of technological processes and production equipment	**72**	6	8	58	exam
Resources of environment pollution	**72**	8	10	54	exam
Engineering protection of environment	**36**	2	4	30	Graded test
Ecological standardization and audit	**36**	4	8	24	graded test
Rational use of natural resources	**36**	4	6	26	graded test
Ecological management	**36**	6	2	28	test
Ecological risks	**36**	4	4	28	test
Ecological control at the enterprise	**72**	6	4	26	exam
Ecological recording and paperwork	**72**	10	12	50	exam
Work on probation					Test
Final assessment (final project)	**36**			36	Defence of final project
All in all:	**540**	56	64	420	Exam - 5 Test - 6

Each discipline of program of professional retraining involves performance of practical work in accordance with methodical instructions; there is a fund of rating means for listeners' assessment. In working programs of subjects (courses) there are particular professional competences in accordance

IOP Conf. Series: Materials Science and Engineering **582** (2019) 012046 doi:10.1088/1757-899X/582/1/012046

with labour functions, as well as knowledge, skills and know-how, which will be mastered by listeners in the process of learning study materials. For example, as a result of mastering the discipline "Legal regulation of ecological safety", the educated should demonstrate the following results of study.
To know:
- ecological legislation of the Russian Federation, basic normative legal acts in the sphere of environment protection;
- production and organization structure of enterprise and possibilities of its development;
- normative and methodical materials on environment protection and ecological safety provision;
- order of keeping paperwork on environment protection and ecological safety provision.
To be able:
- record information of results of production ecological control;
- conduct account of values, characterizing environment state in accordance with requirements of normative legal acts in the sphere of environment protection;
- take into account the specific character of the organization while developing ecological paperwork;
- interact with authorized bodies of executive power in constituent territories of the Russian Federation on elaboration of ecological paperwork.
To master:
- methods of analysis of the basic directions of ecological safety improvement with the consideration of the specific character of the production.

Below there are pattern tests on some subjects.

Variant of the test on the discipline "Legal regulation of ecological safety"

1 The subject matter of the ecological law is:
a) social relations in the sphere of protection, recovery and improvement of environment, prevention and disposal of negative consequences of impact of business and other activities on environment;
б) social relations in the sphere of interaction between the man and the nature;
в) social relations in the sphere of environment protection against the impact of physical, chemical, biological and other values and (or) their combination.

2 The objects of environment protection from wasting, exhausting, degrading, spoiling, eliminating or other negative impacts of business or other activities are:
a) lands, mineral resources, soils;
б) surface and subterranean waters;
в) forests and other vegetation, animals and other organisms and their genetic fund;
г) atmospheric air, ozone layer of the atmosphere and near-Earth environment.
All variants are right.

3 Subject matters of ecological legal relations are:
a) citizens, state bodies;
б) individual persons, legal persons, state bodies;
в) citizens, legal persons, state bodies;
г) state bodies, officials, production enterprises.

4 Which does not belong to the resource of ecological law:
a) Constitution of the RF,
б) FL «Of environment protection»;
в) LC RF;
г) Aqueous and Forestry Codes.

5 Ecological crime is:
a) negative change of environment as a result of its pollution, leading to degradation of natural ecological systems and exhaustion of natural resources;
б) guilty wrongful act, breaking environmental regulation and doing harm to environment and human health.

Listeners are admitted to the final assessment after passing all intermediate assessments.

The final assessment of the trainees of the additional educational program of professional retraining "Ecological safety (in industry)" involves the defence of final project. Each listener has a tutor to assist while solving main, principal questions [10,11].

The topic of the final project is set by the department, preparing specialists in the field of study 20.03.01 "Technosphere safety". The listener has the right to offer (in agreement with the tutor) his own topic, which is essential for the organization, where he works now, he works on probation and where he is going to work in future, or which is appealing to his production or scientific interests.

At the defence of final project the readiness of the listener to perform professional functions, included into the professional standard, is checked; practical experience is assessed; as well as the ability to give reliable reasons and to prove performed research and developed solutions in the process of discussion. The final assessment is done by the final attestation commission, the composition of which is formed by the department of Ecology and life safety, signed by the rector. Considering the results the commission makes a decision of passing/failing final assessment.

According to results of successful passing the program:
- the listener of the courses will possess the competences, necessary for realization of his professional activity in the sphere of provision of ecological safety activity in the organizations of production segments;
- the listener, who has finished the educational program successfully, has got the right to perform new kind of professional activity in the sphere of provision of ecological safety activity in the organizations of production segments;
- the listener, who has finished the educational program successfully and satisfies the requirements of the Professional standard from the point of view of education (higher education – specialist, master's degree program) and experience of practical work (no less than five years in his specialization at leading positions in the sphere of ecological safety) has got the qualification "Specialist on ecological safety (in industry)" (the 7th level of qualification).

The listeners of the first group in 2017, mastering the program of professional retraining and passing final assessment successfully, got the diploma of professional retraining, giving the right for professional activity in the sphere of ecological safety. Nowadays many listeners (having experience of work in the sphere of provision of ecological safety) undergo distant form of education.

Conclusion

The necessity of development of the program "Specialist on ecological safety (in industry)" is conditioned by modern requirements of state policy in the sphere of environment protection. In accordance with the professional standard and unique qualification reference of positions of managers, specialists and employees [4,8] specialists, carrying out work in the sphere of ecological safety must have professional education corresponding to his training directions (specialization). If the specialist on ecological safety has no specific education, he has to get additional professional education on the programs of professional retraining. Otherwise, the worker has no right to take a post in ecological services of enterprises and organizations and must be dismissed. In connection with the fact that the mentioned decree has entered into force, enterprises and organization of the Kurgan region feel the critical need for professional retraining of the specialists of this direction.

References

[1] The bases of state policy in the sphere of ecological development of the Russian Federation for the period to 2030 (ratified by the president of RF dated 30 April 2012) URL: https://base.garant.ru/70169264/ (date of access: 16.10.18).

[2] 2. Federal Law «Of environment protection» dated 10.01.2002 № 7-FL (with changes on 29 July 2018). URL: http://docs.cntd.ru/document/901808297 (date of access: 16.03.19).

[3] Law of the Kurgan region dated 2 October 1998 г. N 163 "Of environment protection of the Kurgan region" (amended and supplemented). URL: http://base.garant.ru/18351134/#ixzz5VV90YUuE (date of access: 18.03.19).

[4] Labour code of RF (LC RF) dated 30.12.2001 № 197-ФЗ (new version of the law dated 27.12.2018). URL: http://www.consultant.ru/document/cons_doc_LAW_34683/ (date of access: 16.03.19).

[5] Decree of Ministry of labour and social security of the Russian Federation dated 31 October 2016 № 591н "Of the statement of professional standard "Specialist on ecological safety (in industry)". URL: http://www.consultant.ru/document/cons_doc (дата обращения 26.03.19)., ratified by of

[6] Federal Law of the Russian Federation dated 29 December 2012 № 273-FL “Of education in the Russian Federation” URL: http://www.consultant.ru/document/cons_doc_LAW_140174/ (date of access: 06.02.19).

[7] Decree of Ministry of education and science of the Russian Federation dated 1 July 2013 № 499 « Of the statement of Order organization and realization of educational activity on additional professional programs». URL: https://base.garant.ru/70440506/ (date of access: 26.02.19).

[8] Decision of Ministry of labour of the Russian Federation dated 21 August 1998 г. № 37 “Of the statement of Qualification reference of positions of managers, specialists and other staff (with changes on 27 March 2018)”. URL: http://docs.cntd.ru/document/58839553 (date of access: 27.03.19).

[9] Decree of Ministry of education and science of the Russian Federation dated 21 March 2016 № 246 (new version of the law dated 13.07.2017) “Of the statement of federal state educational standard of higher education in the field of study 20.03.01 “Technosphere safety” (bachelor degree level)». URL: http://www.consultant.ru/document/cons_doc_LAW_197236/ (date of access: 7.02.19).

[10] Smirnova N.K. Professional retraining of specialists on labour protection. // Ecology. Risk. Security. Materials of the IV All-Russian scientific practical with face-to-face and virtual international participation (29-30 October 2015): Collection of research works: – Kurgan: Publishing house of Kurgan state university, 2016. – p.100-102.

[11] Smirnova N.K., Belyakin S.K. The prior character of specialist training in the sphere of technosphere safety. // Science of XXI century: technologies, management, safety [Electronic edition]: collection of materials of the I international scientific practical conference (Kurgan, 26 September 2017) - Kurgan: Publishing house of Kurgan state university, 2017. – P.164-174.

IOP Conf. Series: Materials Science and Engineering **582** (2019) 012047 doi:10.1088/1757-899X/582/1/012047

Investigation of Effect of Electron Beam Surface Treatment of Titanium Alloy Ingots on Structure and Properties of Melted Metal

S V Akhonin [a], **O M Pikulin** [b]

E.O. Paton Electric Welding Institute of the National Academy of Sciences of Ukraine, 11 Kazymyr Malevych St., Kyiv, 03150, Ukraine

E-mail: [a] akhonin.sv@gmail.com, [b] om.pikulin@gmail.com

Abstract. Based on accumulated experience of application of concentrated power sources for elimination of defects of ingot surface layer it is shown that remelting of surface layer of treated ingot provokes evaporation of alloying elements with high vapor pressure form a liquid pool, i.e. change of composition of treated layer in relation to ingot base. This can indicate the possible change of metal structure since it is determined by content of alloying elements.

Ingots of wrought titanium alloys VT3-1, VT-6, VT20, VT22 of 400 and 600 mm diameter and up to 2 m length made by EBM were melted by electron beam for investigation of effect of electron beam treatment of a surface layer on composition, structure and properties of metal of the treated layer. The investigations of mass concentration of alloying elements, penetration depth, macro and microstructure of melted metal layer of treated ingots were carried out.

The results of carried works showed that electron beam treatment allows efficient elimination of defects of the surface layer of titanium alloy ingots to 12 mm depth, content of alloying elements corresponds to grade composition, reduction is observed in content of alloying elements, in which vapor pressure is higher than in the alloy base, and respectively rise of content of alloying elements with vapor pressure is lower than in the alloy base. There is insignificant increase of content of interstitial impurities, i.e. oxygen and nitrogen, which do not have obvious effect on mechanical properties of treated metal. It is determined that macrostructure of metal in the treated layer of titanium alloy ingots is characterized with absence of pores and discontinuities, crystals close to equiaxial, the melted layer is formed by finer crystals in comparison with the rest part of the ingot, elongated in solidification direction, i.e. to ingot center. Solidification rates of more than 50 mm/min in the melted layer of titanium alloy ingots provoke two times refining of α-colonies and α-lamellas of ingot grains in comparison with the ingot base.

1. Introduction

Titanium is one of the most important of modern structural materials and in current industrial production the semi-finished products of titanium alloys take an important place [1, 2]. Titanium alloys having high indices of mechanical properties, high corrosion and high-temperature resistance are perspective structural materials, which have found application in aircraft, space [3, 4], marine engineering [5], power machine building, oil-and-gas and chemical branches [5-7].

Titanium is alloyed with aluminum, vanadium, chromium, molybdenum and some other elements in order to get set mechanical properties. Addition of alloying elements in predetermined combinations and amounts allows varying the properties titanium alloys in the wide limits [8-9]. Sufficiently wide interval of complex of mechanical properties based on titanium permits selecting the alloys, which in the best way satisfy the technical requirements for this specific application. Comparatively small density in combination with sufficiently high strength characteristics provides higher specific strength in a wide temperature range to titanium alloys in comparison with aluminum alloys, steels and high-temperature nickel alloys [8]. This circumstance has a decisive role, determining preference of application of titanium alloys, and reduction of prime cost of titanium alloy ingots as initial part for manufacture of semi-finished products simultaneously with rise of their quality is a relevant problem.

Today vacuum arc and electron beam melting as well as plasma-arc remeleting are sufficiently effectively used in production of titanium alloys. Electron beam melting (EBM) is the most efficient method of vacuum metallurgy for manufacture of alloys including refractory and high-reactive, with ultra-low content of gases, volatile additives and non-metallic inclusions [10, 11]. However, a series of reasons caused by metallurgical and technological peculiarities in process of ingot melting can provoke on their surface the defects in form of corrugations, cracks and longitudinal band due to metal pouring (see Figure 1). It is virtually impossible to eliminate such type of defects at current level of production, and this in turn complicates further hot treatment of ingots and billets resulting in hot cracking.

The required quality of ingots and billets surface using the traditional technology is reached by means of removal of surface layer by machining. It is necessary to note that mechanical treatment provokes local metal heating in the place of contact with a cutter and, naturally, chips oxidation. High requirements to cleanness of initial charge materials imposes a series of limitations of secondary used chips for ingot manufacture, that in turn leads to irretrievable losses of metal up to 20-35% depending on ingot size [12, 13].

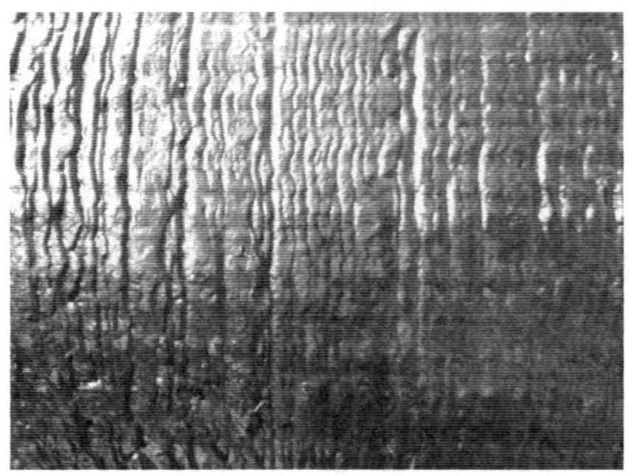

Figure 1. Appearance of VT30 alloy ingot surface produced by EBM method

Treatment of surfaces of ingots and billets with concentrated power sources from point of view of material saving and reduction of technological operations is an effective solution for this problem. Up to the moment there is a huge experience of application of concentrated power sources, namely electron beam and plasma arc for elimination of defects of ingot surface layer [14-16]. The peculiarity of methods of treatment of ingot surface based on application of concentrated power sources is the fact that remelting of surface layer of the treated ingot takes place in vacuum or gas-shielded medium that results in evaporation of alloying elements with high pressure vapor from a liquid pool, i.e. change of composition of the treated layer in relation to ingot base. This can indicate possible change of metal structure since it is determined by content of alloying elements.

At modern stage of development of production an important conditions of its improvement is development of high-performance methods and means for modeling and control of produced parts. Large attention is given to application of modern methods of mathematical and computer modeling, which have become a powerful tool for investigation and understanding of the processes taking place in a liquid metal pool [17-19]. At the present time the models were developed for prediction of depth of the melted layer and its composition.

Taking into account mentioned above the researches on investigation of metal quality of the treated layer of titanium alloy ingots using concentrated power sources are of large interest. The positive results of such works allow reducing the metal loss in form of off-standard wastes (chips) and precious alloying elements.

2. Aim of the work

Aim of the present work lies in investigation of effect of electron beam treatment of the surface layer of titanium alloys VT3-1, VT6, VT20, VT22 ingots on metal quality of treated layer, i.e. its composition, structure and properties.

3. Materials and procedure of investigation

The ingots of wrought titanium alloys VT3-1, VT6, VT20 and VT22 of 400 and 600 mm diameter and up to 2 m length produced by EBM (see Figure 2) were melted for investigation of effect of electron beam treatment of the surface layer on composition, structure and properties of metal of the treated layer. These alloys have sufficiently high content of base alloying element for titanium alloys, namely Al, which in turn has high vapor pressure that results in its evaporation from the treated metal. There are also different classes of phase

IOP Conf. Series: Materials Science and Engineering **582** (2019) 012047 doi:10.1088/1757-899X/582/1/012047

composition, i.e. VT3-1, VT-6 α+β, VT22 - α+β, on composition close to critical, VT20 – pseudo-α that allows deeper investigation of effect of change of the treated metal composition on its structure.

Figure 2. Ingot of VT6 titanium alloy before EBT

EBT (electron beam treatment) process of the ingots was carried out using a technological scheme, at which electron beam is fixed and ingot is rotated around its axis (see Figure 3).

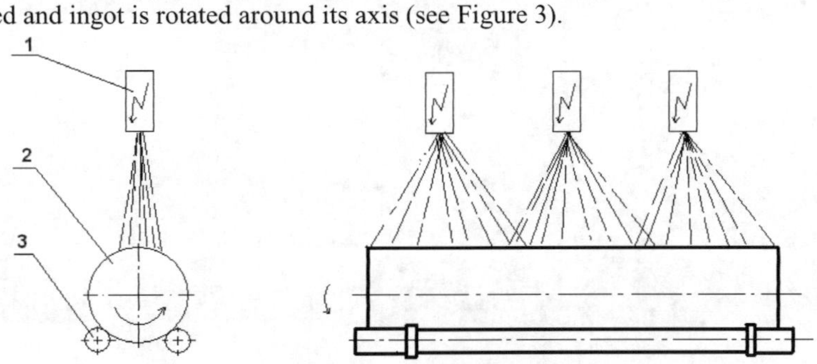

Figure 3. Scheme of electron beam treatment of titanium alloy ingot: 1 – electron beam gun; 2 – ingot; 3 – ingot rotating mechanism rolls

The surface of produced ingots was subjected to electron beam treatment in UE-185 unit with linear melting rate 54 mm/min and specific power of heating 6-7 W/mm^2 (see Figure 4).

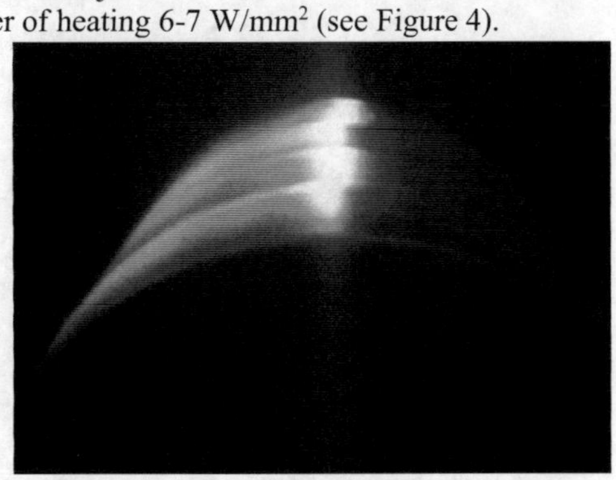

Figure 4. Process of ingot melting

Investigation of penetration depth and structure of metal of the melted layer of titanium alloy ingots were carried out on transverse templates cut out of the treated ingots and for investigation of change of metal composition of the melted layer before and after treatment the probes were taken in form of chips and cut-off samples from the treated layer on ingot length.

Content of chemical elements was determined by atomic absorption method on GOST 19863.1-91 on Saturn-1 device. Content of oxygen and nitrogen was determined using a method of melting of analyzed trial sample in vacuum in a graphite crucible, heating of which was carried out by passing of current of 900-1000 A power, on devices of RO-316, TN-114, RH-3 type of LECO Company (USA).

Metallographic examinations of the samples were performed on optical microscope Neophot-32. The value of grain was determined using a procedure described in "VIAM Instruction No. 1054-76". Microstructure of the samples was examined after detection in the standard chemical agents of $HF:HNO_3:H_2O=(1:1:3)$. The dimensions of structure elements were determined on a procedure developed for titanium alloys [20]. For this the analyzed structures were compared with the reference ones for evaluation of value of α-colonies and for determination of thickness of α-lamellas at corresponding magnification.

The microhardness tests were carried out by indentation method on GOST 9450-76.

4. Investigation results

The experimental evaluation of penetration depth of the surface layer of ingots was performed on transverse templates and made 9-12 mm (see Figure 5). A side surface of ingots after treatment with electron beam acquires a flat microrelief, has smooth mirror-like appearance without obvious defects appearing in process of melting in the surface layer (see Figure 6). Roughness of the surface lied in the Rz20 - Rz80 range at waviness of surface equal to 0.2 – 0.6 mm.

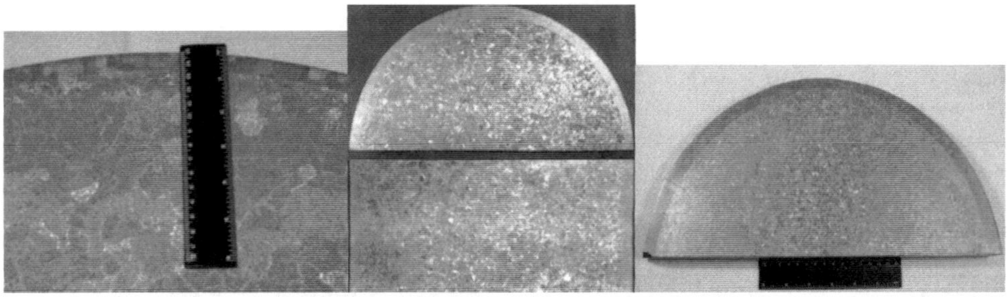

a b c

Figure 5. Penetration depth of surface layer: a - VT3-1, b - VT6, c - VT22

Figure 6. Appearance of side surface of ingot of VT20 titanium alloy melted with electron beam

The results of investigation of mass concentration of the alloying elements in the melted metal layer of ingots showed that their content corresponds to grade composition, decrease of content of aluminum and chromium, in which vapor pressure is higher than in the alloy base, and rise of content of vanadium, molybdenum and niobium with vapor pressure lower than in the alloy base (see Table 1) are respectively observed. It is also determined that there is insignificant rise of content of detrimental impurities, namely oxygen and nitrogen, at that values of their mass concentration are within the limits of standards' requirements.

Table 1. Mass concentration of alloying elements in layer of melted metal of titanium alloy ingots

Alloy	Sampling place	Al	V	Zr	Cr	Mo	O	N
VT6	GOST 19807-91	5.3–6.8	3.5–5.3	≤ 0.3	-	-	≤ 0.20	≤ 0.05
	Initial	6.21	3.77	-	-	-	0.050	0.026
	After EBT	5.98	3.89	-	-	-	0.057	0.030
VT3-1	GOST 19807-91	5.5–7.0	-	≤ 0.5	0.8-2.0	2.0-3.0	≤ 0.15	≤ 0.05
	Initial	6.21	-	-	1.26	2.64	0.014	0.013
	After EBT	5.64	-	-	1.13	2.60	0.032	0.021
VT20	GOST 19807-91	5.5–7.0	0.8–2.3	1.4–2.5	-	0.5–1.8	≤ 0.15	≤ 0.05
	Initial	6.9	2.05	1.81	-	1.57	0.066	0.011
	After EBT	6.48	2.14	1.78	-	1.63	0.078	0.019
VT22	GOST 19807-91	4.4–5.7	4.0–5.5	≤ 0.3	0.5–1.5	4.0–5.5	≤ 0.18	≤ 0.05
	Initial	5.24	4.73		1.45	4.04	0.050	0.010
	After EBT	5.05	4.95		1.14	4.26	0.059	0.012

Metallographic analysis of metal of the melted ingots was carried out for detection of structural changes taking place in metal as a result of thermal effect of electron beam on a side surface and ingot base.

Metallographic examinations were carried out on the samples taken from the transverse templates, which were cut of the ingot. Metal macrostructure is characterized with absence of pores and discontinuities, crystals close to equaxial, the melted layer is formed by more dense crystals in comparison with the rest part of the ingot, elongated in solidification direction, i.e. to the ingot center. Grain size determined on ten-point scale of macrostructures of VIAM Instruction No. 1054-76, corresponds to size Nos. 6-7 in the melted layer and size Nos. 8-9 in the heat affected zone and base metal.

Analysis of microstructure of VT6 alloy ingots, melted with electron beam, showed that the microstructure of base sample, cut of the ingot, is characterized with initial grains of β-phase, fringed on the boundaries with fine α - "fringe". The internal volume of the grains consists of the packages of α-lamellas divided with β-phase interlayers. Electron beam melting of the side surface of ingot surface resulted in re-solidification of alloy structure of the ingot melted layer to 12 mm depth. Microstructure of the melted layer is typical for cast metal and has finer structure in comparison with the ingot base that in turn stipulates higher solidification rates of melted layer (see Figure 7).

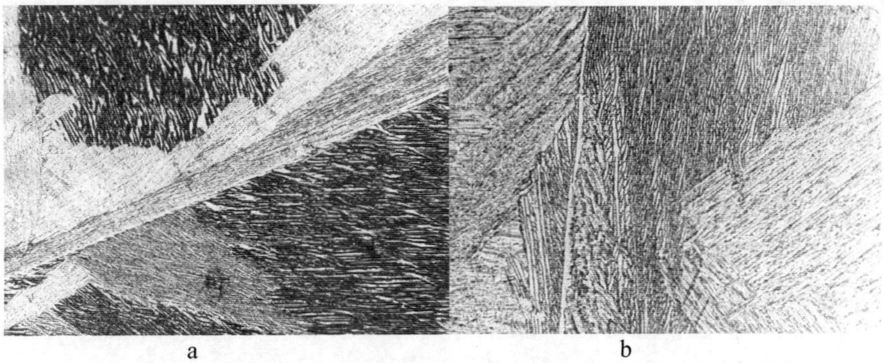

a b

Figure 7. Microstructure of VT6 alloy ingot: a – base, b – melted layer

The sample cut of the ingot of pseudo α-alloy VT20 consists of equiaxial polyedric primary β-grains with lamellar α-phase inside the grain, accumulated in the colonies. Value of the α-colonies in different grains of ingot base metal is somewhat different, since their average size varies from 200 to 300 μm, the average thickness of lamellas of α-phase at that makes around 5 μm. The metal of treated surface layer consists of the colonies of α-phase lamellas, the average size of which varies in 100 – 200 μm limits at α-lamellas thickness 5 μm. i.e. after melting there is refining of the α-colonies at stable thickness of α-phase lamellas. The α-fringe (see Figure 8) is often observed on the boundaries of grains of ingot metal and surface layer.

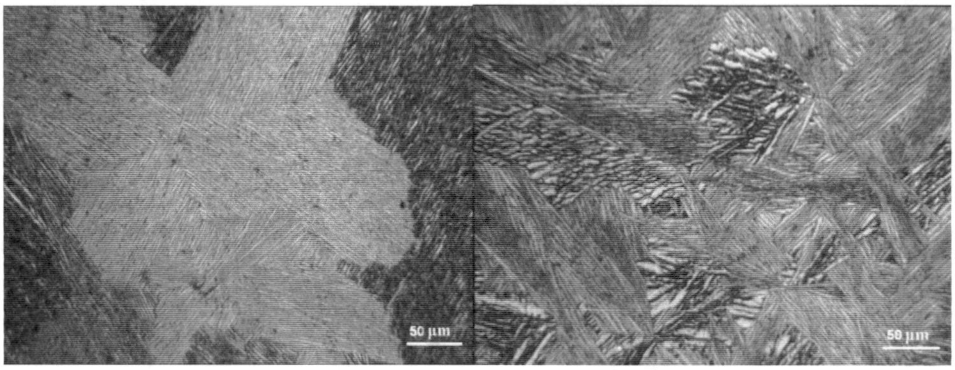

a b

Figure 8. Microstructure of VT20 alloy ingot**:** a – base, b - melted layer

Microstructure of the samples of α+β-alloy of martensite type VT3-1 is presented on Figure 9. Grains of metal of VT3-1 alloy ingot consist of the colonies of lamellar α-phase of preferable size around 300 μm, thickness of α-lamellas makes 5- 7.5 μm. β-phase is located between α-lamellas. The α-fringe is observed on the boundaries of some β-grains. There is refining of size of the α-colonies as well as α-lamellas in the treated surface layer after melting. The average size of the α-colonies varies in 100 -200 μm limits and average thickness of α-lamellas makes 5 μm.

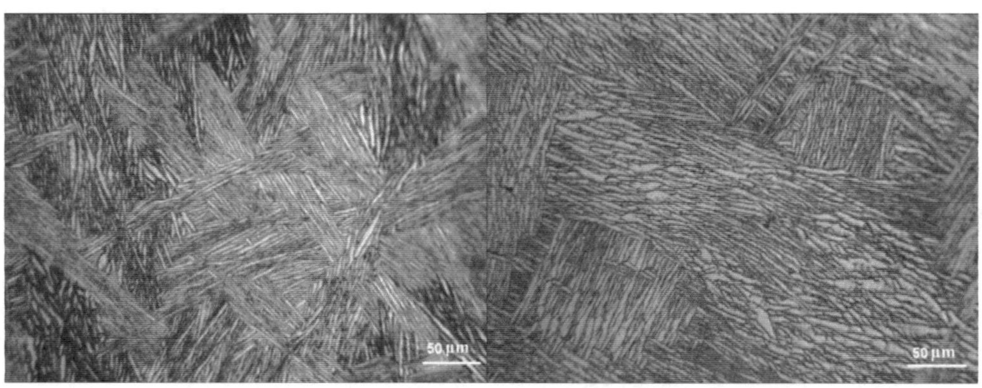

a b

Figure 9. Microstructure of VT3-1 alloy ingot: a – base, b - melted layer

Microstructure of metal of VT22 alloy ingot is presented on Figure 10. In the ingot of VT22 alloy the α-colonies of around 100 μm size, α-lamella thickness makes 2 -3 μm, are located in the β-grain volume. The boundaries of grains are often decorated with the α-fringe, along the boundaries there is precipitation of parallel α-lamellas almost normal to boundary line. β - phase is located in the gaps between the α-phase lamellas, besides, metal structure of VT22 alloy ingot contains locally fixed light areas of β-phase without lamellar α-precipitations or α-phase is at the stage of pre-precipitation. The metal structure of the surface layer after melting is also consists of equiaxial β-grains, however, their size in the spots is less than in the ingot depth. Size of α-colonies and thickness of α-phase lamellas in the surface layers after melting make 50 μm. The boundaries of β-grains in the surface layer also have α-fringe, there are areas of undecayed β-phase.

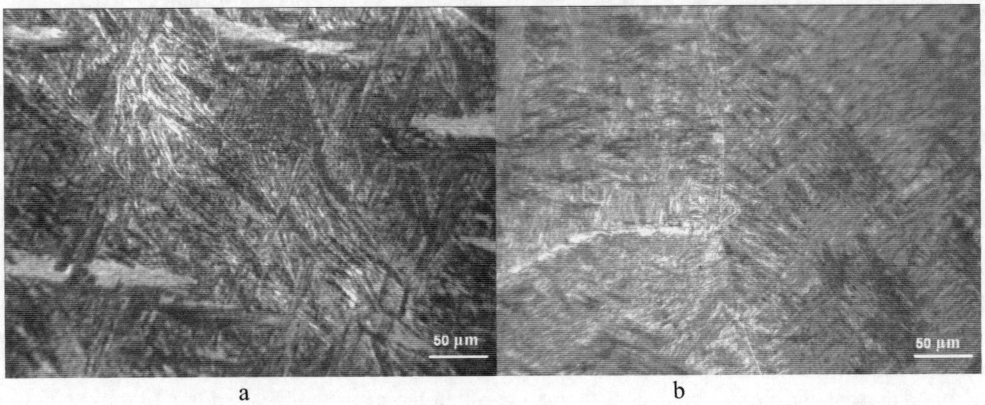

Figure 10. Microstructure of VT22 alloy ingot: a – base, b - melted layer

Carried analysis of the microstructure sample of the melted layer and ingots base showed that solidification rates of more than 50 mm/min provoke in the melted layer of titanium alloy ingots two times refining of the α-colonies and α-lamellas of ingot grains in comparison with ingot base.

Therefore, electron beam melting of the surface of titanium alloy ingot allows eliminating the defects of the surface layers of titanium alloy ingots up to 12 mm depth providing at that typical for the cast metal structure, finer in comparison with the alloy base.

Examination of distribution of microhardness in the zones of transverse template (see Figure 11) was carried out for evaluation of effect of electron beam treatment on mechanical properties of metal of VT3-1 ingot melted layer. Obtained results show that in the melted layer it is insignificantly lower than in the base metal.

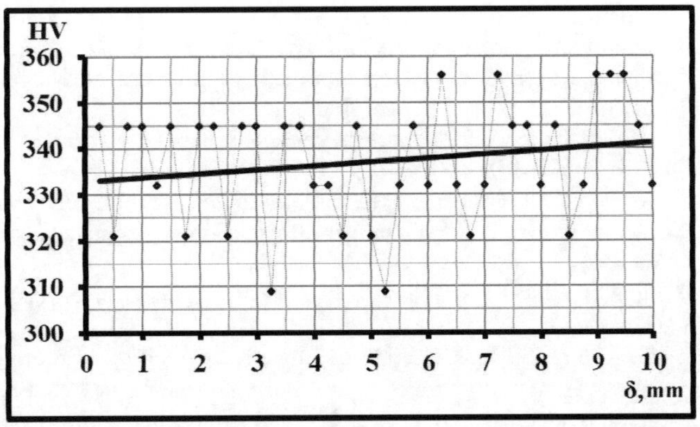

Figure 11. Distribution of microhardness on depth of deposited layer for VT3-1 titanium alloy ingot

From our point of view this is explained by decrease of aluminum content at EBT, which stabilizes and strengths the α-phase in the alloy as well as increase of content of molybdenum, which results in rise of amount of more lamellar β-phase in the alloy. This results also demonstrate that rise of amount of detrimental impurities (oxygen and nitrogen) determined in experimental way does not have negative effect on the properties of metal of the melted layer. Decrease of microhardness of the melted layer at pressure treatment reduces the tendency to cracking.

Thus, electron beam treatment allows effective elimination of the defects of surface layer of titanium alloy ingots at up to 12 mm depth without application of mechanical treatment and get at that high quality of metal of the surface layer with uniform defect-free structure typical for cast metal of finer structure relatively to ingot base.

5. Conclusions

The results of investigation of mass concentration of alloying elements in the metal of melted layer of titanium alloy ingots show that their content corresponds to grade composition, decrease of content of alloying elements, in which vapor pressure is higher than in the alloy base, and, respectively, rise of alloying elements with vapor pressure lower than in the alloy base are observed, insignificant rise of content of detrimental impurities, i.e. oxygen and nitrogen, but at that values of their mass concentration are in the limits of standards requirements.

It was determined based on the results of metallographic examinations that macrostructure of metal in the treated layer of titanium alloy ingots is characterized with absence of pores and discontinuities, crystals close to equiaxial, melted layer is formed by finer crystals in comparison with the rest part of the ingot, elongated in the solidification direction, i.e. to the ingot center.

It is found that at solidification rate more than 50 mm/min there is two times refining of the α-colonies and α-lamellas of ingot grains in the melted layer of titanium alloy ingots in comparison with the ingot base.

The results of investigation of distribution of microhardness on zones of the transverse template show that change of content of alloying elements and insignificant rise of oxygen and nitrogen in the melted layer does not have considerable effect on mechanical properties of the treated metal.

References

[1] Khorev A, Khorev M 2005 Titanium alloys, their application and development prospects *Materialovedenie* **#7** 25-34.

[2] Antipov V 2012 Development strategy of titanium, magnesium, beryllium and aluminum alloys *VIAM In col. of papers Aviation Materials and Technologies: Jubilee Scientific and Technical Collection - Appendix to Aviation Materials and Technologies Journal* 157-167.

[3] Khorev A 2009 Development of structural titanium alloys for the manufacture of parts of aerospace equipment assemblies *Svarochnoe proizvodstvo* #3 13-23.

[4] Khorev A 2013 Fundamental and applied works on titanium alloys for Buran and perspective directions for their development *Aviation Materials and Technologies Journal* **#S1** 10–14.

[5] Khorev A 2007 Theory and practice of development of titanium alloys for structures *Tekhnologiya Mashinostroeniya* **#12** 5-13.

[6] Afanasyev V, Popova M, Samon V 2014 On the creation of new light wrought alloys for space technology *Metallurgy of machinery building* **#5** 21-28.

[7] Sergienko S, Veselovsky A 2001 Electron beam remelting of titanium *Electrometallurgiya* **#11** 33-34.

[8] Ilyin A, Kolachev B, Polkin I 2009 Titanium alloys. Composition, structure, properties. Handbook *Moscow: VILS-MATI* 520p.

[9] Maslenkov S, Maslenkova E 1991 Steel and alloys for high temperatures. Handbook in 2 books. Book 2 *Moscow: Metallurgy* 832 p.

[10] Paton B, Trigub N, Akhonin S, Zhuk G 2006 Electron beam melting of titanium *Kiev: Naukova Dumka* 250 p.

[11] Boyarshinov V, Shalimov Al, Shcherbakov A 1979 Refining remelting of steel and alloys in vacuum *Moscow Metallurgiya* 304 p.

[12] Korbut E, Labunets V 2011 Features of tool wear out in titanium alloys treatment. *Problems of friction and wear* **#55** 83-93.

[13] Illarionov A, Popov A 2014 Technological and operational properties of titanium alloys: Tutorial *Ekaterinburg: Ural University Publishing house* 137 p.

[14] Paton B, Trygub M and Akhonin S 2011 Electron Beam Melting of Titanium, Zirconium and Their Alloys *Kyiv: E.O. Paton Electric Welding Institute, NASU* 216p.

[15] Tikhonovsky A, Trigub N, Mishchenko V 1985 Melting of ingots by flat electron beam *Spetsialnay Elektrometalurgiya* **#59** 71-74.

IOP Conf. Series: Materials Science and Engineering **582** (2019) 012047 doi:10.1088/1757-899X/582/1/012047

[16] Shilov G, Vrzhizhevsky E, Lihobaba A 1993 On the possibility of application of electron beam and plasma-arc heating for treatment of the surface layer of billets with gross defects *Probl. Special. Electrometallurgii* **#3** 58-63.
[17] Akhonin S, Milenin A, Pikulin A 2005 Modeling the processes of evaporation of alloying elements during electron beam melting of the surface of cylindrical ingots of titanium-based alloys *Advances in Electrometallurgy* **#1** 21-25.
[18] Kozlitin D, Tikhonovsky A, Trigub N, Panchenko G, Derecha A 1986 Numerical solution of the problem of non-axisymmetric heating of continuous cylinder with a movable heat source *Probl. Special. Electrometallurgi* **#3** 52-58.
[19] Chinakhov D, Agrenich E. 2008 Computer simulation of thermo-mechanical processes at fusion welding of alloyed steels *Mater. Sci. Forum* **575-578** 833-836.
[20] Borisova E, Bochvar G, Brun M 1980 Titanium alloys. Metallography of titanium alloys. Eds. S Glazunov and B Kolacheva *Moscow Metallurgiya* 464 p.

Laser Welding in Different Spatial Positions of T-joints of Austenitic Steel

A V Bernatskyi[1,a], O M Berdnikova[1,b], I M Klochkov[1,c], V M Sydorets[1,d] and D A Chinakhov[2,e]

[1]E.O. Paton Electric Welding Institute of the National Academy of Sciences of Ukraine, 11 Kazymyr Malevych St., Kyiv, 03150, Ukraine
[2]Yurga Institute of Technology, National Research Tomsk Polytechnic University, 26 Leningradskaya St., Yurga, Kemerovo Region, 652055, Russia

E-mail: [a] avb77@ukr.net, [b] omberdnikova@gmail.com, [c] mts.paton@gmail.com, [d] sydorvn@gmail.com, [e] chinakhov@tpu.ru

Abstract. It has been selected the most industry perspective laser welding technological implementations of the welded T-joints made of fine-sheeted heat resisting steel AISI 321 by identifying the influence of technological parameters on formation quality, level of mechanical properties and structural peculiarities of the welded T-joints obtained by laser welding in different spatial positions.

1. Introduction

Industry of many industry-developed countries has progressing fast in space industry, shipbuilding, instrument manufacturing, sectoral and precision engineering, etc. of late in view of many important factors, including a very important one that of wide introduction of innovative technologies of materials processing at manufacturing complex spatial structures [1-7]. It is done due to both updating of the existing technologies and developing the new ones [8-10].

At the present period of materials processing when manufacturing complex spatial structures there is a demand for application of local methods of processing as the most progressive welding processes, namely laser [4, 6, 7, 11-14], hybrid and combined laser-arc and laser-plasma [4, 8, 15, 16] and electric-beam ones [17-20]. It is related to a substantial reduction of the heat affection on the units which are joined by welding [8, 15, 21]. Due to this the level of residual strength and deformations of both separate units and the welded structures in a whole reduces [21-24].

The biggest challenge of reduction of residual strength and deformations is faced when manufacturing fine-sheet geometrically complicated structures, for example, panels [3, 12, 21, 25], or the closed section units of different configuration which are used in: shipbuilding to manufacture internal ship structures; aircraft industry to manufacture aircraft fuselage and wings [7, 11, 14, 18]; chemical and food industries to manufacture containers and vessels, etc. The mass of those structures is reduced by decreasing the thickness of materials which are the most frequently joined by different methods of welding. And the specified strength characteristics of the structures are achieved by using both new materials with a higher level of the necessary characteristics and certain structural elements.

T-joints are the most accepted forms of structural elements which are used in fine-sheet geometrically complicated structures to render them necessary characteristics of stiffness and strength [1, 6, 11-14, 22-27] when their weight is not very big. When producing T-joints by different methods fusing welding, the most accepted is a technological implementation (TI) shown on Figure 1,a [24], when welding is produced from the side of the stringer under a certain angle. The advantages of this T-joints welding TI include the following: a relatively simple technical equipment; possibility to produce welding in one or two passes under one-side welding; when doing welding in one pass under both-side welding as shown on Figure 1,b [28] (when using two sources of heating at the same time) formation of fillets is guaranteed in the zone of joining skin and stringer and symmetrical spread of heat fields and stresses; possibility to distribute in time and space of sources of heating under both-side welding; etc.

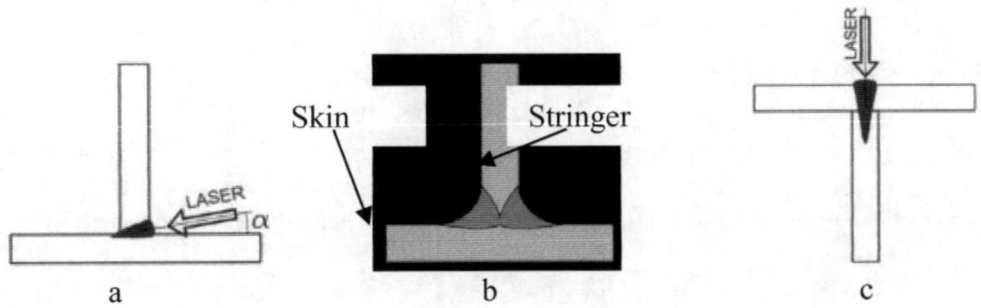

Figure 1. Implementations of laser welded T-joint: a – one-side butt welded joint [24], b – formation of fillets under both-side welding [28], c – I-core T-joint [24]

However there are many structures [3, 12, 21, 25], in which access from the stringer side of the welded T-joint is restricted or impossible. It does not allow use the welding TI shown on Figure 1,a. In that case they use the welding TI shown on Figure 1,c [24]. According to that TI welding is produced from the skin side of the T-joint by a through penetration and melting down of a certain section of the stringer. But in that case there is no possibility for visual control of positioning of the edges of the welded joint when preparing for welding. That complicates and makes more expensive the phase of preparation of the welded T-joints as it needs using additional technological and organizational facilities, like using more precise and, consequently, more expensive technological tools to assemble welded joints.

Besides, problem occurs to ensure necessary strength characteristics of the welded T-joints which are obtained by the welding TI shown on Figure 1,c. For the welded T-joints that work under different loading on the skin of the welded joint from the stringer side, a necessary quality requirement and achievement of the specified long-term strength characteristics is formation of fillets on each side from the stringer. The fillets must have a certain transition radius from the stringer metal to the skin metal as shown on Figure 1,b. If there are no fillets on both sides of the stringer or existence of not welded sections in the zone of stringer and skin joining, creation of defects as undercuts is possible. It may result in big rise of stress in this section of a welded joint, and later, after a long-term exploitation of the welded joint and excessing of a certain stress values, in its destruction.

2. The purpose and objectives of the study

In this connection obtaining of the welded T-joints with a guaranteed formation of fillets with specified radius when producing welding according to the TI shown on Figure 1,c becomes important. This task is more difficult when the units of the welded joints (stringers and skins) are about 1 mm thick. The task may be solved by developing of the welding TI of T-joints as shown on Figure 1,c, using different technological approaches, namely:

- mode of generation of laser radiation;
- changing of the number of passes;
- spatial position;
- position of the focus plate relative to the skin surface.

The objective of the research is the study of the TIs of laser welding of the welded T-joints made of fine-sheeted heat resisting steel AISI 321 and selection of the most innovative solutions for the industry application. It will be based on identification of the technological parameters influence on formation quality, level of mechanical properties and structural peculiarities of the welded T-joints obtained by laser welding in different spatial positions.

3. Research methods and equipment

T-joints were obtained by a laser welding from the samples made of sheet corrosion- and heat-resisting steel AISI 321. To manufacture stringers of the welded T-joints they used AISI 321 sheets 1.2 mm thick, and to manufacture a skin of the welded T-joints – 0.8 mm thick.

For the research Nd:YAG-laser «DY044» produced by «ROFIN-SINAR» (Germany) with the radiation wavelength $\lambda = 1.06$ μm was used.

The samples were fixed in the clamp and welded in the lower and vertical spatial positions. The weld was along the rolling direction of AISI 321 steel which was used as a skin of the welded T-joint. Welding both "on rise" (with moving of the laser head upwards relating to the unmovable clamp with the fixed sample) and "on descent" (with moving of the laser head downwards relating to the unmovable clamp with the fixed sample) was studied.

The quality of the obtained welded T-joints was assessed according to ISO 13919-1:1996 and guaranteed obtaining of fillets of 2…10 mm in radius on both sides of the stringer of the welded joint. To obtain the welds of high – B- or medium – C- quality level (in accordance with ISO 13919-1:1996) the edges of the stringer were mechanically processed (fine milling). Shortly before welding (before fixing in the welding clamp) the samples were degreased by gasoline and later cleaned by acetone. A preliminary heat treatment of the samples before welding was not carried out.

The units of the T-joint were assembled in clamps with edges clamping in 2…5 MPa. The clearance between stringer and skin did not exceed 0.1 mm (the check was done by a tool probe). When assembling the T-joints for a laser welding the tack welds were not used.

The visual control of the obtained welded joints was carried out on the previously mechanically cleaned samples. When carrying out the visual control the combined lighting was used (day and local lighting using filament lamp) – about 1000 lx. Measuring devices and tools was used such as: magnifier glasses 4× and 10×zoom, measuring magnifier 10×zoom with accuracy of 0.1 mm, trammel ShTs140 (class 2), ruler, portable measuring profiler of watch type with accuracy of 0.01 mm, portable microscope and luxmeter U-116. The following parameters were evaluated by visual inspection results: weld width; flakiness; existence and geometrical size of fillets; existence of craters (shrinkage cavities in craters) and other characteristics as provided by ISO 13919-1:1996.

The radiographic (X-ray) test was carried out on the previously mechanically degreased samples. The following devices were used: X-ray machine RAP 150/300; penetrameter – grooved №1 GOST 7512-82; 240×100 mm radiographic film KODAK - AA400; intensifying screens – metal - Pb-0.027. Radiographic mode: U = 80 kV, I = 10 mA, t = 120 s, focal distance – 1000 mm. Number of expositions – 2 (one on each side, on the part of the stringer side of the T-joint). The length of the controlled area of the welded joint during the radiographic test was 220 mm. Based on the radiographic test the appearance of internal defects in the form of cracks, pores, inclusions, and lack of penetration were assessed.

Metal macrograph study of the welded samples was carried out according to ISO 17639:2003 using optical microscopy (microscopes Versamet-2, Neophot-32). The study was carried out on each section (weld – heat affected zone (HAZ) – base metal) of each sample.

Static tensile strength testing of the welded T-joints samples was carried out on servo-hydraulic machine MTS 318.25 according to ISO 6892-1:2016 on five samples cut out from each welded joint.

The testing objective was to identify loading necessary to destroy a sample of the T-joint under static tension and localization of a break in the welded T-joint.

4. Results and Discussion

4.1. Laser welding in the lower spatial position

Initially, a variant of one-pass welding of a T-joint in the lower spatial position in the continuous mode of generation of laser radiation was considered. It is the most simple as regards realization. However, to achieve a stable formation of fillets of a specified radius on both sides of the welded joint under the circumstances of the experiment failed. Due to the variation of the values of the parameters of the technological modes (laser radiation power; welding speed; the position of the focal plane, with respect to the position of the surface of the skin of the T-joint, etc.), a number of technological variants of the process realization were tested. However, any minor deviations (less than 5%) from the above technological process parameters, or deviation of the laser head during welding from the symmetry plane of the stringer of the welded T-joint lead to uneven shaping of fillets on both sides of the welded stringer, or even to the formation of only one fillet on one side of the stringer of welded joint.

Therefore, for laser welding in the lower spatial position, more than ten advanced welding TIs were considered, from which three of the most perspective, distinguished by the parameters of the technological modes, were determined, and the industrial application of each of them makes it possible to obtain a control welded T-joint, which meets the requirements for obtaining welds of the specified geometrical sizes (with guaranteed fillets forming) and high B or medium C quality levels (according to ISO 13919-1: 1996).

In all cases, the welding in the lower spatial position was produced with supply of argon as a shielding gas with different spending in certain areas: in the area of the welding pool – 25 l/min; bottom skin – 5 l/min (in total on both sides of the stringer); in the "tail" of the welding pool – 20 l/min.

Based on the results of visual and radiographic tests, no samples were welded in the lower spatial position had defects, but there was insufficient strengthening of the upper bead of the weld. Despite the identified drawback, all samples welded in the lower spatial position correspond to the "high B" quality standard according to ISO 13919-1: 1996.

4.1.1. Welding technological implementation #1 (TI#1)

In order to guarantee the formation of fillets of specified geometric sizes on each side of the stringer of the welded T-joint, a variant of the TI#1 was considered, according to which the laser welding was produced in two passes, produced one after the other, indented at a specified distance Δ from symmetry plane of the stringer of the welded T-joint. The value of this parameter varied within Δ = 0.2...0.5 mm. The welding was produced in the continuous mode of generation of laser radiation with the power of P from 3.0 to 4.4 kW with the defocusing ΔF = −4...+4 mm, at the welding speed V = 50...300 mm/s.

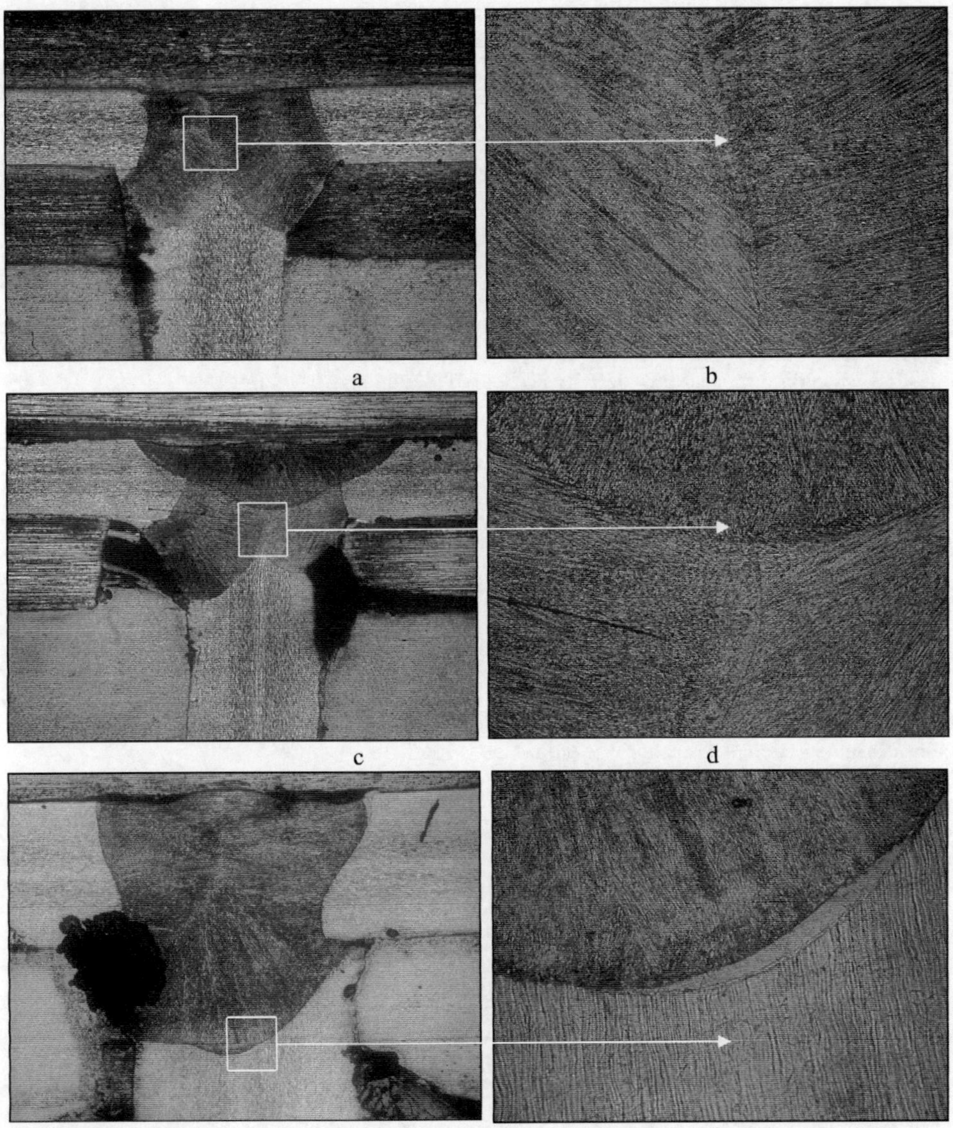

Figure 2. Welded T-joints obtained in the lower spatial position: macrograph, ×25: a – TI#1, c – TI#2, e – TI#3; micrograph: b – zone of overlapping of the welds upon two passes; TI#1, ×320; d – zone of overlapping of the welds upon three passes TI#2, ×156; f – transition zone between weld and stringer TI#3, ×320.

The results, which are better in terms of the formation of welded joints, in particular stability of fillets formation along the weld, were obtained by welding in a continuous mode of generation of laser radiation with a power of 4.4 kW without defocusing ($\Delta F = 0$ mm), at a speed of 100 mm/s indented from the symmetry plane of the stringers by a value of 0.3 mm for each pass.

On Figure 2,a,b it is shown the welded joint produced according to the TI#1. The results of the metal macrograph study of the parameters of the structural features of the welded joint obtained by the TI#1 are given in Table 1.

It is worth noting that according to the results of the mechanical tests on the static tension given in Table 2 it was determined that the welded joints obtained according to the TI#1 ensure a sufficiently high resistance to the destruction force (the average value was calculated based on the results of tests of 5 plates cut off from one sample, without taking into account the maximum and minimum value). Destruction of all samples took place on the base metal of the stringer of the welded T-joint #1 for all experiments carried out. This also indicates the achievement of high strength values of the obtained welded joints.

Producing welding in two passes with a displacement from the symmetry plane of the stringer by 0.3 mm allowed ensure the specified configuration of the welded T-joints and obtain relatively high strength values.

The welded joints obtained according to the TI#1, meet the requirements for obtaining welds of the average quality level C, according to ISO 13919-1: 1996. To the main drawbacks of this welding TI of T-joints made of thin-sheet austenitic steel AISI 321, the following should be considered:

1. the technological complexity of the implementation of the indentation at a specified distance from the symmetry plane of the stringer;
2. existence of a weakened zone in a welded joint with a double-melted metal of weld and an additional HAZ in the specified area (see Figure 2,a);
3. two times lower welding process productivity compared to a one pass welding TI;
4. no strengthening of the upper bead.

Table 1. Structure parameters and mechanical features of the welded T-joints

	Value of structural components (μm)				Microhardness (MPa)		
WTI[a]	$h_1 \times L_1$[b]	$h_2 \times L_2$[c]	D[d]	δ[e]	HV_{WB}[f]	HV_{TZ}[h]	HV_{FL}[g]
#1	4…10×30…120	3…8×30…50	10…20	40…50	2510…2740	2280…2380	2340...2380
#2	2…8×10…100	2…5×20…50	8…20	40…50	2240…2470	2280...2320	2260...2300
#3	5…12×15…100	4…6×30…50	10…20	25…35	2510...2540	2610...2670	2410...2450
#4	10…15×30…120	5…10×30…50	10…20	40…60	2360…2680	2280…2450	2300...2540
#5	8...12×20...100	3…8×30…50	10…20	40…50	2360...2490	2450...2700	2480...2540
#6	15...25×20...100	3…8×30…50	10…20	30…45	2450...2670	2480...2660	2450...2640
#7	10…20×20…100	3…10×20…50	10…20	40…60	2540...2740	2560...2740	2360...2660

[a] WTI – Welding technological implementation.
[b] $h_1 \times L_1$ – Sizes of crystal grains in weld.
[c] $h_2 \times L_2$ – Sizes of crystal grains in transition zone.
[d] D – Size of the austenitic grain from the side of the base metal of stringer in the fusion line.
[e] δ – Width of the transition zone.
[f] HV_{WB} – Microhardness of crystal grains in a weld.
[g] HV_{TZ} – Microhardness of crystal grains in transition zone.
[h] HV_{FL} – Microhardness of austenitic grain from the side of the base metal of stringer in the fusion line.

Table 2. Results of mechanical testing of the welded T-joints

	Q[b] (H)		N[c] (%)	
WTI[a]	Value range	Average range	On the metal of the weld	On the base metal of the stringer
#1	20289...21240	20951	0	100
#2	19590...21830	20460	60	40
#3	18044...19504	18714	100	0
#4	20944...21829	21566	40	60
#5	18500...21882	20883	40	60
#6	15234...18763	17587	100	0
#7	21394...23232	21887	0	100

[a] WTI – Welding technological implementation.
[b] Q – Resistance of the welded joint to the breaking tension.
[c] N – Number of samples that were destroyed in the corresponding location of the welded T-joint.

4.1.2. Welding technological implementation #2 (TI#2).
Unlike the TI#1, laser welding according to the TI#2 included a third pass, except for two main passes, indented from the symmetry plane of the stringer by 0.3 mm (in modes identical to the TI#1 modes) that were produced for a proper formation of fillets on both sides of the welded T- joint. The third pass was produced to melt the surface to ensure a proper formation of the upper bead of skin of the welded joint. The third "smoothing" pass was carried out in the continuous mode of generation of laser radiation in the variation of the parameters of technological modes in the following ranges: P=1.0…3.0 kW; ΔF=+10...+30 mm; V=1…50 mm/s.

Figure 2,c,d shows the welded joint carried out according to the TI#2, according to which the third pass was carried out in the continuous mode of generation of laser radiation with the power of P = 1.5 kW, at defocusing of ΔF = +20 mm, and welding speed 25 mm/s. The position of the focus of laser radiation was deepened with respect to the surface of the sample by 4 mm. Welding was carried out at a speed of 75 mm/s. According to the data of the metal macrograph study, the metal in the fusion zone after welding the formation of austenitic dendritic structure with an increased crystallite size (Figure 2,c,d and Table 1) is characteristic. The microstructure study has also shown that as regards the depth of the transition from melted metal to the weld (in the transition of the skin-stringer) in the size of the structural components the dendritic structure has a uniform character.

An additional pass allows obtain samples with geometric sizes that meet the requirements of high level B of quality (according to ISO 13919-1:1996). However, according to the results of mechanical tests on a static tension, it was found that the samples welded according to the TI#2, showed a large scattering of the values of resistance to the destruction force. Such conclusions are confirmed by the fact that the destruction of 60% of the samples as a result of mechanical tests occurred on the metal weld, and only 40% of the samples were destroyed by the base metal of the stringer of the welded T-joint.

4.1.3. Welding technological implementation #3 (TI#3).
In order to eliminate the negative influence on the structure and mechanical properties of welded joints, as a result of reheating and cooling in the course of additional passes during the formation of welded T-joints in accordance with TIs #1 and #2, laser welding of samples according to the TI#3 was produced in one pass in the pulsed mode of generation of laser radiation. This allowed ensure a uniform formation of fillets on both sides from the stringer of the welded T-joint using more than two times the extended range of parameters of the technological modes, compared with welding in the continuous mode of generation of laser radiation. The parameters were the following ones: pulsed power P_{Max}=3.0…4.4 kW, pulse frequency 100…250 Hz, and fill factor ξ=50…75%. The focus of the laser beam has been deepened on 2.0…4.0 mm relative to the surface of the sample. The welding speed was 50…100 mm/s.

Figure 2,e,f shows the welded joint obtained by the TI#3 in modes that are as close as possible to the optimal, according to the criteria of ensuring the two-sided forming of fillets, the specified sizes and the maximum level of mechanical characteristics of the welded T-joint: at maximum pulsed power P_{Max}=4.4 kW, pulse frequency 250 Hz, and filling factor ξ=75%. The focus of the laser beam has been deepened on 4.0 mm. The welding speed was 75 mm/s.

Summing up the consideration of TIs based on laser welding in the lower spatial position, note the following: according to the results of mechanical tests on a static tension (Table 2) it was established that the samples welded according to the TI#3 ensure a relatively low value of the resistance of the weld joint to the breaking force. We emphasize that the destruction of all samples took place on the metal weld of the welded T-joint TI#3 for all tests carried out. Thus, the features of the joints obtained in accordance with the TI#3 include the unevenness in the formation of fillets on both sides of the welded joint and low level of strength characteristics, which does not give us reason to recommend it for use in industry.

4.2. Laser welding in a vertical spatial position
In case of laser welding produced in a vertical spatial position, the welding modes were considered and the welding variants were investigated as "on rise" (with the laser head moving along the vertical upwards, relative to the unmovable clamp with the fixed sample), and "on descent" (with the moving of the laser head downwards, relative to the unmovable clamp with fixed sample). It should be noted that the general positive tendency of expansion (almost two times) of the range of technological parameters of the modes, at which the stable reception of the specified form and geometrical sizes of the weld of the welded joint with the guaranteed formation of fillets on both sides of the stringer is achieved. It distinguishes welding in the vertical space position from welding in the lower spatial position.

For laser welding in the vertical spatial position, four welding TIs have been identified, the application of each implementation allows obtain the control welded T-joint meeting the requirements for obtaining welds of high B or medium C quality levels (according to ISO 13919-1:1996). They used the same equipment to protect the metal of the welded joint by argon from the influence of the environment, as that was done for welding in the lower spatial position.

According to the results of the visual test it was established that all samples welded in the vertical spatial position correspond to the "high B" quality level, according to ISO 13919-1:1996.

In all samples, welded in a vertical space position, lone pores with dimension not more than 0.5 mm are detected. According to the results of X-ray test it was established that all samples welded in the vertical spatial position correspond to the quality level "C moderate" according to ISO 13919-1:1996.

4.2.1. Welding technological implementation #4 (TI#4).

The TI#4 differed from the TI#1 only by a change of the spatial position from the lower to a vertical one. Welding was carried out downwards. Other parameters of the technological modes were not changed, as in TI#1.

The best results from the point of view of obtaining the geometric sizes of the welded joint and the stability of the forming of fillets were obtained by welding in a continuous mode of generation of laser radiation with a power of 4.4 kW without defocusing $\Delta F = 0$ mm, at a speed of 100 mm/s indented from the symmetry plane of the stringer by 0.3 mm for each pass.

Despite relatively high strengths (see Table 2), the samples obtained under the TI#4 were broken both by the base metal of stringers (60% of samples) and by the metal weld (40%).

The comparison of the samples obtained by TI#1 (Figure 2,a,b) and #4 (Figure 3,a,b) shows a slight increase in strength and microhardness values when the welding position from the lower to the vertical changes. At the same time, it should be noted increasing of instability in the formation of the weld, as well as the multivariance of position of the destruction location of welded joints.

4.2.2. Welding technological implementation #5 (TI#5).

The TI#5 differs from the TI#1 and #4 by the parameters values of the modes, which are the best in view of the formation of the welded joint and stability of the fillets forming, namely by increasing the size of defocusing to $\Delta F = +4$ mm, increased from 100 mm/s to 113 mm/s speed of welding, as well as changing the spatial position from the lower to the vertical one. According to the TI#5 welding was produced upwards. Other parameters of the technological modes were not changed, as in TI#1.

As can be seen from the photos shown in Figure 3,c,d and from Table 1, the change in the focus position relative to the surface of the sample resulted in significant changes in the geometric sizes of the welded joints as compared to the samples obtained according to the TI#1.

Unlike the samples obtained according to the TI#2, the welded T-joints obtained during the welding process according to the TI#5, demonstrate both high level of mechanical characteristics and high stability in the formation of welded joints. This is confirmed by the fact that all welded joints obtained according to the TI#5 were destroyed by the base metal of the stringer, as distinguished to the samples obtained in the lower spatial position according to the TI#2.

According to the results of mechanical tests on a static tension, it was found that the samples welded according to the TI#5, show a big dispersion in the values of resistance to the destruction force. This may indicate a lack of stability in the formation of welded joints.

The conclusions like these are confirmed by the fact that the destruction of most samples occurred on the base metal of the stringer of the welded T-joint.

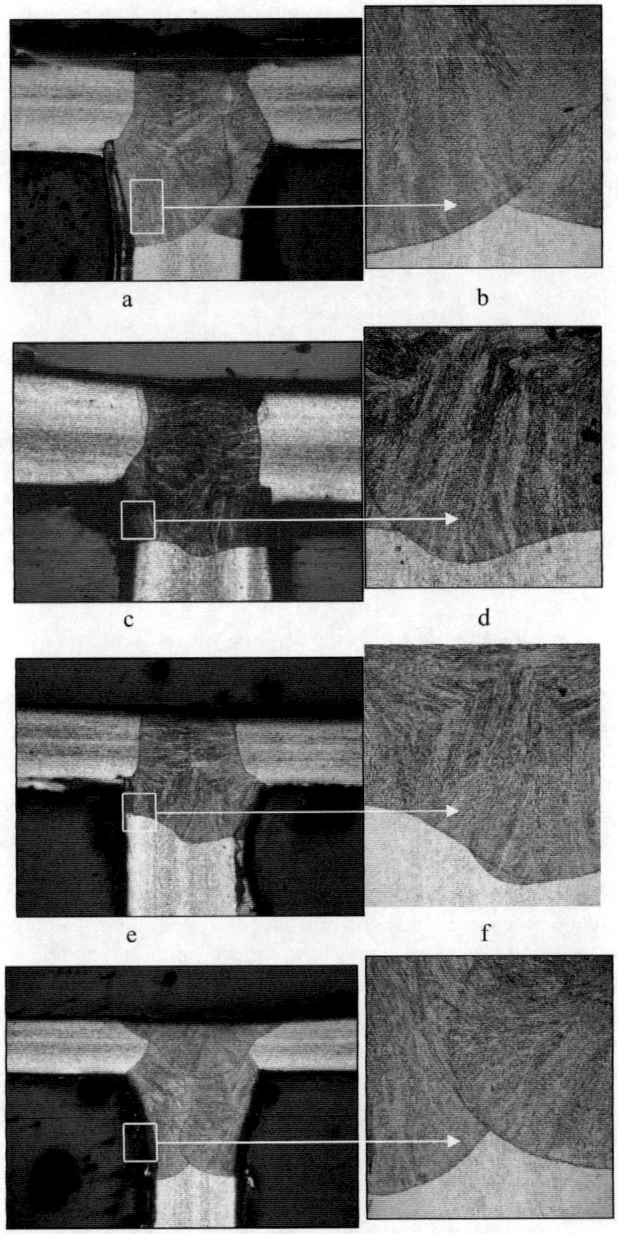

Figure 3. Welded T-joints obtained in vertical spatial position: macrograph, ×30: a – TI#4, c – TI#5, e – TI#6, g – TI#7; micrograph: b – transition zone between weld and stringer TI#4, ×100; d – transition zone between weld and stringer TI#5, ×100; f – central weld zone TI#6, ×100; h – central weld zone TI#7, ×100.

4.2.3. Welding technological implementation #6 (TI#6).

Laser welding of the welded T-joints according to TI#6 was produced in pulse mode, which according to its parameters was identical with the modes of the TI#3, except for the value of defocusing. For TI#6, the optimal value of the focus position was ΔF = 0 mm. Besides, as distinguished to TI#3, the samples according to the TI#6 were welded in vertical position downwards.

The changes like those in the modes resulted in significant changes in the formation of welded joints (Figure 3,e,f), but only lightly affected structural changes (Table 1). At the same time, the lack of stability in formation of the upper bead of the welded joint according to the TI#6 affected the strength (Table 2). According to the results of mechanical tests on a static tension, it was found that all samples welded according to TI#6 were destroyed by a metal weld of the welded joint and showed the lowest strength of all TIs.

4.2.4. Welding technological implementation #7 (TI#7).
In a similar way to the TI#2, laser welding according to TI #7 was produced in two continuous passes "on descent" indented from the symmetry plane of stringer by 0.3 mm with different defocusing (the best value was fixed $\Delta F = 0$ mm) and the next third pass for slight melting of the welding zone at a speed of 25 mm/s and a power of 1.5 kW with a 20 mm of defocusing.

According to the results of mechanical tests of the welded joints obtained according to the TI#7 (Figure 3,g,h), it was defined that all samples were destroyed along the base metal of the stringer. The value of the resistance to the destruction is quite high.

According to the above data, in some cases as relates to the samples welded according to TIs (#2, #4 and #5), there is a destruction both on the base metal and on the metal of weld. This may be related to the lack of stability in the formation of welded joints, and in particular the structural elements as fillets. At the same time, for samples welded in the vertical position, the destruction of the sample on the stringer is more characteristic, while in the samples welded in the lower position this is not found. The results of mechanical tests show that in comparison with samples welded in the lower spatial position, the samples welded in the vertical spatial position less dispersion value is more characteristic. This may be related to a more stable formation of welded joints, as well to a higher average resistance value to the destruction force, which is higher by about 10...20% as compared to the welded joints obtained in the lower spatial position under similar TIs.

5. Conclusions

The studied technological implementations of laser welding of the welded T-joints made of fine-sheet heat resistant steel AISI 321 make it possible to select the most promising solutions for industrial applications.

The general positive tendency of the expansion (almost two times) of the range of technological parameters of the modes is defined, at which a stable obtaining of the specified form and geometrical sizes of the weld of the welded joint with the guaranteed formation of fillets on both sides of the stringer is achieved. This distinguishes the welding in the vertical spatial position from the welding in the lower spatial position.

According to the results of the comprehensive research, it has been defined that the most industry perspective technological implementations of laser welding of the welded T-joints made of fine-sheet heat-resistant steel AISI 321 are the technological implementation #1 (for welding in the lower spatial position) and the technological implementation #7 (for welding in a vertical spatial position). The selection of these technological implementations is justified by the most stable formation of fillets on both sides of the T-joint as well providing the necessary geometrical sizes of the welded joints and high level of mechanical properties.

When choosing between technological implementations #7 and #1, it is recommended to render preference to the implementation #7 since when producing welding in the vertical spatial position a comparatively higher values of mechanical characteristics are achieved.

References

[1] Mittelstädt C, Seefeld T, Woizeschke P and Vollertsen F 2018 Laser welding of hidden T-joints with lateral beam oscillation. *Procedia CIRP* **74** 456-460 doi:10.1016/j.procir.2018.08.151

[2] Paton B, Akhonin S and Prilutsky V 2011 Development of welding technologies in titanium component manufacturing *Proc. of the 12th World Conference on Titanium* (Beijing) vol 2 (Beijing: Science Press) 1585-1591

[3] Kozak J 2009 Selected problems on application of steel sandwich panels to marine structures *Polish Maritime Research* **16** 9-15 doi:10.2478/v10012-008-0050-4

[4] Shelyagin V, Khaskin V, Bernatskyi A, Siora A, Sydorets V and Chinakhov D 2018 Materials Science Forum **927** 64-71 doi:10.4028/www.scientific.net/MSF.927.64

[5] Akhonin S, Belous V, Berezos V and Selin R 2018 Effect of TIG-welding on the structure and mechanical properties of the pseudo-β titanium alloy VT19 welded joints *Mat. Sci. Forum* **927** 112-118 doi:10.4028/www.scientific.net/MSF.927.112

[6] Romanoff J, Remes H, Socha G, Jutila M and Varsta P 2007 The stiffness of laser stake welded T-joints in web-core sandwich structures *Thin-Walled Structures* **45** 453-462 doi:10.1016/j.tws.2007.03.008

[7] Sun X, Shehab E and Mehnen J 2013 Knowledge modelling for laser beam welding in the aircraft industry *Int. J. Adv. Manuf. Technol.* **66** 763-774 doi:10.1007/s00170-012-4364-0

[8] Reisgen U, Krivtsun I, Gerhards B and Zabirov A 2016 Experimental research of hybrid welding processes in combination of gas tungsten arc with CO_2- or Yb:YAG-laser beam *J. Laser Appl.* **28** 022402

doi:10.2351/1.4944096
[9] Chinakhov D, Chinakhova E, Grichin S and Gotovschik Y 2016 Influence of welding with two-jet gas shielding on the shaping of a welding joint *IOP Conf. Series: Mat. Sci. Eng.* **125** 012013 doi:10.1088/1757-899X/125/1/012013
[10] Chinakhov D and Agrenich E 2008 Computer simulation of thermo-mechanical processes at fusion welding of alloyed steels *Mat. Sci. Forum* **575** 833-836 doi:10.4028/www.scientific.net/MSF.575-578.833
[11] Reitemeyer D 2013 Laser welding of large scale stainless steel aircraft structures *Phys. Procedia* **41** 106-111 doi:10.1016/j.procir.2018.08.151
[12] Meng W, Li Z, Huang J, Wu Y, Chen J and Katayama S 2014 The influence of various factors on the geometric profile of laser lap welded T-joints *Int. J. Adv. Manuf. Technol.* **74** 1625-1636 doi:10.1007/s00170-014-6114-y
[13] Yang Z, Tao W, Zhao X, Chen Y and Shi C. 2017 Numerical modelling and experimental verification of thermal characteristics and their correlations with mechanical properties of double-sided laser welded T-joint *Int. J. Adv. Manuf. Technol.* **92** 1609-1618 doi:10.1007/s00170-017-0257-6
[14] Yang Z, Zhao X, Tao W, Jin C, Huang S, Wang Y and Zhang E 2018 Comparative study on successive and simultaneous double-sided laser beam welding of AA6056/AA6156 aluminum alloy T-joints for aircraft fuselage panels *Int. J. Adv. Manuf. Technol.* **97** 845-856
[15] Krivtsun I, Reisgen U, Semenov O and Zabirov A. 2016 Modeling of weld pool phenomena in tungsten inert gas, CO_2-laser and hybrid (TIG+CO_2-laser) welding *J. Laser Appl.* **28** 022406 doi:10.2351/1.4943994
[16] Markashova L, Berdnikova O, Alekseienko T, Bernatskyi A and Sydorets V 2019 Nanostructures in welded joints and their interconnection with operation properties *Advances in thin films, nanostructured materials, and coatings. Lecture notes in mechanical engineering* (Singapore: Springer) doi:10.1007/978-981-13-6133-3_12
[17] Paton B, Nazarenko O, Nesterenkov V, Morozov A, Litvinov V and Kazimir V 2004 Computer control of electron beam welding with multi-coordinate displacements of the gun and workpiece *Avtomaticheskaya Svarka* #5 3-7
[18] Nazarenko O, Nesterenkov V, and Ilyushenko R 2005 Weldability of aircraft aluminum alloys of great thickness in EBW *Avtomaticheskaya Svarka* #8 25-30
[19] Ilyushenko R and Nesterenkov V 2006 Novel technique for joining of thick section difficult-to-weld aluminium alloys *Mat. Sci. Forum* **519-521** 1125-1130 doi:10.4028/www.scientific.net/MSF.519-521.1125
[20] Yerofeyev V, Logvinov R, Nesterenkov V and Mazo A 2014 Formation of the equivalent heat source for calculating strains in structures in electron beam welding *Welding International* **28** 557-561 doi:10.1080/09507116.2013.840042
[21] Kim J, Kang S and Jang B 2012 Prediction of laser welding deformation of sandwich panel in 3D thermal elasto-plastic analysis *Proceedings of the Twenty-second International offshore and polar engineering conference* (Rhodes) 271-277
[22] Piekarska W, Kubiak M and Saternus Z 2013 Numerical simulation of deformations in T-joint welded by the laser beam *Arch. Metall. Mater.* **58** 1391-1396 doi:10.2478/amm-2013-0181
[23] Gallo P, Romanoff J, Frank D, Karttunen A and Remes H 2017 Synthesis of experimental testing and fatigue behavior of laser stake-welded T-joints on medium-high cycle fatigue range *Procedia Structural Integrity* **5** 809-816 doi:10.1016/j.prostr.2017.07.055
[24] Piekarska W, Saternus Z, Sapieta M and Kopas P 2019 The influence of joining technique on the deformation of laser welded T-joints *MATEC Web of Conferences* **254** 02011 doi:10.1051/matecconf/201925402011
[25] Karttunen A, Kanerva M, Frank D, Romanoff J, Remes H, Jelovica J, Bossuyt S and Sarlin E 2017 Fatigue strength of laser-welded foam-filled steel sandwich beams *Mater. Des.* **115** 64-72 doi:10.1016/j.matdes.2016.11.039
[26] Meng W, Li Z, Lu F, Wu Y, Chen J and Katayama S 2014 Porosity formation mechanism and its prevention in laser lap welding for T-joints *J. Mater. Process. Tech.* **214** 1658-1664 doi:10.1016/j.jmatprotec.2014.03.011
[27] Yang Z, Zhao X, Tao W and Jin C 2019 Effects of keyhole status on melt flow and flow-induced porosity formation during double-sided laser welding of AA6056/AA6156 aluminium alloy T-joint *Opt. Laser Technol.* **109** 39-48 doi:10.1016/j.optlastec.2018.07.065
[28] Welding overview *Olympus IMS* https://www.olympus-ims.com/ru/ndt-tutorials/flaw-detection/weld-overview/

IOP Conf. Series: Materials Science and Engineering **582** (2019) 012049 doi:10.1088/1757-899X/582/1/012049

Renewal of Elements and Construction Units of Gas Turbine Engines by Means EBW

V M Nesterenkov[a], Yu V Orsa[a], K S Khripko[a]

E.O. Paton Electric Welding Institute of the National Academy of Sciences of Ukraine, 11 Kazymyr Malevych St., Kyiv, 03150, Ukraine

E-mail: [a] office@paton.kiev.ua

Abstract. Renewal of gas turbine engines (GTE) is very urgent and at the same time highly complicated technical task. A cost effective alternative to new replacement parts is the restoration of the damaged or worn construction units and elements of GTE, including compressor and fan blades, as well as guide cases and hydraulic cylinders of nozzle block. The work contains the results of the development of reliable and efficient methods of gas turbine parts renewal by means electron beam welding. The technology of repair of three types of blade airfoil defects, replacement of damaged guide vanes and restoration of worn hydraulic cylinders was mastered. It is confirmed that electron beam welding is ideal for the solution of aircraft GTE construction units repair problems.

1. Introduction

Electron beam welding is one of the leading technological processes used in the production and repair of structures for aerospace industry. Application of computer numerical control (CNC) tools for EBW brought this technology into the category of leading processes, owing to the capabilities of precision control of both electron beam movement trajectory while welding and its energy, including regulation of the total value of power and shape of this power distribution in space [1, 2]. For many years Paton Electric Welding Institute (PWI) has taken leading positions in development of the advanced technologies and specialized equipment for EBW [3].

Appropriateness of restoration repair of expensive parts of the gas turbine engines (GTE) does not raise doubts [4]. A cost effective alternative to new replacement parts is the repair of the damaged or worn blades – the damaged section is removed and a repair «patch» is electron beam welded in place. In the final step the patch is machined and contoured to the original shape.

The main reason of early replacement of the aircraft GTE in their operation is damage of titanium blades of the fan and compressor as a result of foreign objects coming in the engine [5]. Usually engine operation is allowed at insignificant defects of leading and trailing edges of the blade airfoil without tears. Part of similar (i.e. without tears), but somewhat larger defects can be repaired directly on the engine. Thus, for example, material rising near the nicks is dressed and bends of the blades are eliminated by flattening. Then, the repaired places are polished. Sometimes, removal of the nicks due to smooth rounding of the edge of up to 10–12 mm radius is allowed. In contrast, correction of local damages of the blades, exceeding allowable norms, requires dismantling of the damaged blade and repair under production conditions. Usually such a repair lies in mechanical removal of the defective zone till the boundaries of knowingly undamaged blade metal with further connection (welding, brazing) of a welded-in piece of corresponding size instead of it and with technological tolerance on thickness for acquiring a necessary profile of blade repair section using further mechanical treatment [6–8].

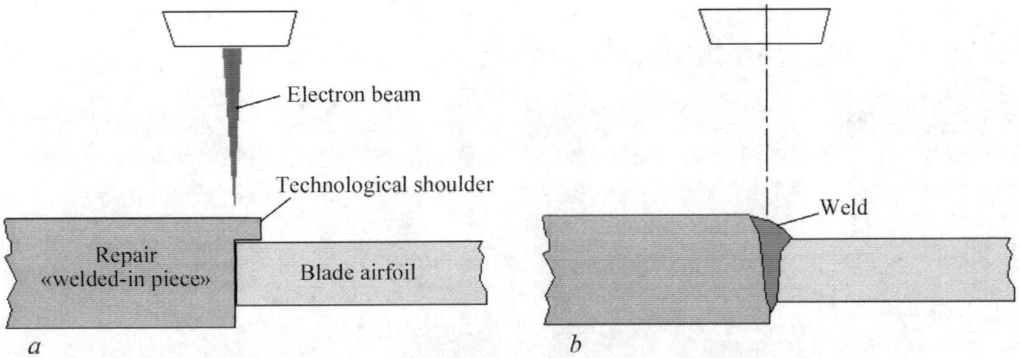

Figure 1. Scheme of repair EBW-joint of «welded-in piece» and blade airfoil:
a – before welding; *b* – after welding

2. EBW as a tool for Renewal and repair of gas turbine engines

Generally, the solution on repair permissibility for each specific defect of the blade is made by the certified repair organization, following the next aspects, i.e. belonging of the blade to low pressure compressor (LPC) or high pressure compressor (HPC), blade type (namely, blade or vane), number of compressor stage, whether it is leading or trailing edge of the airfoil, and at last, evaluating directly the defects sizes, their quantity and location relatively to the places of maxi- mum service stresses (including resonance ones).

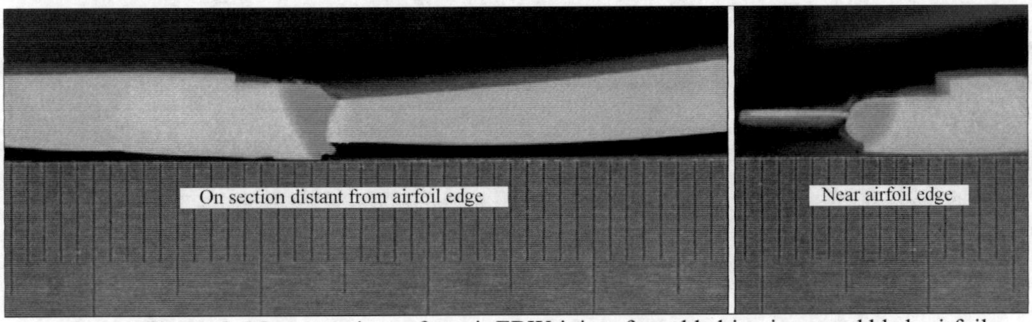

Figure 2. Macro sections of repair EBW-joint of «welded-in piece» and blade airfoil

Our task lied only in development of an efficient and reliable method, which fundamentally will allow such a repair using electron beam welding (EBW) un- der domestic production conditions. The technological methods were mastered on the spot and extended defects of airfoil leading edge of different size titanium blades, namely fan blades and blades of GTE LPC.

The following scheme of repair «welded-in piece» and blade airfoil welded joints was developed. It in principle does not depend on the fact whether it is spot or extended defects (Figure 1).

In all cases the joint is carried out by a single-pass EBW, and smooth (without undercuts) transfer from «welded-in piece» surface to base metal is provided by moderate concentration of the electron beam as well as sufficient quantity of additional metal due to applied structure of welded joint with «overhanging shoulder». Moreover, such a smooth transfer can be reached in the areas with sufficiently larger airfoil thickness as well as in the thinnest places adjacent to the edge (Figure 2).

The repair technology was mastered for three types of airfoil blade defects, i.e. spot damage of the blade angle, spot damage of the edge of blade airfoil main part as well as extended local defects of the edge starting from blade angle.

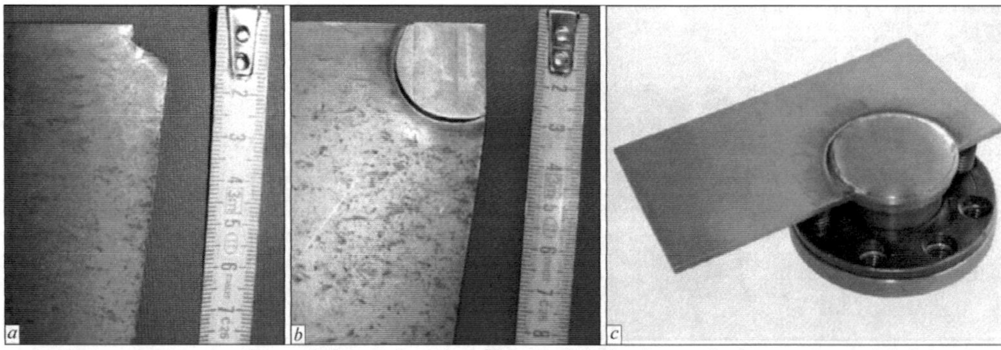

Figure 3. Repair of spot defects of angle of blade airfoil using cylinder repair «welded-in pieces»: *a* – damaged angle of blade before repair; *b* – repaired angle (ends of «welded-in piece» after EBW); *c* – assembly-welding fixture for EBW of cylinder «welded-in pieces» (blade simulator)

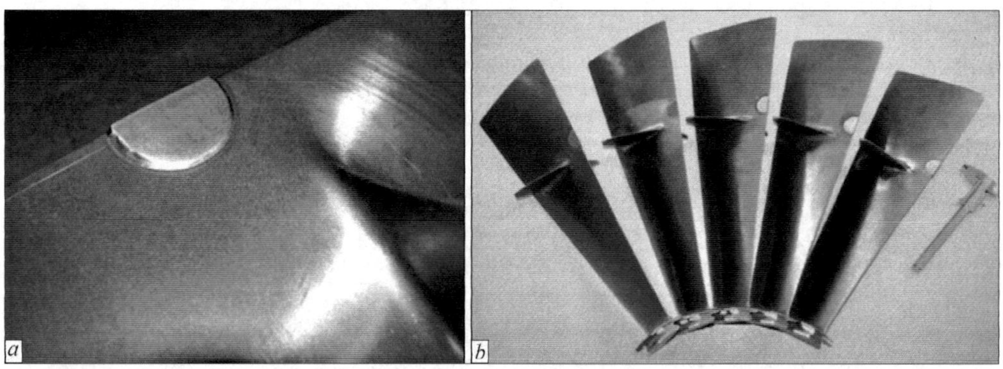

Figure 4. Repair of spot defects of main part of blade airfoil edge using cylinder repair «welded-in pieces»: *a* – cylinder repair «welded-in piece» at airfoil edge (its edge was cut out at initial mechanical treatment after EBW); *b* – pilot batch of GTE LPC blades with cylinder «welded-in pieces» on airfoil blade (before final mechanical treatment)

Repair of both types of spot defects is carried out using cylinder «welded-in pieces» (Figures 3 and 4). There are several dimension types of diameters de- pending on size of airfoil edge defect.

Such a structure is efficient by the fact that matching of diameters of «window» cut out in the blade and «welded-in piece» itself provides fixing and tight contact of butted parts in radial direction. Therefore, sufficiently simple assembly-welding device (Figure 3, c) should only provide pressing of the edge of the cut out «window» to mentioned above «overhanging shoulder» as well as holding the whole assembly in space during performance of electron beam pass along the arc of corresponding circumference.

At the beginning the works were carried on the samples simulating real parts. Then the results were successfully tested on the pilot batches of defective blades, provided by SE «Ivchenko-Progress» (Figure 4, b).

The corresponding extended «welded-in pieces» are used for the extended defects of different areas of the blade airfoil edge. Their form can be changed depending on width and shape of the damaged zone of blade airfoil edge. In particular, it can be a narrow band replacing only damaged part of the airfoil edge (Figure 5, a) as well as «welded-in piece», width of which is very widened to the corner for the case of significant damage of not only the edge, but the angle as well (Figure 5, b).

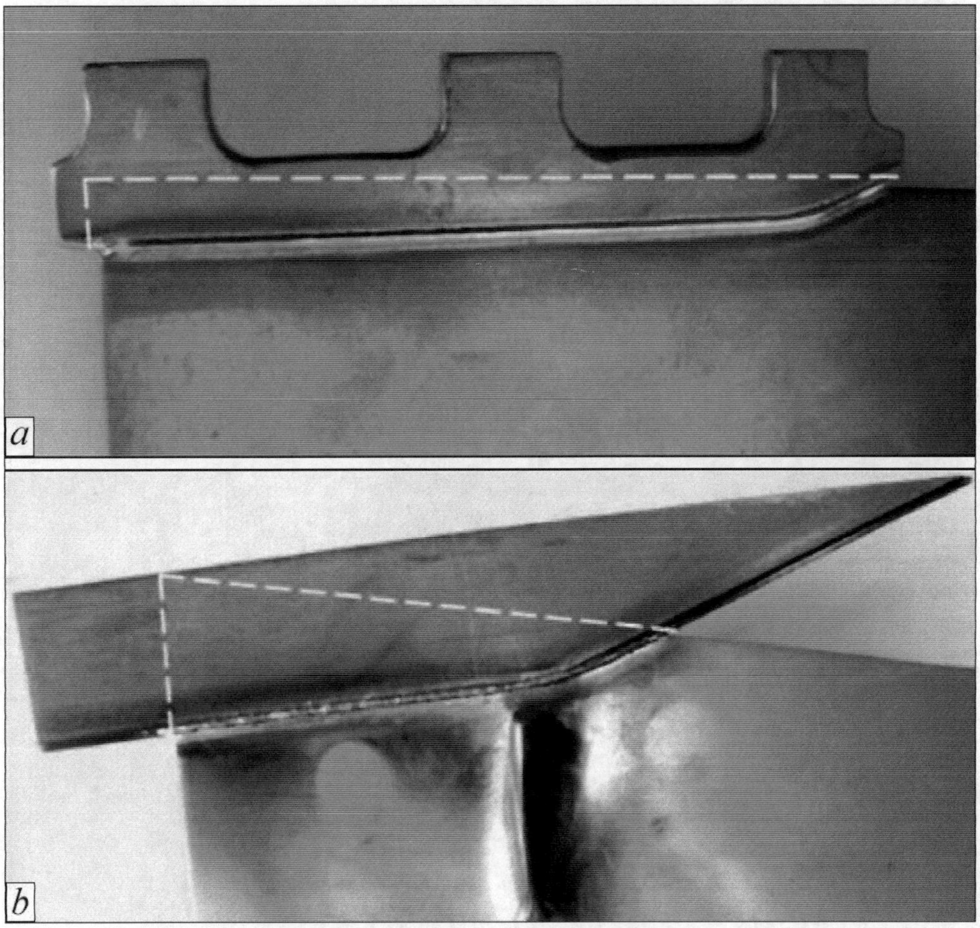

Figure 5. Repair of extended defects of blade airfoil edge: reconstruction of narrow zone adjacent to airfoil edge (*a*); reconstruction of wider zone with very damaged airfoil angle (*b*)

Respectively, uniform pressing of such «welded-in pieces» along the whole length of curved joint with the airfoil requires other more complex assembly-welding fixture (Figure 6).

3. Repair technology

The developed repair technology was tested on the edges of blade airfoil of different sections and, respectively, dimension types in repair of the defects of various size and shape (Figure 7).

In addition to repair of the blade local damages, it is also important task to replace separate elements of permanently assembled units of the gas-turbine engine. Guide cases of 3–8 stator stages of high-pressure compressor in gas-turbine engine consist of semi-rings with a set of brazed in them cantilever blades.

In operation of such engines there are also the cases of appearance of nicks and cracks on the blades as well as their tear out due to local lack of their brazing with the semi-ring wall.

Figure 6. EBW joint of blade airfoil with extended repair «welded-in piece» in assembly-welding fixture

Figure 7. Examples of realization of developed repair technology for different dimension types of the blades (large – fan sections, small – LPC) as well as shapes and extension of airfoil edge defects

Replacement of the defective blades is allowed following «Engine Overhaul Manual». At that the defective blade is removed by milling from semi-ring till its wall, including the brazing filler material holding the blade, and this place is filled with an undamaged donor-blade (Figure 8).

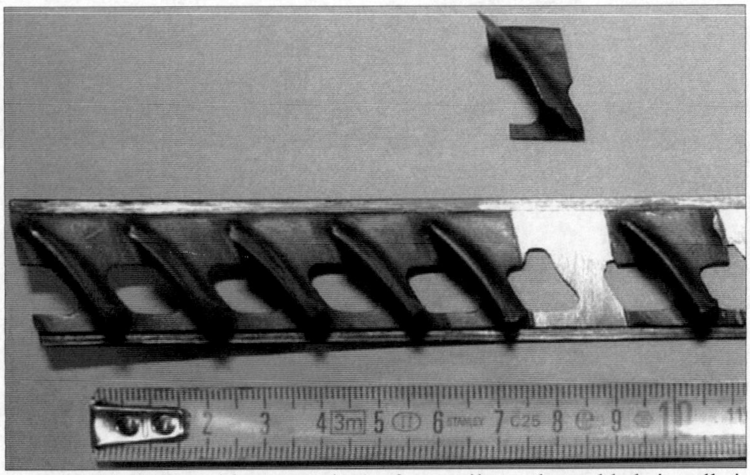

Figure 8. Semi-ring with prepared area for cantilever donor-blade installation

In other words a completely cleaned section is prepared in the semi-ring for installation of the donor-blade. It is obviously impossible to repeat initial brazing process without effecting adjacent still suitable blades. Therefore, a fixing method is necessary having local and concentrated temperature effect on the whole assembly unit. Previous technology for donor-blade fixing assumed application of argon-arc welding with the next dressing of weld reinforcement in order to reconstruct the geometry of end of the guide case semi-ring. The disadvantages of such a technology were significant distortion of product shape, as well as small penetration depth and, as a result, insufficient area of joint zone, remaining after weld reinforcement treatment and, thus, high probability of tear of the replaced blade during engine operation.

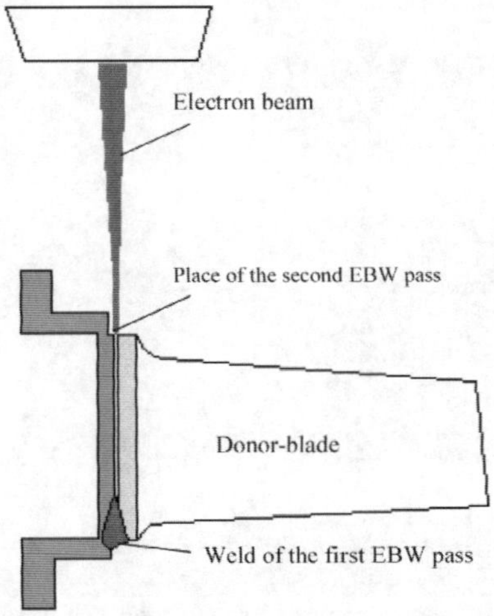

Figure 9. Scheme of double-side EBW of donor-blade with wall of guide case semi-ring

EBW method is ideal in this case allowing receiving the reliable welded joint of sufficient depth in comparison with small heat input in the product being welded.

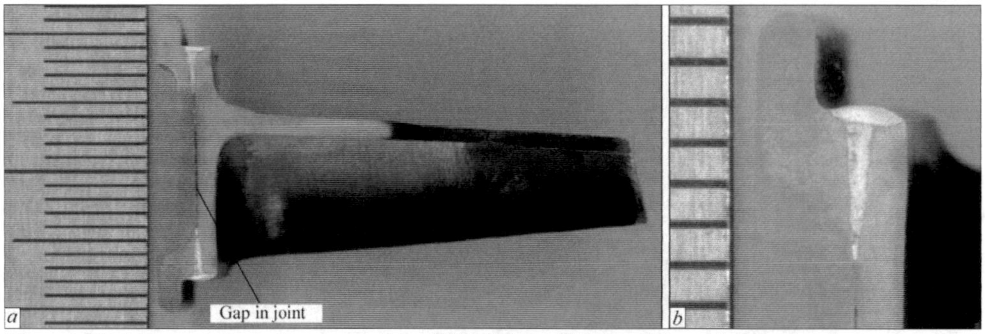

Figure 10. Macrosection of cross section of EB-joint of blade with guide case semi-ring received at up to 0.1 mm gap in joint (*a*) and section area containing the weld itself (*b*)

The next scheme of welded joint of donor-blade and wall of semi-ring guide case was developed (Figure 9). Joining of the blade with wall is carried out using double-side EBW with intermediate turning of the product by 180°. Structural strength of two similar welds is enough for reliable fixing of the blade, which is not inferior to brazed ones.

The main process problem, related, in particular, with the peculiarities of EBW method, was possible gaps in the joint between the semi-ring wall and donor-blade base. It is a result of difficulty of local mechanical removal of the damaged blade and formation of the area for the donor-blade. Such gaps are very critical, since thickness of the wall, to which the blade is welded, makes only 0.5 mm in some places.

The results of the experiments, carried out on the sample-simulators, provided the optimum compromise between the power parameters of the electron beam and possibility to form sufficiently deep weld in a thin-wall joint at gap presence in the butt. At that, pulsed EBW mode was used for improvement of weld formation and reduction of total heat input. This allowed getting quality joining of the parts at local gaps in the joint up to 0.1 mm. Then, selected EBW modes were corrected on real joints of the blades with semi-ring of the guide case (Figures 10 and 11).

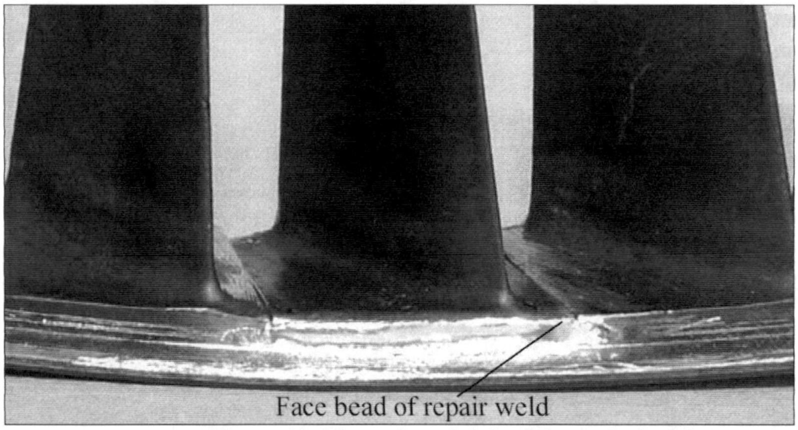

Figure 11. View of repair joint from end of guide case semi-ring

Electron-beam welding also is applied at the renewal of the hydraulic cylinders of the aircraft GTE nozzle block. The repair of welded titanium hydraulic cylinder is caused by the necessity of replacement of the wear groups of seals, that is, the piston seals and the seals of the plunger. To carry out the repair procedure three technological cuts in the places of the previous welded joints are to be done.

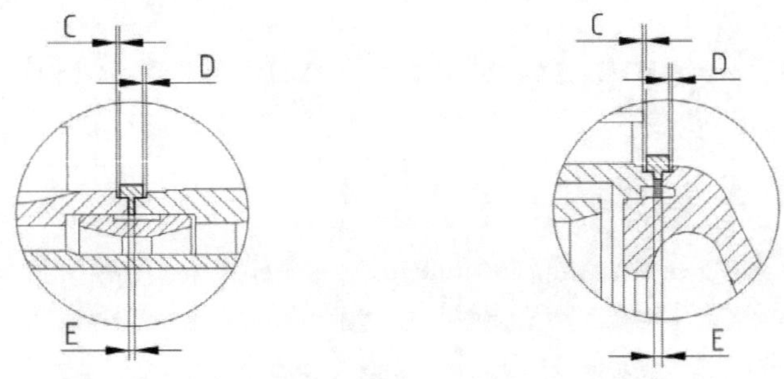

Figure 12. Setup of the hydraulic cylinder repair joints: *a*) repair joint on plunger; *b*) repair joint on cylinder body

This allows taking out the detached parts of the piston-rod from the body of the hydraulic cylinder. After replacing the seals, the hydraulic cylinder requires reassembly. In this case the estimated travel of plunger and the total length of the whole assembly unit are to be ensured. To guarantee the design strength and accuracy of the linear dimensions of the restored hydraulic cylinder its assembly is carried out using electron beam welding. The design setup of welded joints by means electron beam was developed, where the role of the element centering the edges of the connectable parts, as well as an additional filler material is performed by special rings with a T-shaped cross-section (Figure 12).

The thickness of the wall of such a ring "E" is determined taking into account the width of the technological cut and the correction for a small transverse shrinkage of the EB welded seam, and should provide the estimated travel of plunger and, most importantly, the linear dimension of the entire unit. To ensure the reliable contact between the edges of the cylinder parts joined with the T-shaped ring wall, grooves are provided on the front side of the joints. Thus, the necessary "C" and "D" clearances are guaranteed.

The parameters of the welding modes were determined on the basis of the requirements of obtaining the relatively narrow seam with guaranteed full penetration and stable formation of the front and root beads, the absence of undercuts and concaves on the front and back surfaces of the joint [9]. In this case it was necessary to take into account the presence in the repair joint of the design protective elements - the jar sleeve of the plunger and the inner shoulder of the cylinder head, through penetration of which is unacceptable due to the loss of the unit serviceability (Figure 12). The appearance of the GTE nozzle block hydraulic cylinder repair EBW joints is shown on Figure 13.

4. Conclusions

The developed technology has passed successful test in repair of a batch of real GTE nozzle block hydraulic cylinders under industrial conditions.

References

[1] Paton B E, Nazarenko O K, Nesterenkov V M et al. 2004 Computer control of electron beam welding with multi-coordinate displacements of the gun and workpiece *The PatonWelding J.* **#5** 2–5

[2] Akopiants K S, Nesterenkov V M, Nazarenko O K 2002 Electron beam welding of 60 mm thick steels using longitudinal oscillations of beam. *The PatonWelding J.* **#9** 2–4.

[3] Nazarenko O K 2008 Up-to-date equipment of the E.O. Paton Electric Welding Institute for electron beam welding. *The PatonWelding J.* **#10** 31–35.

[4] Beziyazychny V F, Strizhov A N 2010 Increase in efficiency of repair of gas turbine engines on the basis of its localization. *Refer. Book. Ingenerny J. s Prilozheniem* vol 8 p 52–56 [in Russian].

[5] Karpinos B S, Korovin A V, Lobunko A P, Vedischeva M Yu 2014 Operational damages of the gas turbine two-circuit aircraft engines with afterburner. *Vestnik Dvigatelestroenija* **#1** 18-25 [in Russian].

[6] Eliseev Yu S, Krymov V V, Malinovsky K A et al. 2002 *Technology of service, diagnostics and repair of gas turbine engines* (Moscow: Vysshaya Shkola) [in Russian].

[7] Krymov V V, Eliseev Yu S, Zudin K I 2002 *Production of blades of gas turbine engines*. (Moscow: Mashinostroenie) [in Russian].

[8] Moshkin Yu B, Eliseev Yu S, Poklad VA et al 2002 *Method of restoration of blades of turbomachines*. Pat. 2185945 RU, Int. Cl. 8 B 23P6/00 [in Russian].

[9] Zamkov V N, Shevelev A D, Khripko K S 1993 The selection of the mode parameters of EBW of titanium of middle and large thickness. *Avtomaticheskaya Svarka* **#1** 27-32.

IOP Conf. Series: Materials Science and Engineering **582** (2019) 012050 doi:10.1088/1757-899X/582/1/012050

Electron Beam Welding, Heat Treatment and Hardening of Beta-Titanium

S V Akhonin[a], V Yu Belous[b], R V Selin[c]

E.O. Paton Electric Welding Institute of the National Academy of Sciences of Ukraine, 11 Kazymyr Malevych St., Kyiv, 03150, Ukraine

E-mail: [a] akhonin.sv@gmail.com, [b] belousvy@gmail.com, [c] selinrv@gmail.com

Abstract. Analysis of the electron beam welding and different modes of heat treatment such as local heat treatment, furnace annealing, slow speed cooling, water quenching and aging impact on the structure and mechanical properties of pseudo-β titanium alloy VT19 welded joints. Microstructure of obtained welded joints in the states after welding and heat treatment were investigated. Using welded joints microsections approximate amount of β-phase has been obtained. Mechanical properties of the welded joints were analyzed and dependency of tensile strength and amount of β-phase were build.

1. Introduction

Pseudo-β titanium alloys are very important commercial materials to a lot of industries, specifically to aircraft construction and rocket science. One of the main advantages of modern pseudo-β titanium alloys are theirs high manufacturability in comparison to alloys with pseudo-α and (α+β)-structure and high strength properties. These advantages of pseudo-β titanium alloys, that has highlyalloyed VT19 alloy as distinctive representative, make this class of titanium alloys very promising for use in new equipment and modernization of existing [1]. Landing gear design of Boeing 777 uses pseudo-β alloy Ti-10V-2Fe-3Al [2]. Its application allows to reduce weight in comparison with high strength low alloy steels, such as 4340 or 300M. One of the latest successful developments was modern commercial pseudo- β alloy Ti-5Al-5Mo-5V-3Cr or Ti-5553, which found its application in constructions of aircrafts and landing gear of Boeing 787 [3]. Pseudo-β titanium alloys constructions strength, including VT19, determined by strength of welded joints. It also can operate in at higher temperatures, than some of high-strength steels [4]. Welds, obtained by electron beam welding, in comparison with TIG-welding or hybrid TIG+CO_2+Laser welding [5, 6], possesses deeper and much more narrow seams, reduced heat affected zone. Another advantage of EBW is possibility of combining such technological operations, such as welding and heat treatment, which allows to ensure high quality of obtained welds. Therefore, developing technology of electron beam welding and post-welding heat treatment of obtained welds, that allows to ensure phase composition and strength level at 0.9…0.95 of base metal strength, is the main and very important task.

2. The purpose and objectives of the study

Main aims of this work are to determine influence of local heat treatment, furnace annealing and hardening heat treatment regimes on properties of pseudo-β-alloy VT19 welded joints, obtained by electron beam welding. With this purpose research were conducted to determine influence of welding thermal cycle and post-welding heat treatments on structure and properties of welding joints of this alloy.

3. Research methods and materials

Welding of pseudo-β titanium alloy VT19 samples with size 200x100x8 mm was made on this regime: U=60kW, I=120mA, V=7mm/s [7].

Local heat treatment, as electron beam welding itself, in comparison with TIG-welding, performed in vacuum chamber [8]. Scheme of local heat treatment given in work [9]. Width of local heat treatment zone is 20 mm. Power of electron beam in LHT process is near 3 kW and was adjusted to maintain temperature in the treatment area on level of 750°C.

Furnace annealing of welded joints performed on regime that presumes heating up to 750°C and exposure for 1 hour.

Modes of hardening heat treatment are shown in Table 1.

Table 1. Modes of hardening heat treatment of VT19 welded joints

Mode #	Heat treatment	Note
1	Heating to 760°C, slow cooling with speed of 1 °C/s	Heating and cooling performed in vacuum furnace
2	Furnace annealing on mode: heating to 750C and exposing for 1 hour; cooling to 680°C, exposing for 1 hour, cooling to 380°C, exposing for 8 hour, cooling on air; aging at 450°C, exposing for 4 hours, cooling on air	Heating in furnace without protective atmosphere
3	Heating to 760°C, exposure for 1 hour, quenching in water, aging at 450°C for 4 hours	Heating in furnace without protective atmosphere

In all samples, obtained by EBW, as per X-ray examination and structural analysis, such defects as pores, lack of penetration, cracks, non-metallic inclusions are absent.

Structural research were made with the optical microscope "NEOFOT-30", equipped with a digital photography console. Quantification of β-phase in welded joints were conducted via experiments using microsections. For this, the ratio of light and dark areas of the structure, corresponding to β-phase and α-phase respectively, was estimated on the scanned microsections.

Mechanical properties of base metal and welded joints are given in Table 2.

4. Results

4.1. VT19 base metal

The base metal of the VT19 alloy contains equiaxial polyhedral grains with dispersed precipitates of the α-phase, uniformly distributed throughout the grain body (Figure 1, a). The size of α–particle is 1...2 microns and less. The amount of β-phase in base metal in the state after rolling is 44%

The conducted studies of the joints structure in the state after welding led to the conclusion, that in the weld metal and the HAZ, obtained with a welding speed of 7 mm/s, the structure is non-equilibrium and requires the use of heat treatment to obtain a uniform structure. Large, equiaxed polyhedral β-grains predominate in this welded joint. The weld metal consists of practically pure β-phase (see Figure 1, b) with hair-like borders, the amount of β phase is 99%. The strength of welded joints made by EBW in the state after welding is at the level of 91% (Table 2).

4.2. Local heat treatment

In the weld metal made by EBW using LHT (Local Heat Treatment) at 750°C for 10 min, the amount of dispersed excretions of the metastable α'-phase increases (Figure 2, a). The amount of β-phase as a result of the application of preheating has decreased to 53%. The strength of welded joints is at 99% of the strength of the alloy itself.

Thus, the use of local heat treatment makes the structure more uniform across the entire width of the welded joint in comparison with the structure of the samples without using LHT, which allows to ensure uniform strength of the welded joint and mechanical properties at 99% of the strength of the alloy in the condition after rolling. A further increase in the strength of welded joints is limited by the strength of the base metal.

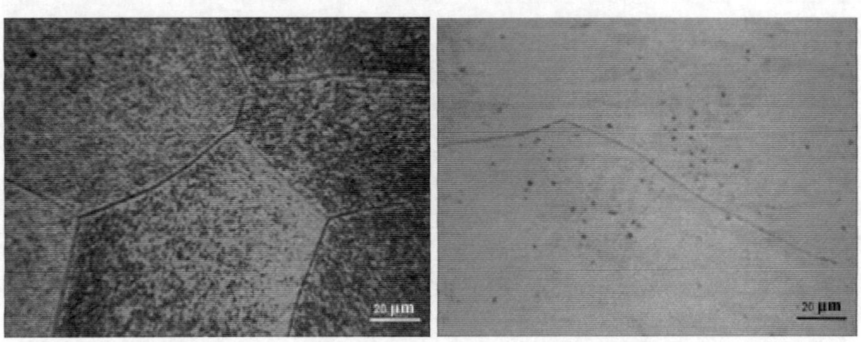

a *b*

Figure 1. Microstructure of VT19 welded joint, obtained by EBW in the state after welding:
a – base metal, x20; b – seam metal, x20;

IOP Conf. Series: Materials Science and Engineering **582** (2019) 012050 doi:10.1088/1757-899X/582/1/012050

4.3. Furnace annealing

Studies of welded joints subjected to furnace annealing at 750°C showed that in this case the weld metal consists of primary β-grains that are elongated in the direction of heat removal and have uniform, two-phase structure (Figure 2, b), consisting of α- and β-phase particles. The α-phase particles have a lamellar morphology, the length of the α-plates is 1…5 μm, with a thickness of 0.5...0.8 μm. The amount of β-phase as a result of furnace annealing is minimal for welded joints and is at 35%. The strength of welded joints in this case is maximum at 105...107% of the alloy strength in the state after rolling. It should be noted that VT19 alloy allows the use of heat treatments at lower temperatures compared to high-strength two-phase titanium alloys, such as VT23, T110 or T120 [10]. Therefore, the annealing temperature is 750°C, which is lower than the LHT temperature of 850°C or the temperature of the recommended vacuum annealing of 900°C for alloy T120.

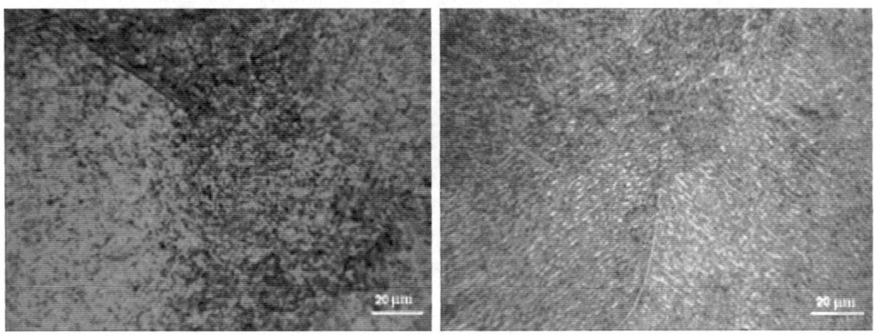

a *b*

Figure 2. Microstructure of VT19 welded joint, obtained by EBW: a – after local heat treatment at 750°C, 10 minutes, x20; b – after furnace annealing at 750°C for 1 hour, x20

4.4. Regulated Step Annealing

The weld metal and HAZ consists of elongated in the direction of the heat sink and equiaxed grains located along the axis of the weld (Figure 3).

It should be noted that the dendritic structure of the electron-beam weld after heat treatment almost did not appear, apparently because cooling of the weld metal of a small volume occurred so quickly that diffusion processes did not find such development as, for example, during TIG-welding. The size α of plates in the weld metal made by EBW is 1...6 μm with a thickness of up to 1 μm. The amount of β-phase in the weld metal was 43%.

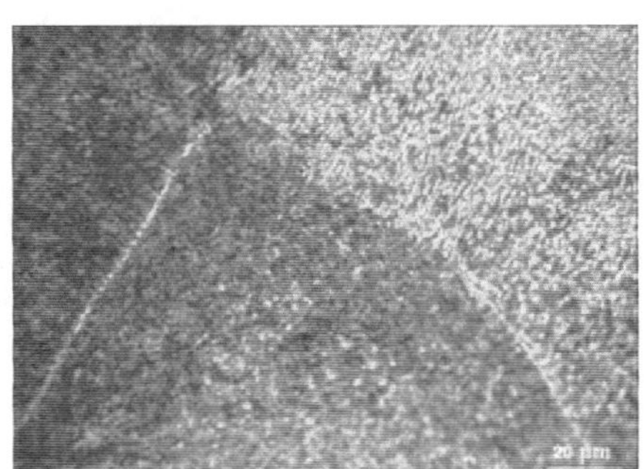

Figure 3. Microstructure of VT19 welded joint, obtained by EBW after regulated step annealing

4.5. Slow cooling with regulated speed

The microstructure of the weld metal of this welded joint is shown in Figure 4. In the upper part of the weld, made by EBW, crystallite accretion occurs along the weld axis, equiaxial or elongated grains grow in the middle and root parts along the axis. Heat treatment in the weld metal causes the decomposition of the β-phase with the release of α- and β-particles of different sizes. Along with finely dispersed decay products with a size

IOP Conf. Series: Materials Science and Engineering **582** (2019) 012050 doi:10.1088/1757-899X/582/1/012050

of less than 1 μm, on the background of the β-matrix there are α-plates 2…5 μm long and less than 1 μm thick, that predominate in the structure.

In the HAZ there are dispersed particles with a size of about 1 μm or less, as well as lamellar α-particles 2…4 μm long and 1…1.5 μm thick with the β-matrix in background. The amount of β-phase in the weld metal after this heat treatment mode was 41%.

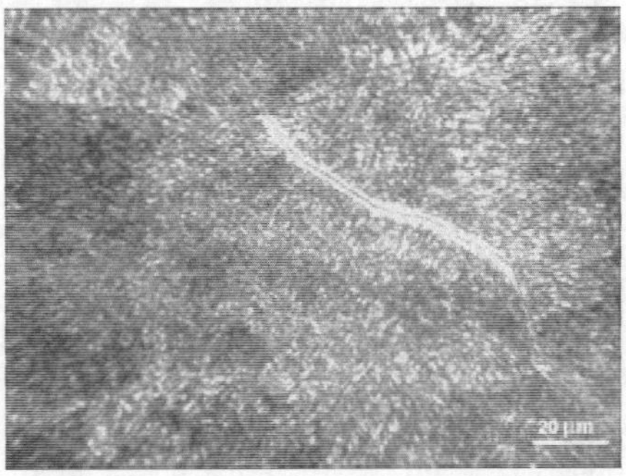

Figure 4. Microstructure of VT19 welded joint, obtained by EBW after slow cooling with regulated speed

4.6. Hardening and aging

Studies of the microstructure of welded joints with hardening heat treatment - quenching and subsequent aging showed that the intragranular structure of the weld metal is finely dispersed, the size of decomposition products most often does not exceed 1 μm (Figure 5). In individual grains of the weld metal, a substructure is observed.

The HAZ metal consists of equiaxed polyhedral grains. In the HAZ, fine-dispersed decomposition products are also formed, mainly with the size of 1...1.5 μm and less, in some grains individual particles reach 2...3 μm. The fine structure in all zones of the welded joint should provide it with high strength. The amount of β-phase in the weld metal was 32%.

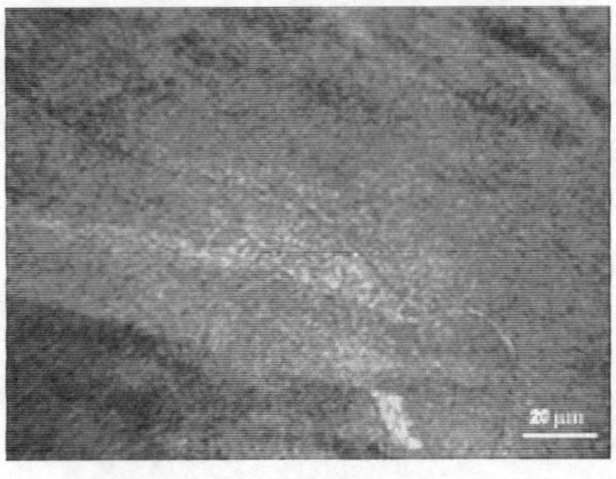

Figure 5. Microstructure of VT19 welded joint, obtained by EBW after water quenching and hardening

5. Discussion

The use of local heat treatment to a welded joint obtained by EBW reduces the amount of metastable β-phase in the weld metal and HAZ, which allows obtaining equal strength welded joints of titanium pseudo-β-alloy VT19. At the same time, in order to obtain a uniform structure in all zones of the welded joint, it is possible to use furnace annealing at 750°C for 1 hour.

Step annealing leads to the formation of larger elements in weld metal and HAZ, so that the size of the α-plates in the weld metal is 1...6 μm with a thickness of up to 1 μm. In the HAZ the α-phases inside the grain have a length of 1...4 μm 1.5 microns. The amount of β-phase in the weld metal after step annealing is 43%, and slow cooling reduces the volume of β-phase in the weld metal to 41%. The smallest amount of β-phase obtained after quenching and aging is at 32%. It is due to the decomposition during aging of quenching structures formed during the quenching process. In the process of aging, the final decay products are the dispersed α- and β-phases in close to the equilibrium state, the formation of which causes dispersion hardening of the alloy [11, 12].

The study of the mechanical properties of titanium pseudo-β VT19 alloy welded joints made by EBW in mode No. 1, which were subjected to such types of thermal hardening as quenching in water and aging, regulated step annealing, slow cooling at a regulated rate of 1°C/s, showed that the highest strength values are welds subjected to quenching in water followed by aging and are at 1.285 MPa (Table 2).

Studies have led to the conclusion that because of quenching and subsequent aging, the finest intragranular structure of the weld metal in which the size of the decomposition products most often does not exceed 1 μm is formed. In the HAZ size is 1...1.5 μm. The fine structure in all zones of the welded joint of the VT19 alloy provides it with high strength reaching σ_B = 1285 MPa, with high impact toughness KCV = 23 J/cm^2. It should be noted that for welded joints of titanium pseudo-β alloy VT19 effective hardening heat treatment is quenching in water with subsequent aging, which ensures the strength of joints at 130% of strength level of the alloy in the rolled state.

Table 2. Mechanical properties of pseudo-β-titanium alloy VT19 base metal and welded joints, obtained by EBW in the state after heat treatments

Mode #	Sample type, Type of heat treatment	Tensile strength σ_B, MPa	Yield strength $\sigma_{0.2}$, MPa	Elongation δ_s, %	Reduction of area Ψ, %	Impact toughness KCV, J/cm^2
1	Base metal, after rolling.	958	887	12	47	27
2	Welded joint, 7 mm/s	876	842	11.3	36.8	29
3	Welded joint, LHT 750°C - 10 minutes.	937	868	5.3	19	20
4	Welded joint, Annealing at 750°C - 1 hour	1026.7	985.7	12.0	31.5	26
5	Welded joint, Regulated step annealing 750°C, 680°C, 380°C, aging 450°C	1047	985	6.0	25.4	28
6	Welded joint, Slow colling with 1°C/s speed	1068	1012	11.3	36	22.5
7	Welded joint, Water quenching and aging at 450°C	1285	1234	4.7	20.6	23

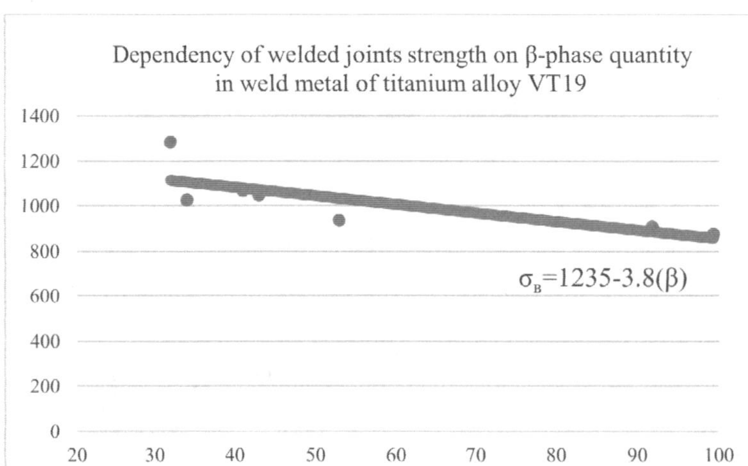

Figure 6. Dependency of welded joints strength on β-phase quantity in weld metal of titanium alloy VT19

Studies of the microstructure and it comparison with mechanical properties of the joints, allowed us to establish an inversely proportional dependence of the titanium pseudo β-VT19 alloy welded joints strength on the amount of β-phase in the weld metal (see Figure 6) in the form σ_B = 1235 - 3.8 (β, %), according to which the minimum strength values σ_B = 881 MPa of welded joints are fixed at the β-phase content of 99%, and the maximum σ_B = 1285 MPa at the β-phase content of 32%.

6. Discussion

EBW in combination with local heat treatment allows to adjust the ratio between α - and β - phases in the weld metal and reducing the β - phase content in the weld metal of the VT19 alloy from 91% to 53%, as well as increasing the strength of the welded joints from 876 MPa to 937 MPa and as a result ensuring equal strength of welded joints to the base metal. To form a homogeneous structure, increase the strength of the base metal and welded joints to the level of 1020 MPa, complete decomposition of the metastable phases in VT19 alloy welding joints, it is necessary to subject them to furnace annealing at 750°C for 1 hour, as a result of which the strength level of the joints increases to 105...107 % compared to the strength of the alloy after rolling. As a result of heat treatment of hardening and subsequent aging, in welded joints of VT19 alloy, made by EBW, the amount of β-phase in the weld metal decreases from 99% in the state after welding to 48% in the state after hardening heat treatment. The smallest intragranular structure of the weld metal formed in the seam. This structure provides the welded joint with high strength, which is 1285 MPa.

References

[1] Khorev A I 2012 Super-strength titanium alloy VT19 *Tekhnologiya Mashinostroeniya* **#6** 2-5.

[2] Kiese J, Wagner L 1996 Fatigue Behavior in Ti-10V-2Fe-3Al: Microstructural Effects on Crack Nucleation and Microcrack Growth, Fatigue '96, Eds. G. Lütjering and H. Nowack (Oxford: Pergamon Press) p 959.

[3] Cotton J, Briggs R, Boyer R, Tamirisakandala S, Russo P, Shchetnikov N, Fanning J C 2015 State of the Art in Beta Titanium Alloys for Airframe Applications *JOM* **67** 1281–1303

[4] Berdnikova O, Sydorets V, Alekseienko T 2014 Structure and properties of laser-welded joints from high-strength steels *Applied Mechanics and Materials* **682** 240-245.

[5] Chinakhov D A, Chinakhova E D, Grichin S V, Gotovschik Y M 2016. Influence of welding with two-jet gas shielding on the shaping of a welding joint. IOP Conf Series: Materials Science and Engineering **125** 012013.

[6] Krivtsun I, Reisgen U, Semenov O, Zabirov A 2016 Modeling of weld pool phenomena in tungsten inert gas, CO_2-laser and hybrid (TIG+CO_2-laser) welding *J Laser Applications* **28**(2) 022406.

[7] Il'Yaschenko D P, Chinakhov D A, Mamadaliev R A 2018 Effect of Inverter Power Source Characteristics on Welding Stability and Heat Affected Zone Dimensions *IOP Conf Series: Earth and Environmental Science* **115** 012041.

[8] Chinakhov D A, Sapozhkov A S 2016 Influence of the gas shielding method upon the properties of the weld joints *AIP Conference Proceedings* **1783** 020026

[9] Akhonin S V, Belous V Yu, Selin R V, Vrzhizhevskiy E L, Petrichenko I K 2018 Electron beam welding and heat treatment of welded joints of pseudo-β-titanium alloy VT19 *Avtomaticheskaya svarka* **#7** 12-17.

[10] Akhonin S V, Belous V Yu, Selin R V, Petrichenko I K, Vrzhizhevskiy E L 2015 Structure and properties of welded joints of two-phase titanium alloys, obtained by EBW and TIG-welding. *Avtomaticheskaya svarka* **#9** 16-19.

[11] Antipov A I, Moiseev V N, Moder N I 1996 Titanium alloy VT35 age hardening. *Metalloved Term Obrab Met* **#12** 22-25.

[12] Gysler A, Terlinde G, Lütjering G 1982 Influence of Grain Size on the Ductility of Age-Hardened Titanium Alloys. *Titanium and Titanium Alloys* Eds Williams J C, Belov A F (Boston: Springer) 1919-1931

IOP Conf. Series: Materials Science and Engineering **582** (2019) 012051 doi:10.1088/1757-899X/582/1/012051

Structure and Properties of Titanium Modified Silicon-carbide at EBM

S V Akhonin[1], V O Berezos[1], A Yu Severyn[1,a], M P Gadzyra[2], Y G Tymoschenko[2], N K Davydchuk[2]

[1] E.O. Paton Electric Welding Institute of the National Academy of Sciences of Ukraine, 11 Kazymyr Malevych St., Kyiv, 03150, Ukraine
[2] I. M. Frantsevich Institute of Problems of Materials Science of the NAS of Ukraine, 3 Krzhyzhanovskogo St., Kyiv, 03142, Ukraine

E-mail: [a] titan.paton@gmail.com

Abstract. Aim of the work is development of technology of production of ingots from sparsely-alloyed titanium alloys using electron beam melting method. To get high mechanical characteristics in titanium alloy, the modifiers of high-dispersion carbide compounds were used as foundry alloys. They in process of electron beam melting provide strengthened with nanoparticles structure of a material suitable for further deformation processing. In order to develop nanosized alloying modifiers for high-strength titanium alloys there were synthesized nanosized powder in form of carbon solution in silicon-carbide. A technology was developed for production of ingots from sparsely-alloyed titanium alloys using electron beam melting method. The works on production of 200 mm diam. titanium ingots with addition of 0.5% and 1.5% of the alloying modifiers based synthesized nanosized silicon-carbide were carried out with the help of determined modes. Hot deformation processing of ingots was performed, 15 mm thick deformed sheets were produced and then subjected to further heat treatment. Structure and properties of obtained material were investigated.

1. Introduction

Titanium alloys are the structural material, which has a lot of unique properties, among which, in this case, it is necessary to outline, first of all, high values of specific strength [1]. Today all over the world there is a stable tendency to increase the portion of high-strength materials, which provide high complex of mechanical and service properties and at that have good weldability [2-5]. The recent investigations show that some titanium alloys in application of special technologies for their production can provide 1600-1800 MPa strength, and in comparison with steel products of the similar strength the one produced from titanium have virtually two times lower weight. Sufficiently cheap high-dispersion compounds of carbides, nitrides and intermetallics can be used as a foundry alloy for development of high strength of titanium. They will provide in the process of melting the ingots with high-dispersion structure strengthened by nanoparticles suitable for further deformation processing.

2. Methodology

To get sparsely-alloyed titanium alloy it was decided to an use alloying element in form of silicon-carbide powder, which was added in necessary proportions to titanium sponge of TG-120 grade. The synthesis of SiC powder is based on reaction of self-propagating high-temperature synthesis (SHS) in specially designed graphite or ceramic crucibles of predetermined volume as well as at necessary temperature mode [6-7]. The appearance of obtained particles is given on Figure 1 and characteristics of produced powder in Table 1.

Figure 1. Appearance of microstructure of silicon carbide particles

Table 1. Characteristics of synthesized SiC powder

Average size of grains, nm	Lattice parameters, nm	Content of impurity phases, %		
		C	Si_3N_4	Si_2N_2O
70	0.43486	0.2	2	3

Among the current methods of special electrometallurgy the electron beam melting (EBM) is one of the efficient methods of vacuum metallurgy and it has found application in research practice and industry for production of alloys, including refractory and high-reactive, with ultra-low content of gases, volatile impurities and non-metallic inclusions [8]. EBM allows in a wide range regulating the ingot melting rate due to independent heat source that in turn permits regulation of time of metal staying in a liquid overheated condition. EBM is the most efficient technology that allows virtually complete elimination of inclusions of high and low density [9]. For experimental evaluation of the characteristics of proposed titanium alloys with addition as modifiers of 0.5% and 1.5% nanosized SiC powder there were carried out the trial melts of batches of 200 mm diam. ingots in electron beam unit UE-121.

The briquettes of rectangular form, which were pressed from TG120 grade titanium sponge of up to 30 mm fraction and addition of 0.5% and 1.5% of synthesized SiC powder (Figure 2) were used as charge materials for melting of a batch of titanium alloy ingot of 200 mm diam. In process of melting the consumable billet was continuously fed in a working zone over hearth, where its melting took place under effect of electron beam heating (Figure 3).

IOP Conf. Series: Materials Science and Engineering **582** (2019) 012051 doi:10.1088/1757-899X/582/1/012051

Figure 2. Pressed briquettes of titanium sponge with addition of nanozised SiC powder.

Figure 3. Prepared for melting fixture and process of melting of 200 mm diam. ingots of titanium alloy modifier with nanosized SiC particles.

Appearance of ingots after melting and next machining is presented on Figure 4.

Figure 4. Appearance of melted and machined 200 mm diam. ingots of titanium alloy with addition of 0.5% and 1.5% of nanosized SiC powder.

Ultrasonic flaw detection using UD4-76 flaw detector was used for determination in the ingots of internal defects in form of non-metallic inclusions, pores and discontinuities. Analysis of macrostructure of metal of produced 200 mm diam. ingots was carried out on transverse templates cut out of the produced ingots (Figures 5, 6).

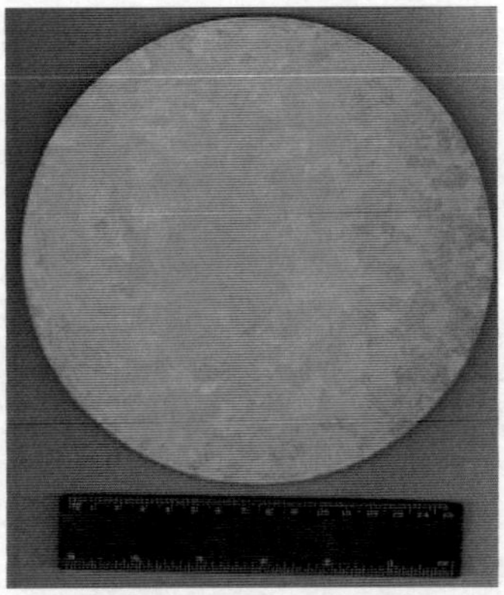

Figure 5. Macrostructure of metal of 200 mm diam. ingot of titanium alloy with addition of 0.5% of SiC

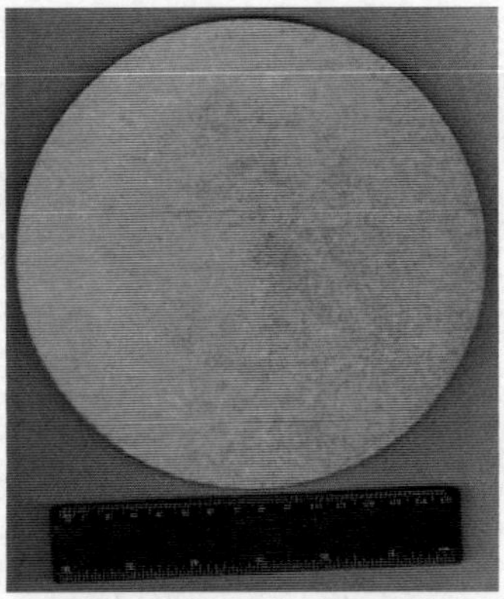

Figure 6. Macrostructure of metal of 200 mm diam. ingot of titanium alloy with addition of 1.5% of SiC

The results of examinations show that a structure of ingot metal is dense, homogeneous, has no zones that have different etching on ingot cross-section. There is no significant difference observed in the structure of a central zone of ingot and peripheral zone. The defects in form of pores, cavities, cracks and non-metallic inclusions were not found. It is determined that increase of content of nanosized SiC in titanium results in significant refinement of macrostructure. Thus, according to GOST 26492-85 [10] the grain size in the metal of the examined ingots of titanium alloys with addition of 0.5 and 1.5% of nanosized SiC powder correspond to sizes 8 and 5, respectively.

The obtained samples of cast metal were used for performance of hardness investigations. Measurement of hardness was carried out on hardness gage TP-73-1 at 10 kg loading on diamond indenter. The measurement of material hardness on the templates in two mutually normal directions showed differences in hardness values not more than 5%. This indicates that the particles of alloying modifier have sufficiently uniform distribution over the whole ingot volume. The values of measurements are given in Table 2.

Table 2. Values of hardness of cast samples of ingots of titanium alloys with addition of SiC-based alloying modifiers

Sample	Hardness HV10.0, GPa	Hardness, HRC
Ti – 0.5%SiC	2.9	29-30
Ti – 1.5%SiC	3.3	32-33

Analysis of data shows that rise of portion of alloying additive of the nanosized alloying modifier results in increase of cast metal hardness. Thus, further rise of portion of the nanosized alloying modifier can be not reasonable due to the fact that increment of hardness level of martial will lead to the difficulties or will make further deformation processing impossible.

Improvement of the mechanical properties of cast material to the necessary level is achieved by further application of technologies of deformation processing and heat treatment. Plastic deformation of the obtained ingots was carried out on a reversing duo rolling mill of Skoda 355/500 grade. Titanium alloy billets with addition of 0.5% and 1.5% SiC were subjected to hot plastic deformation firstly on hydraulic press P-457 with next rolling on a mill at 1050°C temperature without application of shielding atmosphere and shielding coatings. Deformed semi-finished products of 15 mm thickness are presented on Figure 7.

IOP Conf. Series: Materials Science and Engineering **582** (2019) 012051 doi:10.1088/1757-899X/582/1/012051

Figure 7. 15 mm thick plates of titanium obtained by deformation processing from EBM ingots of titanium alloy with addition of 0.5% and 1.5% of SiC

3. Results

The microstructure of titanium alloy doped with 0.5% of SiC at different magnifications is presented on Figure 8. The alloy consists of deformed primary α-grains with considerably deformed boundaries, therefore determination of primary grains is complicated. The material mainly consists of lamellar α phase that makes the colonies of various size. The width of α-phase lamellas makes 5-8 μm and length 50-150 μm. Photos made with large magnification allow observing the finest particles of another phase, size of which is less than 1 μm ,uniformly distributed in the structure (Figure 8).

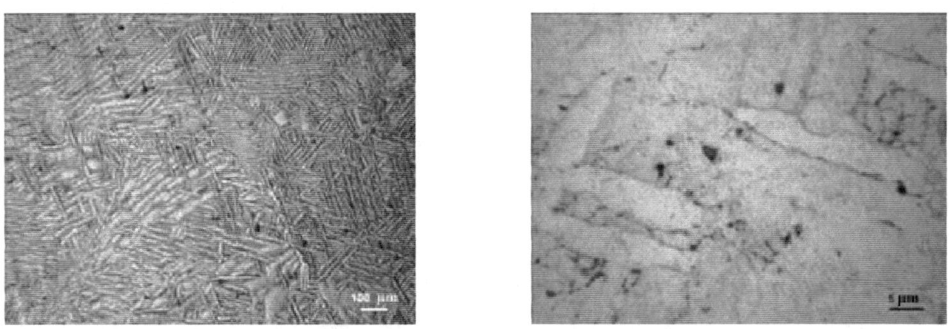

Figure 8. Microstructure of titanium alloy doped with 0.5% of nanosized alloying modifiers based on SiC.

Figure 9 shows the microstructure of titanium alloy doped with 1.5% of SiC. In the same way, as the previous alloy, this titanium alloy consists of deformed primary grains with split boundaries. The main phase constituent of the alloys is lamellar α phase, width of particles of which makes 4-6 μm and length 30-70 μm. i.e. lamella of α phase with addition of 1.5% of SiC is finer than in alloy with 0.5%of SiC. In addition to lamellar α phase in Ti+1.5%SiC alloy it is possible to find in a small amount unknown for us phase with clear boundaries in form of round, oval or elongated particles with 2-4 μm width and 2-12 μm length (Figure 9). To understand the nature of the described phase constituent the further investigations are necessary. In the structure of researched alloy Ti-1.5% SiC there is also presence of a disperse phase of 1 μm size and less. The similar phase constituent was observed in the research alloy Ti+0.5%SiC, however, there it is even smaller and uniformly distributed in the metal volume.

IOP Conf. Series: Materials Science and Engineering **582** (2019) 012051 doi:10.1088/1757-899X/582/1/012051

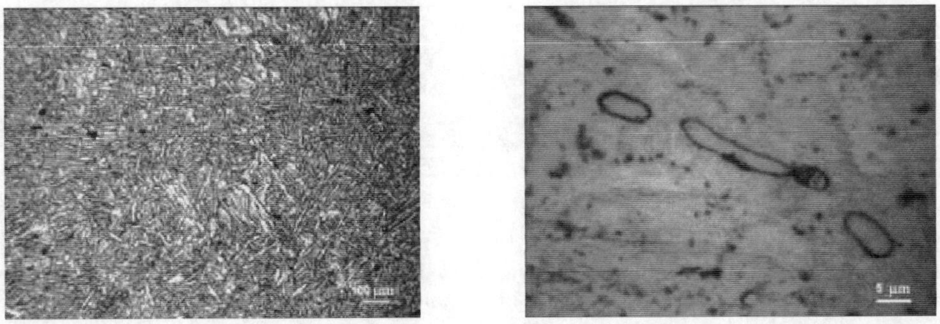

Figure 9. Microstructure of titanium alloy doped with 1.5% of nanosized alloying modifiers based on SiC

Obtained mechanical properties of the designed alloys correspond to structural states that are realized in the process of only hot rolling without additional thermal processing. As it can be seen from Table 3, increase of concentration of nanosized silicon carbide leads to decrease of ductility and rise of strength.

Cold rolling was carried out at maximum 10% level of deformation per one pass. The total level of deformation did not exceed 60%. Cold rolling of hot-rolled titanium alloy results in strength growth and decrease of ductility (Table 3). Such dependence is proved by known dependence of properties on level of stress that is caused by formation of the dislocations in the structure at cold deformation.

Table 3. Mechanical characteristics of hot rolled titanium-based alloys obtained by addition of nanosized carbides (where CR – cold rerolling)

Composition	Yield limit σ_y, MPa	Short-term tensile strength limit , σ_t, MPa	Ductility δ, %	hardness HV30, GPa
Ti+0.5%SiC	689	770	14.3	2.0
Ti+1.5% SiC	745	823	9.4	2.7
Ti+1.5% SiC (CR)	927	1038	7.5	-

Heat treatment was carried out to improve the material properties. It was done in a vacuum furnace by heating to 1200°C temperature. Cooling took place together with a furnace at 100° per hour rate. As can be seen form Table 4, the characteristics of alloys after heat treatment have almost no differences, only insignificant difference in ductility is observed.

Table 4. Characteristics of titanium alloys (compression deformation) after heat treatment at 1200°C temperature.

Composition	Short-term tensile strength limit, σ_t, MPa	Yield limit σ_y, MPa	Ductility δ, %
Ti+0.5% SiC	2173	1515	9.4
Ti+1.5% SiC	2196	1555	8.6

The obtained alloys were also quenched. Heating was carried out in a resistance furnace to 1100°C temperature. The quenching medium was water of 20°C temperature. Quenching was carried out till hardness increase and considerable decrease of ductility (Table 5).

Table 5. Characteristics of titanium alloys (tensile deformation) after quenching

Composition	Short-term tensile strength limit , σ_t, MPa	Yield limit σ_y, MPa	Ductility δ, %	Hardness, HRC
Ti+0.5%SiC	1370	1250	0.8	44-46
Ti+1.5% SiC	1440	1350	0.6	54-56

4. Conclusions

1. In addition of SiC nanosized powder as a modifier the structure of cast titanium after electron beam melting is characterized with high dispersion of grains and uniform distribution of nanosized particles. It is determined that increase of portion of nanosized silicon carbide in titanium results in rise of cast metal hardness.

2. High-temperature plastic deformation of cast titanium ingots after electron beam melting leads to formation of sufficiently high mechanical properties of material.

3. Heat treatment of wrought titanium alloys significantly affects the final mechanical characteristics. Quenching of the obtained metal composites results in considerable increase of strength and significant decrease of ductility. The same effect can be obtained by means of cold rolling due to formation of high dislocation density. The best combination of strength and ductility of metal composite is achieved after heat treatment at 1200°C temperature.

References

[1] *Titanium and Titanium Alloys* 2003 Eds. M. Peters, C. Leyens. (Weinkeim: Wiley)
[2] M.J. Donachie Titanium: A Technical Guide second Edition, ASM International (2004)
[3] G. Lütjering, J.C. Williams, Titanium. Second edition, Berlin Springer (2003)
[4] Shelyagin V, Khaskin V, Bernatskyi A, et al 2018 Multi-Pass Laser and Hybrid Laser-Arc Narrow-Gap Welding of Steel Butt Joints Mat. Sci. Forum 927 64-71,
[5] T. Pasang, Y.Tao, M.Azizi, Welding of titanium alloys, MATEC Web of Conferences 123 , 00001 (2017)
[6] Gadzira M, Gnesin G, Mykhaylyk O, Andreyev O 1998 Synthesis and Structural peculiarities of Nonstoichiometric β-SiC *Diamonds and Related Materials* **7** 1466–1470.
[7] Gadzyra N F, Gnesin G G, Mikhailik A A 2001 Mechanism of Formation of Solid Solution of Carbon in Silicon Carbide *Powder Metallurgy* **#9/10** 15–18.
[8] Tikhonovskiy A.L., Tur A.A. 1984 Refining of metals and alloys by the method of electron beam melting (Kiev: Naukova Dumka)
[9] H. Funagane 2018 Electron beam melting toward inclusion-free titanium alloys *Materials Science and Engineering 424 (2018) 012018*
[10] GOST 26492-85 Rolled rods of titanium and titanium alloys. Technical conditions (Moscow: Izdatelstvo standartov)

IOP Conf. Series: Materials Science and Engineering **582** (2019) 012052 doi:10.1088/1757-899X/582/1/012052

Influence of the Thermal-force Effect on the Process of High-temperature Synthesis of the Ni_3Al Intermetallic Compound

V E Ovcharenko [1,2 a], K O Akimov [1], E N Boyangin [1], Yu Baohai [3]

[1]Institute of Strength Physics and Materials Science of Siberian Branch Russian Academy of Sciences (ISPMS SB RAS), Tomsk, 634055, Russia

[2] Yurga Institute of Technologies (affiliated) National Research Tomsk Polytechnic University, Lenigradskaja, 26, Yurga, Kemerovskaya oblast, Russia

[3]Institute of Metal Research, Chinese Academy of Sciences, Shenyang 110016, China

E-mail: [a] ove45@mail.ru

Abstract. In this work, the intermetallic compound Ni_3Al was obtained by high-temperature synthesis under pressure at various values of the preliminary pressure on the initial powder mixture (3Ni+Al). The study of pressure-time and displacement-time diagrams gave a coherent picture of the synthesis passage over time. It was found that an increase of preliminary pressure leads to a decreasing of the powder compacts porosity. In this regard, the smallest displacement of the press plunger after initiating the synthesis reaction in the powder compact was observed at the highest value of the preliminary pressure on the compact. The role of preliminary pressure on the initial powder mixture in the process of the grain structure formation of the Ni_3Al intermetallic compound synthesized under pressure was determined.

1. Introduction

The Ni_3Al intermetallic compound, related to alloys with high-symmetry L12 superstructure, is of considerable interest in modern materials science. Due to its outstanding properties such as high heat resistance, wear resistance, and anomalous temperature dependence of strength, Ni_3Al has gained widespread use as the basis for many modern high-temperature nickel superalloys used in aircraft, space industry, etc. [1-3]. At present, intermetallic compounds are obtained by various methods: electrolytic recovery of intermetallic crystals from melts, aluminothermic reduction of oxygen compounds, smelting in arc and induction furnaces, etc. Generally, traditional methods of synthesis, such as casting and melting, are not used in the manufacturing of most intermetallic compounds, mainly due to the large differences between the melting points and the densities of the initial components [4,5].

Currently, one of the most effective and versatile methods for producing Ni_3Al intermetallic compounds is the method of self-propagating high-temperature synthesis (SHS), based on the initiation of exothermic reactions of the chemical compounds formation in powder mixtures of the corresponding composition. The SHS method allows one to obtain not only various chemical compounds, including multicomponent compositions but also alloys and composite materials based on them [6]. To the advantages of the SHS method, its necessary include the high reaction rate of synthesis of chemical compounds (from several seconds to several minutes) and the relatively high purity of the final products [7].

The main directions in the development of production technologies and manufacturing of SHS products are theoretical and experimental studies of the kinetics of high-temperature synthesis of materials with a given structural and phase state and the use of SHS technologies in the materials science of products with unique physical and strength properties for extreme operating conditions. The prospects of this area of modern materials science are shown in fig.1 by data on the dynamics of the increase in the number of publications over the years over the past 10 years on the example of SHS known intermetallic compounds, SHS intermetallic compounds Ni_3Al and SHS as the transformation processes of the physicochemical state of disperse media under the conditions of the exothermic reaction to form the target chemical compound.

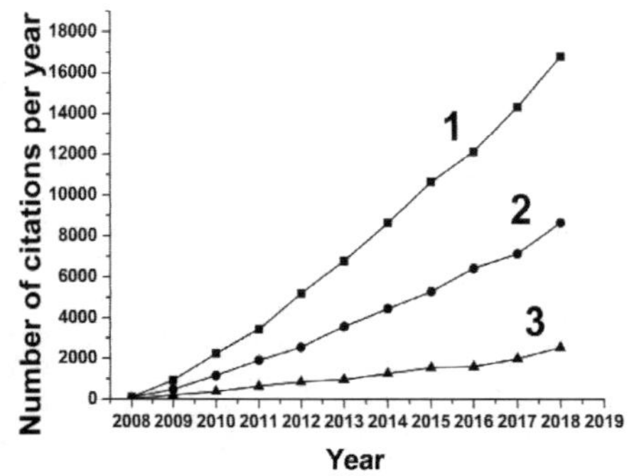

Figure 1. Dynamics of publication activity by year: 1 - SHS intermetallic compounds, 2- SHS intermetallic compounds Ni_3Al, 3 - SHS as a science about the processes of transformation of the physicochemical state of dispersed media under the conditions of the exothermic reaction of the target chemical compound formation (according to the Web of Science for February 2019).

A special place in the SHS technologies of the Ni_3Al intermetallic compound and alloys based on it is occupied by the SHS Ni_3Al technology under pressure, which allows the formation of billets or structural products. High speeds of phase transformations in the conditions of the bulk exothermic reaction of the intermetallic compound synthesis in the initial powder mixture create considerable difficulties in controlling the process of the structural-phase state formation and the strength properties of the synthesized under pressure intermetallic product. One of the ways to solve this problem is to use high-speed recording equipment of the volumetric exothermic reaction processes of intermetallic compound synthesis in the initial powder mixture, which allows registering temporal, temperature and power parameters of the processes of synthesis and compaction of the synthesis product in continuous shooting regime.

The purpose of this work is an experimental study of the influence of the thermal-force effect on the powder mixture of the initial elements on the parameters of the high-temperature synthesis process and the formation of a grain structure in a compact synthesized under pressure from the intermetallic compound Ni_3Al.

2. Material and methods

Samples of Ni_3Al intermetallic compound were synthesized in a powder mixture of nickel with aluminum of stoichiometric composition (nickel particle size of ~ 2.0 μm and aluminum ~ 1.0 μm) in cylindrical steel molds with an inner diameter of 58 mm on a hydraulic press by heating the molds by currents high-frequency till auto-ignition of powder compacting and force compacting of a high-temperature synthesis product. The process of the product's compacting was synchronized in time with the burning process of powder compact using an automatic control system of the preload on the initial powder compact, the heating speed of the powder compact and the digital recording system of the compaction front displacement of the thermo-sensitive mixture with a recording step of 0.25 s. The phase composition of the synthesized intermetallic compound samples was studied on an x-ray diffractometer DRON-7 in $Co_{K\alpha}$ radiation at an accelerating voltage of 35 kV and a current of 20 mA. The relative error in the measurements of displacements and pressures was no more than 2%. The grain size was determined by the method of random secants with averaging over 150 measurements.

3. Results and discussion

In fig. 2a is represented a diagram of a die mold for the high-temperature synthesis of intermetallic compounds under pressure. The performed diffractometric analysis of the obtained samples showed the identity of the phase composition of the samples synthesized at different preliminary pressures to the reference sample of stoichiometric composition (Fig. 2b).

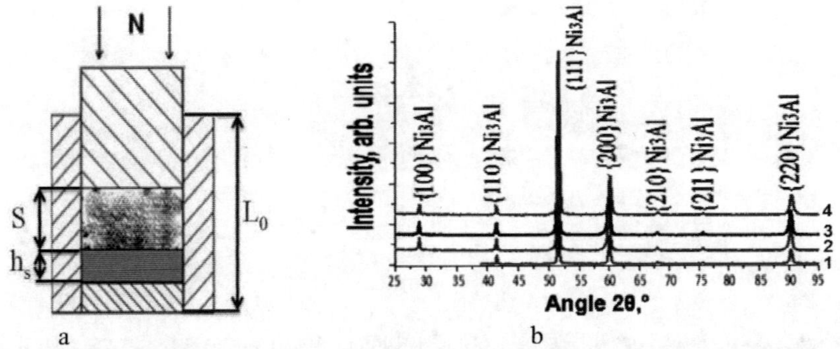

Figure 2. The scheme of the die mold for high-temperature synthesis of intermetallic compounds under pressure (a), diffractograms from samples of intermetallic compound Ni3Al synthesized under pressure at various preliminary pressure values on the original powder compact: 115.1 MPa (1), 76.5 MPa (2), 32.6 MPa (3), Ni_3Al (reference) (4) (b).

At the first stage, studies of the synthesis passage in time were carried out for samples of the intermetallic compound Ni_3Al, depending on the value of the preliminary pressure on the initial powder compact. The high-temperature synthesis of the Ni_3Al intermetallic compound under pressure was implemented with the following sequence of steps of the synthesis process (Fig. 3 a, b):
(1) pre-pressure was applied to the powder mixture (3Ni + Al) in the die mold,
(2) self-ignition of the powder mixture was initiated by heating of the mold by high-frequency currents,
(3) upon ignition of the powder compact, the thermosetting powder mixture was compacted with a predetermined force of the hydraulic press.

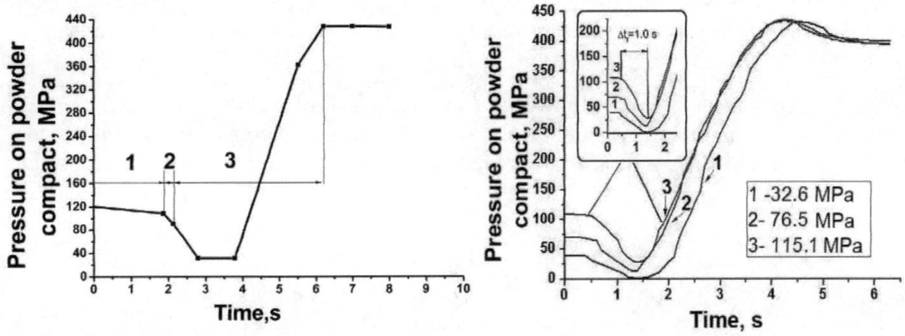

Figure 3. Scheme of the high-temperature synthesis process of the Ni3Al intermetallic compound (a); the pressure-time diagrams of the high-temperature synthesis process of the Ni_3Al intermetallic compound with continuous heating of the powder mixture (3Ni + Al) and various values of the preliminary pressure on the initial powder compact (b), Δt_f – time to reduce pressure to a minimum.

The preliminary pressure on the powder mixture in the die mold ensures the formation of the optimum porosity of the powder compact in the range of 30-40%, in accordance with known conditions of the compact's ignition when heated to the melting point of the aluminum component [8]. From the moment when application synthesis product compacting pressure is applied to the thermo-reacting mixture, a high-precision recording of the hydraulic press plunger linear displacement occurs. In fig. 4 shows the time dependences of the plunger linear movement in the process of the synthesis product compaction at different values of the preload on the initial powder compact.

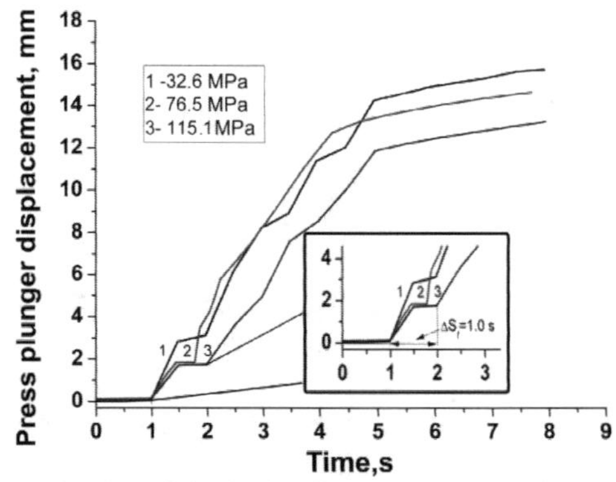

Figure 4. The time dependences of the hydraulic press plunger linear displacement in the process of the synthesis product compaction oat different preload values on the original powder compact: 32.6 MPa (1), 76.5 MPa (2), 115.1 MPa (3).

For shown in Fig. 4 dependencies are characterized by the presence of sites in time, due, apparently, with the formation of a liquid phase in powder compacts, which fills the pores of powder compacts. Similar sites were previously recorded on thermograms of high-temperature synthesis of Ni_3Al intermetallic [9]. Further, as the content of solid-liquid suspension containing intermetallic phases increases in the product of high-temperature synthesis, the process of the high-temperature synthesis product compaction accelerates up to the reaching the final value of the time dependence of the press plunger linear displacement during the formation of an intermetallic compound of a given composition in the entire volume of the synthesized compact. The difference in the size of the sites is due to the difference in the porosity values of the powder compacts - the smallest displacement of the press plunger after initiating the synthesis reaction in the powder compacting occurs at the highest value of the preliminary pressure on the compact at 115.1 MPa, which is caused by the highest compacting density. This regularity determines the nature of the time dependence of the relative compaction of the powder mixture and the dependence of the maximum relative compaction of the powder compact on the value of the preliminary pressure, which was calculated by the formula

$$\delta_i = \frac{S_i}{S+h_s}$$

where δ_i - is the relative compaction of the powder mixture at the current time, S_i - is the compaction of the powder mixture at the current time in the synthesis process, S - is the final compaction of the powder mixture during the synthesis, h_s - is the thickness of the obtained sample of the intermetallic synthesis product.

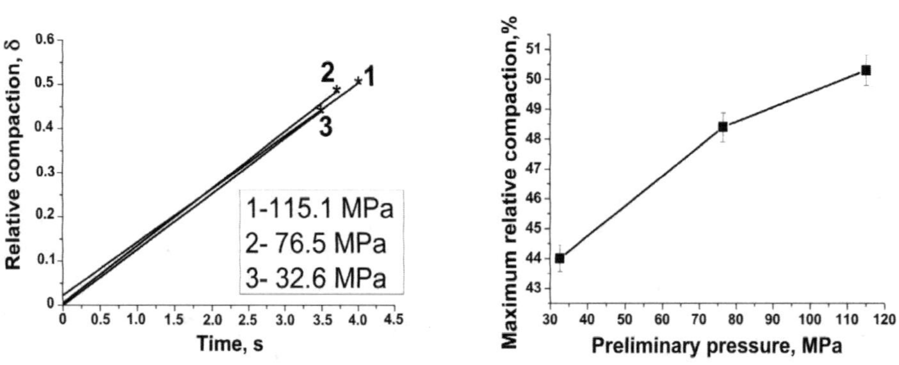

Figure 5. The time dependence of the relative compaction of a powder compact during the high-temperature synthesis of the Ni_3Al intermetallic compound: 115.1 MPa (1), 76.5 MPa (2), 32.6 MPa (3) (a) and the dependence of the maximum relative compaction of the powder compact on the pre-pressure (b).

Based on the data on the time dependencies of the relative compaction at the current time (Fig. 5a), the dependence of the maximum compaction of the original powder compact on the value of the preliminary pressure (Fig. 5b) is constructed, from which it follows that the maximum compaction corresponds to the maximum preload on the compact. This is more clearly seen in the behavior of the speed and acceleration curves of powder compacting: the highest value of the speed of synthesis product compaction corresponds to the minimum amount of preliminary pressure on the powder compact. This is due to the longer duration of the high-temperature synthesis process of an intermetallic compound under pressure at the maximum porosity value of the powder compact (Fig. 6 a, b).

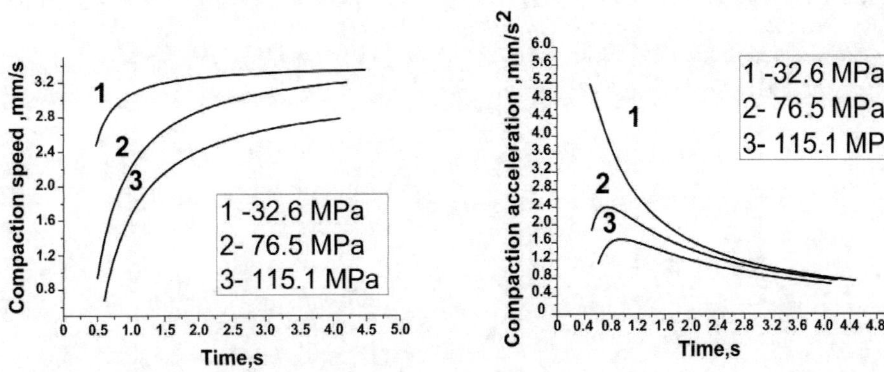

Figure 6. The time dependences of the compaction speed of the powder compact (a) and the compaction acceleration of the compact (b) at different values of the preliminary pressure on the compact.

In other words, in the case of initiation of high-temperature synthesis in a powder compact with the highest porosity, the thermo-reactive powder mixture exists at high temperatures for the longest time, which determines more favorable conditions for grain growth in the intermetallic compound synthesized under pressure. Indeed, as shown in [10], a decrease in the preload on the initial powder compact leads to an increase in the grain size in the intermetallic compound synthesized under pressure. For samples synthesized with a preliminary pressure value of 115.1 MPa, the average grain size was the smallest, and the value of the intermetallic compound ultimate strength was greatest (table 1).

Table 1. The values of the average grain size and ultimate strength for Ni_3Al samples synthesized at different values of preliminary pressure [7].

Preliminary pressure, MPa	32.6	76.5	115.1
Average grain size, μm	10.5 ±0.7	8.7 ±0.5	3.8 ±0.4
Ultimate strength, MPa	336±6	406.3±8	482.35±9

4. Conclusion

The kinetics of the high-temperature synthesis process of the Ni_3Al intermetallic compound in a powder mixture (3Ni + Al) under pressure under the conditions of a volumetric exothermic reaction of the intermetallic compound formation is largely determined by the values of the preload on the original powder compact and the relative compaction. A comprehensive study of the process of shrinking the initial powder pressing throughout the process of an intermetallic compound high-temperature synthesis under pressure makes it possible to identify and evaluate the role of preload on the initial powder mixture and compaction rate in the formation of the grain structure and strength of the intermetallic compound Ni_3Al synthesized under pressure.

Acknowledgments
This work was supported by the Program for Basic Scientific Research at the State Academies of Sciences for 2013–2020 (Project No. 23.2.2).

References

[1] Stoloff N S *et al* 2000 *Intermetallics* **8**(9-11)1313
[2] Sikka V K *et al* 2000 *Intermetallics* **8**(9-11)1329
[3] Jozwik P *et al* 2015 *Materials (Basel)* **8**(5) 2537

IOP Conf. Series: Materials Science and Engineering **582** (2019) 012052 doi:10.1088/1757-899X/582/1/012052

[4] Yeh C L and Sung W Y 2004 *J. Alloys Compd.* **384** 181
[5] Jozwik P and Bojar Z 2007 *Arch. Metall. Mater.* **52 (2)** 321
[6] Naplocha K 2018 *Self-propagating high-temperature synthesis (SHS) of intermetallic matrix composites* (Intermetallic Matrix Composites Properties and Applications) ed R Mitra (Woodhead Publishing) pp 203-220
[7] Borovinskaya I P *et al* 2017 *Concise Encyclopedia of Self-Propagating High-Temperature Synthesis. History, Theory, Technology, and Products* (Amsterdam: Elsevier) p 466
[8] Nayborodenko Y S and Itin V I 1975 *Combust. Explos. Shock Waves* **11(3)** 343
[9] Lapshin O V *et al* 2005 *Combust. Explos. Shock Waves* **41(1)** 64
[10] Ovcharenko V E *et al* 2017 *Mater. Sci. Forum* **906** 95

IOP Conf. Series: Materials Science and Engineering **582** (2019) 012053 doi:10.1088/1757-899X/582/1/012053

The Effect of Electron-pulse Modification of the Surface Layer on the Strength Properties of the Ni_3Al Intermetallic Compound

V E Ovcharenko [1,2a], Yu Baohai [3]

[1]Institute of Strength Physics and Materials Science of Siberian Branch Russian Academy of Sciences (ISPMS SB RAS), Tomsk, 634055, Russia

[2] Yurga Institute of Technologies (affiliated) National Research Tomsk Polytechnic University, Lenigradskaja, 26, Yurga, Kemerovskaya oblast, Russia

[3]Institute of Metal Research, Chinese Academy of Sciences, Shenyang 110016, China

E-mail: [a] ove45@mail.ru

Abstract. In this paper it was shown that pulsed electron irradiation forms in the surface layer of the Ni_3Al intermetallic compound samples a columnar crystal structure oriented perpendicular to the irradiated surface. The dimensions of the crystals of the columnar structure and the depth of the surface layer modification depend on the power density and the duration of the irradiation pulses - with power density increasing , the dispersion of the columnar structure increases, with increasing duration of irradiation pulses, the depth of the surface layer structure modification increases. Modification of the surface layer structure improves the strength properties of Ni_3Al intermetallic compound samples.

1. Introduction

The current state of research in the field of structural-phase states modification of intermetallic alloys can be estimated by the example of creating and improving the most highly loaded parts of gas turbine engines - blades of heat-resistant alloys [1-4]. The development of fourth-generation aircraft engines in the 70s-80s of the last century, connected with the need to increase the operating temperature of the coolant from 1200 to 1500-1550 ° C, required solving the problem of high-temperature protection of the blade surface. Since the 80s of the last century, intensive research has been launched to develop and create high-performance protective coatings from high-temperature gas corrosion and from the formation of thermo-fatigue defects in the form of micro- and macro-cracks in the surface layers, leading to the formation of major cracks on the surface with subsequent destruction of the blades. The solution of the first problem was achieved by creating multicomponent coatings of the type Me-Cr-Al-Y, applied by electron-beam, ion-plasma and vacuum-plasma technologies. Hardening of the surface layers towards thermal fatigue damage was achieved by modifying the structural-phase state to which the nanostructuring of the surface layer relates [5-7]. The physical meaning of this modification is to reduce the scale level of plastic deformation localization in the nanostructured surface layer, leading to a more uniform distribution of elastic stresses in a larger volume of material with external mechanical or temperature effects on the surface. As a result, the energy of nucleation of stress concentrators in the surface layer substantially increases the probability of formation in the surface layer of defects the internal structure decreases. In other words, the nanostructured surface layer exhibits damping properties in relation to the base material under shock mechanical and thermal effects, preventing premature nucleation and propagation of main cracks from the surface into the main bulk of the material. This paper presents a part of the results of the study of the pulsed electron irradiation influence in the micro- and submillisecond range of exposure time on the structural-phase state of the Ni_3Al intermetallic compound surface layer on the strength and ductility of intermetallic compound samples.

2. Material and methods

The Ni_3Al intermetallic compound was synthesized in a powder mixture of nickel (PNK-S1B, particle size 0.98 μm) and aluminum (ASD-4, particle size 1.0 μm) of stoichiometric composition (76.0 at.% Ni – 24.0 at. % Al). The high-temperature synthesis of an intermetallic compound was initiated in the thermal explosion regime of the initial powder mixture in a closed-type die mold reactor by applying pressure to a powder compact preheated to a predetermined temperature.

IOP Conf. Series: Materials Science and Engineering **582** (2019) 012053 doi:10.1088/1757-899X/582/1/012053

Intermetallic samples were irradiated in a pulsed mode by a wide-aperture (covering the entire surface of the sample) electron beam with given values of the pulse duration, the energy density per pulse and the frequency of the pulses at the facility generating irradiation pulses from 20 to 200 μs. Studies of the microstructure of the surface and surface layers of the intermetallic compound before and after pulsed electron-beam irradiation were carried out by methods of optical metallography (Neophot 32), scanning (Quanta 200 3D) electron microscopy with micro X-ray spectral analysis. The microhardness of the intermetallic compound surface was investigated on the device PMT-3. Tests of intermetallic samples before and after irradiation were carried out according to the standard scheme on an Instron-3369 machine with a loading rate of 0.2 mm/min.

3. Results and discussion

In fig. Figure 1 shows the microstructure of the Ni3Al intermetallic compound surface synthesized under pressure in the initial state, prior to electron beam irradiation. The spectrum of the characteristic radiation shows the presence in the analyzed point only nickel and aluminum in the stoichiometric ratio.

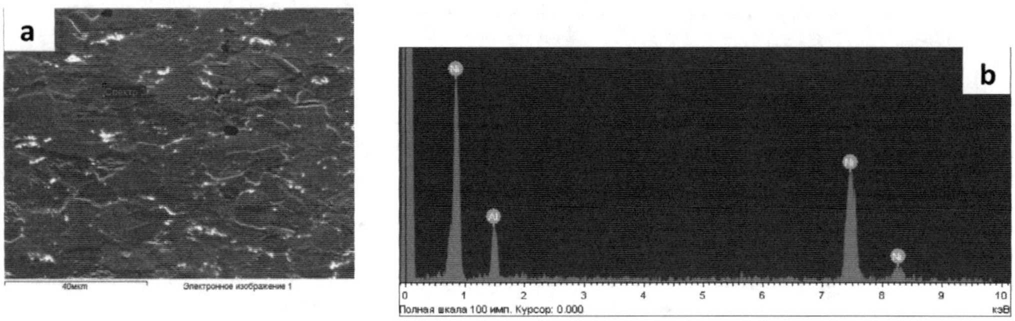

Element	%, weight	%, atomic
Ni, K_α	87.296	75.950
Al, K_α	12.704	24.050
Total	100.00	100.00

Figure 1. The microstructure of the surface (a) and the data of the grains elemental composition of the Ni_3Al intermetallic compound synthesized under pressure (b).

Typical images of the Ni_3Al intermetallic surface microstructure after pulsed electron beam irradiation are shown in Fig. 2.5.

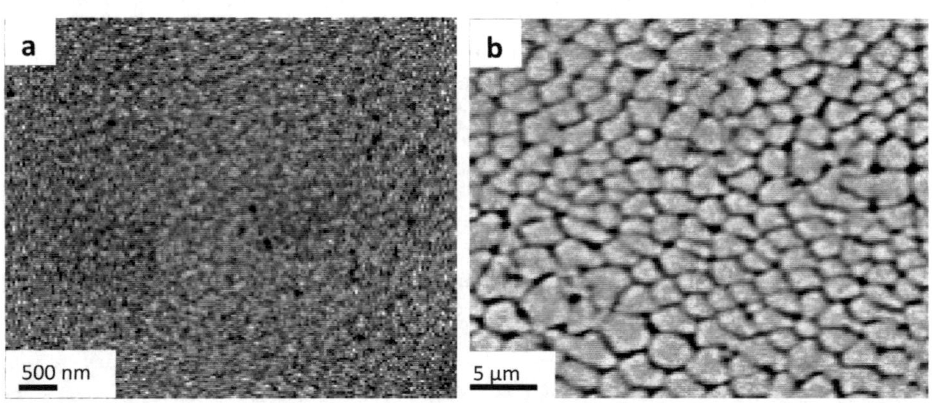

Figure 2. Microstructures of the Ni_3Al intermetallic samples surface after pulsed electron-beam irradiation at an electron beam power density of $6.0 \cdot 10^6$ W / cm^2 (a) and $0.2 \cdot 10^6$ W / cm^2 (b); the number of irradiation pulses 5

The grain size in the modified surface layer largely depends on the parameters of pulsed electron-beam irradiation — as the electron beam power density (W_S) increases, the grain size in the surface layer decreases. If at $W_S = 6{\cdot}10^6$ W/cm^2, the average crystallite size of the surface layer does not exceed 100 nm (Fig. 2, a), then at $W_S = 0.2{\cdot}10^6$ W/ cm^2, the crystallite size varies in the range of 0.5...3.0 μm (Fig. 2, b). On the other hand, the increase in the irradiation power density associated with a decrease in the irradiation pulses duration multiply reduces the depth of modification of the surface layer structure. In fig. 3 shows the structures of the transverse fracture of the intermetallic samples surface layers after irradiation with pulses of 3 μs (a) and 50 μs (b) duration.

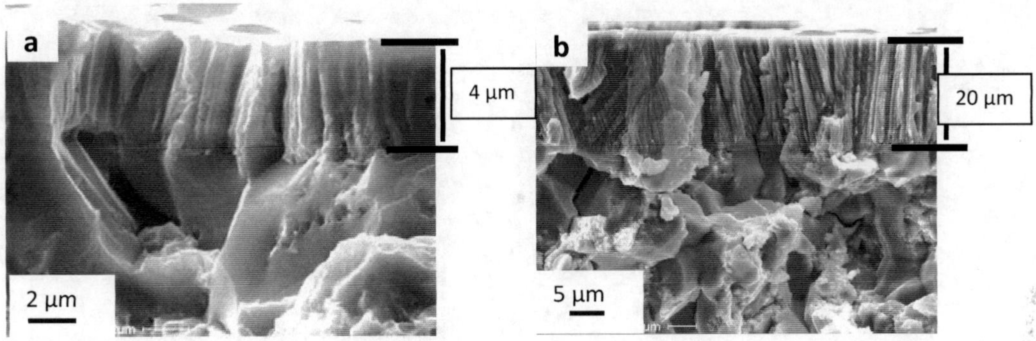

Figure 3. Fractographs of cross-sections of the intermetallic samples surfaces layers after electron irradiation with pulses of 3 μs (15 J/cm^2, 5 pulses) (a) and pulses of 50 μs (20 J/cm^2, 5 pulses) (b).

It can be stated that the cross-sectional structures of the surface layers after pulsed electron irradiation are distinguished by the columnar nature of the structure, consisting of individual needle-like crystallites oriented perpendicular to the irradiation surface. The transverse crystallite size depends on the magnitude of the irradiation power density - with an increase in the irradiation power density the transverse size of the crystallites significantly decreases. At the same time, the longitudinal size of the crystallites (the depth of modification of the surface layer structure) depends on the duration of irradiation pulses - with an increase in the duration of the pulses from 3x to 50 μs, the depth of modification of the surface layer structure increases from 4x to 20 μm.

At the same time, the longitudinal size of the crystallites (the depth of modification of the surface layer structure) depends on the duration of irradiation pulses - with an increase in the duration of the pulses from 3x to 50 μs, the depth of modification of the surface layer structure increases from 4x to 20 μm. In fig. 4 shows the load-elongation and bending stress-strain relations for the Ni_3Al intermetallic compound after pulsed electron irradiation.

Electron irradiation by pulses with a duration of 3 μs leads to a noticeable demonstration of plasticity during bending of intermetallic samples (plastic manifestation is noted in Fig. 4c and 4d). Of interest is the superposition of intermetallic samples irradiation with short and longer irradiation pulses. Figure 5 shows the microstructures of the cross-section of the Ni3Al intermetallic sample surface layer after irradiation with pulses of 50 μs duration and subsequent irradiation with pulses of 3 μs duration.

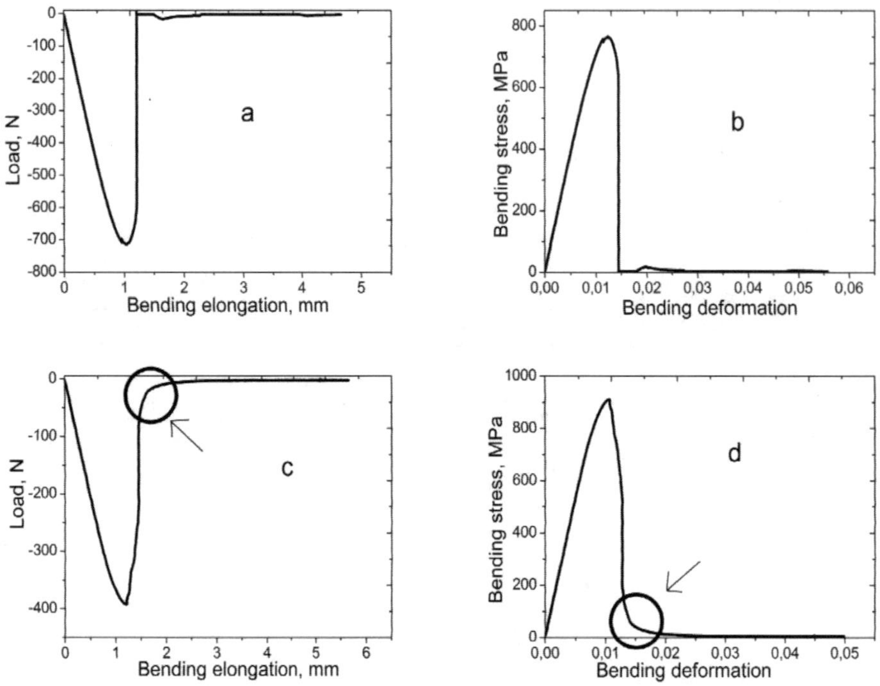

Figure 4. Dependences "load - elongation" (a, c) and "stress-strain" (b, d) during bending of Ni_3Al intermetallic samples in the initial state (a, b) and after pulsed electron irradiation according to the 5 J/cm^2 mode, 3 μs, 5 pulses, (b, d).

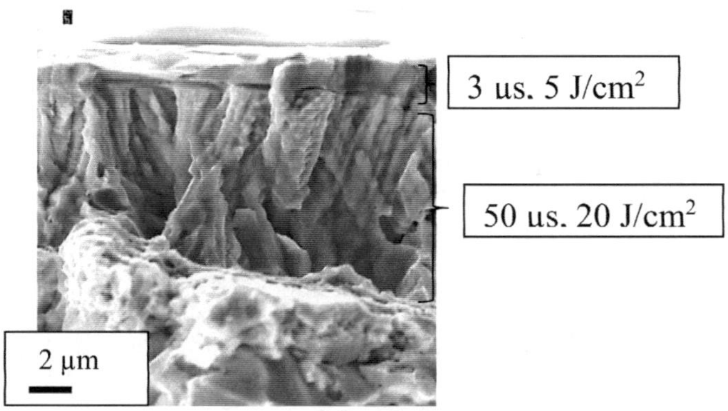

Figure 5. The microstructure of the fracture of the intermetallic sample surface layer after double pulsed electron irradiation - pulses with a duration of 50 μs + pulses with a duration of 3 μs.

The microstructure of the cross-section of the intermetallic compound surface layer after twofold irradiation consists of a thin upper layer and a more voluminous layer of the columnar structure. Tests of samples with a two-layer structure of the surface layer under tension showed a noticeable increase in the strength of the samples (Fig. 6).

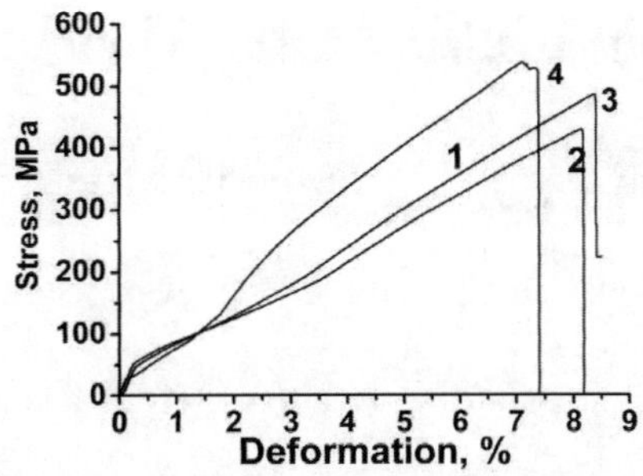

Figure 6. Extension curves of Ni_3Al intermetallic compound samples in the initial state (1), after irradiation with pulses of 50 μs (20 J / cm^2, 3 pulses) (2), pulses of 3 μs (5 J/cm^2, 5 pulses) (3) and after irradiation with pulses of 50 μs (20 J/cm2, 3 pulses) with subsequent irradiation with pulses of 3 μs (5 J/cm^2, 5 pulses) (4).

It can be stated that the formation in the surface layer of a two-layer structure with a gradient dispersion of the structure increases the ultimate strength compared with its value for samples of the intermetallic compound after single irradiation.

4. Conclusions

1. Pulsed electron irradiation drastically modifies the structure of the Ni3Al intermetallic compound samples surface layer - a directionally crystallized structure in the form of columnar intermetallic grains oriented perpendicular to the irradiation surface is formed in the surface layer.

2. With an increase in the duration of irradiation pulses, the thickness of the layer with a modified structure increases.

3. As the power density of the electron irradiation increases, the cross-sectional dimensions of the columnar structure grains decrease to nanoscale level.

4. The formation in the surface layer of a gradient fibrous structure consisting of a thin layer of highly dispersed (nanoscale) structure on a thicker layer of directionally crystallized fibrous structure, by double electron irradiation, initially with a lower power density in the electron beam, increases the ultimate strength of the intermetallic compound samples, including the number of samples after a single electron beam irradiation.

Acknowledgments
This work was supported by the Program for Basic Scientific Research at the State Academies of Sciences for 2013–2020 (Project No. 23.2.1).

References

[1] Kablov E.N. 2001 *Cast blades of gas turbine engines (alloys, technology, coatings)* (Moscow: MISIS) p 632
[2] Stoloff N S *et al* 2000 *Intermetallics* **8**(9-11)1313
[3] Sikka V K *et al* 2000 *Intermetallics* **8**(9-11)1329
[4] Jozwik P *et al* 2015 *Materials (Basel)* **8**(5) 2537
[5] Schulson E.M. *et al* 1985 *Acta Metallurgica* **33** 1587
[6] Antolak-Dudka A. *et al* 2013 *Intermetallics* **42** 41
[7]. Kwai S. Chan. 1990 *Scripta Metallurgica et Materialla* **24** 1725
[8] Polkowski W. *et al* 2015 *Mater. Lett.* **139** 46
[9] Ovcharenko V.E. *et al* 2015 *Physics of the Solid State* **57**(7) 1293
[10] Antolak-Dudka A. *et al* 2013 *Intermetallics* **42** 41

About Rational Using of the Controlling Longitudinal and Transverse Magnetic Fields at arc Welding and Surfacing

Alexander D Razmyshlyaev [1, a], Marina V Ahieieva [2, b], Elena V Lavrova [1,c]

[1]State High Educational Organization "Priazovsky State Technical University", Mariupol (Ukraine),
[2]Donbass State Machine-building Academy, Kramatorsk (Ukraine)

E-mail: [a]razmyshljaev@rambler.ru, [b]marinamykyta@rambler.ru, [c]lavrova.e.v.pstu.edu@gmail.com

Abstract. The use of the controlling longitudinal magnetic fields (LMF) and transverse magnetic fields (TMF) at submerged arc welding and surfacing by a wire is shown. Application both LMF and TMF at the induction optimal values of the magnetic fields allows increasing the electrode wire melting productivity (melting coefficient) at arc welding and surfacing on 20...30 %. However there are no publications comparing the efficiency of the controlling LMF and TMF use at submerged arc welding and surfacing by a wire. A depth and area of parent metal penetration zone considerably decrease at the use of both ferromagnetic and non-magnetic basic and welding materials at submerged arc welding and surfacing by a wire with act of LMF and TMF. The methods of induction components calculation that generate input devices of LMF and TMF in the welding bath zone are briefly described. This comparative analysis allows establishing the areas of rational use of controlling LMF and TMF at arc welding and surfacing, taking into account the magnetic properties of basic and welding materials.

Introduction

The increasing of arc welding and surfacing efficiency is an actual task in the field of welding science. There are ways to influence the stability of melting and electrode metal transfer, as well as the weld formation, such as welding with a consumable electrode and the gas dynamic effect of the two-jet at the welding [1,2]. As shown in [3] welding efficiency development and reduction of its harmful effect on human organisms are important.

The use of longitudinal magnetic fields (LMF) and transverse magnetic fields (TMF) during submerged arc welding and surfacing by wire allows to improve the efficiency of these processes. However, there are no publications in which a comparison of the effectiveness of the use of the controlling of the LMF and TMF during submerged arc welding or surfacing by wire was made.

Data on the effectiveness of the LMF use for arc welding and surfacing are summarized in [4-6]. It is established that when action of LMF in the arc welding (surfacing) process, the melting coefficient of wires (α_m) with a diameter of 3; 4; 5 mm increases by 20...30 %, if they are made of ferromagnetic materials. If the wire is made of a non-ferromagnetic material, then α_m does not increase at surfacing (welding) [4].

Data on the efficiency of the TMF use in submerged arc welding (surfacing) with wire is partially given in [7]. It is shown that during welding (surfacing) with act of TMF, the electrode wire melting coefficient also increases (by 20 ... 30% when action of constant TMF), the penetration depth of the base metal decreases. However, there is no comparative analysis of the advantages that are achieved at arc welding or surfacing using LMF, or TMF. Such an analysis is necessary in order to make a rational choice of such controlling (external) magnetic fields by technologists in the field of welding production, taking into account the capabilities of each of these magnetic fields, taking into account the features of the welded (surfaced) construction.

IOP Conf. Series: Materials Science and Engineering **582** (2019) 012054 doi:10.1088/1757-899X/582/1/012054

The aim of the work is to perform a comparative analysis of technological advantages achieved when using the controlling LMF and TMF at submerged arc welding and surfacing by wire.

Experiments and discussion

The input device (ID) of the LMF is a solenoid with a ferromagnetic core (1) and a winding (coil) – (2) (Fig. 1), which are installed coaxially with the welding wire (4). The presence of a core of ferromagnetic material, and the electrode and products of non-magnetic materials, as established in [8], increases the longitudinal component of induction - B_z at the plate surface 2 times compared with the option when there is no ferrocore. In [9], it was shown that if a product is made of a ferromagnetic material, then the induction under the electrode end (with a gap $\Delta = 5$ mm) rises still 2 times. If not only the product-plate, but also the electrode wire are ferromagnetics, then the induction B_z under the electrode end (with $\Delta =$ 5 mm) rises 7 times and more.

Thus, the process of submerged arc welding and surfacing by wire with the action of the LMF is expedient to use if the product and the electrode wire are from a ferromagnetic material. It is expedient to use LMF for single-pass submerged arc welding with wire butt joints of plates from ferromagnetic materials with a thickness of up to 6...8 mm (on a lining). If the thickness of plates from ferromagnetic materials is 12 mm (with single V-butt weld), then LMF is not expedient for welding, because at a depth of 10...12 mm in the butt (and in the weld pool) induction B_z decreases by 6...10 times compared with its value at the surface of the plates to be welded (in the butt zone, i.e, single V-butt weld) [7]. This does not occur when welding butt joints from non-magnetic materials, the induction B_z in this zone decreases slightly.

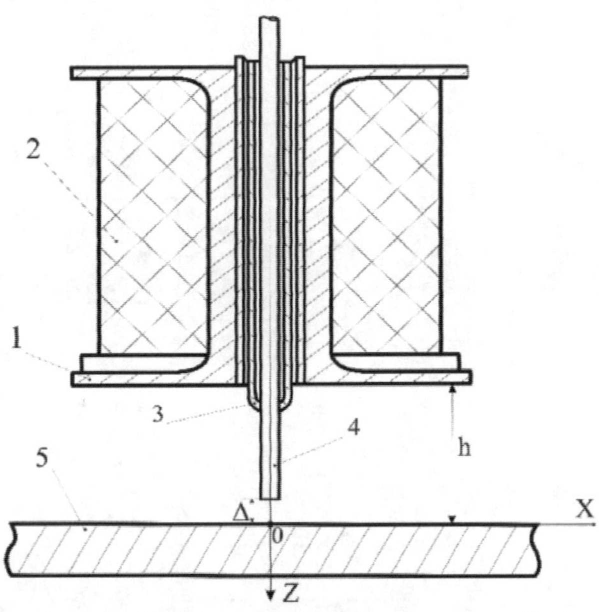

Fig. 1 Device for creating a longitudinal magnetic field: 1 - core; 2 - coil; 3 - welding tip; 4– electrode wire; 5 - product

The diameter (outer) of the ferrocore of the ID LMF should not exceed 50 mm [9], since its further increase reduces the level B_z at the surface of both the ferromagnetic and non-ferromagnetic plates.

At submerged arc welding and surfacing with a wire ID TMF is located above the weld pool surface (above the welded or surfaced plate). In this case, the ID TMF rods are located either along the axis of the weld (Fig. 2) or across the axis of the weld (Fig. 3). The ID TMF contain rods (1) with a cross section of 6...12 cm^2 from electrotechnical steel sheets and a jumper (2) of the same material and cross-section (it is possible without a jumper). Coils from copper insulated wire with a diameter of 1.2...2.0 mm with the number of turns $W = 100...200$ are placed on the rods. The ID TMF plane passes through the axis of the electrode 5. It is possible to regulate the distance from the lower rods ends to the surface of the plates (4) - the parameter h, and the distance between the rods - the parameter a. In all studies, induction at the surface of the plates was measured when the distance $\Delta = 5$ mm (measured before welding or surfacing).

When welding (surfacing) with action both LMF and TMF, the increase of α_m electrode wires (up to 30%) is due to the forced removal (due to the action of LMF, TMF) electrode drops from the melting electrode end. The mechanism of LMF action on the removal of electrode droplets from the electrodes ends during welding is described in detail in [4], and the TMF action is in [5]. It should be noted that, unlike the LMF action, during welding (surfacing) with TMF action, α_m increases when using wires and from non-magnetic materials. With the location of the ID TMF rods along the axis of the weld and the presence of a product from ferromagnetic material in the liquid metal zone, there is no transverse induction component, but mainly the longitudinal component of induction B_z. If a product is from a non-magnetic material, then at this arrangement of the ID TMF rods it is possible to avoid the presence of a longitudinal induction component (B_z), if the significant distance between the rods a (about 90 mm). Thus, when welding and surfacing with the action of such a magnetic field in the zone of the molten metal of the weld bath, two components of this field are present, and it is not clear which of the induction components induces the flow of the bath molten metal.

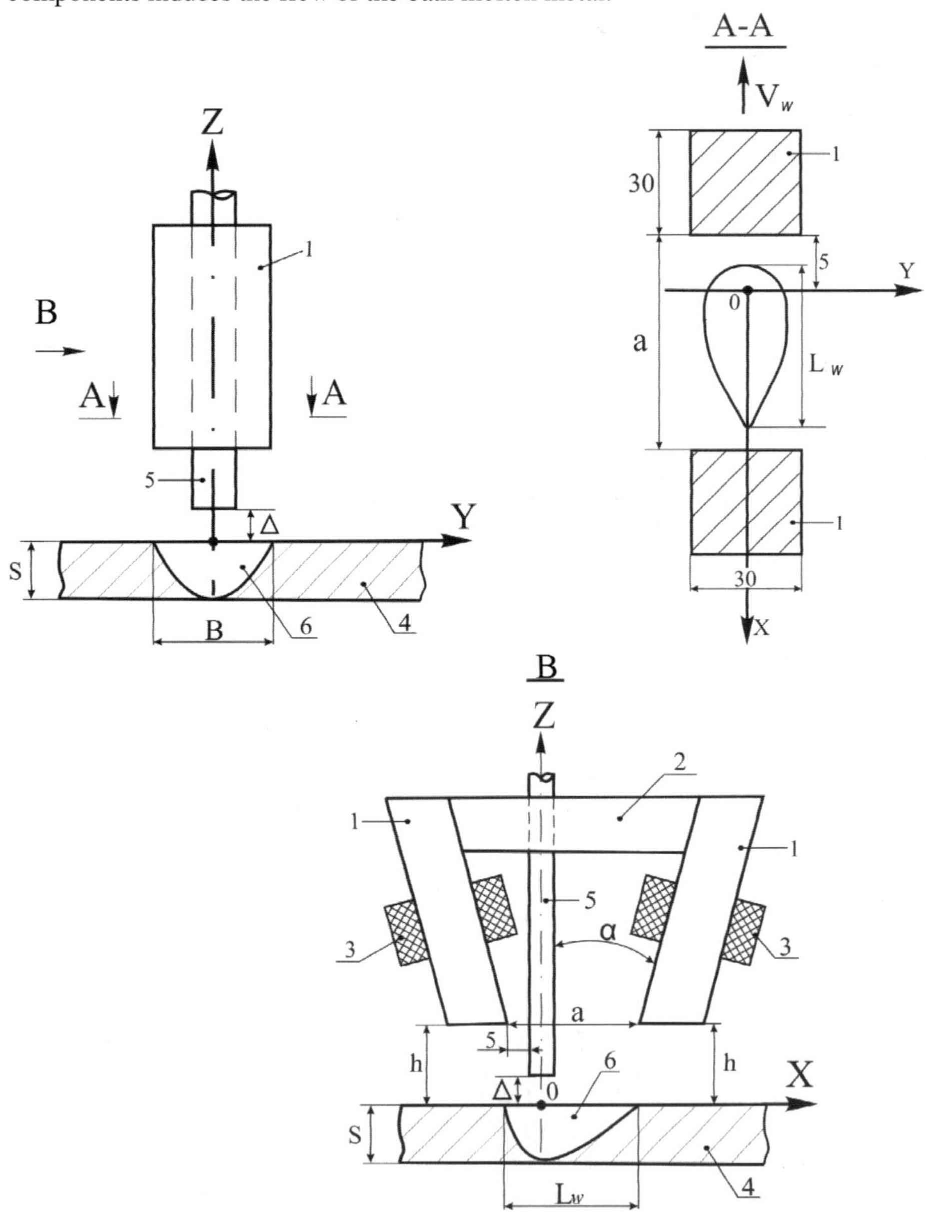

Fig. 2 The location of the ID TMF rods along the weld axis - *OX*: 1 - rods; 2 - jumper; 3 - coils; 4 - plates (base metal); 5 - electrode; 6 - welding bath

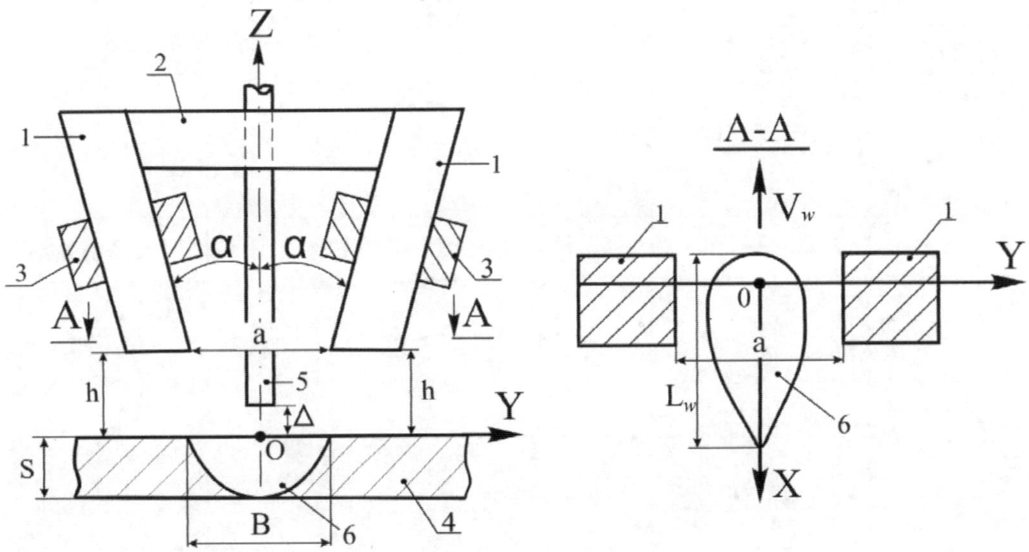

Fig. 3 The scheme of the ID TMF rods across the weld axis

With use to single-pass welding of butt joints of plates made of steel VMSt3sp with a thickness of 12...16 mm (with single V-butt weld (Fig. 4) with a gap of 2 mm according to GOST 8713-79), it was stated that for parameters a = 30 mm, h = 10 mm, direct current in the coils I_c = 16 A ($I{\cdot}W$ = 1600) in the grooving creates a level of induction B_z = 20...30 mT (and more), sufficient to induce the flow of liquid metal in the weld bath (Fig. 5).

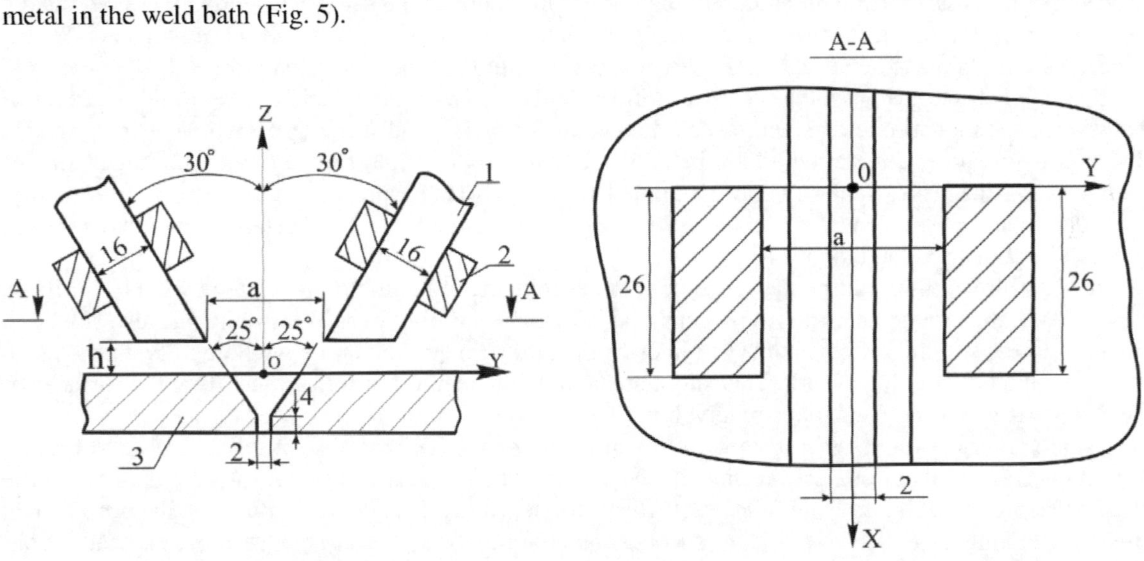

Fig. 4 The scheme of placing of the ID TMF rods: 1 - rods; 2 - coils; 3 - plates

Moreover, at the root of the grooving (curve 2 in Fig. 5), the transverse induction component B_y, is larger than that at the plates surface (curve 1 in Fig. 5). However, if there is a backup weld at the root of the grooving, the magnetic flux closes through it and in the grooving the value of B_y will approach zero. Thus, the transverse placing of the ID TMF rods only for single-pass welding of plates from ferromagnetic materials is expedient to use. If the product (butt joint of plates) from non-magnetic steels, then in the molten metal of the bath (in single V-butt weld) induction B_y is observed in the presence of any number of previously made welds (beads) (there is no shunt ferromagnetic jumper). In this case, the TMF use for welding is expedient.

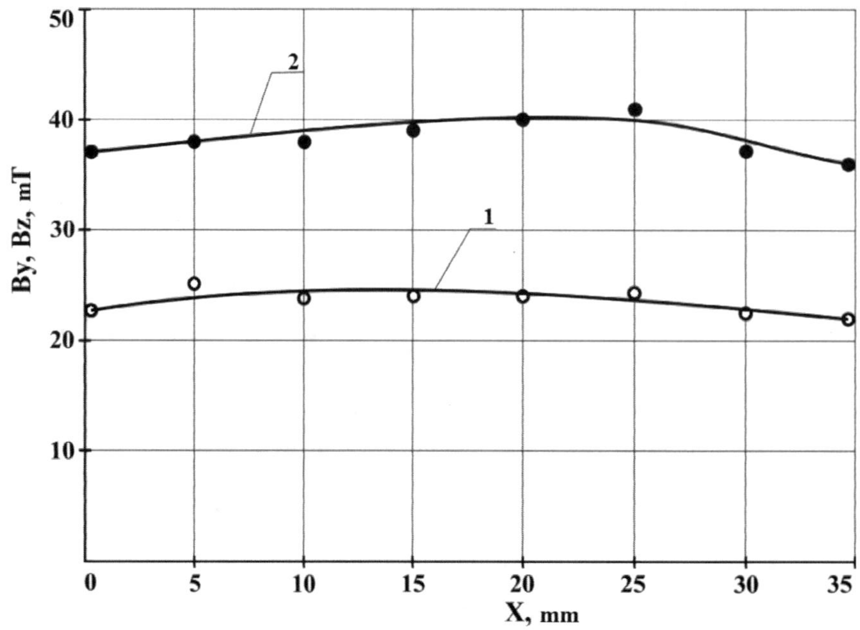

Fig. 5. The induction distribution of B_y along the axis OX (Y = 0): 1 - at the surface of the plates; 2 - at the grooving root

At welding and surfacing with the action of LMF and TMF in the molten metal of the bath flows which affect the process of crystallization of the metal are induced [10-14]. Currently, there are many hypotheses about the mechanism of grinding the structure of the welds with the act of magnetic fields. The question of how to achieve a certain level of the longitudinal induction component of B_z in the weld bath area when the LMF action is set forth in [4]. Difficult was the question of defining the level of transverse the induction component even when using surfacing and welding of butt joints of products (plates) from non-magnetic materials (in Figs. 2, 3). It was possible to use the method of analogies between the magnetic field structure generated by the ID TMF and the electrostatic field structure generated by analogical sizes with charged bodies. At the same time, electrostatic equations are used, which simplifies the calculations [7].

More difficult was the solution to the problem of defining the induction B_y generated by the ID TMF (in Fig. 2, 3), if the product is made from a ferromagnetic material. However, this problem is also solved in relation to the welding of butt joints of plates from ferromagnetic materials with single V-butt weld It also uses the laws of electrostatics, but supplemented by the method of reflections (from the surfaces of ferromagnetic plates). These data are given in [7].

The important question is what material to manufacture of the ID TMF ferrocore (according to Fig. 1). Applying a set of plates of electrotechnical steel is difficult. Thus satisfactory results are obtained if the ID LMF core is machined from mild steel (for example, St. 3). It is better if after this the core is heat treated (heating to 650 °C with slow subsequent cooling). The ID TMF rods are best made from electrotechnical steel plates (0.35 mm thick, or 0.5 mm thick). However, it is permissible to manufacture them from low carbon steels.

In conclusion, it should be noted that, for arc welding and surfacing ID LMF must provide in the electrode drop zone, the welding arc and in the head part of the molten metal of the weld bath the longitudinal induction component B_z = 25...30 mT. Analogically ID TMF should provide the same level of transverse induction component B_y (B_x) in these zones. This will increase the efficiency of the magnetic fields use in arc welding or surfacing.

Conclusions

1. The use of longitudinal and transverse magnetic fields in arc welding and surfacing allow improve the electrode wire melting productivity, reduce the penetration depth of the base metal, and grind the structural components of the welds (beads).

2. Comparative analysis allowed us to state the area of rational use of controlling magnetic fields (LMF and TMF) at submerged arc welding and surfacing by wire, taking into account the magnetic properties of welding and basic materials.

References

[1] Saraev Yu.N., Chinakhov D.A., Ilyashchenko D.I., Kiselev A.S., Gordynets A.S. Investigation of the stability of melting and electrode metal transfer in consumable electrode arc welding using power sources with different dynamic characteristics. Welding international. Vol. 31, No 10 (2017), pp. 784–790. DOI: 10.1080/09507116.2017.1343977

[2] Chinakhov D.A., Chinakhova E.D., Gotovschik Y.M., Grichin S.V. Influence of welding with two-jet gas shielding on the shaping of a welding joint. IOP Conf. Series: Materials Science and Engineering 125 (2016) 012013. doi:10.1088/1757-899X/125/1/012013

[3] Chinakhov D.A., Il'yaschenko D.P., Gotovschik Y.M. MAW Productivity Development and Reduction of its Harmful Effect on Human Organisms. Applied Mechanics and Materials Vol. 682 (2014), pp. 122-126. www.scientific.net/AMM.682.122

[4] Razmyshlyaev A.D., Mironova M.V. Magnetic control of formation of rollers and welds at arc surfacing and welding. Mariupol: PSTU (2009), pp. 242.

[5] Boldyrev A. M., Birzhev V. A., Martynenko A. I. Examination of the effect of the alternating axial magnetic field on the process of melting of electrode wire**.** Welding International. Vol. 23, No 3 (2009), pp. 223-224.

[6] Chen Shu-jun, Wang Jun, Wang Hui-xia, Lu Zhen-yang, Yin Shu-yan, Bai Shao-jun. The principle of rotating droplet transfer under the action of a longitudinal magnetic field. Trans. China Welding Institute. Vol. 26, No 3 (2005), pp. 58-62, ISSN 0253-360X.

[7] Razmyshlyaev A.D., Vydmysh P.A., Ahieieva M.V. Automatic submerged arc welding with the action of external magnetic field. Mariupol: PSTU (2017), pp. 209.

[8] Chernysh V.P., Kuznetsov V.D., Briskman A.N., Shelenkov G.M. Welding with electromagnetic stirring. Kiev: Tekhnika (1983), pp. 127.

[9] Razmyshlyaev A.D. Magnetic control of welds formation at arc welding. Mariupol: PSTU (2000), pp. 245.

[10] Ryzhov R.N., Semenyuk V.S., Titov A.A. Features of welds formation and crystallization in TIG welding with arc deflection by magnetic field. Avtomaticheskaya svarka. No 4 (2004), pp. 17-20. ISSN 0005-111X.

[11] Ryzhov, R. N. The effect of pulsed electromagnetic influences on the formation and crystallization of welds. Avtomaticheskaya svarka. No 2 (2007), pp. 56-58. ISSN 0005-111X.

[12] Villafuerte J. C., Kerr H. W. Electromagnetic stirring and grain refinement in stainless steel GTA welds. Welding journal. No 1. Vol. 69 (1990). pp. 1-13.

[13] Liu Zheng-jun, Chen Hong, Liu Chen, Liu Duo. The influence of magnetic field control on the microstructure and properties of the surfaced layer at plasma surfacing. Trans. China Welding Institute. Vol. 26, No 8 (2005), pp. 16-18. ISSN 0253-360X.

[14] Malinovski-Brodnicka M., G. den Ouden, Vink W. J. P. Effect of electromagnetic stirring on GTA welds in austenitic stainless steel. Welding journal. No 2. Vol 69 (1990), pp. 52-59.

IOP Conf. Series: Materials Science and Engineering **582** (2019) 012055 doi:10.1088/1757-899X/582/1/012055

Development of the Specialized Contact Highly Productive Scanning Sensor

A V Lavrinovich [a], M A Bolotov, N V Ruzanov, E V Kudashov

Samara National Research University, Moskovskoye shosse, 34, Samara, Samara region, 443086, Russian Federation

E-mail: [a] alexackpro@gmail.com

Abstract. The article is devoted to the development of specialized high-contact scanning probe, intended to equip the technology of measuring and processing equipment. Conducted a literature review allowed to identify the main types of design solutions for contact measurement sensors on the basis of which was formed by generalizing the decision tree by type layouts. The first part of the selection of parameters is performed using a sensor structure in a simulation environment ANSYS. After selecting the design performed sensor fabrication including the use of additive technology for the production of the main structural elements of the sensor. Experimental studies to identify the calibration curves and the homogeneity of the developed sensitivity of the sensor fields.

Introduction

The conditions and Industry 4.0 IoT are important sensors and sensors of high accuracy, enabling to perform dimensional control parts, workpieces, assembly units, as well as to collect the processes of data errors and accidents occurring in production. Sensors are needed for the creation of high accuracy measurement systems, the demand for which is increasing in various industries and spheres of human activity. Particularly relevant in this matter production GTE blades having complex surfaces of double curvature. Control surfaces of double curvature in the production process is carried out using high-precision measuring systems, which include the coordinate measuring machine (CMM). The accuracy and reliability of the coordinate measuring machine depends on the design used sensor. An important condition for extensive use of measuring sensors in the production cost is relatively low. Proceeding from aggressive media and hard operating conditions, the sensors must meet high standards moisture-dustproof protection, have a relatively high accuracy, and also have optimum weight and size characteristics for easy operation and maintenance. One of the most difficult tasks is difficult to process control profile details. The relevance of the development of specialized high-contact scanning probes is high in the aviation industry and general engineering. Appearances developed measuring sensors are shown in Fig. 1, as: uniaxial not symmetrical measuring sensor used: uniaxial symmetric measuring sensor.

Literature review

A review of the articles in the direction of the development of contact scanning probes. Encoders have a significant influence on measurement errors of coordinate measuring machines [1]. Level measurement sensors perfection offers substantial opportunity to improve the accuracy and performance measurement of geometrical parameters on a coordinate measuring machine [2]. Increasing technological processing characteristics of critical parts, including aircraft engines can be achieved through the use of inline measuring sensors including optical laser machines [3, 4]. The sensitivity and reliability of the contact sensors depends upon their design and materials used in them.

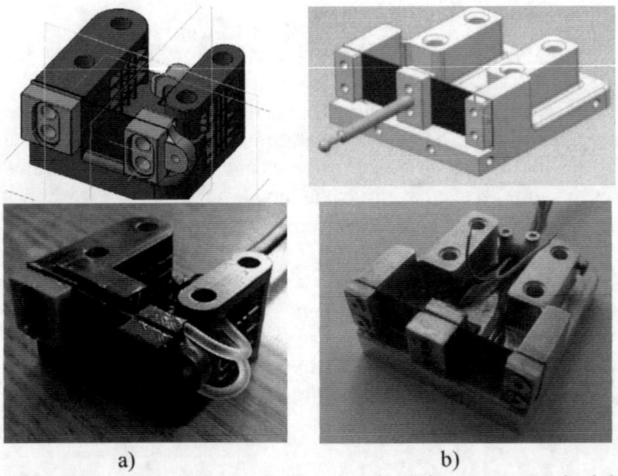

Figure 1. The external kinds of measurement sensors, a) uniaxial not symmetrical, b) uniaxial symmetric

The use of structures with a large number of sensors ramemory location in all directions of deformation, providing enough information on the vector and the unit of force applied to the measuring probe, but such designs have a significant disadvantage of a large limb of the probe during the measurement. The use of smaller amounts of the axes and the sensors reduces the limb at a force probe, but introduces complexity in analyzing data from the sensor. Three axial sensors located in one plane sensing elements are of non-linearity associated with the vector of applied forces. This requires the development of sophisticated mathematical models to compensate for sensor measurement errors with similar designs.

Analysis known designs allowed to form generalizes decision tree represented in Figure 2. The solutions Generalizing tree predictive positive and negative characteristics or properties of this design the contact sensor. With this tree may make certain changes in the already pre-selected design, to reduce measurement errors and approaching the measuring sensor characteristics to the required values. An integral part of measuring sensors are sensors which, depending on the physical principle implemented steps geometrical parameter measured is converted into an electrical value. The design of sensors may be provided as own sensors (SCHE) directly sensing a force or other effects in the measurement of the geometrical parameter or additional sensing elements (DCHE) which indirectly perceive phenomena arising in the measurement of the geometrical parameters and converts them into an electrical magnitude. Application DCHE involves the use of extra transmission elements or mechanisms provided in the sensor design. To implement scan cycle involving continuous contact tip of the measuring sensor with the measured control surface must perform its spatial position. The control problem is maintaining the predetermined mode parameters arising in the contact measurement probe tip and measured surface, to which the nominal pressure force, limits this pressure change, the nominal slip velocity, the allowable slip speed

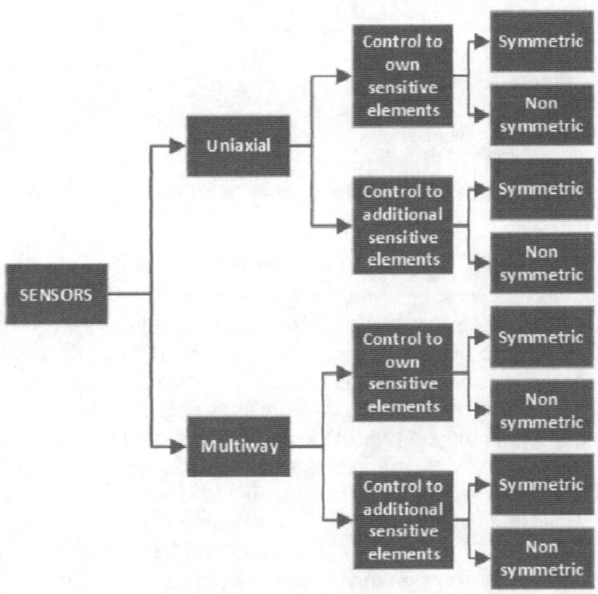

Figure 2. Generalizing the decision tree

IOP Conf. Series: Materials Science and Engineering **582** (2019) 012055 doi:10.1088/1757-899X/582/1/012055

limits harmful phenomena arising during probing, such as vibration. The control can be performed based on data from DCHE or SCHE.

Depending on the form of the integral sensitivity zone sensors can be isolate having symmetrical and balanced characteristics. For the purpose of hanging the measurement accuracy, elimination of nonlinearities occurring during the conversion of the measured geometrical parameter, it is desirable to use sensors with symmetrical integral sensitivity zone. It should be noted that the various tasks and areas of application can tolerate sensors not symmetrical integral sensitivity zone, which often reduces the cost of the sensor.

The work will be considered the task of developing single-axis sensor for process control difficult profile details. Selecting new sensor arrangement execute based on the results of experiments conducted with the use of uniaxial not symmetric sensor developed previously [6]. According to the results of experimental studies, it was decided to apply the principle of symmetry in the design of the new sensor. The issue is not fully resolved and requires further research.

Theoretical part

Finite-element model, which can be used for sensor selection and design of rational geometrical dimensions and for the specified requirements developed in MATLAB and ANSYS systems. Scheme designed finite element model of the sensor shown in Figure 3.

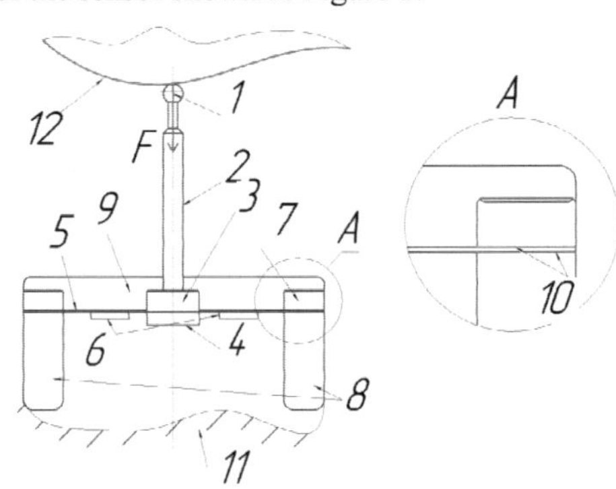

Figure 3. Scheme of a finite element model of the sensor

During the measurement process, the point of contact of the ball 1 and the measured surface 12 acts reaction force F, through which the leg of the probe 2 is transmitted to the tool holder 3 and the clamp 4. This createsan effortin the elastic element 5, such as a carbon plate, resulting in its deformation. On its surface pasted sensitive elements 6 in the form of strain gauges, the contacts of which are defined in the finite element model (FEM) type - Bonded, or converting a force or other effects occurring during the measurement into an electrical magnitude. The elastic member 7 is pressed against the hold-downs to the vertical supports 8 provided on the base of the measuring sensor 9. The contact surface 10 between the supports, the elastic member and the clamps are set in the finite element model type - Bonded. measuring sensor in the base of the FEM is assigned to limit displacement in space.

The finite element model can account of different materials: D16 aluminum, steel, carbon fiber. Mathematical model of the elastic element of the material, which acts as carbon fiber, has been verified by means of tests of prototypes made from the same material.

Performing research to develop sensor under the specified requirements conducted in accordance with the flowchart shown.

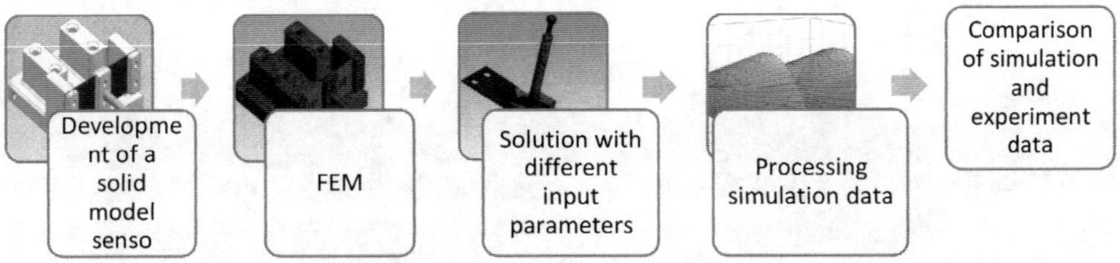

Figure 4. Flowchart perform studies

Studies two measuring sensor arrangement were examined with a different arrangement of the plate and carbon fiber probe shown respectively in Figures 5.a and 5.b.

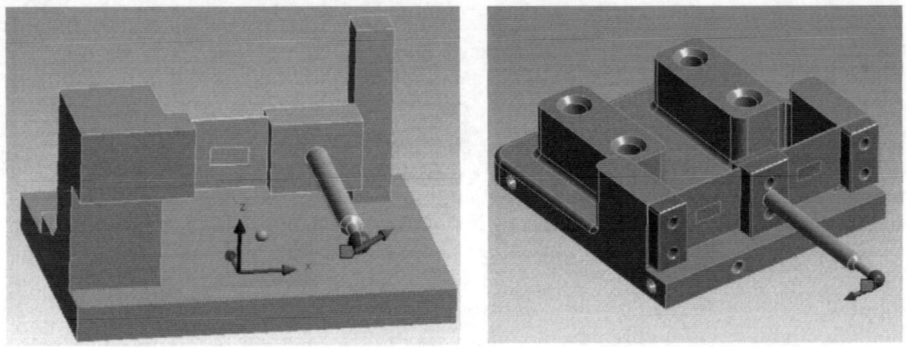

a) b)

Figure 5. a) uniaxial not symmetrical sensor, b) uniaxial symmetric sensor

Arrangements for data in the system ANSYS simulation was conducted impact forces measured on the surface of the measuring tip. Particular attention was paid to the task model material carbon fiber plate. We considered various ways of specifying the parameters and the force vector applied to the tip of the probe.

The developed finite element model makes it possible to perform the calculation for a single experiment with the given parameters. must perform a series of experiments at different input parameters for constructing stray fields developed sensor sensitivity. This calculation has been automated with the use of MATLAB system, the system loads the appropriate settings for the current experiment, launched a calculation, and then performs the export of the results to a file.

Research

The developed sensor is designed to solve the problems associated with partial automation of industrial processes such as the dimensional control parts, workpieces, assembly units. The basis of measurement is the principle surface contact method. Control of the geometry of the measuring object is performed by determining the coordinates of the touch surface to be measured and the probe. The tip of the probe is made of synthetic ruby, having high hardness and wear resistance, thus ensuring high stability of operation and durability of the probe.

Analysis turbomachine blade geometry limits of variation shown applying a force to the measuring tip angle: $\alpha\varepsilon$ [0; 360], $\beta\varepsilon$ [60; 120]. The simulation was calculated change in the strain gauge length arising from a carbon fiber stiffener plates and the change in coordinates of the probe center. Field sensitivity dispersion for sensors, with the first version and new designs are shown in Figures 6a and 6b, respectively.

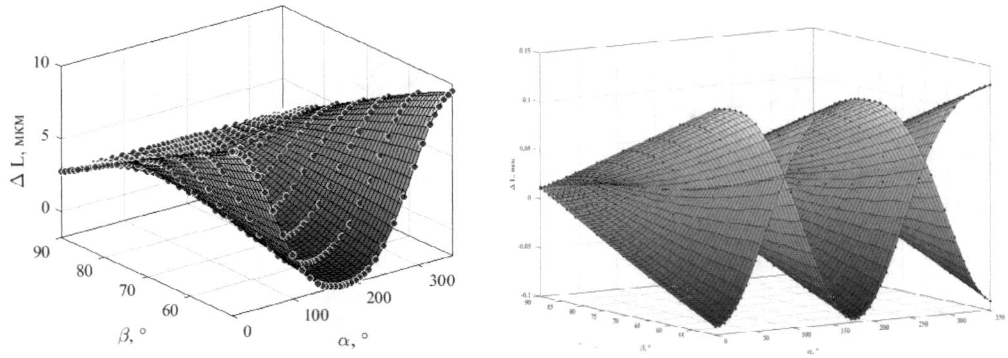

a) b)

Picture 6. Field sensitivity dispersion: a) uniaxial sensor is not symmetrical; b) uniaxial symmetric sensor

According to the calculation results of the analysis of the data revealed a strong change in sensor sensitivity. It was found that the introduction of symmetry in the sensor design has resulted in a dramatic increase of hardness sensitive element and consequently reduce the sensitivity of the sensor. But desensitization compensated by the presence of large amounts of data collected from the sensor, which allows to estimate and compensate for errors with increased accuracy.

Experimental studies on the measurement of the geometric pattern with known geometry, sensor calibration and measurement repeated patterns was performed to determine the accuracy of the developed sensors. We designed the sensor there is a high repeatability of measurement results.

Conclusions

During the study, it was designed with the basic measurement sensor sensors strain gauges. To improve the accuracy and reliability of the measuring sensor calculated mathematical model, two principal variants flowmeter sensor arrangements have been developed and analyzed. According to the results in the ANSYS simulation system layout has been selected for the developed sensor, allowing us to obtain a greater amount of data needed to analyze errors and improve measurement accuracy. After the analysis of the fields were obtained simulation dispersions integral sensitivity of the sensor. According to the results of practical experiments were constructed a graph of change of the strain gage resistance from vector applied to the measuring bulb strength. It can be concluded, developed that perform the tasks sensor, characterized by a relatively low cost, small weight and size characteristics, the accuracy of measurement reference element is 3-4 m. Measuring sensor will be used for process control high geometry gas turbine engine blades.

Further studies are planned in the direction of the development of high-contact scanning probe. Will be developed issues of increasing measurement accuracy and improving the integrated area sensor sensitivity due to the use of data obtained during calibration of the mathematical function for generating the sensor measurement error compensation. Using the functions of compensation errors will significantly speed up the operation of the sensor and to increase its accuracy.

References

[1] AF Rezchikov, AV Kochetkov, and OV Zakharov. Mathematical models for estimating the degree of influence of major factors on performance and accuracy of coordinate measuring machines. MATEC Web Conf. 2017. Pp. 1054

[2] OV Zakharov, AF Balaev, and AV Kochetkov. Modeling Optimal Path of Touch Sensor of Coordinate Measuring Machine Based on Traveling Salesman Problem Solution. Procedia Engineering. 2017, Pp. 1458-1463

[3] SP Murzin, Formation of structures in materials by laser treatment to enhance the performance characteristics of aircraft engine parts, CO, 2016, Volume 40, Issue 3, 353-359

[4] Microstructuring the surface of silicon carbide ceramic by laser action for reducing friction losses in rolling bearings Murzin, SP, Balyakin, VB 2017 Optics and Laser Technology

[5] Development of a touch probing system for gaging the compressor blades of the GTE Ruzanov, NV, Bolotov, MA, Pechenin, VA, Matek, ER 2018 Key Engineering Materials 769 KEM, p. 242-249
[6]Development of a coordinate measuring machine (CMM) touch probe using a multi-axis force sensor Jae-jun Park1, Kihwan Kwon2 and Nahmgyoo Cho Meas. Sci. Technol. 17 (2006) 2380-2386
[7] Probing-error compensation using 5 degree of freedom force / moment sensor for coordinate measuring machine Minho Lee1 and Nahm-Gyoo Cho 2,3 Published July 26, 2013 • 2013 IOP Publishing Ltd
[8] Three-dimensional touch probe using three fibre optic displacement sensors Takaaki Oiwa and Hiroshi Nishitani Published 4 November 2003 IOP Publishing Ltd
[9] Investigation on calibration methods for multi-axis, linear and redundant force sensors CM Oddo1, P Valdastri2, L Beccai3, S Roccella3, MC Carrozza3 and P Dario2,3 Published January 19, 2007 • 2007 IOP Publishing Ltd
[10] Development of a touch probe based on five-dimensional force / torque transducer for coordinate measuring machine (CMM) Author links open overlay panel Qiaokang Lianga Dan Zhangbd Yaonan Wanga Yunjian Gec

Institute of Physics
Dirac House, Temple Back
Bristol BS1 6BE UK

ISSN: 1757-8981
ISBN 978-1-5108-9813-4

9 781510 898134

CURRAN ASSOCIATES INC.
proceedings.com